▼ Trilobites of New York

TRILOBITES

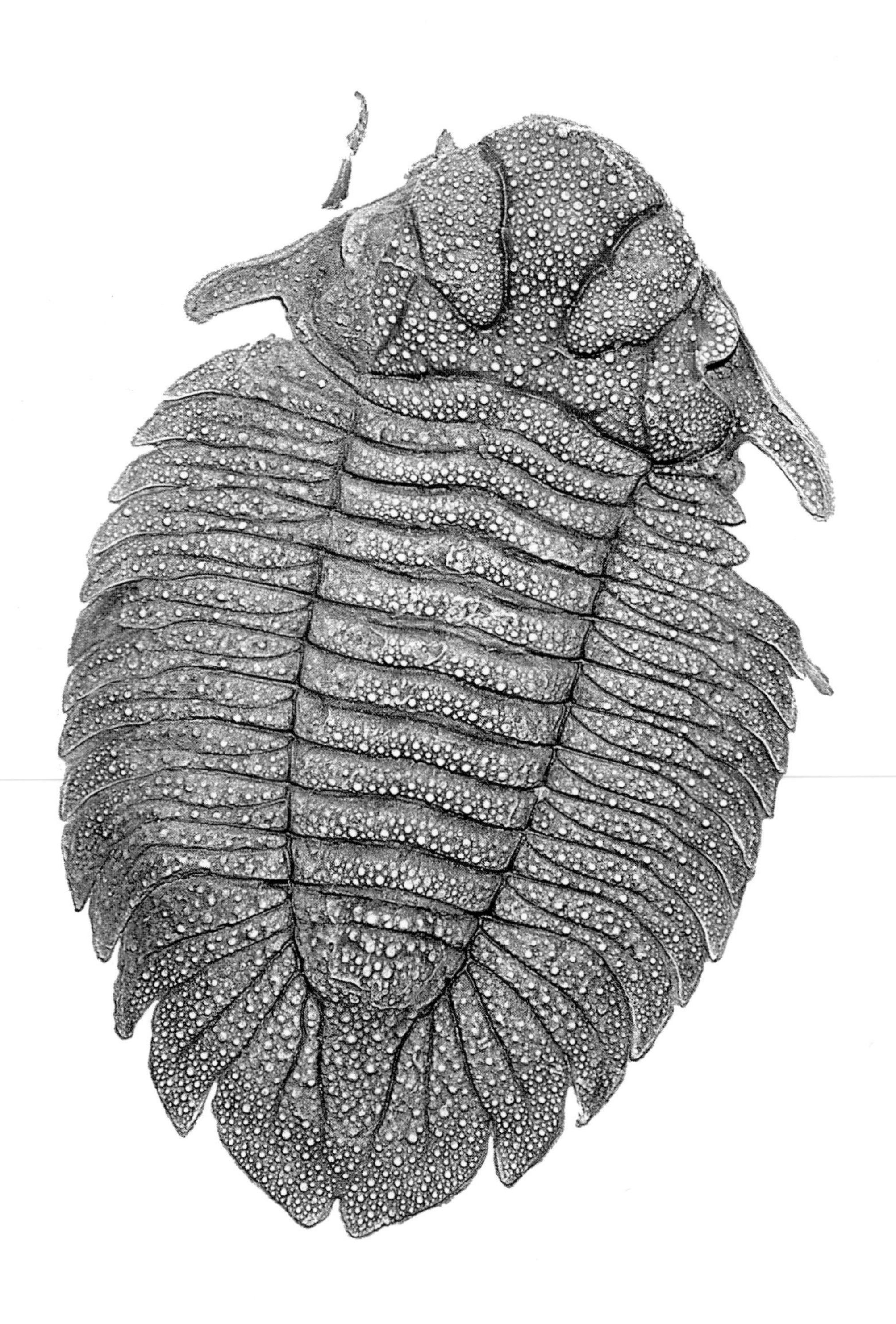

of NEW YORK

AN ILLUSTRATED GUIDE

Thomas E. Whiteley
Gerald J. Kloc
Carlton E. Brett

with a foreword by Rolf Ludvigsen

Published in cooperation with the
Paleontological Research Institution, Ithaca, New York

COMSTOCK PUBLISHING ASSOCIATES, *a division of*
CORNELL UNIVERSITY PRESS

Ithaca and London

First published 2002 by Cornell University Press

Published in cooperation with the Paleontological Research Institution, Ithaca, New York

Printed in the United States of America

Library of Congress Cataloging-in-Publication Data

Whiteley, Thomas E. (Thomas Edward), 1932–
Trilobites of New York : an illustrated guide / Thomas E. Whiteley, Gerald J. Kloc, and Carlton E. Brett.
p. cm.
Includes bibliographical references and index.
ISBN 0-8014-3969-9 (acid-free paper)
1. Trilobites—New York (State) I. Kloc, Gerald J. II. Brett, Carlton E. (Carlton Elliot) III. Paleontological Research Institution (Ithaca, N.Y.) IV. Title.

QE821 .W397 2002
565′.39′09747—dc21

2001054767

Cornell University Press strives to use environmentally responsible suppliers and materials to the fullest extent possible in the publishing of its books. Such materials include vegetable-based, low-VOC inks and acid-free papers that are recycled, totally chlorine-free, or partly composed of nonwood fibers. For further information, visit our website at www.cornellpress.cornell.edu.

Cloth printing 10 9 8 7 6 5 4 3 2 1

Contents

Text Figures vii
Text Tables ix
Foreword xi
Preface xv

Chapter 1
BACKGROUND INFORMATION 1
Historical Notes 1
Trilobite Names 2

Chapter 2
THE BIOLOGY OF TRILOBITES 4
Exoskeleton 4
Ontogeny 13
Soft Body Parts 17
Life-Mode 21
Evolution and Cladistics 30

Chapter 3
TAPHONOMY 32
Death, Decay, and Disarticulation 32
Transport and Reorientation 35
Fragmentation and Biased Preservation 36
Abrasion, Corrosion, and Encrustation 37
Fossil Diagenesis: Geochemical Processing of Potential Fossils 37
Trilobite Taphofacies 40
Trilobite Lagerstätten 40

Chapter 4
THE PALEOZOIC GEOLOGY OF NEW YORK 43
Overview 43
Cambrian Period 46
Ordovician Period 56
Silurian Period 81
Devonian Period 95

Chapter 5
THE TRILOBITES 117
Order Agnostida 117
Order Redlichiida 118
Order Corynexochida 118
Order Lichida 124
Order Phacopida 128
Order Proetida 148
Order Asaphida 157
Order Ptychopariida 163

Appendix A: Trilobites and Their Environments 168
Appendix B: The Photography 170
Glossary 172
References 177
Trilobite Index 191
Index 201
Plates 204

Text Figures

2.1 Trilobite structure using *Eldredgeops rana* 5
2.2 Trilobite structure using *Kettneraspis tuberculata* 6
2.3 The structure of the trilobite cephalon using *Calymene* species 8
2.4 Trilobite eyes 9
2.5 Cephalic sutures 9
2.6 Exoskeletal pits or circular perforations 11
2.7 Ventral anatomy of the exoskeleton 12
2.8 Ontogeny of the trilobite 14
2.9 Trilobite exoskeletons with attached fauna or injury 18
2.10 Trilobite appendage reconstruction and nomenclature 19
2.11 Ventral anatomy and appendages 20
2.12 Internal anatomy of the trilobite 22
2.13 Trilobite shapes and functions 23
2.14 Trilobite traces 25
2.15 Trilobite injuries 27
2.16 *Cryptolithus*, the most-studied trilobite genus found in New York 28
3.1 Fossil assemblages reflecting various conditions and timing of burial 34
3.2 Conditions for the formation of pyritized, well-preserved fossils 39
4.1 Time scale for Earth history and for the Paleozoic rocks in New York 44
4.2 Precambrian Grenville (1.0 BP) metamorphic/igneous rocks 46
4.3 Map showing the extent of the Grenville belt in eastern North America 47
4.4 Paleomagnetic reconstruction of the supercontinent Rodinia 48
4.5 Position of the Laurentian plate and neighboring Baltica and Avalonia terranes from the Cambrian to the Devonian 49
4.6 Cambrian rocks in New York 50
4.7 New York in the Cambrian 52
4.8 Close-up of Upper Cambrian Potsdam Sandstone 53
4.9 Upper Cambrian limestones 54
4.10 Simplified stratigraphy of the Cambrian-Ordovician rocks in the Taconic allochthon 55
4.11 Lower Cambrian allochthonous beds of the low Taconic Mountains 56
4.12 Stratigraphic chart of the Ordovician exposures in New York 57
4.13A–C New York during the Early and early Middle Ordovician 58
4.13D–F New York during the Middle Ordovician 59
4.14 Details of the stratigraphy of the Chazy Group in northeastern New York 63
4.15 Flute casts on Austin Glen graywacke 65
4.16 Black River limestones 66
4.17 Middle Ordovician Black River Group 67
4.18 Close-up of sharp Black River/Trenton (Watertown-Napanee) contact 68
4.19 Close-up of storm beds in the Middle Ordovician Kings Falls Limestone 69

4.20 Maps of New York during late Middle Ordovician times 70
4.21 Middle Trenton at Trenton Falls 72
4.22 Middle Ordovician Trenton Group limestone 73
4.23 Trenton at Trenton Falls 74
4.24 Middle Ordovician Dolgeville and Utica rocks 77
4.25 New York maps during the late Middle Ordovician and the Upper Ordovician 79
4.26 Ordovician/Silurian unconformities 81
4.27 Stratigraphic chart of the Silurian rocks in New York 82
4.28 Ordovician-Silurian (Cherokee) unconformity/Ordovician-Lower Silurian succession 84
4.29 Maps of New York during Wenlock times 87
4.30 Irondequoit-Rochester bioherm at the upper part of the Lower Silurian Clinton Group 88
4.31 Mid Silurian successions 91
4.32 Silurian carbonates 92
4.33 Domal stromatolites 94
4.34 Composite stratigraphic chart for the northern and central parts of the Appalachian Basin 96
4.35 New York during the Early Devonian 97
4.36 Upper Silurian/Lower Devonian rocks 99
4.37 New York during the early Middle Devonian 103
4.38 Upper Silurian/Onondaga Limestone 104
4.39 Middle Devonian rocks 106
4.40 New York during the Middle Devonian Hamilton deposition 108
4.41 Devonian successions 113
4.42 Middle Devonian/Upper Devonian successions 114
4.43 Stratigraphy of the Upper Devonian and the upper Middle Devonian of New York 115
5.1 Lichids from the lower and lowest Middle Devonian of New York 125
5.2 Key cephalic differences between Dalmanitidae and Synphoriidae 136
5.3 Trilobites of the families Dalmanitidae and Synphoriidae 138
5.4 Lower Devonian and lower Middle Devonian phacopins 143
5.5 Features of New York *Cryptolithus* 164

Text Tables

5.1 Trilobites of the suborder Agnostina 118
5.2 Trilobites of the suborder Eodiscina 119
5.3 Defining features of the New York asteropygins 129
5.4 Features of the genera *Bellacartwrightia* and *Greenops* 130
5.5 Features of the Lower Devonian phacopids from New York 141
5.6 Features of the Middle Devonian phacopids of New York 142
5.7 Features of the genus *Odontocephalus* of New York 147
5.8 Features of the described proetids of New York 153

Foreword

Tom Whiteley, an accomplished amateur paleontologist, has taken the lead in compiling a much-needed popular account of the trilobites of New York. Sumptuously illustrated with generous photographs of complete specimens of New York trilobites, this book is more than a regional field guide. It also testifies to Gerry Kloc's expertise in preparation and Carlton Brett's keen insight about the rocks and complex facies of the state. In essence, the book reprises the work of Charles Walcott, another accomplished amateur paleontologist of a century and a quarter ago.

The technical literature on trilobites is vast. It is dispersed through numerous paleontological journals, in specialized books, and as century-old monographs. Even a dedicated trilobite paleontologist with access to a university library would have difficulty retrieving all of it. The popular literature on trilobites is a lot easier to access, if only because there is so little of it. Few professional paleontologists have considered it important to write field guides to fossils along the lines of the popular regional guides to flowers, mushrooms, trees, insects, and birds that are available in every bookstore across America. Fewer still have written books exploring the natural history of fossils.

Trilobites of New York is a lavishly illustrated bestiary of New York trilobites in which one can sense the spirit of Samuel Latham Mitchill's monograph, *The Fishes of New York*, published in 1815. Both books are basically encyclopedic and scientific in their approach, but each includes snippets of practical information. Mitchill advised fishermen that the blackfish will run in the spring when the dogwood blossoms open. Whiteley suggests that a 35-mm TIFF color image of a trilobite is about 25 megabytes in size and should be stored on a CD-ROM. Times change!

Here is a summation of nearly two centuries of discovery and study of New York trilobites by a succession of paleontologists and as an exposition of the natural history of these fascinating fossils. The book will be welcomed especially by American paleontologists, professional as well as amateur, and by anyone who delights in the exquisite beauty of these ancient fossils.

Professional paleontologists working on New York fossils have always been greatly outnumbered by amateur paleontologists and avocational fossil collectors. Although often (and mistakenly) dismissed as dilettantes, amateurs (the root is the Latin *amator*, for "lover") are those who pursue an activity out of interest instead of financial gain. Amateurs, like professionals, tend to specialize. In New York, as in Ontario, Ohio, Indiana, Oklahoma, and Utah—regions that have a lot of fossiliferous Lower Paleozoic outcrop—many amateurs and fossil collectors concentrate their efforts on trilobites. They have collected and prepared trilobite specimens that, in quality and completeness, rival those described by academic paleontologists. The extent of popular interest in trilobites is exemplified by hundreds of websites on the Internet, almost all of them hosted by amateurs.

The photographs in this publication reveal why trilobites have long been considered to be among the most desirable and curious of fossils. A late Paleolithic hunter in what-is-now central France carried one around his neck as a pendant suspended from a leather thong. Ancient Chinese philosophers called them "stone silkworms" and "batstones." The Pahvant Ute tribe of western Utah knew them as *timpe khanitza pachavee*, meaning "little water bug like stone house." Strung as amulets, they were thought to possess magical powers. In 1698, Edward Lhwyd, curator of the Ashmolean Museum in Oxford, included in the pages of the *Philosophical Transactions of the Royal Society* a plate with a finely lined oval fossil from the shales exposed near Llandeilo in southern Wales. He thought it was a "flat fish." In nearby areas in Wales suffused with the legend of King Arthur, such fossils are called "stone butterflies" and are widely believed to have been entombed in rock by spells cast by the wizard Merlin.

European naturalists were fascinated by these fossils. Although they described and figured them in considerable detail, they had great difficulty in determining what they were and how they should be classified. The great naturalist Carolus Linnaeus applied the term *Entomolithus paradoxus* ("paradoxical stone insect") to such fossils from Sweden; others thought they were fossil crabs or possibly weird mollusks.

With their evocative and disparate names, these stony objects were corralled in 1771 when the German naturalist Johann Ernst Immanuel Walch published the third volume of *Der Naturgeschichte der Versteinerungen* [*Natural History of Petrifactions*]. In it he proposed a collective name derived from the most obvious feature—their three-lobed appearance. Eventually naturalists studying fossils (who were now beginning to be called *pale-*

ontologists) accepted that trilobites comprise a distinct group of fossil arthropods.

By the early years of the nineteenth century, trilobites of many different types had been documented from Britain, Germany, Scandinavia, and Bohemia. Then, as now, they attracted attention because of their peculiar shape, their striking ornamentation, and their great age. Among the most ancient of fossils, they were found mainly in the sectors of Earth history soon to be named the Cambrian and Silurian systems.

Trilobites are the most lifelike of fossils—many well-preserved specimens belie their great antiquity and seem almost ready to arch their bodies, peer about with their compound eyes, and crawl forward as if to continue a journey that was interrupted hundreds of millions of years ago. A trilobite is an ancient arthropod, but it is certainly not a lesser arthropod.

Trilobite discoveries in the New World followed closely on those in the Old. By 1832 so many different kinds of trilobites had been collected that Jacob Green, a professor at Jefferson Medical College in Philadelphia, had sufficient material to write a 93-page monograph detailing all the species then known. He might have titled it *Trilobites of New York* instead of *A Monograph of the Trilobites of North America*, because of the 32 species he dealt with, all but 7 came from that state.

New York was home to the first American school of geology. In 1824 Stephen Van Rensselaer provided the funding that allowed Amos Eaton to start the Rensselaer School in the town of Troy. This school left a solid mark on early American geology and paleontology. It graduated a remarkable contingent of geologists—one that effectively dominated American geological surveys—along with a few paleontologists who ensured that New York remained the paleontological heartland of North America for the rest of the nineteenth century. One of the Rensselaer graduates was sure that he had found a living trilobite.

As a member of the United States "Exploring Expedition of 1830," James Eights of Albany was the first American scientist to study the marine animals, landforms, and geology of Antarctica and its surrounding islands. Among his discoveries from the shallow seas around the bleak South Shetland Islands was a peculiar creature that he named *Brongniartia trilobitoides* and illustrated alongside its presumed relative, *Brongniartia boltoni*. *B. boltoni* was a large fossil trilobite that had been described from ancient Silurian shales scooped up by tarriers digging the Erie Canal near Rochester. And if *B. boltoni* was a stony extinct trilobite, then surely the lively trilobitoides must be living members of that ancient clan. The Antarctic animal, however, turned out to be a crustacean—an isopod of the genus *Serolis*. But even today *Serolis trilobitoides* (Eights) is cited as a textbook example of the convergent evolution of isopods and trilobites.

New York paleontology entered a new state-sanctioned phase in the early 1840s when James Hall, the state paleontologist and the best-known graduate of the Rensselaer School, received support from the state legislature to prepare a single volume on the fossils of the state. Hall had greater ambitions, and entirely due to his stubborn determination and ability to browbeat legislators, this volume became the first of no fewer than 13 quarto volumes of the monograph series *Palaeontology of New York*. The series comprised thousands of pages and many hundreds of plates published over the next half century. To pursue his work, Hall amassed huge fossil collections at his laboratory along the Beaverkill in Albany. Of Hall it can truly be said that he never met a fossil he didn't covet. He hired a succession of assistants to collect, prepare, describe, and draw the specimens. Hall himself described many trilobites from the state in his *Palaeontology*, but independently one of his assistants made the class Trilobita his own.

Charles Doolittle Walcott was a young man of 26 when he started to work for Hall in 1876. Born at New York Mills in the Mohawk Valley, he had little formal education but a wealth of practical knowledge about fossils. A few years previously he had sold a collection of trilobites he had compiled from Trenton Falls on West Canada Creek to the Museum of Comparative Zoology at Harvard University for $5000. At night, after he had finished his work for Professor Hall, Walcott polished sections of tightly enrolled specimens of the trilobite *Ceraurus* in an attempt to determine the structure of its infolded limbs. This was enormously difficult—akin to attempting the restoration of an orchid by slicing serially through a rolled-up flower—and, not surprisingly, resulted in inaccurately reconstructed trilobite limbs. Walcott also collected from localities in New York where few trilobites had been known before. He made large collections from deformed Lower Cambrian rocks of the Taconic Mountains and from Upper Cambrian limestones near Saratoga Springs. However, Walcott chafed at the treatment he received from the mercurial Hall, who after a few years dumped him for a younger and more compliant assistant. Walcott benefited by gaining a position in the newly formed U.S. Geological Survey, then became director of the survey, and rose to become the secretary of the Smithsonian Institution. Walch might have named it, but it was Walcott who conceived the trilobite taxon. He was the first to suggest that the class Trilobita (or, as some paleontologists now favor, phylum Trilobita) was a group of arthropods quite distinct from crustaceans.

New York had to relinquish its primacy in matters trilobitic in the early years of the twentieth century as the research focus shifted to other parts of the continent—to the Great Basin of Nevada and Utah, to Newfoundland, to Virginia, to the Upper Mississippi Valley, and to the southern Canadian Rockies. In the 1960s a new crop of paleontologists applying modern paleontological ideas sparked renewed interest in the New York trilobites that had been described a century earlier. Among these scientists were Franco Rasetti, the paleontologist/nuclear physicist from Johns Hopkins University who focused on Cambrian trilobites from the tortured rock of the Taconic region; Harry Whittington, the Woodwardian professor of geology at Cambridge University, who restored the anatomy of pyritized *Triarthrus* from Upper Ordovician shales near Rome; and Niles Eldredge, the paleontologist at the American Museum of Natural History in New York City, whose work ensured the centrality of Middle Devonian *Phacops* (now *Eldredgeops*) in the new evolutionary model of punctuated equilibria.

So the authors of this book are to be congratulated for bringing us full circle as they update and enhance the story of the trilobites of New York, bringing new visions and fresh perspective to these wonderful creatures.

Rolf Ludvigsen, *Head*
Denman Institute for Research on Trilobites
Denman Island, British Columbia, Canada

Preface

Scientists estimate that life on Earth may have begun as early as three *billion* years ago. For much of its history, however, life was confined to single-celled bacteria. Stromatolites, layered mounds of sediment trapped by mats of blue-green cyanobacteria, are the predominant fossils for nearly two billion years of geologic history. Finally, in the late Precambrian Ediacarian Period, about 570 million years ago, enigmatic soft-bodied forms of multicellular life first appeared as impressions in sandstone.

About 543 million years ago life made a further profound change in direction. A sudden burst of new organisms with hard skeletons in the fossil record has been called the "Cambrian Explosion." The earliest fossil skeletons were simple tubes, probably made by worms; shell material clearly made by complex living creatures such as mollusks appeared about 540 million years ago. Then, beginning about 520 million years ago, highly sophisticated skeletons of trilobites, early representatives of Earth's most abundant complex animals—the Phylum Arthropoda—appeared in marine strata worldwide. The trilobites not only appeared dramatically in the fossil record but for millions of years they dominated it.

Trilobites are the quintessential archaic marine animals. Few if any other invertebrate fossils have attracted more attention from paleontologists and fossil collectors than these ancient arthropods, distant relatives of today's crustaceans and insects. Paleontologists have learned a great deal about trilobites because they were ubiquitous in the oceans and seas of the early Paleozoic Era and because they possessed readily preserved hard skeletons. In the mid to late 1800s lithographed images of trilobites became symbolic of the rapidly developing field of paleontology in New York State as well as in England, two hotbeds of early research by serious amateurs and professional scientists when interest in the nascent field of geology was first beginning to burgeon. The beautifully preserved, segmented exoskeletons of trilobites—in shades of saddle brown and blue gray to black—are truly spectacular objects, but perhaps above all it is the well-developed, commonly compound eyes of trilobites that have made them attractive to paleontologists and lay persons alike. Trilobites were certainly among the first organisms to form relatively clear images of their world.

New York State is and has long been a magnet for trilobite hunters. Historically, New York was of central importance in the study of Paleozoic fossils, and New York's trilobites were among the first illustrated fossils in North America. New York strata are the source of many specimens accepted worldwide as the best of their kind. These fossil remains are actively sought, studied, and traded. With its extensive shale deposits, New York is a particularly rich source of trilobites, many of which are shown for the first time in this volume. Many outstanding localities in New York State, from the majestic Ordovician limestone bluffs of Trenton Falls to the Silurian beds in the great gorge of Niagara River to the Devonian shale cliffs of Lake Erie, continue to yield abundant and spectacular trilobite fossils. New York strata have also yielded more trilobites with preserved appendages and other "soft parts" than almost any other region of the world. The rarity and aesthetic beauty of complete outstretched or enrolled trilobites gives trilobite fossils special value to collectors. Spectacular, ornate trilobites from New York, ranging from a few millimeters to nearly a half-meter in length, are featured in museums all over the world; some extraordinary examples are prized by collectors and have been sold for thousands of dollars.

Yet despite the fame of New York State's trilobites, no recent text has attempted to document comprehensively these remarkable fossils. With a little effort one can find trilobites in New York State rocks ranging in age from the time of their earliest occurrence in the Early Cambrian up to their last time of major abundance in the Devonian Period, about 370 million years ago. Thus, although New York strata do not document the entire evolutionary history of trilobites, the abundant, high-quality material available in this area offers a rare opportunity to discover and study these intriguing representatives of early life history.

Trilobites of New York is intended to be a nearly complete compilation of the trilobite species found in New York: a review of the biology of the trilobite; insight into trilobite preservation in the rocks; a short course on the Paleozoic geology of New York, emphasizing trilobite-bearing strata; and a collection of high-quality images of representative New York trilobites. The book is not and was never intended to be a field guide or identification matrix to trilobites. As such, there is no specific locality information, although many readers will find the photographs very useful in identification and in differentiating similar species.

This work started more than 20 years ago as an attempt by Tom Whiteley to compile illustrations of the trilobites found in New York. Although New York has a history of trilobite discovery and research since 1824, references on trilobites are scattered and often not available except in the libraries of large universities. The only collective works on New York trilobites were the classic volumes by James Hall, and the last of these was published in 1888. It soon became apparent that the New York Paleozoic exposures are too varied for one person to really understand all the trilobites and their locations. Hence, Gerald Kloc became involved with this project for his knowledge of the Silurian and Devonian exposures and their trilobites and for his contacts in the amateur community.

As in all research programs, background literature is an essential starting point. There are a few texts on trilobites that provide more in-depth information on the animal itself than we include. The works of Johnson (1985), Levi-Setti (1975), Ludvigsen (1979b), and Whittington (1992) are good references for additional reading. Every trilobite publication has references, and these references lead to other publications, which in turn lead to more references and so on. Hundreds of publications were examined, and the relevant ones were put into a database. Fieldwork was also carried out in the more promising exposures. However, this fieldwork resulting from literature surveys was limited and nowhere near the hours and days of work spent by many professionals and amateurs in the field collecting each individual specimen. Specimens of the quality illustrated in this book are uncommon, even rare. A number of the trilobite specimens are unique in their quality of preservation and preparation, and very few like them exist anywhere.

The trilobite collections in a number of major northeastern United States museums were examined carefully, and specimens that were unusual or of high quality were photographed and the accompanying information recorded. These museum collections represent the efforts of dozens of individuals over a period of more than 150 years. A number of amateur collectors made their specimens available for photography, which was helpful as the best and most complete material is often not in a museum. In a few cases research paleontologists made their photographs of uncommon material available for reproduction. The photographic procedures are provided in Appendix B, but in general the photographs were taken in a museum or in a laboratory environment. Often the specimens were whitened with ammonium chloride to bring out detail. The images were then scanned into a digital file, and all of the final preparation of images was done on computer. No information was added or subtracted from the digital image at any time. Of the thousands of photographs, only about 200 could be selected for the book. Selection was made on the basis of quality, rarity, and representation of the material present in the New York rocks.

Anyone who collects fossils of any kind understands that finding a choice specimen is only part of the process. Most trilobites are embedded in a matrix of shale or limestone, and to really appreciate their quality one must prepare them by removing the stone from the part of the trilobite that is to be displayed. Most of the illustrated specimens were prepared or "touched up" by Gerald Kloc. Even some museum specimens were worked on, with the museum's permission, to bring out the details concealed by matrix.

As work on this project progressed, it became clear that to compile data and images on trilobites was not enough. A listing of nearly 500 separate species, while interesting to a few specialists, is not very helpful to the collector or the student. To be really useful, we needed to include information on why the trilobites are found where they are and how their preservation comes about. To this end, Carlton Brett provided a review of trilobite taphonomy, as well as an overview of the geological history of the New York Paleozoic.

The remaining issue concerned the intended audience, or who is expected to read the book. Dr. Warren Allmon of the Paleontological Research Institution suggested that high-school earth science teachers represented the right level for content, as this level would provide information useful to the collector, teacher, and student. Dr. Allmon also made the first contacts with Cornell University Press.

As already mentioned, it was necessary in the course of this work to visit many of the major natural history museums in the northeast United States. The American Museum of Natural History, Museum of Comparative Zoology (Harvard), Peabody Museum (Yale), New York State Museum, National Museum of Natural History (Smithsonian), Rochester Museum and Science Center, Royal Ontario Museum, and Paleontological Research Institution were visited, some many times, and their collections carefully examined. The cooperation of the museums' management and their collections managers in particular was unreserved. Without their help, this work would not have been possible.

The cooperation and assistance of Fred Collier, Jan Thompson, Ed Landing, Niles Eldredge, Janet Waddington, Tim White, Wendy Taylor, Paul Krohn, Fredrick Shaw, Stephan Westrop, and George McIntosh were all important to this work. Fred Collier, in particular, while at the United States National Museum, greatly influenced the early direction of these efforts with his professionalism and enthusiasm.

We gratefully acknowledge the collectors who made their specimens available for photography and in some instances donated these specimens to a museum. They are William Pinch, Kent Smith, Lee Tutt, Paul Krohn, James Scatterday, Gregory Jennings, Fred Barber, Kym Pocius, Sam Insalaco, Kevin Brett, Steve Pavelsky, Tod Clements, Douglas DeRosear, Fred Wessman, Gordon Baird, and William Kirchgasser.

Rolf Ludvigsen, Nigel Hughes, and George McIntosh read early drafts of the book and made many valuable suggestions. Warren Allmon made the first contacts with Peter Prescott, science editor of Cornell University Press, who was instrumental in giving the book focus and helped turn what was a collection of information and pictures into something publishable. Alyssa Sandoval, Lou Robinson, and Candace Akins also provided valuable assistance. To all we express our gratitude.

Thomas E. Whiteley
Gerald J. Kloc
Carlton E. Brett

▼ Trilobites of New York

1 Background Information

Historical Notes

E. Lhwyd provided the first record of trilobites in the literature in 1698, with the publication of plates depicting two Welch trilobites, identified as fish. In 1771, J. I. Walch originated the use of the name "trilobite" as a distinct class of animal. L. D. Herrmann, however, used the term *trilobus* as part of the name for a trilobite fossil, as early as 1711. (For the very early trilobite references, see the publications by H. Burmeister (1843, 1846).)

In 1822, C. Stokes was the first to describe North American trilobites, with *Asaphus* (now *Isotelus*) *platycephalus* from Canada. J. E. DeKay provided the first unequivocal description of a New York trilobite, *Isotelus gigas* from Trenton Falls (north of Utica, New York), in 1824. This report was followed by that of *Arctinurus boltoni* by Bigsby in 1825. Of the 40 trilobites described in the classic works by the Philadelphia physician J. Green in 1832 and 1835, most were from New York.

New York State took an early leadership role in North American Paleozoic invertebrate paleontology, in part due to the number of lower Paleozoic exposures within the state and also to the history of the state itself. The early 1800s saw a general expansion westward within the United States. New York participated both by pressing settlement into the rich farmlands of western New York and by aggressively seeking to become the communication route to the nation's Midwest. Roads, canals, and permanent construction were all part of these goals, and all needed building stone to succeed. Limestone was the ideal material both for buildings and for the cement and mortar to hold them together. Thus, small and large limestone quarries became common along the Hudson-Mohawk River corridor. The state government also was concerned about its knowledge of the natural treasures contained within its borders. In 1818 the Lyceum of Natural History of New York was established in New York City, and in 1823 the Albany Lyceum of Natural History was founded. *Isotelus gigas* was described in the *Journal of the Lyceum of Natural History* (New York). The *Isotelus* fossils from the Trenton Limestone of central New York, particularly Trenton Falls, were long known and collected for sale by the local residents and became part of many early natural history collections. The *Arctinurus* specimen described by Bigsby was first found during the digging of the Erie Canal locks in what is now Lockport, New York.

In 1836, New York began a general natural history survey of the state, including its geology and mineralogy, and at the same time formed the New York Geological Survey. That same year the first state paleontologist was named, T. A. Conrad, a Philadelphia conchologist.

For the geological survey the state was divided into four districts, and the results from each district were published as separate volumes. Starting in 1842, the first of these was published. The *Geology of the Fourth District of New York* (1843) was the beginning of the career of New York's second and most well-known state paleontologist, James Hall. Hall was also the principal author of the eight-volume *Palaeontology of New York*, issued between 1847 and 1894. Volumes 1, 2, 3, and 7 (written with J. M. Clarke) contain significant trilobite information and are the primary references for early trilobite work in New York. Hall had a number of assistants who began their careers with him: F. B. Meek, F. V. Hayden (future director of the United States Geological Survey), C. A. White (future state paleontologist for Iowa), W. A. Gabb, R. P. Whitfield, C. Calloway, C. D. Walcott (future director of the United States Geological Survey and the secretary of the Smithsonian Institution), C. E. Beecher (future professor at Yale University), and J. M. Clarke (future state paleontologist for New York). Hall was a difficult man to work with and consequently had a high turnover in assistants. It is claimed that Hall took on some of his assistants to gain access to their personal fossil collections (Yochelson, 1987).

The turn of the century introduced additional important contributors to the knowledge of New York trilobites, such as R. Ruedemann and P. Raymond. By the early twentieth century, New York was no longer a major area for new trilobite discoveries, as the focus had shifted westward with the general expansion of the United States. However, significant contributions are still made today, for example, in the understanding of some less well-described areas such as the Middle Ordovician Chazy Group and the Cambrian in eastern New York. The general shift in emphasis from discovery to understanding still keeps New York trilobites in the limelight. In later chapters we will point out the importance of New York trilobite beds in our understanding of trilobite biology and fossil preservation.

Trilobite Names

Trilobites are named using the rules of zoological nomenclature published in English and French in the *International Code of Zoological Nomenclature* (ICZN). This code is used worldwide by all scientists irrespective of the language of publication. In dealing with the names of trilobites (or any other kind of organism), it helps to understand the basic rules, how names come about, and how they can change over time.

For example, and as mentioned already, in 1824 J. E. Dekay first reported *Isotelus gigas* Dekay, 1824, a name that remains unchanged to this day. The first name, with a capital first letter, *Isotelus*, is the **genus** name and refers to a group of animals in which similar characteristics indicate a close evolutionary relationship. Genus names must be original and not used for any other grouping of fossil or living animal. The second name, *gigas*, which is not capitalized, is the **species** name and ideally should refer only to a coherent group of interbreeding populations. Species names do not have to be unique, except within the same genus. There can be only one *gigas* within the genus *Isotelus*, but species names can and do reoccur in different genera (plural of *genus*). Accordingly there are seven different New York trilobite genera with the species name *trentonensis*. The proper name, genus and species, is italicized in print. In descriptive literature the name of the describing author (Dekay in our example) and sometimes the date of publication (1824 in our example) follow the name. Changes to the genus name are not uncommon, and in these cases the name of original author and date are given in parentheses.

The **etymology** of trilobite names often refers to a **morphological** feature. *Isotelus* means "similar" (*iso-*) "end" or "tail" (*-telus*), referring to the similarity between the head and tail of this species, and *gigas* means "large" or "giant," referring to the large size of this species compared to most trilobites. Names are usually derived from Latin, ancient Greek, latinized ancient Greek, or Indo-European and are selected by the describing author. Names are often latinized, and the gender of the species follows that of the genus. In other words, if the genus name is changed, the ending on the species name must sometimes change to agree with it in gender.

Genera, that are considered to be closely related on the basis of shared characteristics, are grouped into **families.** Because families are larger and more distinctive, it is often easier to determine the family to which a new trilobite belongs than to determine its genus. New genera are constantly being erected, and trilobite species are frequently moved around as descriptive methodology becomes more sophisticated and new classification standards are adopted.

Families are further collected into **orders**, again based on inferred evolutionary relationships. There are currently eight orders in the **class** Trilobita of the **phylum** Arthropoda (Kaesler, 1997). There are additional taxonomic relationships rarely used herein, such as superfamily and suborder. Family names always have the suffix -idae; superfamily names, -acae; suborder, -ina; and order, -ida.

The common New York trilobite *Eldredgeops rana* is taxonomically described as follows:

phylum
Arthropoda
class
Trilobita
order
Phacopida
family
Phacopidae
genus
Eldredgeops
species
rana

To take another, somewhat more complex example, *Eldredgeops rana* (Green, 1832) was first named *Calymene bufo* var. (short for *variety*) *rana* by J. Green in 1832. (Green in the same publication had described *Calymene bufo* from a poorly preserved phacopid specimen in a float boulder. The condition of the fossil was too poor to use for determining clear relationships and the name subsequently was abandoned.) In 1860 Emmons changed the species to *Phacops bufo* because of the greater similarity of the genus to the European genus *Phacops* than to *Calymene.* Hall (1861) first called the common trilobite from the Hamilton shales and limestones, *Phacops rana.* Green's name and publication date now appear in parentheses because of the change in his original genus designation. This story is further complicated by the assignment of some new phacopids to subspecies of *Phacops rana* such as *Phacops rana milleri* Stewart, 1927 and *Phacops rana rana* (Green, 1832). (The designation of subspecies is not often used with trilobites because strictly speaking the term implies geographically separated populations that could interbreed given the opportunity. This is nearly impossible to determine from fossils, and most authors prefer the single species names such as *Phacops rana* and *Phacops milleri.*) Struve (1990) re-examined the New World "*Phacops*" species and found them significantly different from the type species *Phacops latifrons* that was originally described from the Devonian of Germany. (The *type species* is the single species used to describe and define the genus.) He erected a new genus, *Eldredgeops*, with *Eldredgeops milleri* as the type species. The very familiar former *Phacops rana* is thus now properly referred to as *Eldredgeops rana*.

Species names are never changed by later authors, not even to correct spelling errors. The exceptions to this are if the name has already been used for a closely related animal in the same genus or if the same animal has been named by another person in an earlier publication. In almost all cases, priority is with the name given by the first author.

The number of trilobite species and genera is constantly increasing due to both new finds and redescriptions of previously collected material. There are also differing approaches to the concept of species. Some authors view speciation on the basis of small external changes and tend to propose new species based on these differences. Others view many small external differences as within the normal intraspecies variation and include a wider variety of specimens within a single species. The concept of species is not unambiguous in extant animals, and in the case of fossils morphological features are usually all there is available for evaluation. Statistical evaluation of fossils using measurements of key features is often currently used to define intraspecies variability, and the comparison of derived (uniquely shared) characteristics between closely related species is used to determine their evolutionary relationship. **Systematics** is the study of the similarities and differences in organisms and their related species. A. B. Smith (1994) presented an excellent in-depth review of systematics for the fossil record.

The rules of zoological nomenclature now require that an author, when describing a new species, must designate a specimen that clearly exemplifies this new species. This specimen should then be deposited in an appropriate public collection, such as a museum. This one specimen is called the **holotype**. Other specimens of a reference group (the "type series" or *hypodigm*) from which the holotype was chosen are called **paratypes**. In the 1800s and early 1900s authors often illustrated their new specimens but failed to designate a single type specimen and its repository. Later authors in referring to these specimens, provided they could be found, consider the members of the type series to be **syntypes**. The term **cotype** is also seen in collections; it is a synonym for syntype or paratype and its use is discouraged by the ICZN. Should an author need to choose a single specimen from the designated syntypes to be the single species representative, then this specimen becomes a **lectotype**. Other specimens in the original series become **paralectotypes**, and the term *syntype* can no longer be applied. If the species has no original type specimens that can be found and a sufficient taxonomic purpose is present, an author may designate a specimen to represent the type for the species. This specimen is called a **neotype**. If at all possible, the neotype should be from the same location and horizon as the originally described material. The description and name must be publicly issued as a permanent scientific record and available in multiple, identical copies. All these rules are thoroughly spelled out in the ICZN (1985).

Other type designations are commonly used in collections but do not bear ICZN recommendation. **Hypotype** is a specimen that was referred to, usually in publications, to extend or correct the knowledge of a species. **Topotype** refers to a specimen from the type locality, and **plesiotype** refers to specimens very close to the type. **Plastotype** is an artificial cast of the original type.

A difficulty one often encounters with trilobites, as well as other fossils, is that species were originally named when only a partial specimen was available. There are a number of instances where, for example, the pygidium of an uncommon trilobite bore one name and the cephalon a different one. Also in the 1800s trilobites were often identified with the same names as species from other locations but of the same geological age. A number of New York trilobites, for example, were given the same names as species from the Midwest, particularly Ohio, as well as some from Europe. Some of these names are only now being corrected as careful studies are made.

Another interesting situation is that early in the twentieth century, scientists going through museum collections saw differences in specimens and gave them new names, or listed them as subspecies, by noting the name on a label and leaving it with the specimen. These "museum label" names are commonly encountered with the Ordovician trilobites at the National Museum of Natural History (Smithsonian or USNM). Museum label names are not recognized by the ICZN and have no priority or recognition, except when subsequent authors choose to use them to name specimens. Later authors sometimes recognized these unpublished names with the designation "MS" (for "manuscript"). The official date for the name is when it is published, however, not when the museum label was made. An example of such a name is *Isotelus walcotti* Ulrich in C. D. Walcott, 1918. In this case E. O. Ulrich saw differences between the specimens from the New York Trenton Limestone, which were being called *Isotelus iowensis*, and the authentic trilobite from Iowa. He named the New York species *I. walcotti* on a museum label, and C. D. Walcott recognized the name in a subsequent publication in 1918. There is no specific rule or protocol for the recognition of museum label names, and it is solely to the discretion of the publishing author whether the name is recognized or not. Walcott chose to recognize Ulrich; thus it is appropriate, but not required, to add his name to the final formal trilobite name.

2 The Biology of Trilobites

Trilobites are the earliest unambiguous arthropods found in the fossil record, evidently because they were the first arthropods to develop the **mineralized** skeleton necessary for frequent preservation. Modern arthropods have organic **exoskeletons** that are often strengthened with minerals such as calcite. In contrast, trilobites had an exoskeleton composed primarily of calcite, although it undoubtedly was modified with organic materials or other minerals. The appendages, legs, and antennae were probably an organic material like the chitin of extant arthropods, which is only preserved when replaced by minerals under special conditions (see Chapter 3). None of the organic skeletal materials have survived directly in the fossil record. The word **arthropod** means "jointed foot or leg," and the phylum Arthropoda includes trilobites, crabs, lobsters, spiders, horseshoe crabs, centipedes, millipedes, and the largest of all living animal groups, the insects. This large group includes animals with strong external skeletons that must be shed periodically to enable physical growth. Many also go through **ontogenetic** phases in their growth, which can include free-ranging near-microscopic larvae.

This chapter is divided into four parts: the external skeleton or exoskeleton, the growth phases or ontogeny, the appendages and internal anatomy (the unmineralized or soft body parts), and the mode of life of trilobites. This order roughly corresponds to the level of knowledge about the biology of trilobites, as most is known about the exoskeleton and least is known about the life-mode. The last part, life-mode, is inferred from knowledge of trilobites in the fossil record and observations of living arthropods. All of the illustrations were chosen from trilobite genera found in New York.

Exoskeleton

The word *trilobite*, freely translated from Latin, means "having the nature of three lobes." The name refers to the three lengthwise, lateral parts or lobes of the trilobite body (Figure 2.1A), not the three parts making up the body—the **cephalon** or head (Figure 2.1D), **thorax** (Figure 2.1E), and **pygidium** or tail (Figure 2.1F). The central of the three lobes is referred to as the *axial lobe* (Figure 2.1B) and the side lobes of the thorax and pygidium, as the *pleural lobes* (Figure 2.2C).

Trilobites and other arthropods probably evolved from annelid wormlike ancestors. This is reflected in their segmented body and the observation that each segment, whether fused together or jointed, carries a pair of appendages. The segments in the front of the body are fused to form the cephalon, and those in the rear of the body are fused to form the pygidium. The central segments forming the thorax or trunk were joined by flexible tissue that enabled many trilobites to flex inward to the point of enrolling (Plate 109), but probably did not allow for much, if any, sideway flexing. Some trilobites are found arched or flexed dorsally, which indicates the flexibility of the trilobite articulation. For a more detailed account of the elements of trilobite articulation, see the publications by Whittington (1992), Bergström (1973), Levi-Setti (1975, 1993), Moore (1959), and Kaesler (1997). The work by Moore (1959), *Treatise on Invertebrate Paleontology*, Part O, Arthropoda 1, often is referred to simply as the ***Treatise***. Kaesler's publication (1997) is the first of a three-part revision in progress but only the first part has been released.

Four terms are used repeatedly in describing trilobites. These are **dorsal**, meaning the uppermost surface of the body in the animal's life position; **ventral**, the lower surface of the body;

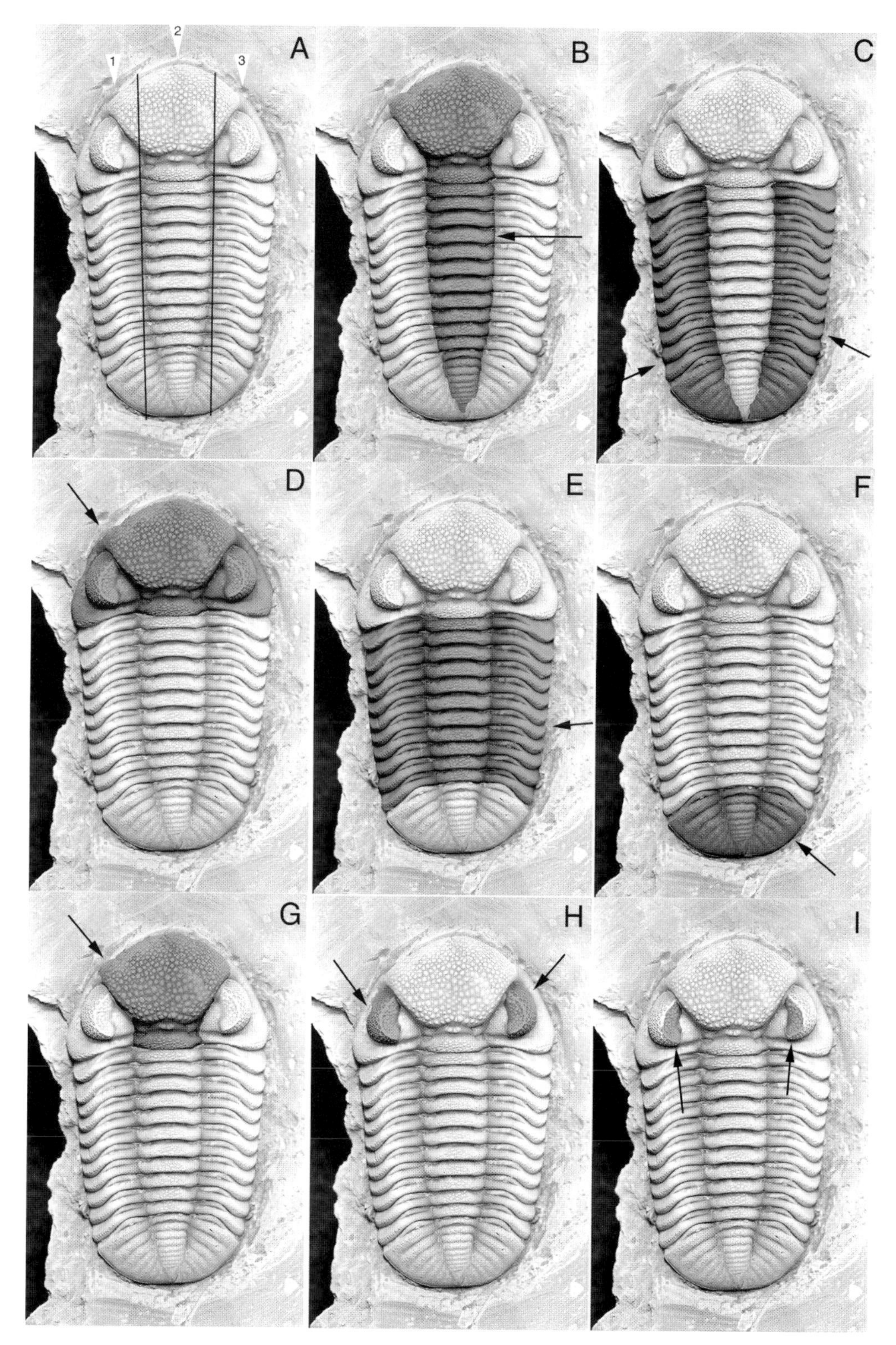

FIGURE 2.1. Trilobite structure using *Eldredgeops rana* (PRI 49656, whitened). A. The three lobes from which the name trilobite was derived. B. The axial lobe. The axial lobe is subdivided into the thoracic axis and the pygidial axis. C. The pleural lobes. These may be further defined as thoracic pleurae and pygidial pleurae. D. The cephalon or head of the trilobite. E. The thorax. F. The pygidium or tail. G. The glabella. H. The eyes. I. The palpebral lobes on top of the eyes.

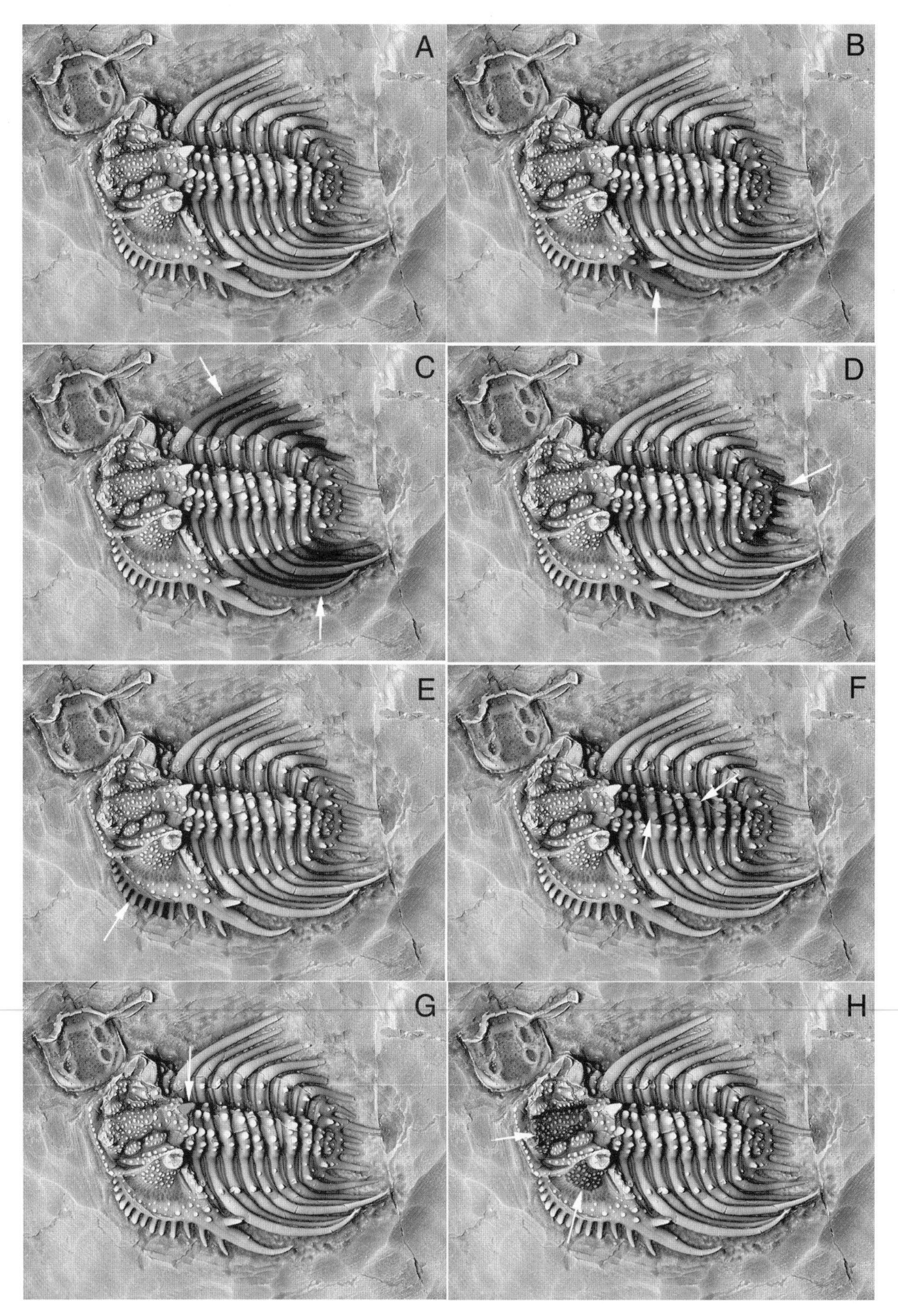

FIGURE 2.2. Trilobite structure using *Kettneraspis tuberculata* (GJK collection, whitened). A. The exoskeleton without the right free cheek. B. The genal spine extending off the free cheek (arrow). C. Lateral thoracic pleural spines (arrows). D. Pygidial spines (arrow). E. Anterior cephalic spines (arrow). F. Axial nodes, raised areas on the thoracic axis (arrows). G. Axial occipital node (arrow). H. Pustules, small, randomly scattered raised areas (arrow).

anterior, toward the front; and **posterior**, toward or at the rear of the body part being described. *Dorsal* and *ventral* are absolute terms in the sense that the dorsal and ventral surfaces are the same regardless of the part being described or its orientation. Anterior and posterior relate to the particular part or parts being described. Thus, a **suture**, or inflexible joining, might be anterior to one body part and posterior to another.

The hard mineral exoskeleton, also called the **cuticle**, is a complex structure. The external surface is variously ornamented with **ridges**, **terrace lines**, **nodes** (Figure 2.2F, G), **pustules** (Figure 2.2H), **tubercles**, and **spines** (Figure 2.2B, C, D). Terrace lines are common on the trilobite exoskeleton and are described in some detail by Miller (1975). Nodes are discrete, rounded, raised areas on the dorsal exoskeleton, which are usually bilaterally symmetrical (Plates 45 and 134). Pustules, on the other hand, are raised areas that are more frequent on the surface and also tend to be more randomly distributed (Plates 24 and 25). Large pustules are often called *tubercles* (Plate 27). There commonly are channels from the surface of these structures to the interior, and it is believed that these channels served for sensory capability, enabling the trilobite to sense currents and chemical changes to the environment (Figure 2.6). Some of these perforations also may have been follicles for sensory hairs. Dorsal spines are the physically extended equivalents to nodes (Plates 46 and 128). Spines, which extend or radiate from the edges of body parts, will be discussed later.

The cephalon is the most complex and the most important part of the trilobite to be understood by the student or collector because it is often the most easily recognized evidence for trilobites in the rocks. The center of the cephalon is the **glabella** (Figure 2.1G, 2.3E), a raised portion with characteristics often important to the recognition of trilobite species. Although the glabella is treated here as separate from the **occipital ring** (Figure 2.3F, the lobe labeled LO), the terminology used in the *Treatise*, most trilobite workers today include the occipital ring as part of the glabella (this latter protocol is used in the Kaesler (1997). Laterally outward from the glabella are the cheeks or **genae** (singular, *gena*). These areas usually have sutures (Figure 2.5) that separate them into free cheeks (Figure 2.3C) and fixed cheeks (Figure 2.3D), called **librigenae** and **fixigenae**, respectively. The librigenae usually separate from the cephalic area on molting, and the fixigenae remain permanently attached. The glabella, together with the fixagenae, is called the **cranidium** (Figure 2.3B). Many trilobite species, particularly in the Cambrian, are differentiated on the basis of details in their cranidia.

The posterior part of the glabella, the occipital ring, is a raised portion almost always separated from the main body of the glabella by a groove. Some occipital rings have a central prominently raised area, node, or spine that is very characteristic for particular genera or species. Spines are often overlooked because of the ease with which they are broken off and lost during the process of removing the trilobite from the rock.

The glabella is the dorsal covering of the stomach of the trilobite. It usually has lateral grooves in its surface known as **lateral glabellar furrows** (Figure 2.3G). The furrows rarely cross the surface completely but can be deep, forming prominent areas on the glabella called **lateral glabellar lobes** (Figure 2.3F) or simply, glabellar lobes. Because of the use of furrows and lobes in trilobite identification, they are designated starting from the occipital ring and occipital furrow. The furrow or **sulcus** separating the occipital ring from the glabella is labeled **SO**; the next most anterior glabellar furrow is **S1**, and so on. (The *O* in *SO* and *LO* refers to "occipital" and is not a zero.) The lobes are similarly designated, with the occipital ring called **LO**, the lobe directly anterior to it **L1**, and so forth. It is not always easy to distinguish the most anterior lateral lobes because the furrows can be very faint; thus, the most anterior portion of the glabella is called **La**, for anterior glabellar lobe. The furrow separating the glabella from the cheek area (Figure 2.3H) is the *glabellar furrow*. On many trilobites the edge of the cephalon is distinctive (Figure 2.3I) and is called the *border*.

The cheeks generally bear the eyes of the trilobite (Figure 2.1H). (There are eyeless, presumably blind, trilobites but these are an exception and will be noted in the specific descriptions.) The eye is often prominently raised, actually being on a stalk, in a few species. The area between the eye and the glabella is called the **palpebral area** (Figure 2.1I) and can have its own furrows and lobes.

The surface of the eye has multiple lenses to form images, and this surface can be very distinctive. Eyes with a closely packed optical structure and a smooth, continuous outer surface (cornea) are called **holochroal** (Figure 2.4A). Although they have a smooth appearance, these are compound eyes, similar to those in some modern insects. Eyes with discrete individual lenses, each with its own corneal surface, are called **schizochroal** (Figure 2.4B). In addition, the lenses in schizochroal eyes are separated by a thick interlensal **sclera**. There is a third type of eye, resembling the schizochroal eye, found in the family Pagetidae, called **abathochroal** (Jell 1975). Abathochroal eyes lack the deep interlensar scleral projection and the intrascleral membrane of schizochroal eyes.

Holochroal eyes are found in the majority of trilobites, while the schizochroal eyes are found only in the suborder Phacopina, which arose in the Ordovician and disappeared by the end of the Devonian. Phacopins are well represented in New York, and this type of eye structure is often observed. The number of optical elements in schizochroal eyes is usually less, often far less, than that in holochroal eyes.

In a series of elegant experiments, Towe (1973) demonstrated that the eyes in at least two trilobites were single crystal calcite oriented to give high-quality imaging. Towe mounted the eye surface of *Eldredgeops rana* and an *Isotelus* species and demonstrated that each facet of a holochroal eye (*Isotelus* species) and each lens of a schizochroal eye (*E. rana*) gave an

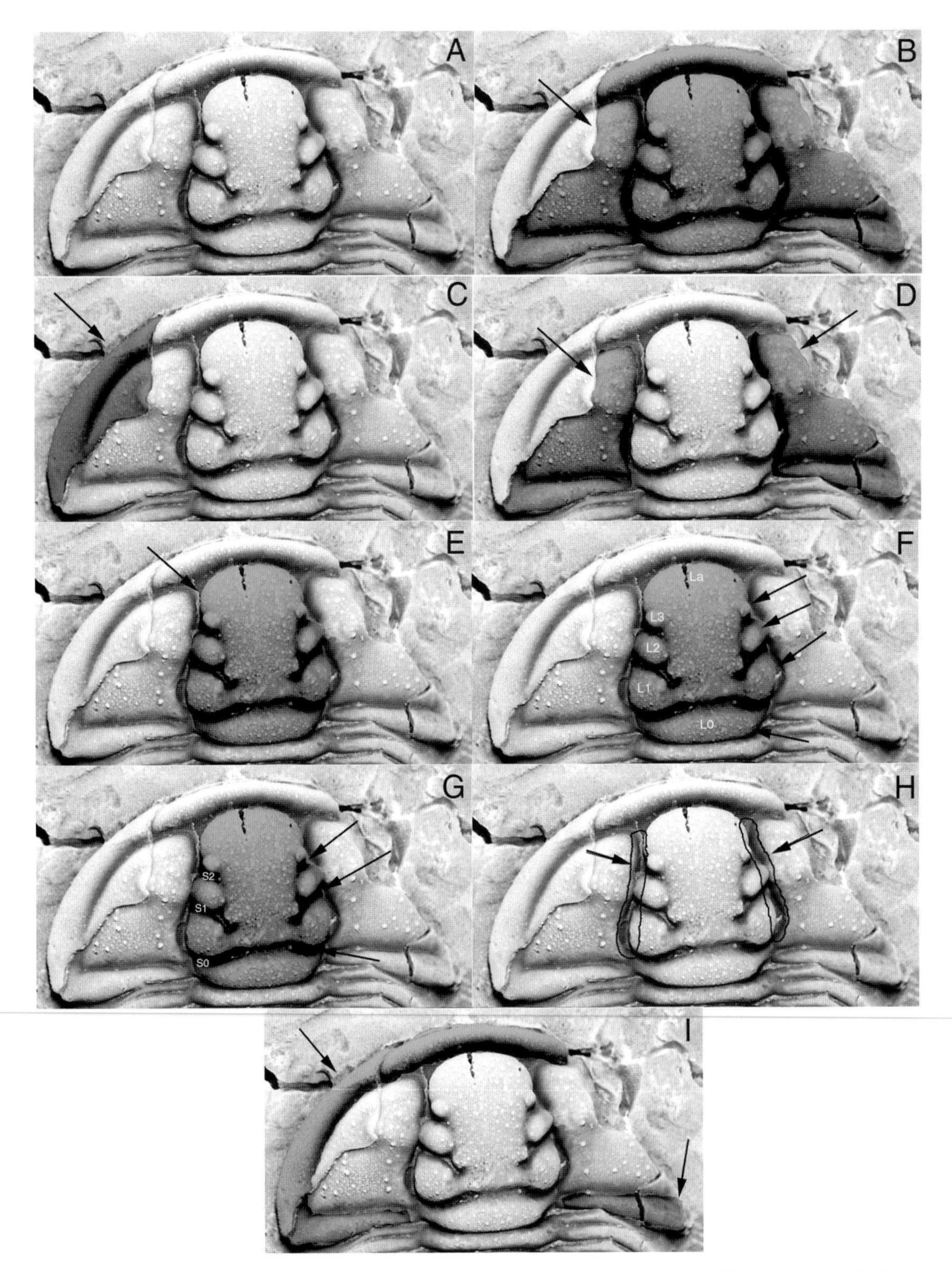

FIGURE 2.3. The structure of the trilobite cephalon using *Calymene* species (S. Insalaco collection, whitened) (cephalon only). A. Cephalon lacking the right free cheek. B. The cranidium (arrow). C. Free cheek (arrow). D. Fixed cheeks (arrows). E. The glabella (arrow). F. The glabella, with the glabellar lobes numbered (arrows). G. The glabella, with the lateral glabellar furrows numbered (arrows). H. The glabellar furrows (outlined, with arrows). I. The cephalic border (arrows).

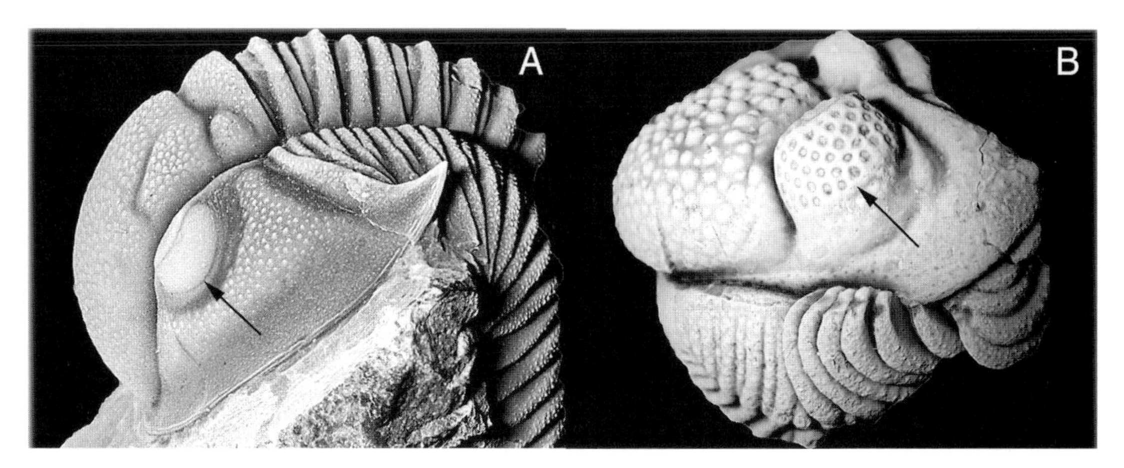

FIGURE 2.4. Trilobite eyes. A. *Monodechenella macrocephala* (G. Jennings collection, whitened) with holochroal eyes (arrow). B. *Viaphacops bombifrons* (GJK collection, whitened) with schizochroal eyes (arrow).

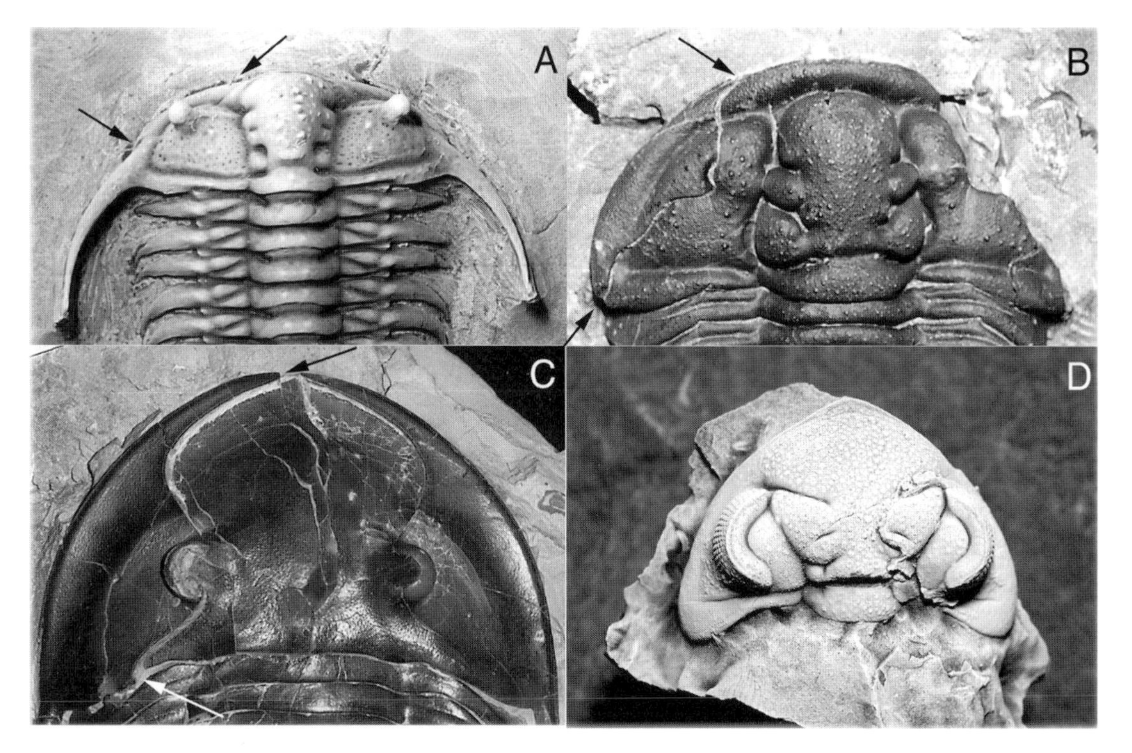

FIGURE 2.5. Cephalic sutures. A. *Ceraurus pleurexanthemus* (GJK collection, whitened) showing proparian cephalic sutures (arrows), both ends emerging anterior to the genal angle. B. *Calymene niagarensis* (S. Insalaco collection) with gonatoparian cephalic sutures (arrows), the posterior end emerging at the genal angle. One free cheek has been lost on the right side. C. *Isotelus maximus* (PRI 49651) with opisthoparian cephalic sutures (arrows), where the posterior end emerges along the rear of the cephalon. D. *Calyptaulax callicephalus* (GJK collection, whitened), a trilobite in which the facial sutures are fused and do not separate upon molting. This is common in the order Phacopida.

inverted image when viewed from the rear, just as a simple glass lens does.

Stürmer and Bergström (1973) studied the internal structure of the eyes of *Phacops* specimens from the Hunsrück slates in Germany. Some of the soft tissue of the trilobites was replaced by the mineral pyrite (i.e., it was pyritized), and X-ray photographs showed clear evidence of optical fibers extending from the lens into the central cephalon. One asteropyge in their study also had similar fibers extending from the lens. These structures were considered comparable to similar structures in extant arthropods. Structures of this kind were reported previously in asteropyges but not confirmed until this study.

On many trilobites there is a medial, glabellar node on the **meraspid** (juvenile) that generally disappears by the **holaspid** (adult) phase or early in the holaspid growth. This node is interpreted as having a visual or light-sensing capability (Ruedemann 1916b, Jell 1975). *Cryptolithus* is a common Ordovician trilobite genus in New York that lacked normal eyes but retained this median glabellar node into the mature holaspis (Plates 163 to 167). A review of the evolution of trilobite eyes with references for detailed reading is given by Clarkson (1975). More will be said about eye function later in this chapter when the possible modes of life of trilobites are discussed.

In post-Cambrian and most Cambrian trilobites there is a suture running across the upper edge of the eye, separating the lens surface from the palpebral area. This suture or separation continues to the edge of the cephalon in both directions and results in a portion of the cheek area (librigena) that can separate from the cranidium. This **facial** or cephalic **suture** separates the free cheek (librigena) from the fixed cheek (fixigena).

In most Cambrian trilobites an additional suture runs below the eye and joins the facial suture, forming the circumocular suture so that the visual surface separates on molting and is lost. It is also common in articulated calymenids for the visual surface to be missing. This absence suggests that a suture surrounded the visual surface or that it was weakly mineralized, if at all.

The anterior margin of the cephalon is rounded and the posterior margin is less curved laterally, resulting in the formation of a corner at the posterolateral extremity known as the **genal angle**. Depending on the species, this structure varies from a blunt well-rounded angle to a **genal spine** (Figure 2.2B) that extends back along the body. One end of the facial suture crosses the genal area and emerges on the anterior margin of the cephalon, and the other end emerges either in front or behind the genal angle. If both ends of the sutures emerge anterior to the genal angle, it is known as a **proparian** (Fig 2.5A) suture; if one end emerges on the posterior cephalic margin, the suture is called **opisthoparian** (Figure 2.5C). As one might expect, there are trilobites where the suture emerges precisely on, or very near, the genal angle, and that condition is called **gonatoparian** (Figure 2.5B). In some trilobites the cephalic suture (Figure 2.5D) is fused and does not open on molting.

The area immediately in front of the glabella varies from nearly nonexistent to a broad border platform or brim. In some families, particularly the calymenids (Figure 2.3I), the shape of this area is important to the identification of genera.

Sometimes this anterior border has a lateral furrow. The area between this preglabellar furrow and the glabellar furrow is the **preglabellar field** (Plates 139 and 142). The area between the preglabellar groove and the anterior margin is the **anterior cephalic border.**

The thorax is divided into a number of segments that form a highly flexible part of the exoskeleton. The number of **thoracic segments** can range from zero in the unusual Cambrian family Naroidae to 40 or more. Each segment has a central or axial portion and a lateral part on each side of the body called the **pleura** (plural, *pleurae*). The usually grooved pleurae also may extend laterally beyond the body into short rounded extensions called **lappets** (Plates 50 to 56) or even into more extended and pointed **pleural spines** (Figure 2.2C). The distinction between lappets and spines is qualitative, and in some cases it is not clear whether the extensions should be termed *long lappets* or *short spines* (Plates 46 and 48). Other structures can increase flexibility and sometimes tight enrolling, but these are beyond the scope of this book. These structures are covered in detail in the work by Bergström (1973).

The most posterior part of the trilobite is the pygidium (Figure 2.1F). The pygidium, similar to the thorax, has segments with axial and pleural portions. However, the pygidial segments are totally fused so there is no flexibility among the parts. The number of axial segments, the number of pleurae, and the general shape of the pygidium are often diagnostic to species. Pygidia are common in the trilobite fossil record, and these features are very important to the identification of trilobites. The pygidium can have **marginal lappets** or **marginal spines** (Figure 2.2D) as extensions of the pleurae, and sometimes has a central posteriorly directed spine called the **terminal axial spine** (Plate 88). In addition, there is sometimes a narrow featureless area around the margin of the pygidium simply called the **border.** In some genera of trilobites, the pygidium is almost featureless, and when found separate, it looks like nothing more than a dark thumbnail on or in the rock. This finding should not be overlooked as an indicator of the presence of particular species.

On the surface of many trilobites, there are pitted areas that often penetrate the exoskeleton. The exoskeleton of the large trilobite *Dipleura dekayi* is literally covered with pits, seen in Figure 2.6A as small white specks. Figure 2.6B shows a broken area of this trilobite, with the pits as complete perforations through the cuticle. *Isotelus gigas* is similarly covered with pits (Figure 2.6D). Pits such as on these two trilobites may have served a sensory purpose and may have contained sensory hairs. A different kind of pit is seen on the cephalic border of *Cryptolithus* species (Figure 2.6C). These are also perforations but go completely through the border area. Their role is unknown, but they may have served to sense water currents or movement.

The exoskeleton covering the entire dorsal surface of the trilobite sometimes curls under at the edges of the cephalon and pygidium to form the **doublure** (Figure 2.7H, I), a flat terrace around the ventral edges of the cephalon and pygidium. In addition to the doublure, there are two important pieces of the exoskeleton on the ventral side of the cephalon. The **rostral plate,** found on many but not all trilobites, is a continuation of the doublure but is separated from the rest of the cephalon by sutures; it is located directly under the front central part of the cephalon (Plate 65 shows a displaced rostral plate immediately in front of the cephalon). Under the approximate center of the cephalon is

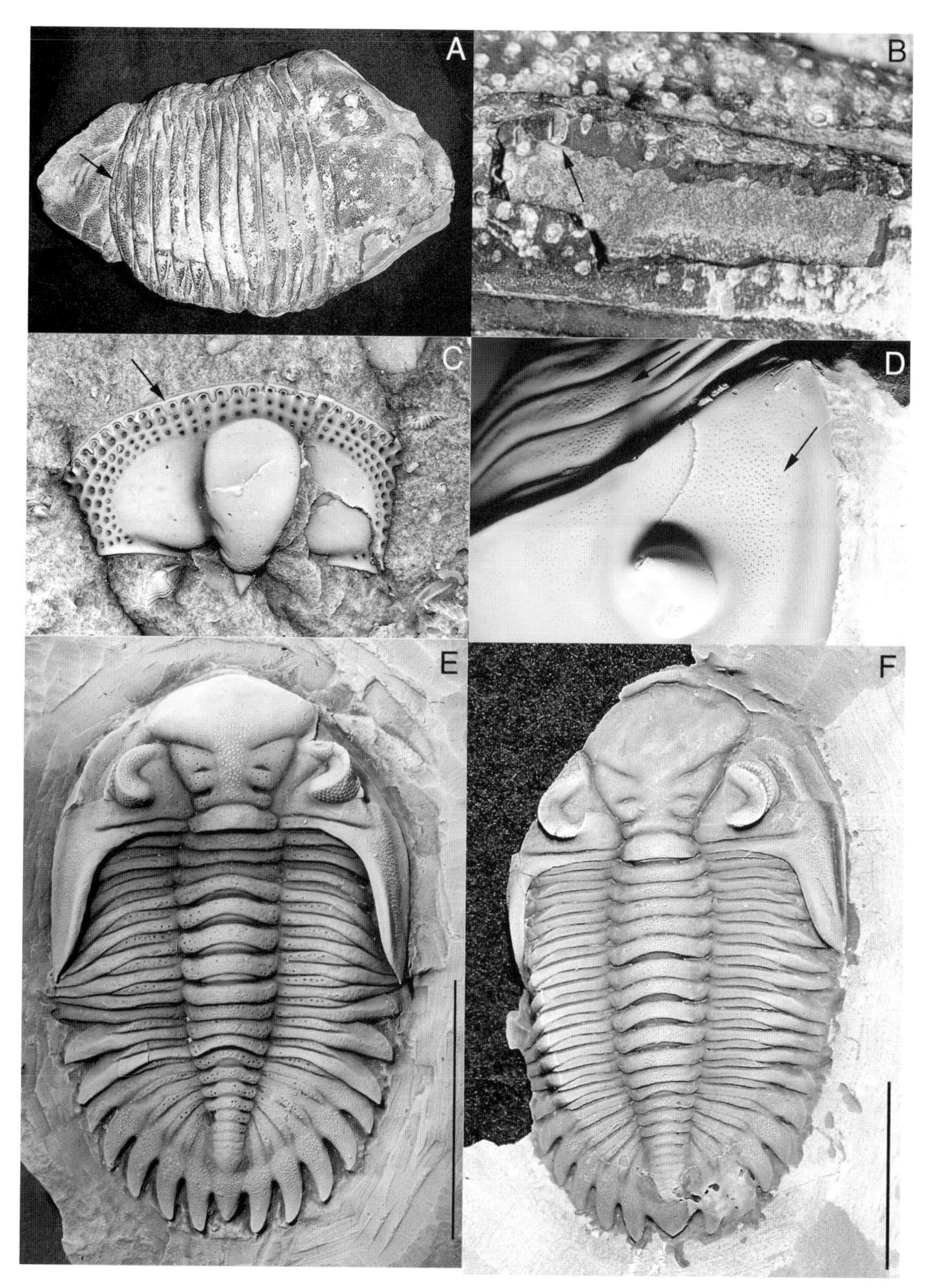

FIGURE 2.6. Exoskeletal pits or circular perforations. A. *Dipleura dekayi* (PRI 49629). White specks over the body are sediment-filled pits. The arrow points to the area shown in B. B. Close-up of *Dipleura* pits, showing that they extend through the exoskeleton (arrow). C. *Cryptolithus lorettensis* (PRI 49657, whitened), showing pits on the cephalic border (arrow). These are actually perforations that extend all the way through the border. D. *Isotelus gigas* (TEW collection, whitened). The exoskeleton is heavily pitted (arrow), a diagnostic character of this species. E. *Greenops grabaui* (F. Barber collection, whitened), showing rows of circular perforations characteristic of the New York astropygids. F. *Greenops?* species (GJK collection, whitened). Unnamed, new?, species of *Greenops* in which the circular perforations are degenerate.

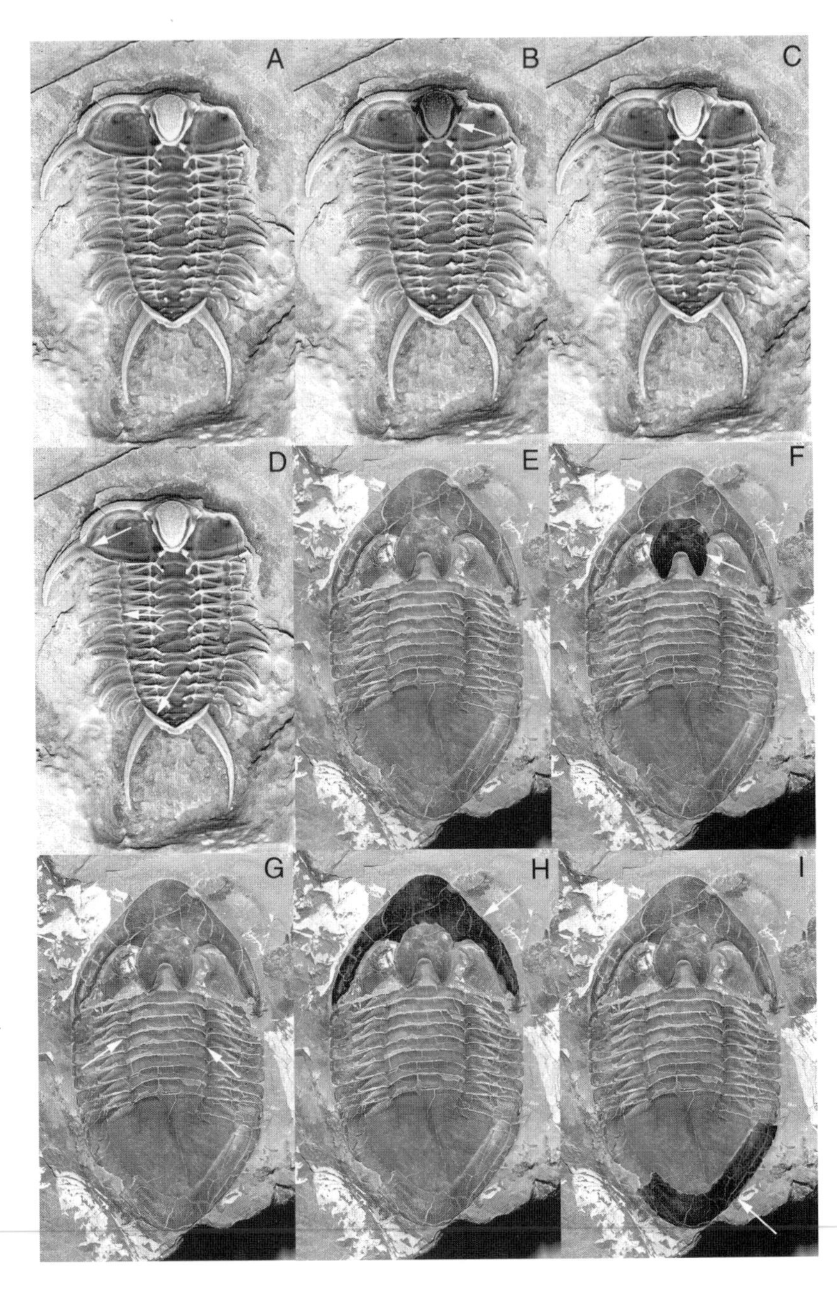

FIGURE 2.7. Ventral anatomy of the exoskeleton. A–D. *Ceraurus pleurexanthemus* (PRI 49658, whitened), prepared to show the ventral exoskeleton. B. The hypostome (arrow). C. Two of the apodemes (arrows). The apodemes are arranged along the ventral surface under the axial furrow. There is a pair of apodemes for each pair of cephalic appendages, two for each pair of thoracic appendages (which also correspond to the number of thoracic segments), and apodemes associated with the pygidial appendages. The often large numbers of pygidial appendages are not as well reflected in prominent apodemes. D. The arrows point out the entrance into the hollow exoskeletal genal spines, pleural spines, and pygidial spines. E–I. *Isotelus* species (Kevin Brett collection), ventral exoskeletal anatomy. A specimen from Canada. F. The hypostome (arrow), the anterior margin and "wings" of which are under the doublure. G. The apodemes (arrows) of *Isotelus*, which are far less prominent than those of *Ceraurus*. Given that these represent muscle attachment, the shape must represent their use and consequently the life-mode of the trilobite. H. The cephalic doublure (arrow). I. The pygidial doublure (arrow), which is incomplete in this specimen.

a rounded plate called the **hypostome** (plural, **hypostoma**) (Figure 2.7B, F). The mouth was at the rear of the hypostome.

The hypostome is a more important feature. For some species it is very robust, distinctive, and a not uncommon part of the fossil record. (In at least one rare Ordovician trilobite, *Hypodicranotus*, the presence of its hypostome is a very good indicator, and one of the only indicators, of its stratigraphic range.) The significance of the hypostome will be pointed out for the individual species. The hypostome is directly under the glabella, and together they form an envelope covering and protecting the stomach. The mouth of the trilobite was at the posterior central notch of the hypostome (Plates 37, 47, 77, 117, 153, and 157).

The attachment of the hypostome is proposed to have significance for the high-order classification of trilobites (Fortey [1990a] using observations made by Fortey and Chatterton [1988]). **Natent** hypostoma are those separated from the cephalic doublure, or rostral plate when present, by a gap and are displaced or absent in most trilobite specimens. **Conterminant** hypostoma are closely joined to the cephalic doublure or rostral plate and consequently are more likely to be present on specimens (Figure 2.7B, F). In both of the above cases, the anterior margin of the glabella is directly above the anterior margin of the hypostome. The third type of hypostome, **impendent**, is when the anterior margin of the hypostome is not close to the anterior margin of the glabella and the anterior margin of the glabella coincides with the cephalic margin. Examples of trilobites with the different types of hypostome attachment are as follows: natent—some proetids; conterminant—*I. gigas* (Plates 153 and 155) and *Ceraurus pleurexanthemus* (Plate 77); and impendent—*E. rana*.

Under the lateral sides of the ventral axial region of the thoracic segments, glabella, and pygidium are thickened areas called **apodemes** (Figure 2.7C, G) or sometimes **appendifers**. The apodemes are believed to have been points of muscle attachment and may provide evidence for the lifestyle of the trilobite. On some trilobites, such as *Ceraurus pleurexanthemus* (Plate 77), the apodemes extend down from the ventral surfaces of the segments to form prominent ridges or posts, yet in others, as illaenids, the apodemal area is fairly smooth. The long apodemes suggest good leverage for the attached muscles and a high level of limb mobility, perhaps for swimming, rapid crawling, or digging.

Ontogeny

Ontogeny is the biological life cycle of an animal; for the trilobite this would be from the presumed egg to the smallest larvae and the various intermediate stages, to the end of its life cycle. Considerably more is known about the adult phase of the trilobite life cycle because the vast preponderance of the fossil record is composed of the pieces of the exoskeleton representing late growth stages. Careful workers have found, however, a significant amount of information related to trilobites' earlier growth stages. Chatterton and Speyer (1997) provide an extensive and current review of the present state of knowledge of trilobite ontogeny.

Trilobites most likely began life as an egg, which was either laid outside the body or hatched within the animal, although there is no unequivocal evidence of trilobite eggs. C. D. Walcott (1877c), in a study of remarkably well-preserved trilobites in New York, found spherical objects within the cross sections of some trilobites, which he identified as eggs. This and earlier reports about trilobite eggs by Barrande in 1852 apparently have not been investigated completely.

The well-defined phases of trilobite growth are labeled the **protaspid**, meraspid, and holaspid phases. Fortey and Morris (1978) described some small exoskeletons from the Lower Ordovician as a pre-protaspid phase of trilobite called **phaselus**. Chatterton and Speyer (1997) also found phaseluses in beds with silicified trilobite remains. It is still not clear whether these are from trilobites or another fossil arthropod.

The first well-defined phases in growth of the trilobite, after the presumed egg and the phaselus, is the protaspid phase. (The name *protaspis* (plural, *protaspides*) was given to the individual **silicified** exoskeletons found in or on rocks by C. Beecher (1893a, 1893b).) Protaspides are difficult to find because of their small size. The larger ones are about the size of the "o" in the printed word "protaspis," and most are between that and half that size. One needs very sharp eyes to spot them among the normal fossil debris, or to spend hours with a stereomicroscope scanning likely surfaces. Most protaspides that have been studied are those in which their exoskeletons have been replaced by silica, since silicified fossils survive acidic dissolution of a limestone or shaly matrix. One then uses the microscope to pick out the important material from the insoluble residue, including protaspides. The advantage of this procedure is that in silicified material, very fine details, including spines and in some cases hypostoma, are often preserved.

The shape and other physical features of the protaspis are unique for specific fossil families and genera, and for this reason the protaspides are used for taxonomic assignments and confirmation (Chatterton and Speyer 1997). Protaspides are assigned to specific trilobites and growth series through association with pieces of the more mature fossil in the same debris and through the prior knowledge of protaspis-adult relationships. Using the presence of mature trilobites, however, is not always a secure way to assign protaspis-adult relationships.

It is probable that the pre-protaspid and protaspid phases of the trilobite growth cycle served to disperse the trilobites in their environment, similar to the situation with many modern marine Crustacea. Some protaspides were **benthic** (bottom dwelling) and others **planktic** (in the plankton), drifting in the Paleozoic seas. Because of their differing life-mode and preservation, some protaspides may never be found or never be associated with their later growth and adult phases.

Parts B and C in Figure 2.8 show two protaspides of trilobites from the Lower Devonian of New York. The one in B is a lichid (family Lichidae) and the one in C is a phacopid (family Phacopidae). They are magnified to the same scale and each is about a millimeter in actual length. At this phase of growth, morphological features, which represent a future body part in the adult, are given the *proto*-prefix. Thus, the protaspis has a **protocephalon**, described similarly to the cephalon of the adults, and a **protopygidium**. Trilobites also had growth stages *within* the protaspid phase. This is observed by increases in the number of lateral lobes on the glabellar axis and the general size (Figure 2.8A). The protocephalon and protopygidium, however, are fused and remain together in the molted protaspid exoskeleton, although the free cheeks and the hypostome may be detached. Each successive molt is called an **instar** (Figure 2.8A). Whether or not there is a direct correlation between the number of instars and the number of morphologically defined phases is not clear.

When the trilobite larvae clearly begin to display the separation of the protocephalon from the protopygidium, they are referred to as being in the meraspid phase (Figure 2.8E). The segments develop from the anterior of this **transitory pygidium**. (Once the segmentation is definite, the protopygidium is no longer "proto" but is still not the final pygidium, as the thoracic segments are being formed anterior to this new pygidial

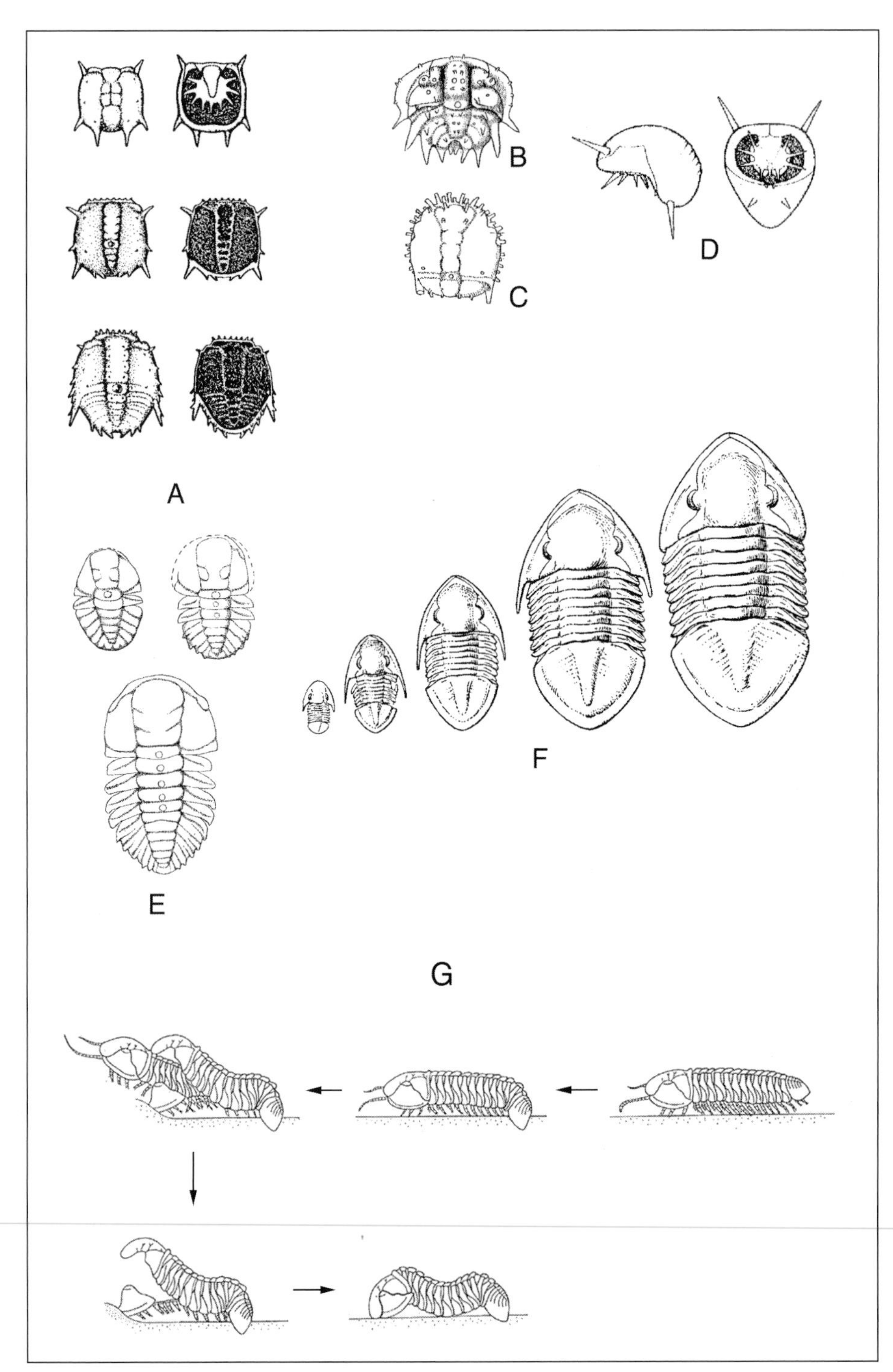

FIGURE 2.8. Ontogeny of the trilobite. A. *Flexicalymene senaria* protaspides from the Ordovician of New York. These silicified specimens were prepared and reported on by Chatterton et al. (1990). Reproduced with permission. B. Possibly a lichid protaspid from the work of Beecher (1893a) and reproduced by Whittington (1957). C. A phacopid protaspid from the same source as B. D. The protaspid of *Isotelus gigas* reported on by Chatterton and Speyer (1990). Reproduced with permission. The shape precludes this protaspid from being a bottom dweller. It probably was planktic, living and drifting near the surface of the sea. E. *Triarthrus* meraspid instars of degree 1, 2, and 4 from Whittington (1957). This is part of a nearly complete growth series collected and reported on by Walcott (1918). The meraspides are reproduced at about eight times their life size. F. A growth series of *Isotelus gigas* holaspides from Raymond (1914). The holaspids are natural size. Young *Isotelus* holaspides have prominent genal spines, which are lost, in New York specimens, when they reach about 50 mm long. G. A molting sequence for calymenid trilobites proposed by Mikulic and Kluessendorf (2001). The trilobite on the right first pushes its pygidium into the surface as an anchor, the cephalic sutures open, and the animal crawls forward and out. The cephalic parts held in position by the ventral integument fall back together, leaving a molt with the cephalon and pygidium curved downward and the thorax in a concave curve due to the pushing up of the cranidium during the process.

structure—hence, the name *transitory pygidium*.) In trilobites the early meraspides show a line of separation from the cephalon on the anterior margin of the transitory pygidium but no thoracic segments. This is referred to as a "degree 0 meraspis." During successive molts, as the segments detach from the transitory pygidium and can be considered separate, the meraspis grows in size and degree number. Each detached thoracic segment is counted, and this number is the degree assigned to the meraspis. The fact that segments are added from the anterior side of the transitory pygidium is ascertained by the growth of trilobites with pleural or axial spines on specific thoracic segments. The spine first appears on a segment adjacent to the transitory pygidium, and each succeeding segment is added behind it until the adult number of segments is reached.

There are usually profound changes in the shape of the cephalon during the meraspid phase. When the meraspid trilobite molted, the cephalon and transitory pygidium separated, and thus they are found as distinct elements, unlike the one-part protaspis molts. The meraspis can grow until it gains the number of free thoracic segments found in the adult, at which time it is called a *holaspis*.

Figure 2.8E illustrates a limited growth series of *Triarthrus eatoni*, part of the only nearly complete series known from a New York trilobite. The meraspides appear to be missing their free cheeks and are probably molts. Meraspids are also known for *Elliptocephala asaphoides* from the Lower Cambrian of New York (Ford 1877, 1878), but articulated holaspids are rare. Parts D and F of Figure 2.8 illustrate the morphological changes in *Isotelus gigas* from the protaspid (Figure 2.8D) through the early holaspid (Figure 2.8F). It is surprising, given the number of adult trilobite remains found in New York, that so few examples of growth series have been recorded. Either the exoskeleton on the juvenile forms of most trilobites is too fragile and not easily preserved in the fossil record, or the juveniles did not occupy areas where their remains could be readily preserved or found.

The trilobite exoskeleton did not necessarily become fixed in its physical structure at the early holaspid phase. The achievement of the holaspid phase generally only means that no more thoracic segments were added during continued growth. There are, however, variations in the number of thoracic segments in the holaspis of a very few species; in *Aulacopleura koninki*, from the Silurian of Bohemia, thoracic segments were added after the holaspid phase was reached (Hughes and Chapman 1995). For all other trilobites the segment number was stable. Trilobites did not reach maturity, or at least their final body proportions, until they grew substantially from the first holaspis. In some trilobites such as *Eldredgeops* species, the growing holaspis changed very little and the small ones looked essentially like the adults. *Isotelus gigas*, on the other hand, possessed long genal spines in meraspids and early holaspides (Figure 2.8F) and did not lose the spines until it reached about one-third the size of a full mature specimen. *Isotelus* tergites or molt remains from specimens longer than 125 mm (5 inches) are common in the Middle Ordovician Trenton age rocks of New York.

As in all arthropods, growth in trilobites required that the exoskeleton be shed or molted at regular intervals. In modern arthropods, molting (**ecdysis**) begins by the formation of a new cuticle or shell beneath the current one, separation of these shells by a space filled with molting fluid, and resorption of much of the old cuticle to provide the base chemicals to finish the new one. When this process is complete, the old shell splits and the animal emerges with a soft cuticle in place. By swelling this soft cuticle through the intake of liquids, the new body rapidly becomes larger. The new exoskeleton then hardens and the animal has grown one more increment. Because of the resorption of some of the old cuticle, the molted shell is significantly thinner and more fragile than it was on the animal. This process also lowers the energy requirements of growth because the resorbed chemicals are available for the new exoskeleton.

Trilobite molts, on the other hand, are robust and at least as thick as molts attributed to living animals that died and were preserved. This information indicates that trilobites did not resorb a significant amount of the minerals of the old exoskeleton and must have emerged from the molting process with a rather thin cuticle that was little more than a soft template within which the new mineralization took place. Very thin and compressed or wrinkled trilobite fossils are known and usually are attributed to the exoskeletal remains or impressions of recently molted or "soft-shelled" individuals. There is at least one example of a trilobite preserved in the Burgess Shale beds that is completely unmineralized and believed to be a very recently molted individual (Whittington 1985). This finding suggests that trilobites were vulnerable to predators and external trauma for an extended time and that building the new exoskeleton required significant energy from the animal. It also suggests that as trilobites gained size, they became increasingly vulnerable during molting and that the number of trilobites that might grow to an exceptional size for the species was severely limited. Early trilobite predators such as *Anomalocaris* species have been identified in the Middle Cambrian, and the number of potential predators such as cephalopods and fish increased throughout the Paleozoic.

There are many possible strategies for trilobite molting, and the cephalic sutures play a part in most of them. Henningsmoen (1975), McNamara and Rudkin (1984), Speyer (1985, 1990b, 1990c), and Whittington (1992, 1997) discussed specific molting strategies in detail. Disarticulated exoskeletons are common fossils in many New York rocks. Since most of these parts are attributable to molts, significant information is gained from their examination. For example, in *Eldredgeops rana*, the very common trilobite of the Middle Devonian of New York, complete cephala and pygidia are the parts most often found. This evidence shows that the facial sutures were fused and that the connections between the cephalon and thorax and between the thorax and pygidium were opened or weakened during the molting process.

The animal generally emerged through the split between the cephalon and thorax.

One often-illustrated set of molt remains is the upright thorax and pygidium, with the inverted cephalon just in front of it. In this case the molting animal must have pushed forward after the cephalon and thorax parted, with the forward margin of the cephalon pushed down in the sediment. This action would pivot the cephalon molt over upside down, and the animal could then emerge with the thorax and pygidium almost intact and right side up. That the pygidium is usually found separate suggests that the connection is weakened during the molting process and sometimes separates from the thorax during or shortly after ecdysis. There are other arrangements, besides the one just discussed, of the molt remains of *E. rana* found in western New York. S. E. Speyer (1990c), who has studied these molt remains, summed it up well: "Trilobites, like modern arthropods, displayed a variety of moult behaviors which vary according to ecological considerations (e.g., substrate consistency) and individual convenience."

Most other trilobites, however, lose their free cheeks during the molting process. The most common fossil remains of *I. gigas*, for example, are cranidia, free cheeks, hypostoma, and pygidia. Separated whole cephala of *Isotelus*, or articulated specimens without their free cheeks are very rarely found in the fossil record. *I. gigas* evidently molted by the facial sutures opening and the free cheeks separating, with the suture between the cranidium and the doublure opening, possibly along with a break in the connection between the thorax and the cephalon. These breaks permitted ecdysis by the trilobite moving straightforward. This molting strategy is important because *I. gigas* in the early phase had quite long genal spines, and it had to be able to free them to molt properly. The most efficient way to do this was to emerge in a forward direction through the opened sutures in the cephalon.

Spines on the trilobites were hollow, and the tissue inside had to be withdrawn during molting (Figure 2.7D). Any molting strategy of an individual trilobite must accommodate the physical shape of the animal. Since the newly molted animal was unmineralized and soft, there must have been some strategy to provide some protection while the new exoskeleton became suitably mineralized and hardened.

It is not always possible to distinguish the molted parts from the disarticulated remains of a dead trilobite. A complete exoskeleton, with free cheeks, is almost always the fossil of the carcass of a trilobite. As in any rule, however, there are exceptions. Trinucleids and harpids do not have dorsal facial sutures. The dorsal cheeks and preglabellar areas are separated from the ventral doublure by a suture that runs parallel to the horizontal plane. In other words, there is a suture separating the dorsal surface from the ventral surface, and the suture line is around the edge of the cephalon. Molting occurs by the opening of this suture and the trilobite emerging forward. In trinucleids the genal spine is on the lower **lamella.** An articulated specimen of *Cryptolithus* without the genal spines can be reasonably assumed to be a molt (Plates 163 and 165). In some extant arthropods like the horseshoe crab, however, which uses this same molting technique, the suture can reseal after molting and the molted exoskeleton remains whole. Thus, when one finds a whole, articulated *Cryptolithus* specimen with its genal spines, one cannot say with absolute certainty it is not a molt.

Most authors agree on the generalities of trilobite molting, but some unanswered questions are rarely discussed. The roles of the ventral, unscleritized integument and the scleritized appendages have been largely ignored (but see Whittington 1992). It is not surprising that there is little preserved evidence for the ventral membrane; there is so little soft tissue evidence from the fossil record, and only one of the known soft tissue preservation sites contains any significant ventral anatomical information beyond the appendages (Walcott 1881, 1918). In the trilobites in which the loss of the free cheeks is an important first step in ecdysis, the inversion of these parts helps demonstrate the actual process (McNamara and Rudkin 1984). Brief mention is made of the possibility that the free cheeks may still have been attached to the ventral membrane and would have been inverted as they came away from the animal. McNamara and Rudkin as do others, explain the inversion of the cephalon in molt remains as evidence that the trilobite pushed its cephalon down in front while arching its thorax to break away the cephalon at the cephalothorax suture. As the trilobite continued to move forward, the cephalon inverted. Many trilobites are found with the inverted cephalon under the thorax. Many of these also have long genal spines.

Some trilobite genal spines are totally enclosed except for the area at the genal angle. For this mechanism to take place, the soft spines must be dragged from their exoskeleton prior to total inversion of the free cheeks. Such a mechanism is illustrated by Whittington (1992, Figure 9). Not illustrated is the final struggle of the animal during the ecdysis process, which drags the dorsal cuticle forward over the now-inverted free cheeks.

Based on the observation that calymenid trilobites from the Silurian of Wisconsin and Illinois are often found with the cephalon and pygidium curled somewhat down, with a distinct concave sway to the thorax, Mikulic and Kluessendorf (2001) propose that the trilobite molted by pushing its pygidium into the substrate to anchor it, by the sutures between the free cheeks and cranidium as well as the anterior cephalic suture opening, and by the animal crawling forward, leaving the old exoskeleton behind. They further propose that since the upper and lower portions of the cephalon are held in place by the ventral integument, after the molting process the parts fall back into place, leaving a molt that may be indistinguishable from a carcass (Figure 2.8G). The same process is seen in extant horseshoe crabs, which leave behind an intact molted exoskeleton.

Growth was rapid during the early holaspid phase and could be expected to slow as the animal reached maturity. In a few rare cases, long periods between molts of larger individual species have been inferred. Tetreault (1992), Kloc (1993, 1997), and Brandt (1996) observed epizoans (encrusting animals) on whole, articulated exoskeletons of several species of trilobites. Such encrusters must have been on the living animal, as articulated trilobites would not remain whole unless buried a very short time after death and epizoans would have little opportunity to become attached to the buried carcass. These observations indicate either a terminal molt, after which there is little growth in the animal with no further molting, or long intervals between molts of the mature species. Tetreault further observed that brachiopods on the exoskeleton of *Arctinurus boltoni* were in four distinct size classes. From this he deduced, assuming these brachiopods spawned once a year, that the largest brachiopods were 4 years old and that in mature *Arctinurus* animals molting was terminal or occurred at as much as 4-year intervals.

Many collectors find populations of trilobites that are significantly larger than the norm. Excellent specimens of *E. rana* 5 to 6.4 cm (2 to 2.5 inches) long and *I. gigas* 13 to 15 cm (5 to 6 inches) long are not uncommon. Specimens of *E. rana* of 10 to 13 cm (4 to 5 inches) and *I. gigas* of 30.5 cm (12 inches) and larger are found, but rarely. The largest reported articulated trilobite is an Ordovician asaphid from the Arctic. It is 72 cm (28 inches) long. This indicates that some trilobites continued growth throughout life, as do extant lobsters, and could achieve an exceptional size in favorable environments. Raymond (1931), describing an unusually large hypostome of an *Isotelus* species from the Ordovician Chazy limestones, estimated the length of the trilobite at 61 to 64 cm (24 to 26 inches). Estimates of sizes of other New York trilobites, from molt remains or partial specimens, listed by Raymond are as follows:

Basilicus whittingtoni	~12 inches	(30 cm)
Isotelus "giganteus"	~24–26 inches	(61–64 cm)
Isotelus gigas	~17 inches	(43 cm)
Isotelus maximus	~18–19 inches	(46–48 cm)
Terataspis grandis	~20–24 inches	(51–61 cm)
Trimerus major	~15–16 inches	(38–41 cm)
Coronura myrmecophorus	~15 inches	(38 cm)

Soft Body Parts

The unmineralized or soft parts of the trilobite body are very rarely preserved in the fossil record. Allison and Briggs (1993) made a listing of sites of exceptional fossil preservation, called by the German name **Konservat-Lagerstätten**. They recognized 19 marine sites worldwide in the Paleozoic, where soft body fossils are preserved. Nine of these sites are in the United States, and six of them yield trilobites. Only one site in the United States in their listing has significant trilobite appendage and other soft body information: Beecher's Trilobite Bed in New York. In fact, most of what we know about the soft parts of trilobites comes from only five localities worldwide, including the really remarkable new Cambrian sites in China (Shu et al. 1995). Two of these five sites are in New York and one of them, the Walcott-Rust Quarry, was not included in Briggs and Allison's list. Information on soft or weakly skeletonized body parts is very rare, and because of this these data are generalized to a wide range of trilobites. The rarity of this soft-bodied information is exemplified by the remarkably preserved biota of the Burgess Shale in British Columbia. Most of what is known of soft-bodied animals in the Cambrian initially came from these beds, yet of the 22 species of trilobites known from the Burgess Shale beds, apparently only 4, so far, have yielded appendage information and in one of these it is from a single specimen. When you are reading through the following descriptions, remember that all the soft tissue data come from a very few sites and only a bare handful of trilobite species.

An indirect relationship of trilobites to the annelid worms was introduced earlier. This evolutionary trail is supported by the multisegmented body of the trilobite and the observation that each recognizable segment bears a pair of appendages. This observation includes the cephalon and pygidium, in which the segments are fused. Careful studies, by C. D. Walcott (1876, 1881, 1918, 1921), of the specimens he had available established that the appendages are **biramous** (Figure 2.10A). In other words, each individual appendage is divided into two parts, one part for walking, the **endopod** (Figure 2.10D), and one part, the **exopod** (Figure 2.10B), possibly an apparatus similar to the gill of fishes, for breathing.

The only trilobite appendages that are not biramous are the most anterior, which are modified into **antennae** (Figure 2.11A, B, C). C. E. Beecher (1893c, 1894a, 1894b, 1896), after years of study on meticulously prepared *Triarthrus eatoni* from Beecher's Trilobite Bed, published the most famous, and most often reproduced, illustration of the trilobite ventral anatomy (Figure 2.11A). Beecher, possibly following Walcott's lead, gave *T. eatoni* an extra set of appendages under the cephalon, but this does not take away from his remarkable achievement.

The works of Walcott and of Beecher have been modified and augmented by a number of later workers, particularly Raymond (1920a), Størmer (1939, 1951), Stürmer (1970), Bergström (1969, 1972, 1990), Bergström and Brassel (1984), Cisne (1975, 1981), Whittington (1980, 1992), and Whittington and Almond (1987).

Biramous appendages are also the rule in extant crustaceans. In the trilobites all the appendages, with the exception of the antennae, are very similar, differing primarily in size. All but one of the trilobites recently studied have three pairs of biramous appendages in the cephalon, a pair for each thoracic segment, and multiple pairs in the pygidial section. Four pairs of biramous cephalic appendages reportedly were found in one trilobite (Bergström and Brassel 1984), but some workers think this finding is questionable. In the pygidium the segmentation has to be inferred. For example, the number of appendage pairs under

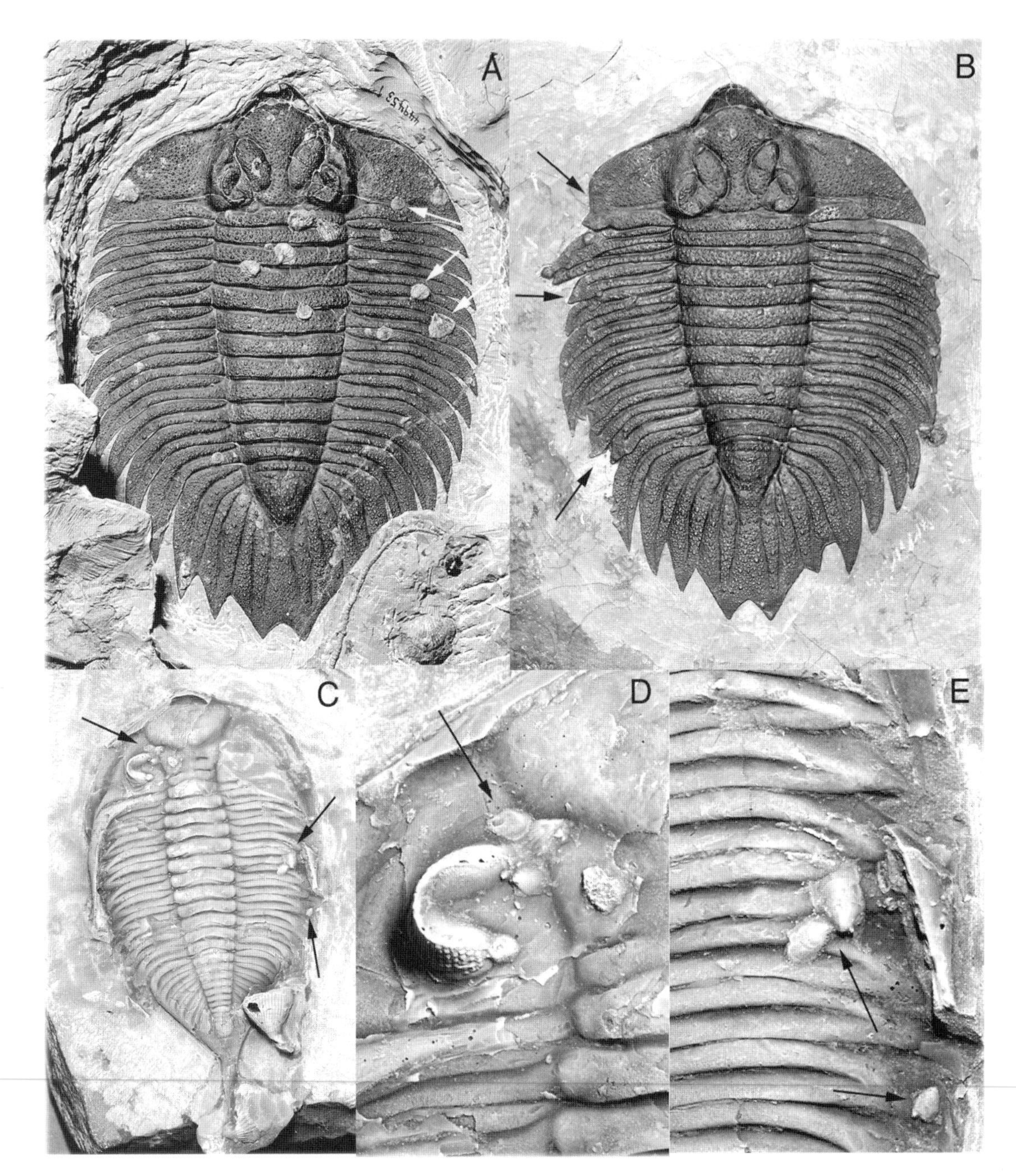

FIGURE 2.9. Trilobite exoskeletons with attached fauna or injury. A. *Arctinurus boltoni* (USNM 449453). This trilobite has brachiopods of different sizes (arrows) attached, indicating the length of time between molts of mature specimens. *Arctinurus* specimens with attached brachiopods are not rare in the Rochester Shale beds, where these came from. It is unlikely that they settled on the exoskeleton after death because the exoskeleton would have to have been buried to remain articulated. B. The same species of trilobite as in A with healed injuries to the exoskeleton. Three areas (arrows) have been damaged and been through at least one molt (PRI 42095). C. *Dalmanites limulurus* (F. Barber collection, whitened). This trilobite has attached brachiopods (arrows). This brachiopod attachment is very unlikely to have occurred postmortem for the same reasons. D. The same trilobite as in C, with the eye area enlarged to better show the brachiopods (arrow). E. The same trilobite, with the thoracic area enlarged to show the brachiopods (arrows).

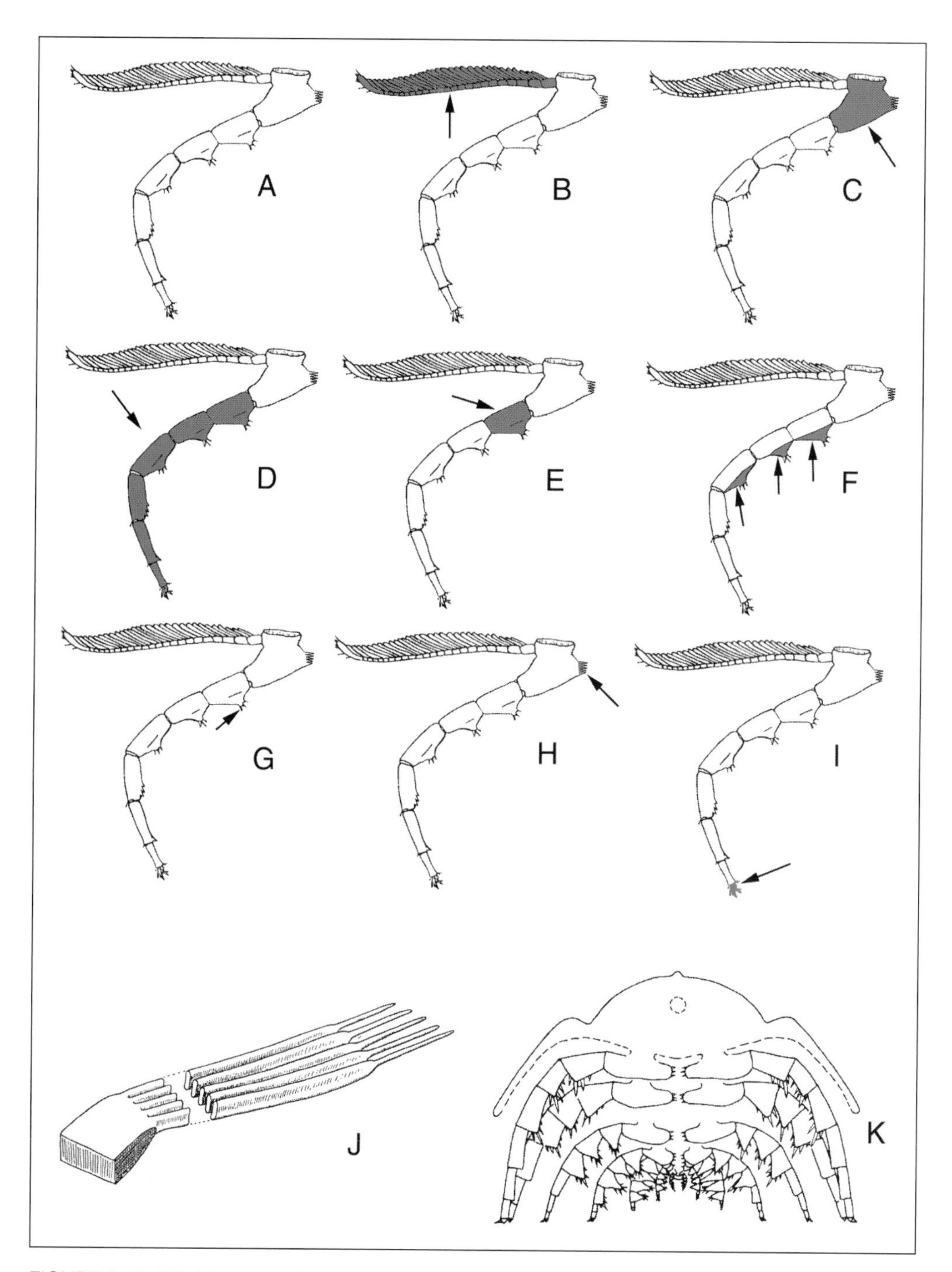

FIGURE 2.10. Trilobite appendage reconstruction and nomenclature. A–I. *Triarthrus eatoni* biramous appendage after Størmer (1939). B. The exite or brachial appendage (arrow) with the comblike structures. C. The basis (arrow), also called the *coxite* in early literature. D. The walking leg or telepodite (arrow). This leg has seven segments, including the foot, in all trilobites where the appendages have been studied. E. A podomere or individual segment (arrow) of the walking leg. F. Endites (arrows), small triangular inwarding-facing projections on the podomere. These were possibly used to help transport food along to the mouth at the rear of the hypostome. G. Setae (arrow), hairlike projections on the exites. H. Gnathobases (arrow) are sharp projections on the basis that may have been used to masticate, to reduce the size of food particles. I. The foot (arrow) with its setae. J. A reconstruction of the filaments of the exopod of *Ceraurus pleurexanthemus* by Størmer (1939). K. A partial axial view, looking toward the rear, of *Triarthrus* as reconstructed by Whittington and Almond (1987, p. 42, Fig. 43). Reproduced with permission.

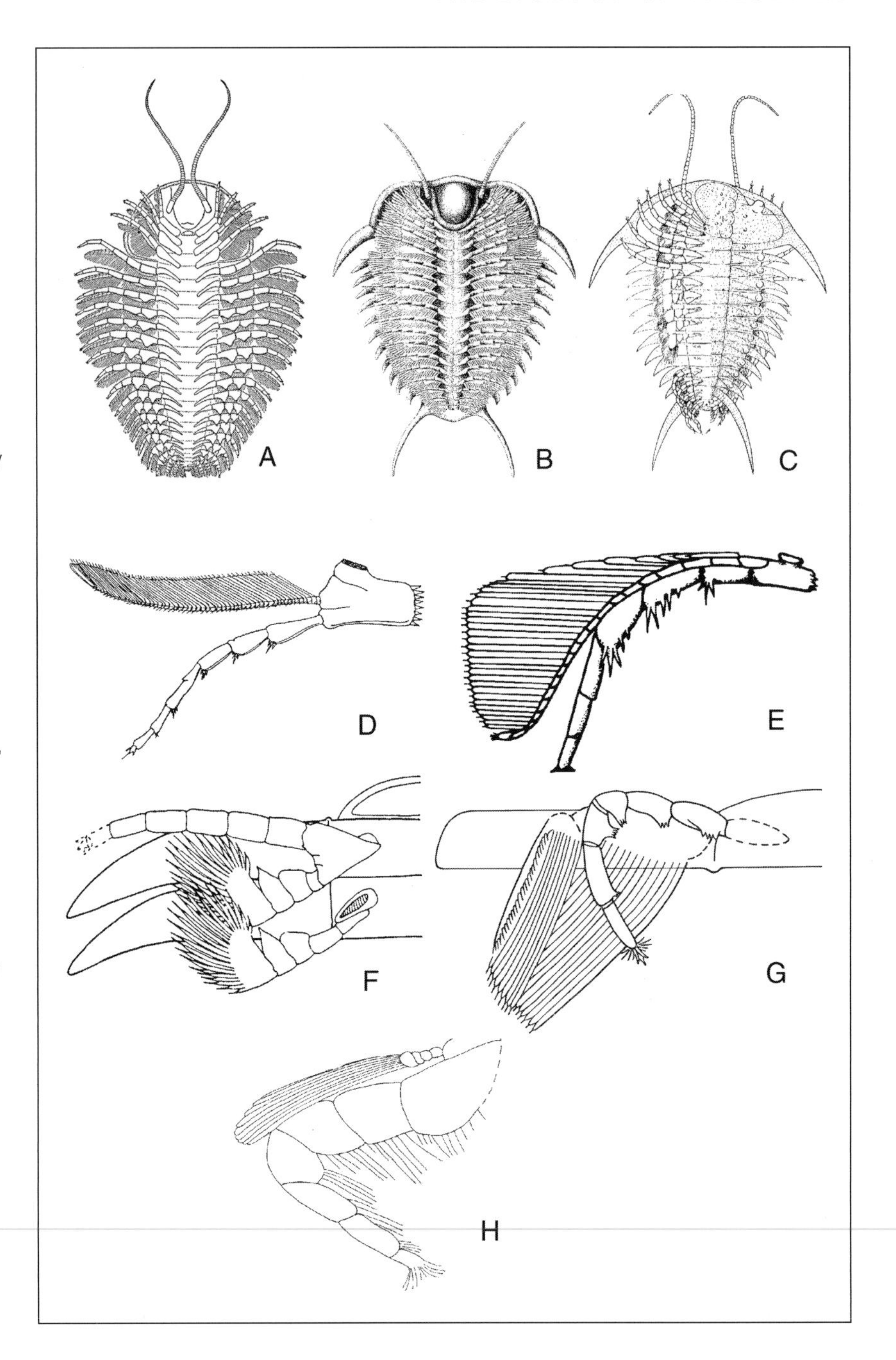

FIGURE 2.11. Ventral anatomy and appendages. A. *Triarthrus eatoni*, ventral anatomy. The first essentially correct reconstruction of a trilobite's ventral surface and appendages. After Beecher (1896). B. *Ceraurus pleurexanthemus*, ventral anatomy. This reconstruction from Raymond (1920a) was from cross sections made by Walcott in the late 1900s. C. *Ceraurus pleurexanthemus*, ventral anatomy. This reconstruction by Størmer (1951) was from specimens collected by Walcott and uniquely prepared. D. *Triarthrus eatoni*, appendage structure developed by Cisne (1975, p. 49, Fig. 3) from high-resolution radiographs of pyritized specimens. Reproduced from Fossils and Strata, www.tandf.no/fossils, by J. L. Cisne, 1975, vol. 4, 45–63, by permission of Taylor and Francis AS. E. *Triarthrus eatoni*, appendage structure developed by Whittington and Almond (1987, p. 31, Fig. 41), from direct observation of very carefully prepared specimens. Reproduced with permission. F. *Ceraurus pleurexanthemus*, appendage structure drawn by Bergström (1972) from information developed by Størmer (1939, 1951). Reproduced with permission. G. *Cryptolithus bellulus*, appendage structure developed by Bergström (1972, 1973) from pyritized specimens prepared by Beecher. Reproduced with permission. H. *Phacops* cf. *P. ferdinandi*, appendage structure developed by Bergström (1969) using the radiographs of pyritized specimens from the Hunsrück Shale in Germany. It is assumed that the phacopid trilobites of New York will have similar structures. Reproduced with permission.

the pygidium in *T. eatoni* is significantly more than the number of axial furrows on the dorsal surface of the pygidium, showing a weak relationship between segments and axial furrows.

A more detailed look at an individual trilobite appendage illustrates that it is a complex structure. The attachment to the ventral surface of the trilobite body is through a part called the **basis** (Figure 2.10C). (For a current view on appendage nomenclature, see the article by Ramsköld and Edgecombe (1996).) The exopod or outer branch is attached to the basis, which in turn is attached to the ventral membrane of the trilobite. Beneath the exopod, and also attached to the basis, is the walking leg, or endopod. Until the work of Cisne (1975, 1981), all reconstructions of the appendages included a precoxa from which the exopod extended (Størmer 1939, Figure 1, p. 155). This view is incorrect, based on all the material recently examined; both the exopod and the walking leg are attached to the same apparatus, the basis. The exopod has a series of thin, flattened filaments extending from it, giving it a feather- or comb-like appearance, and it is carried up under the pleurae. Only in *T. eatoni* are the exopods known to be long enough to extend well out from the lateral edge of the dorsal exoskeleton. The large surface area of the small filaments led most authors to believe that they have served for breathing, as an external gill. Bergström (1969) argued that the filaments are too small to support an effective circulatory system, and instead they may have served as a filter of food, as a means to circulate water over gill membranes on the ventral surface, or possibly as a swimming function. In all the studies the exopods are drawn with the filaments lateral and posterolateral. Størmer (1939) actually found the filaments of *Ceraurus pleurexanthemus* pointed forward, but he rotated them in his reconstruction. Bergström (1969) believed that they were pointed forward in life (Figure 2.11F). Størmer (1939) reconstructed the filaments of the exopod of *C. pleurexanthemus* (Figure 2.10J), illustrating the high surface area that supports their use as a brachial organ.

The endopod is multiply jointed with distinct sections called **podomeres** (Figure 2.10E). There is general agreement that there are seven podomeres on all trilobites examined. On some of the podomeres, starting with the ones closest to the basis, are projections, **endites** (Figure 2.10F), with hairlike **setae** (Figure 2.10G) near or on their tip. On *Triarthrus*, and probably most other genera, the last podomere is a footlike tip to the walking leg (Figure 2.10I).

The most thoroughly examined trilobite appendages are those of *T. eatoni* and *C. pleurexanthemus*, both from the Ordovician of New York. There are more specimens available of these trilobites with preserved appendages than there are of any others. Figure 2.11A is the Beecher reconstruction of *T. eatoni* and parts B and C of Figure 2.11 are reconstructions of *C. pleurexanthemus* by Raymond (1920a) and Størmer (1951). Størmer's figure is modified to show the dorsal anatomy on the right and the ventral on the left. Parts D and E of Figure 2.11 are reconstructions of the legs of *T. eatoni* by Cisne (1981) and by Whittington and Almond (1987), respectively. Parts F, G, and H are drawings of the legs of *Ceraurus*, *Cryptolithus bellulus*, and *Phacops* cf. *P. ferdinandi*, all as reconstructed by Bergström (1969).

Figure 2.10K shows *T. eatoni* as reconstructed by Whittington and Almond (1987). The view is from about the midline of the trilobite looking to the rear and illustrates the orientation of the walking legs and their parts. The bases are shown with toothlike **adaxial** projections or **gnathobases** (Figure 2.10H). Food was probably collected, masticated, or pulled apart, and passed along forward to the mouth at the posterior of the hypostome by use of the endopods and bases. Trilobites so equipped would be effective bottom feeders, both as predators and as scavengers. Since this kind of information on appendages is known for so few trilobites, it is difficult to extrapolate too far as to the life-mode of the others.

Pyritized trilobites from New York and Germany that were studied by high-definition X-ray photography provide much of what we know about the internal anatomy (Størmer 1939; Stürmer and Bergström 1973; Cisne 1975, 1981). Their stomach (Figure 2.12A, label c), crop, or foregut is located in the glabella. Along with the stomach in the cephalon are organs called *hepatopancreatic organs* (Figure 2.12A, label h), probably the equivalent of a liver, under the genal areas.

The gut (Figure 2.12A, B, C, label g) passes from the stomach through the axial region of the thorax and terminates at the anus just under the posterior area of the pygidium (Plate 78). The position of the gut is known from several specimens of trilobites because the ingested sediment often survives in place (Figure 2.12D, label g) and has a different texture or color than the surrounding stone (Raymond 1920a; Cisne 1975; Whittington 1993; Whiteley et al. 1993; Brett et al. 1999). It is assumed that some form of circulatory and nervous systems also occupied the axial interior region.

Small rodlike structures seen in radiographs (Cisne 1981) and in unusually well-preserved specimens (Whittington 1993) are believed to represent muscles, and reconstruction of the musculature related to the appendages has been proposed (Figure 2.12A, B, C, label m). Some specimens of *E. rana* show symmetrical dark spots on the exoskeleton (Babcock 1982). These are interpreted as muscle attachment areas.

Two soft-bodied arthropods, *Naraoia compacta* and *N. spinifer*, originally discovered from the Burgess Shale, are now regarded as trilobites and assigned to the family Naraoiidae. The Naraoiidae now includes five genera (Fortey and Theron 1995). Two of these are Ordovician and one survived to Late Ordovician. If, as some authors believe, the heavier exoskeleton of post-Cambrian trilobites is an evolutionary response to more advanced predators, then it is unlikely that soft-bodied trilobites survived past the Ordovician. None of these genera are known from New York, but given the special conditions necessary for their preservation, this does not prove that they were not present.

Life-Mode

The following discussion of life-mode is based almost exclusively on circumstantial evidence. As such, it is highly interpretive. Fortey (1985) pointed out that using the same body of knowledge, trilobites in the family Agnostidae have been hypothesized to be pelagic, benthic, parasitic, and epifaunal, possibly attached to algal strands. The fossil record does not often permit clear, unambiguous conclusions. However, one might assume that "form follows function" and that a trilobite's morphology is often a good indicator of its life habits.

Most paleontologists contend, by analogy with crustaceans, that for many trilobites the protaspid phase was planktic. That is to say that the larvae floated and drifted in the sea, ensuring a wide dispersal of the species.

Trilobites that were primarily benthic as adults probably settled to the sea bottom sometime during the meraspid phase. **Pelagic**, or free-swimming trilobites, may never have left the open seas during their transformation to adults. All through these changes, the immature trilobite was very vulnerable to predators and environmental stress, as the molting process left the animal

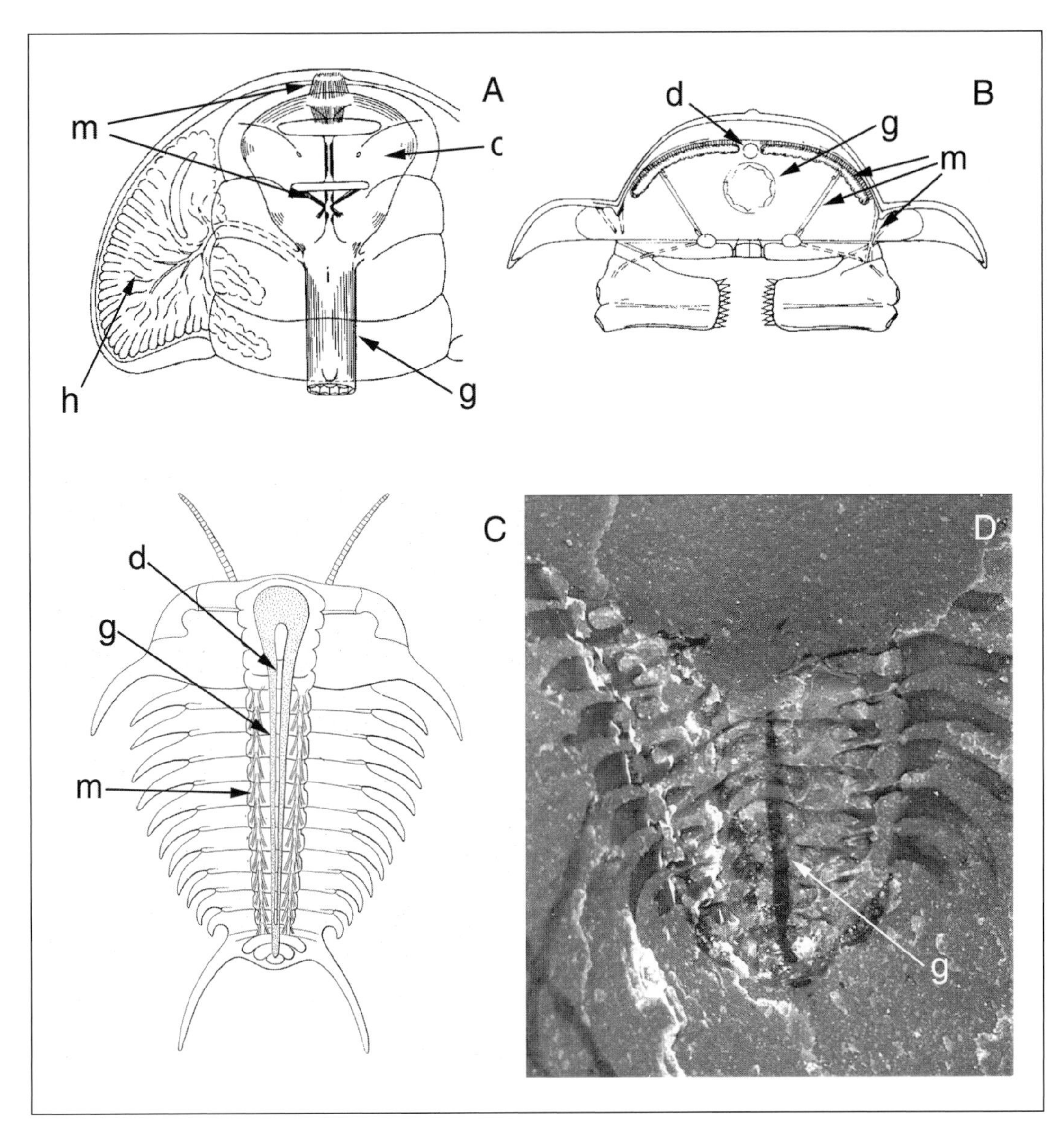

FIGURE 2.12. Internal anatomy of the trilobite. A. Internal organs of the cephalon of *Triarthrus* determined by Cisne (1975, p. 55, Fig. 9): m, muscles; c, crop or stomach; g, gut; h, hepaticopancreatic organ. Reproduced from Fossils and Strata, www.tandf.no/fossils, by J. L. Cisne, 1975, vol. 4, 45–63, by permission of Taylor and Francis AS. B. Cross section of the thorax with the internal organs (Cisne 1975, p. 53, Fig. 7): d, dorsal vessel or "heart"; m, muscles; g, gut. Reproduced from Fossil and Strata, www.tandf.no/fossils, by J. L. Cisne, 1975, vol. 4, 45–63, by permission of Taylor and Francis AS. C. Reconstruction of some internal anatomy of *Ceraurus pleurexanthemus* by Raymond (1920a): d, dorsal vessel or heart; g, gut; m, muscles. D. Specimen of *C. pleurexanthemus* that cleaved to show the gut (g) as a dark ferruginous stain (MCZ 111716).

with little in the way of defense for a period of time and the energy burden of continually building new exoskeletons was considerable. In extant arthropods Clarkson (1979) noted that 80% to 90% of mortality comes during the molting process. The more common trilobites had to produce large numbers of larvae in order for significant numbers to reach maturity.

The **metamorphic** change in an asaphid trilobite, *Isotelus*, from the planktic protaspid (Figure 2.8D) to the benthic adult (Figure 2.8F) is well documented. Both forms are well suited to their respective modes of life. The change to a bottom-dwelling (benthic) form comes at the metamorphosis from protaspis to the meraspis, which is shaped like the adult but with long genal spines. As mentioned earlier, the long genal spines last well into the early juvenile holaspis.

Some trilobite species are widely dispersed throughout the world, suggesting that they were planktic or pelagic as adults and

FIGURE 2.13. Trilobite shapes and functions. A. *Isotelus gigas* (MCZ 311). This trilobite has a smooth shape suitable for shallow plowing of the surface muds for feeding. B. *Hypodicranotus striatulus* (MCZ 100986). The streamlined shape and the 180-degree visual capability suggest a trilobite that might have been a good swimmer and was perhaps pelagic. C. *Achatella achates* (PRI 49659). This trilobite has a fairly flat body with eyes raised well above the rest of the cephalon. In analogy with bottom dwellers with raised eyes, one might expect this trilobite to rest on the bottom, with the body just under the surface of the substrate and the eyes above it. This position is a defense against predators and possibly a means of lying in wait for prey. D. *Cryptolithus bellulus* (PRI 49654). This trilobite has a prominent, robust cephalon compared to the light exoskeletal material on the thorax and pygidium. This feature and other evidence (Campbell 1975) suggest a sedentary lifestyle and filter feeding habit. E. *Triarthrus eatoni* (TEW collection). The appendage, including the exite or brachial branch, extends well beyond the edge of the thoracic shield. This configuration aids the trilobite to survive in the dysoxic, deep-water conditions suggested by the dark shales in which they are found.

moved freely throughout the seas. Planktic larval forms would also serve to disperse species but in a more limited manner. Dispersal into new areas by the normally benthic trilobites is expected to be followed by **speciation**, so one would not expect the species identity to be maintained if dispersal was into areas not previously occupied by the species.

There are wide variations in the size of trilobite eyes, the number of lenses, and the angle of vision. These variations are certainly some indication of the life-mode, but modern analogy is often necessary to come up with suggestions. Animals that bury themselves shallowly in the bottom sediment have eyes that are raised above the plane of the head, enabling them to see when they are slightly buried. Some trilobites have such raised eyes (e.g., *Achatella achates* (Figure 2.13C) and *Dalmanites* species), and it is reasonable to suggest that they too buried themselves under a thin layer of sediment.

Body shape suggests the life-mode of many trilobites. Fortey (1985) defined three morphologies of pelagic species: (1) large-eyed, epipelagic, slow-swimming trilobites; (2) pelagic, streamlined, faster swimmers; and (3) possible swimming *Irvingella* types, remopleuridids, and progenetic types. The fast-swimming trilobites have rounded streamlined shapes, which promoted buoyancy and low drag while swimming. These shapes are not often found among New York trilobites, but the Middle Ordovician remopleurid *Hypodicranotus striatulus* has the right shape and is considered pelagic (Figure 3.13B, Plate 160).

Vaulted, smooth exoskeletons such as that on *I. gigas* (Figure 2.13A, Plate 150), *Dipleura dekayi* (Plate 97), and *Trimerus del-*

phinocephalus (Plate 99) were well designed to plow through the upper sediment layers in search of food. *T. eatoni*, with its thin exoskeleton and outer branches that extend beyond the pleurae, was unsuited for shallow, turbulent water and for plowing in sediment and was better designed for surface scavenging in deeper, less oxygenated environments (Figure 2.13E, Plate 172).

It has been proposed that to maximize visual effectiveness, the plane of the upper and lower edges of the eyes should be parallel to the substrate (Plates 5 and 6). In many outstretched trilobites it is apparent that the eyes are parallel to the substrate. However, many species of illaenids, when outstretched, have eyes that are angled upward and posteriorly. Westrop (1983) and Bergström (1973), among others, argued that these trilobites were **infaunal**, burying themselves backward into a soft bottom with their cephalon on the surface at an angle to the **thoracopygidium**.

In this attitude the eye base is parallel to the surface and gives the maximum all-around vision. These infaunal trilobites fed and breathed by the exchange of water on the buried ventral anatomy. This exchange of water was the result of the movement of the appendages and an upstream orientation of the burrow. There are other trilobites with this life-mode, which Westrop termed "illaenimorphs." Whittington (1997b) rebutted this view on the basis of the high flexibility of the thorax in illaenids and that they are more suited to crawling around the bottom and over irregular objects than living or resting primarily in burrows.

Westrop also pointed out that on some of the illaenimorphs there is a median tubercle on the glabella, midway between the palpebral lobes on the sagittal line. This tubercle is also a thin spot and is characterized as having possible light-sensing properties. It is the highest point when the body is in a normal life position, thus covering any blind spots of the conventional eyes. Ruedemann (1916b) found a significant number trilobites with such median tubercles, even nominally blind trilobites such as in the genus *Cryptolithus*. It is reasonable to conjecture that the tubercles had a light-sensing utility and played a role in the trilobite's life-mode. Such light sensing tubercles probably could sense movement but not resolve objects.

In modern arthropods larger eyes are found on nocturnal species or those adapted to low daylight levels. Animals with a wide visual angle need it to watch for predators. Very large eyes and a wide visual angle are seen on some pelagic trilobites (Figure 2.13B). Fortey (1985) considered the large-eyed trilobites as **epipelagic** and slow swimmers.

Compound eyes permit insects to be highly aware of movement but do not necessarily provide high visual acuity. Reduction in eye size has been noted for trilobite genera that moved to deeper water through time. Blind trilobites, such as in the genus *Cryptolithus* (Figure 2.13D), may have burrowed into sediment where sight would have been less important than other sensory capabilities.

Most trilobites were benthic. They passed most of their life on or in the uppermost part of the sediment layer. Some **trace fossils** such as tracks, burrows, and distinctive pits preserved in the fossil record are attributed to trilobites, as supported by the rare find of a trilobite at the end of a trackway or in one of the burrows (Figure 2.14A). Two common traces—***Cruziana*** and ***Rusophycus***—are generally preserved as molds (fillings) on the basal contact of sandstones or carbonate beds in shales. One type of deep, inscribed, horizontal track or furrow, *Cruziana*, shows "V-like" scratch patterns made by the dactyls (claws) of the trilobite and a central groove corresponding to an axial ridge of debris pushed up by the trilobite (Figure 2.14B). The bilobed burrows, or resting pits also showing V-shaped scratches, are called *Rusophycus*, and the trackways consisting of small "footprints" on the substrate are sometimes known as ***Diplichnites*** (see Bromley 1990, p. 161), although Osgood (1970) did not hold this term in high regard.

Trace fossils attributed to various burrowing animals (worms?) have been found ending in *Rusophycus*, suggesting the trilobite had attacked another burrower. These fossils indicate that many trilobites walked around on the bottom and dug into the sediment for both food and resting places (Hall 1852, Plate 9, Figure 1).

In the case of *Cruziana* the direction of travel for the trilobite is toward the open end of the V-shaped appendage traces. Most *Rusophycus* traces form a V, with the gape end or anterior being wider than the posterior, suggesting that the trilobite rested (or hunted) with the cephalon toward the "gape" end. There is little argument that *Rusophycus* traces are most readily explained as trilobite hunting or resting pits, but the same is not true for *Cruziana*. Whittington (1980), based on his in-depth studies of *Olenoides serratus*, believed that trilobite appendages do not readily allow for the type of traces represented by *Cruziana* and that some other animal, perhaps not even an arthropod, may be responsible. However, many *Cruziana* traces end in *Rusophycus*, and since the latter are unambiguously trilobite, this reference is questionable.

In New York, traces attributed to trilobites are common in the Silurian Clinton Group, particularly on the base of sandstone layers near the village of Clinton, Oneida County. Osgood and Drennen (1975) provided a good description of these traces and their literature. In other strata they are far less well known, either because conditions were not right for their preservation or because little effort has been made to find and identify them in appropriate strata.

Fortey and Owen (1999) proposed that trilobite feeding habits can be related to the position and attachment of the hypostome and other physical characteristics. Trilobites that are considered predatory (i.e., they fed off macrofauna such as worms) had a conterminent hypostome fixed or strongly supported at the anteroventral cephalon. The hypostome provided a strong base for the appendages so the trilobite could manipulate and masticate the prey. Food was passed forward along the ventral median by the bases to the mouth at the posterior of the hypostome. Examples of this type of hypostome attachment in an *Isotelus* species may be seen in Figure 2.7E and Plates 153, 155, and 157.

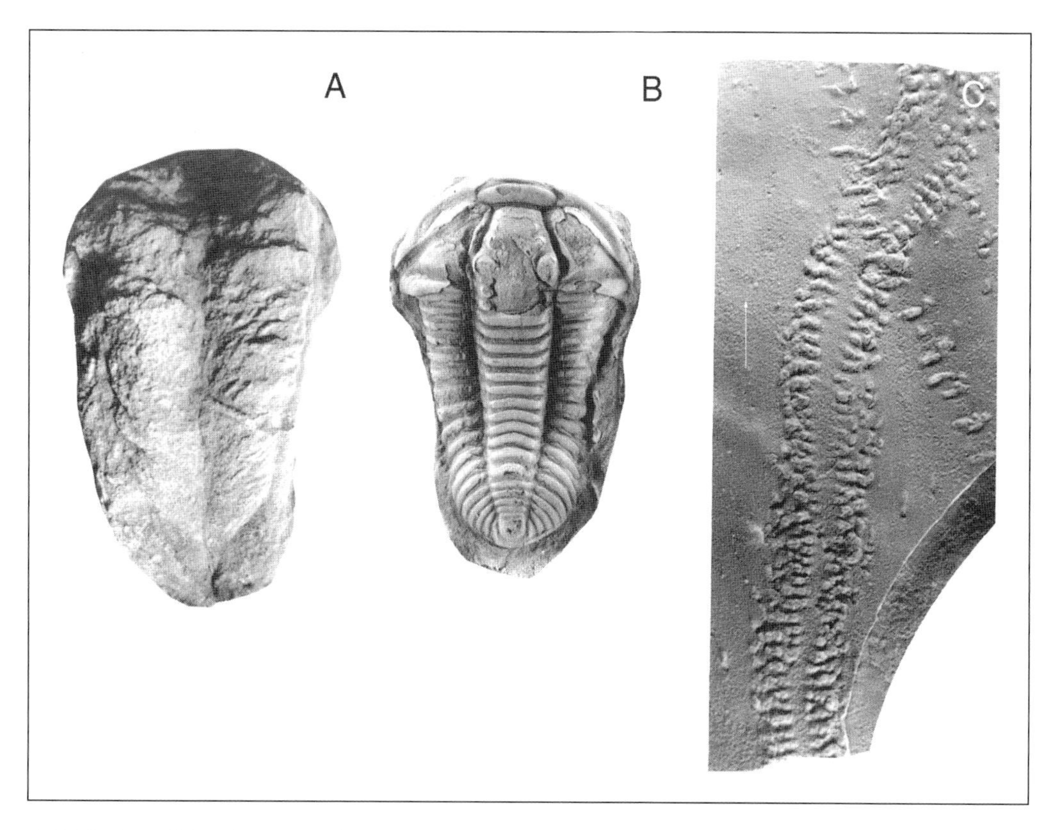

FIGURE 2.14. Trilobite traces. A. *Rusophycus pudicum* (UCM 37574). Sandstone deposits over trilobite-rich bottom muds often have convex, slightly V-shaped traces on their lower surface. These traces are known as *Rusophycus* and have been long regarded as trilobite "resting traces." Osgood (1970) reported on a remarkable *Rusophycus* that had the trilobite responsible for it still in place. B. *Flexicalymene meeki*. This trilobite was found on the *Rusophycus* in A. The trilobite is about 46 mm long and is from the Upper Ordovician of Ohio. C. *Trachomatichnus numerosum* (UCM 37695). A trilobite walking trace attributed to *Cryptolithus* (Osgood 1970). The illustration is life size. All figures reproduced with permission.

Impendent hypostoma that are also attached to the doublure also suggest a predatory habit. *Bellacartwrightia* species (Plate 47) and *Calyptaulax callicephalus* (Plate 117) show this mode of hypostome attachment.

Trilobites whose hypostoma were not strongly attached to the cephalon (i.e., natent) are considered to be particle feeders. They relied on much smaller food particles swept up from the bottom, and the rigid hypostome was unnecessary. These trilobites tend to be smaller than the predatory ones because their food sources were not as rich and concentrated. The genera *Harpidella* (Plate 128) and *Triarthrus* (Plates 170 to 174) represent this group of trilobites.

The last feeding mode to consider here is filter-chamber feeding. This type of feeding is typical of trilobites with reduced mobility and relatively large cephala. These trilobites settled in a position and stirred up the sediment immediately under them, and then filtered out the minute food particles contained in the top layer of sediment. Movement occurred only after the food supply was exhausted. The genus *Cryptolithus* (Plates 163 to 167) is the example used, and the evidence is the relatively small (weak) thorax and thoracic appendages compared to the cephalon, as well as trace fossils clearly attributable to *Cryptolithus* resting (and feeding) sites.

Trilobites are found in a variety of environments, from fairly shallow waters near shore, to reefs, continental shelves and slopes, and moderately deep basins. The observation has been made that trilobites from the shallower areas with more wave turbulence have thicker exoskeletons (Fortey and Wilmot 1991). These thicker exoskeletons are possibly an evolutionary response to the greater environmental energy. As mentioned earlier, exoskeletons are generally thicker in post-Cambrian trilobites, which may also signal the rise of better-developed predation, another form of environmental stress.

Many trilobites were gregarious, at least at some point during their life cycles. The large numbers of death and molt assemblages, well illustrated by *E. rana* in New York, are no statistical accidents. It is not uncommon, within a number of different trilobite species, for a large number of individuals to be found in local "pockets" on the same bedding plane or horizon, indicating some form of group behavior (Plates 59, 89, 102, 128, 146, 147, and 152). All this suggests that it was common for many trilobites to congregate for breeding or molting, or just because it was their normal life-mode to be together in "schools" (Speyer and Brett 1985, Speyer 1990a).

Trilobites such as phacopids and calymenids were capable of very tight enrollment, literally into a ball. This position was undoubtedly a defense mechanism resorted to in times of stress. If the stress was an undersea sediment flow too large to get out of, the trilobite would be entombed in the tightly enrolled position. Most of the post-Cambrian trilobites in New York could enroll, and the frequency of this position versus the open position is perhaps an indication that it was a common reaction to stress for the species.

Some benthic trilobites undoubtedly could swim. Modern horseshoe crabs, a normally benthic species, are good swimmers. They also swim upside down, and it has been shown that the hydrodynamics of their swimming works best in this attitude (Fisher 1975). It is reasonable to assume that some normally benthic trilobites could swim also and possibly quite well. Some likely swam upside down. (H. Burmeister first proposed this in 1843.) This mode of swimming may not have been involved in their food gathering, but they could move from place to place and evade danger by swimming. In one trilobite bed, the result of a burial event, at least 98% of the trilobites with a wide variety of sizes are found buried upside down (Whiteley et al. 1993; Brett et al. 1999). This observation applies to both the lower surface and internal to the limestone. For more on this, see Chapter 3.

Spininess in trilobites has invoked a number of explanations, most of them probably correct for one species or another. There seems to be little need to invoke just a single explanation, any more than there is one explanation for spininess in modern species of arthropods. Spines can be a defense mechanism against predators, provide support on a soft substrate, assist the animal in burying itself when necessary, and sensory devices. The Lower Devonian stratum in New York (and the Devonian strata of Oklahoma and Morocco) has trilobites with elaborate spines, some curling up and over their thorax. These spines of the genus *Dicranurus* commonly carried algae and other encrusting organisms on them, possibly as a defense mechanism (Kloc 1993, 1997). The spines provided good attachments and presumably helped break up the visual body lines to provide camouflage. Some modern arthropods do exactly this.

One can assume, based on modern analogy, that trilobites were subject to injury, parasitism, and predation, as reflected in their preserved parts. Owen (1985) extensively reviewed trilobite abnormalities. He listed three general types of trilobite abnormalities: injury, genetic or embryological malfunction, and pathological abnormalities.

Not uncommonly, trilobites are found with malformations that are ascribed to healed injuries (Figure 2.9B). Most of these injuries are seen on the pleura either by asymmetry or by healed damage. These malformations can come about in several ways, but damage due to problems in molting and damage caused by actual attack by a predator were probably the most common. The mechanism of a defect is not always clear, but healed punctures and crescent-shaped malformations are readily attributed to predation. Figure 2.15 illustrates four examples of exoskeletons with clear indication of predation. Panels A and B show "bite" marks out of trilobites from the Rochester Shale that have been through at least one molt, as shown by the broken edges of the thoracic segments being rounded to a new termination. Panel C shows the unusual situation of *Dalmanites limulurus* with the pygidial spine missing (compare to Figure 2.15A) and the damage healed over. Panels D through F show a calymenid from the Rochester Shale with evidence of boring on its exoskeleton. This finding is particularly interesting because crinoids from the Rochester Shale have been described with similar boring marks (Brett 1978, 1985). Signor and Brett (1984) and Pratt (1998) summarized the possible predators in the Paleozoic. *Anomalocaris* is a well-documented predator of the Cambrian, and many of the circular scars on trilobites are attributed to the circular mouth of *Anomalocaris*. Cephalopods became a predation factor in the Ordovician, and fishes became prominent in the Devonian.

Babcock (1993) has collected information related to the concept of behavioral asymmetry. This means that when a trilobite was attacked by a predator, its response was not random, but there was a preference to move in a manner that resulted in most damage to the right side of the animal. Conversely, preferred damage to the right side of the trilobite could suggest the attack strategy of the predator.

Genetic or embryological abnormalities cannot be easily diagnosed in fossils. Pathological abnormalities are due to disease and parasites. Disease is impossible to document in fossils, but parasitic attack is seen by the presence of worm-shaped borings and gall-like swellings. It is more difficult to determine if borings were made on the living animal (versus on the exuviae or postmortem) than to recognize a healed injury. Swellings are known from a number of different trilobite families, and some can be diagnosed as the result of parasites based on the direct presence of worm-like structures.

McNamara and Rudkin (1984) documented the death of a partially molted *Pseudogygites latimarginatus*. It seems likely that the animal was overcome and buried while molting.

Often trilobites are found in association with other fossils in rapid-burial deposits. If there is little indication of their being transported any distance, then this can be taken as evidence of their ecology. An example is the odontopleurid *Meadowtownella trentonensis* from the Trenton Group. This small, spiny trilobite is often found whole on fossil hash layers, particularly with branched bryozoans, and also in burial event deposits with the cystoid *Cheirocrinus* and fenestrate bryozoans. *M. trentonensis* was probably a scavenger on bottoms with animal debris accumulations and also in bryozoan thickets with their attached

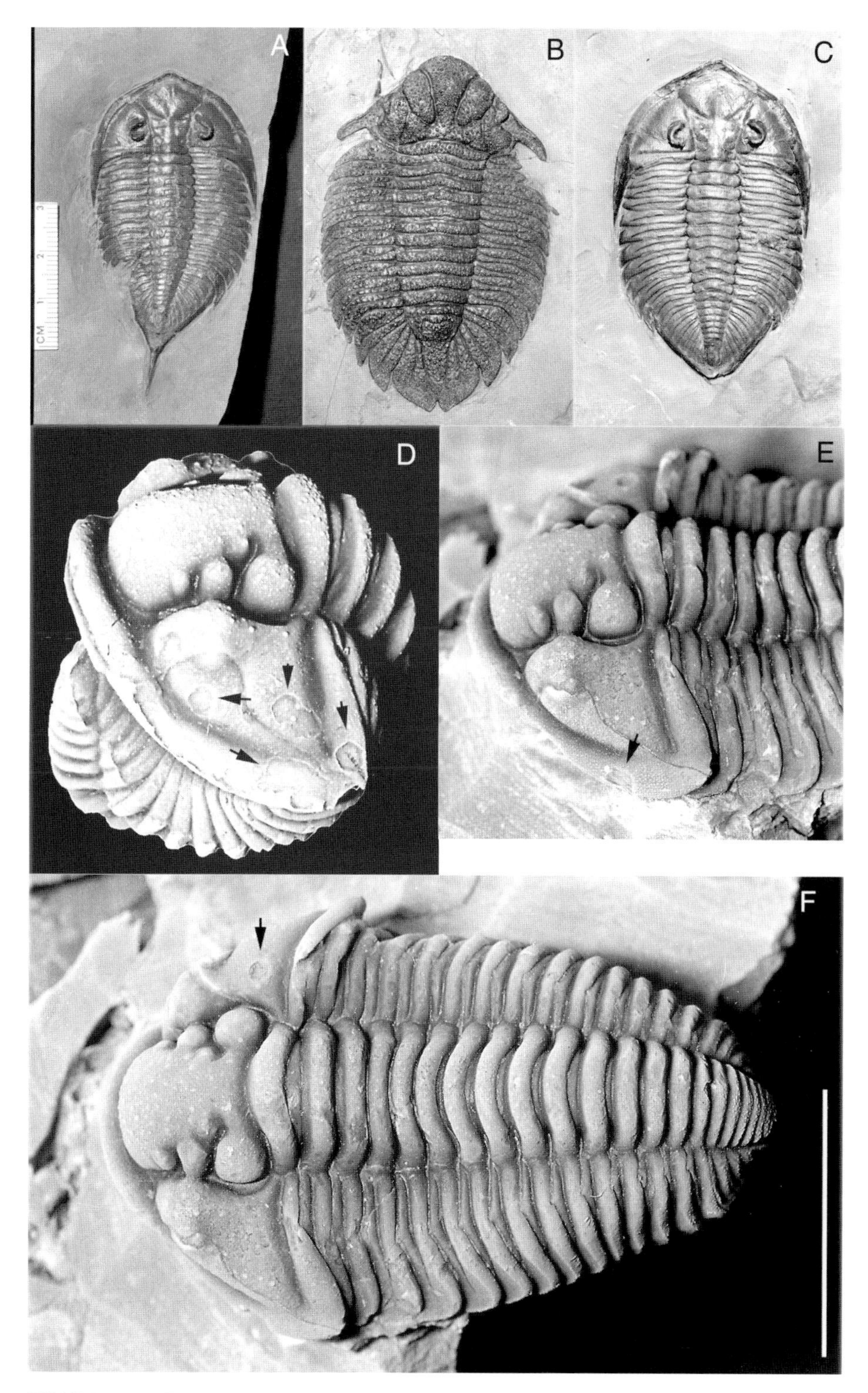

FIGURE 2.15. Trilobite injuries. A. *Dalmanites limulurus* with a semicircular portion of the thorax missing. The damage has "healed" in that the edges of the damage are rounded, indicating the trilobite has molted at least once since the injury (K. Smith collection). B. *Dicranopeltis nereus* with an injury to the thorax. The trilobite has molted at least once since the injury (K. Smith collection). C. *Dalmanites limulurus* with the pygidial spine missing. The end of the pygidium is rounded, indicating one molt since the loss of the spine. (K. Smith collection). D–F. *Calymene* species, showing circular "borings" (arrows) similar to those of *Tremichnus* on several different species of crinoids from the Rochester Shale reported by Brett (1985) (NYSM 16796).

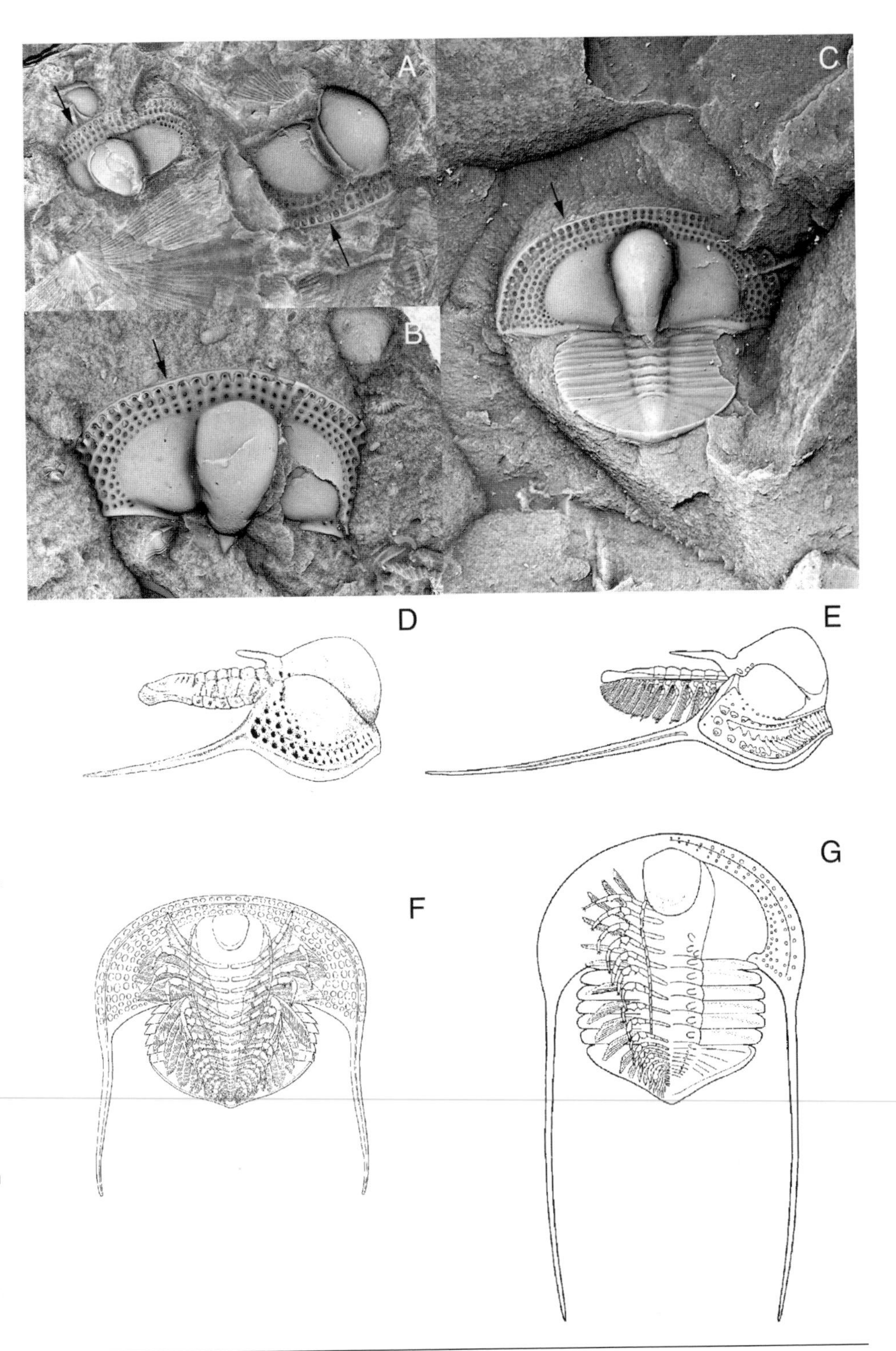

FIGURE 2.16. *Cryptolithus*, the most-studied trilobite genus found in New York. A. *Cryptolithus tessellatus* (TEW collection, whitened). This trilobite, in New York, is found only in the Middle Ordovician Sugar River Formation. It is recognized by the three rows of pits anterior to the cheek area (arrow). B. *Cryptolithus lorettensis* (PRI 49657, whitened). In New York this trilobite is found only in the upper Sugar River Formation. It differs from *C. tessellatus* in that there are four rows of pits anterior to the cheek (arrow). The first approximately nine radial abaxial pit rows are well aligned. C. *Cryptolithus bellulus* (PRI 49654, latex pull, whitened) from the Upper Ordovician Lorraine Group in New York. This trilobite differs from *C. lorettensis* primarily in the poor radial alignment of the first four abaxial pit rows. D. A drawing of a silicified specimen of *C. tessellatus* by Campbell (1975). Reproduced with permission. E. A drawing of the same specimen by Bergström (1972) with the appendages drawn in. The appendage information is from Beecher specimens of *C. bellulus*. Reproduced with permission. F. Raymond's (1920a) reconstruction of the ventral anatomy of *C. bellulus* using specimens prepared by Beecher. G. Bergström's (1972) reconstruction of the same specimens. Reproduced with permission.

cystoids. A further possibility is that this trilobite fed off living bryozoa.

Trilobites of the genus *Cryptolithus* are among the best understood of the New York trilobite genera (Figure 2.16). There are several reasons for this. The genus is widely distributed geographically and geologically. Wide distribution means that it is preserved in a wide variety of environments and the possibility of new information on the lifestyle and biology is enhanced. *Cryptolithus* specimens with appendages are preserved in Beecher's Trilobite Bed, which enables detailed observations of the ventral anatomy (Beecher 1895a, Raymond 1920a (Figure 2.16F), Bergström 1972 (Figure 2.16G)). At least one whole, three-dimensional specimen, preserved in silica, is known. Figure 2.16D is a drawing of the specimen by Campbell (1975), and Figure 2.16E is a drawing of the same specimen by Bergström with the legs included. Walking and digging trace fossils unequivocally assigned to *Cryptolithus* are known from Kentucky. The *Cryptolithus* protaspid is sometimes found silicified and is easily characterized. The unique cephalon with its heavily pitted brim is readily recognized in stratigraphic samples, and it is impossible to mistake it for other genera. A number of authors have studied the distribution of *Cryptolithus* so there is a large database of information.

In New York there are three distinct species or "morphs" of *Cryptolithus* (Whittington 1968, Shaw and Lespérance 1994): *C. tessellatus* (Figure 2.16A), *C. lorettensis* (Figure 2.16B), both from the Middle Ordovician Sugar River Formation, and *C. bellulus* (Figure 2.16C) from the Upper Ordovician Lorraine Group, all in New York. They are distinguished primarily by the rows of pits or the pit arrangement, or both, so that it is justified to assume that any biological information from one species is essentially the same for all of them.

Trinucleids arose as a family in Europe and migrated to North America during the Middle Ordovician (Whittington 1968). The pelagic protaspis (Chatterton et al. 1994) ensured that once in the North American epicontinental sea, the trilobite spread rapidly wherever currents took the protaspis. Molts from protaspides and cephala of what is probably the meraspid phase are commonly found in the deep-water Frankfort Shales including the Beecher's Trilobite Beds. The holaspid is blind except for the possible eye spot or visual receptor centered on the median line of the glabella. There is a wide brim on all but the posterior of the cephalon. This brim is bilaminar, separating into an upper and lower lamella through a plane parallel to the ventral plane of the trilobite. The brim is perforated with holes, commonly referred to as *pits*, that pass through both lamellae. The pits are in circumradially arranged rows for the first three or four circumferal rows. There are a number of speculations as to the function of the pits. The most current thinking is that they were sensory devices, perhaps to determine current direction (Campbell 1975).

The long genal spines are on the lower lamella. Cephala are found with and without genal spines, as are near-whole articulated specimens. The trilobite molted by moving forward through a gap between the lamella, and this gap possibly closed after molting (a mechanism observed in extant horseshoe crabs). Alternatively, the lower lamella became detached during the molting process. Consequently, it is difficult to know if a completely whole articulated specimen with the genal spines intact is a molt or an animal killed and buried. The absence of genal spines and lower lamella, however, ensures it is a molt.

The thorax and pygidium are much reduced compared to the cephalon. There are six thoracic segments and a small triangular pygidium. A nearly whole silicified *Cryptolithus* species was discovered by Whittington (1959). Photographs of this specimen reproduced by Campbell (1975) reveal that the ventral plane of the thoracopygidium was well above the apparent plane of the ventral surface of the cephalon. In order to move about on the bottom and to burrow, the trilobite must have had long, strong, ventral walking legs or have developed other modes of movement. Campbell (1975) explored this in detail, but in summary *Cryptolithus* species could not have been very active crawlers or deep burrowers (see Figure 2.16G). Osgood (1970, Plate 58, Figure 1 and 2) figured the resting trace fossil *Rusophycus cryptolithi*, which because of size and the impressions of genal spines could only have been made by *Cryptolithus* animals. These traces indicate that the trilobite sat in a shallow depression, facing into the current, and swept particles of detritus from the bottom into its mouth.

The mouth of *Cryptolithus* species is at the rear of the small hypostome in the ventral part of the cephalon. The stomach occupies the glabella, and digestive capability extends out into the genal area. The alimentary track lay along the axis of the thoracopygidium and occupied about 20% of the width of the axial rings. The anus is at the posterior point of the pygidium.

Fortey and Owens (1999) described *Cryptolithus* species as filter-chamber feeders, meaning that the resting trilobites stirred the sediment directly under them, using the cavity between the cephalon and the substrate, and filtered out the food particles from the rest of the suspended sediment. Fortey and Owens also suggested that the cephalic perforations provided channels for water to flow out of this cavity while it was swept forward by the appendages. This flow brought the food particles forward to the mouth and kept oxygenated water flowing over the exopods.

As in all specimens of trilobites with preserved ventral appendages, the biramous limbs of *Cryptolithus* included walking legs, endopods, and outer branches, exopods, which may have served as a breathing organ as well as provided some ability to sweep the surface under the trilobite. The exopods do not extend beyond the border of the thoracic segments. There are three pairs of appendages under the cephalon and one pair under each thoracic segment. Under the pygidium there are a large number of much-reduced appendages, reflecting the number of fused segments in the pygidium.

Assuming the relationship of the appendages to the exoskeleton is correct, it is hard to see members of the genus *Cryptolithus* as active surface crawlers.

On the whole, one can assume that trilobites initially occupied many of the environmental niches that are occupied today by modern marine arthropods. The ecological niche for the individual species enabled trilobites to become part of a paleocommunity. Within this community the trilobites shared the local environment with a variety of other animals and plant life. Wherever one finds the necessary associations within the rocks, one can expect to find the other members of the community. This is no different than what is seen in extant, and undisturbed, communities today. Some trilobites, such as the illaenids, are found in a fairly narrow range of environmental conditions, and some like in the genus *Isotelus* are found in a wide variety of communities from shallow nearshore areas to deep ramp-basin transition areas. Fortey (1975) defined communities in the northern Europe Ordovician. Two of his communities, the cheirurid-illaenid and the olenid, fit much of the New York Middle Ordovician very well.

The geographic restrictions of specific trilobite families and genera also are used to identify paleobiogeographic provinces. Whittington (1961b), Whittington and Hughes (1972), and Ross (1975), for example, used this approach to define the early positions of the paleocontinents.

The early evolutionary success of trilobites was probably due to their development of a mineralized exoskeleton and their wide geographic dispersal. The number of trilobite genera increased from their first appearance until the Late Cambrian and then decreased continuously through the Paleozoic. This is graphically shown in Ludvigsen (1979b, Figure 9). A number of explanations for this have been proposed. The number of potential predators continually increased during these same times. It is likely that the increased predation, increased number of competitors for environmental niches, and the burden of molting to a completely defenseless individual all played a part in the reduced success of trilobites through time. Trilobites disappeared completely at the end of the Permian.

Evolution and Cladistics

Fortey and Owens (1997) reviewed the evolutionary history of trilobites. This review should be considered the latest in what will be an ongoing series of arguments.

The phylogeny or evolution of trilobites is known from the first calcified remains in the Lower Cambrian rocks to the last trilobites in the Late Permian. It is generally believed that trilobites as a class are monophyletic; that is they are from a common ancestor (Ramsköld and Edgecombe 1991). The earliest known trilobites are from the suborder Olenellina found exclusively in the Lower Cambrian. These trilobites lack facial sutures, which is considered a primitive characteristic (Fortey and Whittington 1989). In New York *Elliptocephala asaphoides* is a member of this early group. Facial sutures first appear in the suborder Redlichiina, also in the Lower Cambrian. It is probable that there are ancestral trilobites that did not have mineralized exoskeletons, perhaps back into the late Precambrian. There is no compelling reason to believe that all trilobites sprang from olenellin ancestral stock.

From these ancestors evolved the large number of trilobite orders, families, and genera that are recognized today. Evolution, however, is not something that proceeds smoothly with a calendar-like precision. For evolutionary change to occur, at least two major factors must be in place. First, any genetic change or mutation must offer a competitive advantage to the species. Noncompetitive genetic changes will soon disappear, as the animal is unable to compete for the opportunity to pass the changes along to its offspring. Next, the new trait should occur in an isolated small population so that it will not be "lost" in a much, much larger gene pool. Since mutation occurs relatively frequently, a beneficial mutation will happen at somewhat regular intervals, statistically. In a stable environment, however, with many others of the same species, this change may not become fixed. A better opportunity for a new mutation to be passed along is when a group, for some reason, is occupying an environment where there is little outbreeding competition or there are new environmental niches to occupy. Essentially populations are stable until a change takes place in an isolated environment, and after this change (or changes), the new species becomes competitive with the former stable species and displaces it or in some cases occupies a different available niche. Thus, evolution is not a process of continual small changes but one of relative stability followed by abrupt step changes. Eldredge and Gould (1972) used the term **punctuated equilibria** to reflect their observation that evolution is not a smooth transition from one species to the other but rather is composed of periods of relative stability punctuated by periods of rapid change.

The observation that some fossil communities show remarkable stability, sometimes over millions of years, has led to some in-depth investigations. In New York some Middle Devonian fossil communities, particularly in **dysaerobic** environments, show this stability. As a result of these and other studies, Brett and Baird (1992, 1995) coined the term **coordinated stasis** for this form of evolutionary stability (see also Kammer et al. 1986, Morris et al. 1995, Brett et al. 1996). Similarly in the Cambrian, for example, there is evidence that populations of trilobites in the shallower water environments underwent a number of extinctions, and the area where they once were was repopulated with trilobites from deeper water. These episodic extinctions and repopulations, called **biomeres** (Palmer 1965, 1984), are recorded from the Cambrian of North America (see also Edgecomb 1992).

Relationships of trilobites at all taxonomic levels are currently being revised using **cladistic** methodology. Cladistics as used in paleontology is the grouping of taxa by their shared physical characteristics. Close relationships are based on **synapomorphies** or "shared derived" characteristics. Traits may be divided into primitive (plesiomorphic) or derived (apomorphic). For trilobites the biramous appendages are **plesiomorphic** or "shared primitive" characteristics, meaning they are primitive to the group (i.e., they represent a common ancestral trait of arthropods not exclusive to trilobites) and may be used only to compare trilobites with other arthropods. Schizochroal eyes are only found in the suborder Phacopina, and thus this characteristic forms a derived characteristic or **apomorphy**. Another example involves the trilobite hypostome. Fortey (1990a) developed the argument that hypostome attachment, natent versus conterminant versus impendent, is an indicator of trilobite phylogeny. Using these and other relationships, Fortey (1990a, Figure 19) constructed a phylogeny/hypostome attachment chart for trilobites, which summarizes the current state of knowledge.

By selecting a significant number of derived characteristics and evaluating their presence or absence, one can group related trilobites by the timing of the appearance of the selected derived characteristics. With a large number of characteristics, the groupings are best done with computer programs designed for cladistics that search for the closest or most **parsimonious** fit. Cladistics is clarifying many relationships among trilobites. For a more in-depth review, see the works by Fortey (1990b, 2001) and Novacek and Wheeler (1992).

3 Taphonomy

Taphonomy is the study of processes that influence the preservation of potential fossils. This field encompasses the disciplines of *biostratinomy*, the study of processes affecting organism remains or traces prior to their final burial, and *fossil diagenesis*, the investigation of phenomena affecting potential fossils after burial. Recently, taphonomy has developed both as a means of assessing bias in the fossil record and more positively, as a critical tool for paleoenvironmental analysis. Taphonomic analyses and fossil geochemistry provide valuable information on the physical and chemical parameters of environments. The assumptions of uniformitarianism are applicable at this level because the physical and chemical properties of the skeletons of organisms have probably been invariant through geologic time, despite the nonuniformity imparted by evolution. The taphonomic features of trilobites (e.g., articulation of delicate skeletal elements) may provide unambiguous evidence for episodic sedimentation; conversely, highly corroded fossil material provides a distinctive signature of long-term condensation.

Articulated specimens of trilobites, while generally uncommon, may be abundant in some beds. Moreover, the totally mineralized body parts of trilobites commonly have been preserved nearly unchanged for at least 520 million years. However, preservational features of fossil specimens can be used to decipher the taphonomic history of these specimens and provide useful clues as to original environment.

The structural complexity of the trilobite exoskeleton and the process of molting of the exoskeleton during growth increase the number of fossil parts that can be generated by a single individual. Trilobite skeletal parts can be found in a majority of fossiliferous localities in New York, and well-preserved cephala and pygidia are generally identifiable to species. With most organisms, like bivalves or gastropods in which the skeletal growth is by accretion, the fossil is evidence of the death of an individual. Conversely, with trilobites, the presence of body parts is not a simple indicator of population density.

Death, Decay, and Disarticulation

Death

The taphonomic history of most organisms generally begins with their death, although in arthropods, including trilobites, molted exuviae also become a part of the preserved record. Death of organisms may involve gradual normal mortality. However, many of the spectacular Lagerstätten occurrences involve mass mortality of many individuals and species due to environmental crises (Brett and Seilacher 1991). These crises may include storm and seismic shock events, volcanic eruptions such as those at Pompeii, overturn of the water column, and anoxia. However, only the mortality events associated with episodes of burial will be recorded as such. Most bodies decay rapidly after death, with a resultant loss of soft parts.

Scavengers of all sorts are also part of the normal fauna, and a dead animal can be expected to become a food source for a wide variety of other animals, which in Paleozoic seas included other trilobites. The result of this is that a dead trilobite exposed on the seafloor can be expected to become disarticulated and the body parts scattered within a relatively short period of time. Consequently, very shortly after death the animal must be buried at least deeply enough to physically inhibit disarticulation. Lowered oxygen inhibits scavenging and also favors the preservation of intact skeletons.

Soft Tissue Decay

The most destructive process to affect the bodies of organisms is the decay of soft parts. The most volatile parts are tissues, such as internal organs and muscle, and as a result such tissues are very rarely encountered as fossils. Where they are preserved, it is usually (but not always: Butterfield 1990) the result of early diagenetic mineralization (Allison 1988b).

Anoxia historically has been viewed as a prerequisite for the preservation of such tissues (e.g., Whittington 1971b). However, the removal of environmental oxygen does not prevent microbial activity (Berner 1981b; Allison 1988a; Allison and Briggs 1991a). Microbes simply utilize a variety of alternative oxidants for carbon degradation. In fact, it has been suggested that anoxia may retard the decay process by only a factor of two or three (Canfield and Raiswell 1991a). More importantly, anoxia prevents scavenging and favors early mineralization. The microbial reactions that are involved in anaerobic decay also generate a series of reactive elements, which in some circumstances can go on to produce early diagenetic minerals. These minerals, in turn, may preserve the decaying tissues themselves. The minerals most frequently associated with soft-part preservation are pyrite (Allison 1988a), phosphate (Allison 1988b; Briggs and Kear 1993), and carbonates. The activity of anaerobic bacteria is a necessity for the formation of all three. Formation of organic-clay complexes or clay coatings may also be important in soft-part preservation (Butterfield 1990).

Most well-preserved benthic fossils are preserved approximately at their life sites, so bottom-water anoxia can be ruled out as a preservational factor. Allison's (1990) calculations demonstrate that most organism bodies become anoxic microenvironments internally during early phases of decay. Inhibition of scavenging in these environments may prolong the association of skeletal elements. However, even under conditions of anaerobiosis or anoxia, bacterial decay of ligaments is rapid and the slightest currents will serve to disarray pieces (Allison 1988a, 1990).

Not surprisingly, most cases of soft-part preservation among trilobites are associated with dysoxic mudrock facies, typically dark, slightly organic-rich shales. The Burgess Shale (Middle Cambrian, British Columbia), Frankfort Shale (Beecher's Trilobite Bed, Upper Ordovician, New York), and the Hunsrück Shale (Lower Devonian, Germany) are a few of the most notable examples (Allison and Briggs 1991a, 1991b, 1993; see below).

Articulated Remains

Skeletons of organisms in which the skeleton is composed of multiple elements weakly bound together by ligaments or musculature, such as trilobites, are only rarely preserved intact. Modern experimental studies indicate that the degradation of soft tissues in arthropods occurs within a period of a few hours after death, while destruction of ligaments ensues in weeks to months. As such, skeletons may be completely disarticulated within a period of a few months or less (Plotnick 1986, Allison 1988a).

A variety of articulated trilobite remains are found. Speyer and Brett (1986) recognized three basic categories: (1) partially articulated exoskeletons, (2) molt remains or exuviae, and (3) completely articulated skeletons representing carcasses (Figure 3.1). Partially articulated remains, such as groups of thoracic segments, are of indeterminate origin and may represent either partially decayed remains of carcasses or molt parts.

Exuviae in many trilobites are recognizable by the absence of specific structures. For most trilobites molting involves the shedding of free cheeks. Therefore, articulated exoskeletons with cranidia lacking free cheeks are almost certainly exuviae. Phacopids had fused facial sutures and shed the entire cephalic shield in molting. Thus, articulated thoracopygidia ("headless trilobites") suggest molt parts. Molt ensembles are groups of closely associated molt parts, such as free cheeks lying in close proximity to articulated remains with cranidia, or cephalic shields closely associated with thoracopygidia in phacopids.

Carcasses are represented by completely articulated exoskeletons, with the cephala intact and articulated (i.e., free cheeks are intact). This category may be subdivided further into outstretched or prone specimens and in some trilobite species, partially enrolled and fully enrolled individuals.

Organisms with multielement skeletons, including trilobites, are particularly sensitive indicators of rapid and permanent burial. Well-preserved, articulated trilobites typically occur on certain bedding planes within mudrocks that would not be recognizable as event-beds by other sedimentological means. Because such skeletons cannot be reworked, the occurrence of even a single intact specimen of a trilobite is an excellent indicator that the enclosing sediment accumulated rapidly and was not subsequently disturbed. The occurrence of large numbers of completely articulated trilobites provides dramatic evidence for a population of organisms that was abruptly wiped out. Conversely, the occurrence of well-preserved molt ensembles need not imply any mortality but rather indicates burial under relatively low-energy conditions that prevented scattering of parts. Certain widely traceable levels in the Middle Devonian Hamilton Group are characterized by abundant articulated molt ensembles but few, if any, complete trilobites (that would represent carcasses). Such findings may reflect rapid accumulation of thin layers of sediment that did not kill or smother living trilobites but were sufficient to preserve molts (Speyer 1987).

Many assemblages of well-preserved trilobites are also demonstrably in situ (buried in their living sites). A particularly sensitive indicator is the occurrence of trilobite molt ensembles, that is, associated, disarticulated molt parts. It is virtually impossible for different disarticulated portions of the skeleton to be transported any distance and still remain associated. The hydrodynamic properties of whole exuviae versus free cheeks would

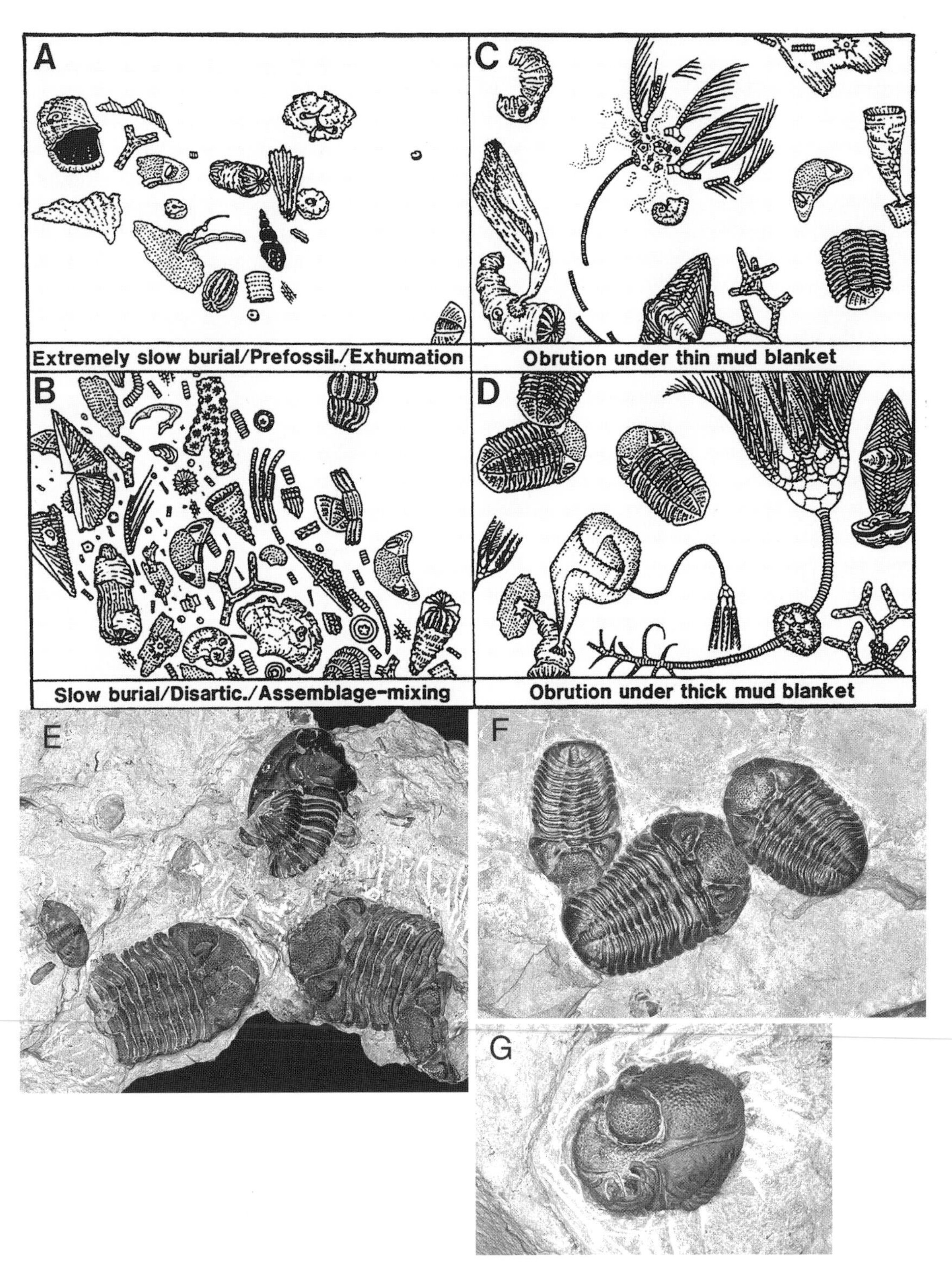

FIGURE 3.1. Fossil assemblages reflecting various conditions and timing of burial. A. Low rates of sedimentation and slow burial along with reworking of the prefossilized hard parts. Sclerites are broken and scattered. B. Slow burial with disarticulation and mixing. C. Obrution: burial, under a thin blanket of sediment, with some disarticulation and infaunal scavenging. D. Obrution: sudden burial under a thick blanket of sediment, resulting in good preservation of articulated fossils and reduced infaunal scavenging. A–D from Brett and Baird (1993). E. Trilobites that exhibit some degree of disarticulation because they were not buried deeply enough or soon enough after mortality, PRI 49661. F. A group of well-preserved trilobites as a result of rapid, relatively deep burial, PRI 49662. G. A trilobite that is tightly coiled and well preserved, PRI 49663. Trilobites coiled under stress, probably associated with the burial event.

be so different that it is extremely unlikely they could be transported and yet end up together. Molt ensembles thus constitute proof of life activity (i.e., molting) by trilobites in the precise site of burial.

Other dramatic examples of in situ trilobites are the very rare specimens of articulated skeletons directly associated with their trace fossils, such as famous specimens of *Flexicalymene* attached to *Rusophycus* (see Figure 2.14B). The occurrence of beds of enrolled trilobites also suggests a behavioral response to stressed conditions in which trilobites may have burrowed into the sediment and enrolled themselves, only to be buried in place and apparently unable to escape from mud blanketing.

Speyer and Brett (1985) also recognized species-segregated clusters of fully articulated skeletons ("body clusters"; Plates 59, 102, 147, and 152) and molt ensembles ("molt clusters") involving at least three species of trilobites on single bedding planes in the Devonian Hamilton Group. These trilobite clusters appear to represent preserved behavioral patterns. Using the argument that the molt ensembles could not have been transported any distance, Speyer and Brett inferred that these represented species-segregated aggregations of trilobites that molted en mass. Speyer and Brett further pointed out that synchronized molting typically serves as a prelude to mating in modern peracarid crustaceans (e.g., marine isopods) and that several species also may time their reproductive cycles to a common external stimuli, such as lunar cycles. This analog suggests that trilobite clusters may represent preserved mating "orgies."

Transport and Reorientation

Trilobite skeletons can be sensitive indicators of hydrodynamic conditions in the depositional environment. Under low-energy environments, typical of mudrocks, organism remains may be buried in situ. One might use the argument that well-articulated fossils such as trilobites must not have been transported and that their occurrence therefore indicates quiet water environments, but this inference must be made cautiously. Aside from obvious cases where this is not so (e.g., in which these skeletons are found at the bases of turbidites or even within skeletal grainstone deposits), there is experimental evidence that if organisms are transported within the first few hours following death, their remains may stay articulated. Allison (1986) demonstrated in tumbling barrel experiments that arthropods such as shrimp, which provide possible analogs to trilobites, may be potentially moved even tens of kilometers without disarticulating, provided that this happens within the first few hours following their demise.

Trilobite fossils also may provide evidence for reworking or transport of skeletons that would otherwise remain unsuspected. It is not unusual in marine mudrocks to observe concentrations of shells in localized pockets, typically associated with evidence for current scour. Such accumulations commonly display convex-downward or lateral-oblique positions along linear features that may be interpreted as indicators of storm-generated scour and fill.

Many skeletal elements, including articulated trilobites, as well as their cranidia and pygidia, have approximately concavo-convex dish-shapes. Random or preferred convex-up or convex-down orientations may be observed in different trilobite assemblages, and each has distinct implications for burial conditions. Instances of random or nearly one-to-one ratios of convex-up and -down orientations occur primarily in heavily bioturbated sediments, where, in addition, skeletal elements may display lateral or edgewise orientations. The mixing of skeletal components into sediment may account for the randomizing of orientation. However, such orientations are unstable under hydrodynamic currents and hence, are probably one of the better indicators of bioturbation or some type of protective concentration trap on the substrate.

Concavo-convex trilobite skeletal elements and whole bodies, like many shells, commonly display preferred convex-up or convex-down orientations. For example, cephala and pygidial shields in a preferred convex-up position are probably most typical of concentrated trilobite beds. Flume studies (e.g., Hesselbo 1987, Lask 1993) showed that even gentle currents will affect trilobite remains resting on the seafloor in such a way that they flip to a hydrodynamically stable orientation, at which point currents will glide over their streamlined, convex-up surfaces. The effect is particularly evident in areas of muddy substrate because of the impedance (i.e., frictional drag).

Hence, the occurrence of abundant convex-up trilobite cephala or pygidia on bedding planes may provide evidence for reworking of skeletal material under slightly current-agitated conditions. Beds of this sort represent skeletal material that was processed by one or more storm-generated current events. Excellent examples are found in large assemblages of *Flexicalymene* and *Isotelus* pygidial and cephalic shields on the bases of hummocky laminated calcisiltite beds, and pavements of *Pseudogygites* from the Ordovician Collingwood Shale (Brett, in prep.).

Beds of preferentially concave-upward trilobite tagma are not as common. In settings where the molt parts are suspended briefly and allowed to resettle from suspension, they will almost invariably settle concave-upward (Lask 1993). Such reorientation would occur in areas very close to storm wave-base in which the rather gentle storm-generated waves would lift skeletal materials temporarily off the bottom and allow them to free-fall back to the substrate. Rapid burial following this stirring would incorporate such shells as a basal pavement of a possibly graded mudstone layer. Excellent examples of this mode of burial are seen in pavements of *Triarthrus* cranidia in Ordovician dark shales. Counts of many bedding planes from the Ordovician Collingwood Shale of Ontario show a predominance of concave-upward orientations of the cranidia and of tiny ostracode valves, even on bedding planes in which larger skeletons are random or convex-upward (C. Brett, unpublished observation, 1998). We suggest that under certain conditions only the smallest, lightweight skeletons were suspended and then resettled—typically concave-upward.

Stacks of nested or shingled fossils apparently occur where densely packed skeletal remains were affected by oscillatory, storm-generated waves or currents, and provide an indication of deposition well within storm wave-base. A similar mode of preservation in shell accumulations involves concavo-convex shells bundled together in nested groupings, either convex-upward or -downward and often both in the same layers. These, too, seem to reflect the effects of settling and concentration of skeletal remains during turbulence events, and they are commonly associated with minor sediment grading. An intriguing example consists of masses of nested cephalic and pygidial shields of bumastid trilobites from Silurian bioherms, such as those in the Silurian Irondequoit Limestone. These may represent molt parts that were concentrated in sheltered "pockets" or scours on bioherm surfaces.

A particularly intriguing and still enigmatic aspect of orientation in trilobites is the predominantly concave-upward (dorsal shield down) orientation of articulated trilobites. Many occurrences of clusters of trilobites display this phenomenon, including mass occurrences of *Ceraurus pleurexanthemus* from the Ordovician Trenton Limestone of New York (Brett et al. 1999), beds of *Dalmanites limulurus* in the Silurian Rochester Shale (Taylor and Brett 1996), and aggregations of *Eldredgeops rana* from the famous trilobite beds of the Lake Erie region (Speyer and Brett 1985). Several explanations for such orientations have been put forth. Raymond (1920a) believed that these represented examples of trilobites buried in a life position, postulating that these animals swam upside down, like horseshoe crabs, and perished in this position. However, most later authors have considered this unlikely and argued for a postmortem reorientation of carcasses. Once again, if the concavo-convex carcasses of trilobites were suspended and resettled freely, they would very likely assume this position. Several features associated with the dorsal-downward trilobites may bear on this issue. For example, Speyer and Brett (1985) noted that some of the "inverted" trilobites were headless and thus, likely molted thoracopygidia. Such specimens obviously were not "swimming," and yet they are found alongside complete specimens that are also predominantly convex-downward. Even among these latter specimens there may be signs of incipient disarticulation (e.g., cephalon or pygidium is rotated slightly away from thorax; thoracic segments are very slightly pulled apart). Such observations suggest that the trilobites represent carcasses that had undergone a very slight amount of decay prior to burial. In all of the above-mentioned cases, there is also a vaguely to strongly preferred long-axis orientation (e.g., see Brett et al. 1999). This would also seem to imply that the trilobites reflect carcasses or molts that were transported, if only slightly, by currents after death. As noted already, however, currents usually have the effect of flipping concavo-convex elements to a preferred convex-upward position. If trilobite remains were lifted into a current slightly and then resettled, they might still assume this position. Also, the generation of decay gasses within the body cavities of trilobites might give them slight buoyancy and make them more subject to the lifting and settling required to invert the carcasses. All such inferences seem to point to a brief interlude between mortality and final burial.

A few occurrences of predominantly convex-upward outstretched trilobites are also known; for example, groupings of *Eldredgeops milleri* from the Silica Shale typically show this orientation. These are commonly associated with perfectly enrolled trilobites and they may represent examples of more nearly instantaneous burial with little disturbance.

Speyer (1987) and Brandt (1985) also considered the possibility of preferential orientation of enrolled trilobites but came to different conclusions. Brandt reported essentially random orientations in some mass occurrences of enrolled *Flexicalymene meeki* from Upper Ordovician shales of the Cincinnati, Ohio, region. In contrast, Speyer recognized preferred and species-specific orientations in Devonian enrolled trilobites from New York. *Greenops* specimens were most commonly found in a cephalon-up position, while *Eldredgeops rana* more commonly showed cephalon-lateral or -downward positions. Speyer suggested that these findings recorded different modes of pre-enrollment behavior by the trilobites.

The preferred azimuthal (compass bearing) orientation of elongate skeletons has been the subject of numerous studies, including both observational and experimental studies (see Kidwell and Bosence 1991, for summary). From such studies, it has become common knowledge that elongate particles typically orient themselves selectively within a current. Elongate objects that do not roll, such as the trilobite carcasses mentioned earlier, commonly will be aligned parallel to the direction of the current. Such may be the case with consistently aligned specimens of *C. pleurexanthemus* reported by Brett et al. (1999). The occurrence of some aligned trilobites in elongated "windrows," such as in the Silurian Rochester Shale, may represent accumulation in very minor scours or gutters (Whiteley and Smith 2001).

Fragmentation and Biased Preservation

Trilobite skeletal elements are variably affected by biotic and abiotic destructive agents postmortem. The degree to which the skeletons are affected is a function not only of the delicacy of their original construction but also of residence time on the seafloor.

Once elements are disarticulated, they may be acted on by various hydrodynamic processes that serve to sort, fragment, abrade, or corrode them. Hydrodynamic size sorting is uncommon in most mudrocks, for which generally low-background environmental energies are characteristic. However, size sorting may occur on the scale of a single graded bed as a result of turbulent events that briefly suspend skeletal particles and sediment and allow them to resettle.

Biased ratios of skeletal parts, for which original proportions are known to have been one-to-one, may provide evidence of sorting or, more frequently in the case of mudrocks, of preferential destruction. For example, it is a common observation that certain portions of trilobite skeletons are preferentially preserved, whereas the others tend to be fragmented. This results in a selective bias that is not related to size but rather to relative robustness of the skeletal parts. For example, the rather robust pygidia of dalmanitids and asaphids are more commonly preserved intact than are cephala. Pygidia-cephala ratios up to ten-to-one were found in some spectacular bedding plane pavements of *Pseudogygites latimarginatus* from the Collingwood Shale in Ontario (C. Brett, unpublished data). Conversely, for the small trilobite *Triarthrus eatoni*, from the same beds the small, fragile pygidia are very rare, and some bedding planes show nothing but cranidia ("head shales"). Beds with approximately one-to-one ratios of cephala and pygidia may indicate rather rapid and intact burial, whereas those that show a strongly biased ratio probably represent time-averaged accumulations in which rather long exposure times on the sea bottom prevailed. During these intervals, minor physical disturbances, such as multiple storms or even biotic disturbances, fragmented the less rigid skeletal parts preferentially.

Abrasion, Corrosion, and Encrustation

It is often difficult to distinguish between trilobite skeletons that have been physically abraded and those that have been corroded by biogeochemical processes. As a generalization most trilobite remains are too fragile to withstand prolonged abrasion, and this fragility may account for the rarity of trilobite skeletons in some nearshore, sandy environments in which trace fossils indicate that trilobites were common. It is well known that clay-sized sediment is ineffective as an abrasive agent. Hence, truly abraded fossils are rare in mudrocks and, if found, might indicate a much more complex history to the deposit in which shells were transported into a quieter water environment by a turbulent event.

Corroded trilobite remains tend to occur in offshore, low-energy environments, and bioerosion, likewise, tends to predominate over physical abrasion in these offshore settings (Kidwell and Bosence 1991; Parsons and Brett 1991). Many apparently abraded shells in mudrocks probably have been chemically etched or acted on by microboring organisms.

Even in life, trilobite exoskeletons may become encrusted with epibiontic organisms, such as bryozoans and even brachiopods (Tetreault 1992; Kloc 1993; Taylor and Brett 1996). Postmortem trilobite remains may be encrusted both externally and internally. Internal encrustation provides an excellent indication that skeletons have lain disarticulated for a period of time on the sea bottom. The extent of encrustation may provide an indicator of exposure time, as well. Conversely, some shells in condensed deposits show few, if any, epibionts. Microborings of forms such as endolithic algae may be recognized and may provide particularly useful indicators of exposure and, in some cases, of relative depth (Golubic et al. 1975; Vogel et al. 1987). Endolithic algae of various sorts, for example, are confined to various portions of the euphotic to upper dysphotic zone. Algal microborings in *Dicranurus* specimens have been used to suggest that these trilobites lived in the euphotic zone. Kloc (1997) suggested that abundant encrusters on the cephalic spines of these trilobites may have settled during the life of the trilobite and served as camouflage against visual predators.

Fossil Diagenesis: Geochemical Processing of Potential Fossils

Early diagenetic phenomena comprise the physicochemical processes that act on organism remains primarily after burial. Diagenetic features of fossils may provide information regarding the geochemistry of bottom waters and the upper "taphonomically active zone" (TAZ) of the sediment column. Diagenetic features of note include evidence for early dissolution, compaction, and mineralization of fossils.

The relative timing of dissolution is commonly recorded in skeletons. Trilobite exoskeletons were impregnated with calcite and thus are relatively resistant to dissolution and are preservable; however, they still may show evidence of early dissolution. Trilobite skeletons that are dissolved prior to compaction may leave no record but, in many cases, may be preserved as plastically deformed molds (Seilacher et al. 1985). Such preservation would indicate undersaturation with respect to calcite in the upper sediments and possibly low pH conditions. Conversely, many fossil skeletons and their molds display mosaic fracture patterns on their surfaces; these are particularly prominent on large shields such as the cephala and pygidia of *Isotelus* species. Such skeletons remained hard in early phases of compaction, which caused brittle fracture. (In rare instances, both plastically deformed "ghosted" specimens and brittly fractured (or unfractured) well-calcified specimens occur on the same bedding planes. Such associations have been used to suggest the presence of "soft-shelled" (immediately postmolt) and intermolt individuals (Speyer 1987).)

Early diagenetic minerals such as siderite, calcite, and pyrite generally form as a result of the action of anaerobic bacteria and are partly composed of their respiratory by-products (Allison 1988a). They may provide valuable information on sediment geochemistry and rates of burial.

One of the best understood of the early diagenetic minerals is pyrite, iron disulfide (Berner 1981a; Canfield and Raiswell 1991a, 1991b). Pyrite is common in many marine mudrocks and is commonly associated with fossil trilobite remains. The ferrous iron required in pyrite formation is available in terrigenous sediments, and dissolved sulfate is abundant in marine water (but not in fresh water). Under aerobic decay of organic matter, the major respiratory by-products of organisms are water (H_2O) and carbon dioxide (CO_2). However, under anaerobic conditions particular types of bacteria, referred to as *sulfate-reducing bacteria*, use sulfate (SO_4^{--}) as an oxygen donor for metabolizing organic matter, and produce hydrogen sulfide (H_2S) and bicarbonate (HCO_3^-) as by-products. Iron reduction, mediated by a second group of anaerobic bacteria, can generate ferrous ions, which in turn may react with the H_2S generated by sulfate reduction, to produce the precursors of pyrite (Canfield and Raiswell 1991b). The presence of pyrite shows that the sediment was anoxic but does not necessarily demonstrate that the overlying water column was anoxic. Pyrite can form either very early or relatively late in the burial history of a sediment (Hudson 1982; Brett and Baird 1986; Allison 1988a, 1988b; Canfield and Raiswell 1991a).

Under conditions of high organic-matter production, anoxic conditions may extend into the water column; in such euxinic settings H_2S is in excess, and any iron that is introduced into the system is pyritized as it is deposited. Thus, pyrite tends to be distributed evenly in the sediment as disseminated tiny crystal aggregates called *framboids*; it is not concentrated around any organism remains that may settle into these settings. Under oxic bottom-water conditions, however, organic material is not distributed so uniformly because much of it is degraded aerobically at or near the sediment-water interface, and the sediment becomes anoxic but nonsulfidic, in Berner's (1981a) terminology. Thus, pyritization tends to occur more locally within the nonsulfidic sediment, particularly in the vicinity of anaerobically decaying organic matter. As a result of local sulfate reduction, sulfide is liberated around this decomposing organic material. Dissolved iron will react at the site of sulfate reduction so that pyrite is restricted to anoxic organic-rich microenvironments within a broadly dysoxic, low-organic setting. Well-preserved pyritized fossils, including trilobites tend to occur in bioturbated gray mudstones and thus are indicators of bottom-water oxygenation (Brett and Baird 1986; Allison and Brett 1995). The earliest phases of pyrite tend to be fine-grained fillings of cavities, such as the interiors of enrolled trilobites. Later generations of larger crystalline "overpyrite" may nucleate on existing pyritic cores. In this way, pyritic nodules may form (Figure 3.2).

There are a number of Middle Devonian fossil beds in New York wherein large amounts of pyrite are found associated with the fossils (Dick and Brett 1986). The best known of these is the Alden Pyrite Beds in the Ledyard Shale. The Ledyard Shale is a dark-gray, generally poorly fossiliferous shale. However, a horizon in the lower Ledyard in western New York is rich in pyrite nodules and pyritized molds of fossils of all sorts. Most common are the fossils that had calcitic or aragonitic shells in life, such as pelecypods, brachiopods, nautiloids, ammonoids, and trilobites. The pyrite nodules have fossils at their core, which indicates that the original pyrite, formed from the organic decay, acted as a nucleus for over-pyrite precipitation. *Greenops grabaui* is the most common trilobite in these beds, and some specimens are partially pyritized and have a solid pyritic core and some coiled specimens form the nucleus of a pyrite nodule. These beds appear to represent the rapid burial of organism bodies in an otherwise low-organic sediment. The high concentration of pyritic material around burrows and fossils indicates nucleation around local centers of anaerobic decay associated with buried organic matter.

Anaerobic decay processes, including sulfate reduction, also generate HCO_3^- (bicarbonate), which may initiate the precipitation of calcite or siderite concretions around decaying organic matter. Where trilobites or other fossils are enclosed within carbonate concretions, they are nearly always three-dimensional. This proves that the concretions formed early and before organism remains could be compacted by overburden pressure.

It is less common for phosphatic nodules to form because phosphorus is only present in very small quantities in seawater. However, the anaerobic decay of organic matter does liberate phosphate-bearing compounds to solution. Also, phosphate can become adsorbed to ferrous hydroxides in the sediment. As these ferrous hydroxides are buried and pass through the anoxic-oxic boundary, they are reduced. This process also liberates phosphates to pore-water. Dissolved phosphate may be released back to the water column if anoxia persists to the sediment-water interface. However, if a micro-oxidized zone exists in the upper sediment, then the phosphates may be reprecipitated, especially around phosphatic skeletal nuclei, such as the chitinophosphatic material forming arthropod skeletons (Swirydczuk et al. 1981; Berner 1981a; Allison 1988b). Trilobites are not uncommonly phosphatized. Phosphatization of trilobites occurs primarily under conditions of slow sedimentation. Thus, for example, in the Middle Devonian Hamilton Group, phosphatized internal molds of enrolled trilobites may occur at minor disconformities.

If the sedimentation rate is high, then the time spent by any particular sediment layer at this micro-oxidized interface will be low, and the phosphorus concentration in pore-water will be increased only slightly. Conversely if the sedimentation rate is low, then the time spent at the interface will be high. Thus, a large proportion of the adsorbed phosphorus compounds will be concentrated at one layer in the sediment. Such concentration can increase pore-water levels of phosphorus so that phosphate minerals can precipitate. These minerals may replace organic remains or form concretions. Thus, the occurrence of phosphatic fossil molds or concretions is nearly always an indicator of low rates of sedimentation.

Other forms of diagenetic modification of trilobite material are uncommon to absent in New York. There are a few references to silica replacement of exoskeletal material in trilobite protaspids but no observations of phosphate replacement or carbonized specimens.

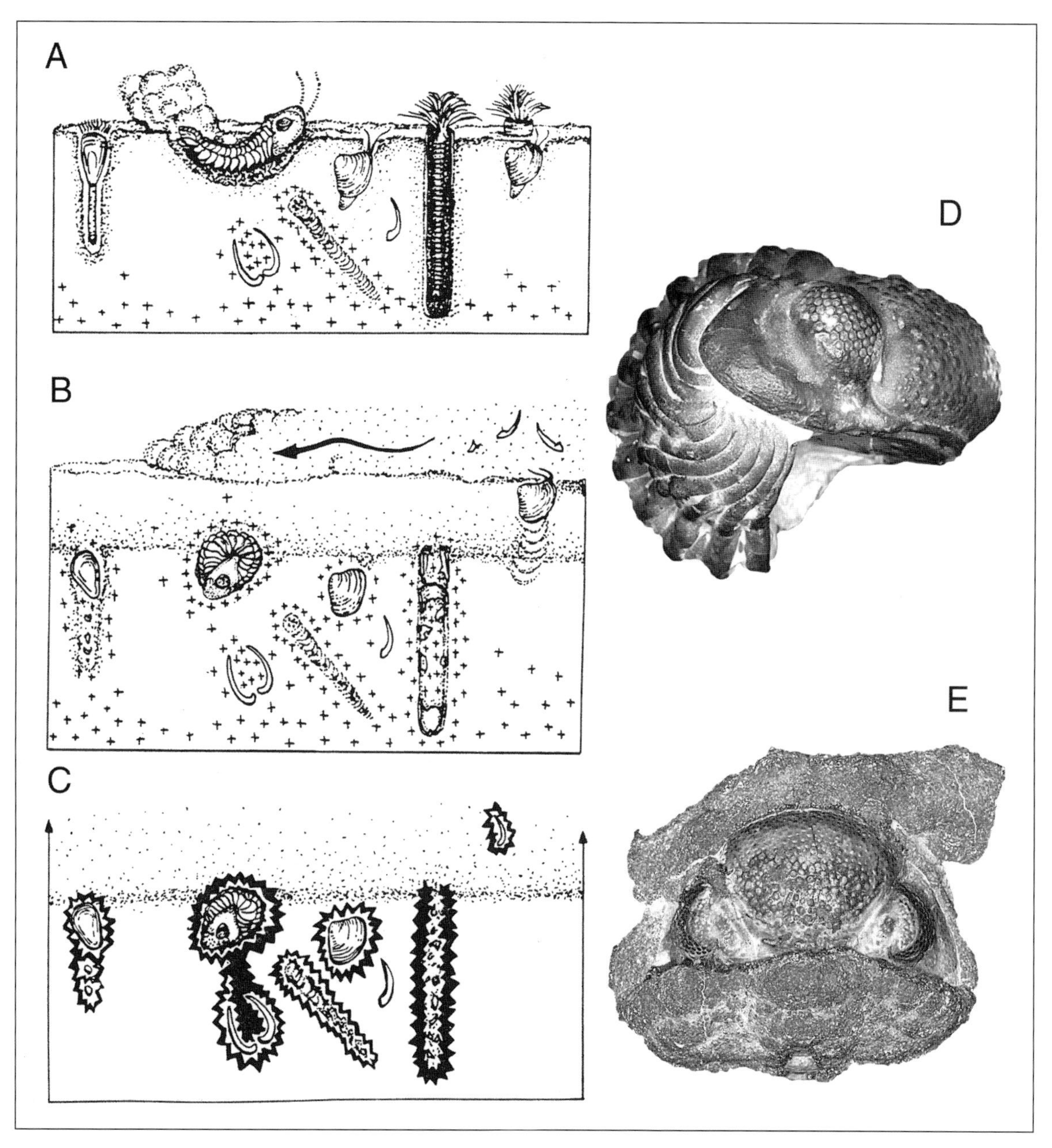

FIGURE 3.2. Conditions for the formation of pyritized, well-preserved fossils. A. Fauna on a bottom with poor oxygenation. There is low diversity, and the bottom is anoxic a short distance beneath the surface. The plus signs (+) indicate anoxic conditions. B. Rapid burial by sediment and the reduced oxygen levels of the water raise the anoxic level to the newly buried organisms. Anoxic bacteria in metabolizing organic matter of the trilobites also reduce the sulfate in the pore water, resulting in sulfide ions that then react with reduced (Fe^{2+}) iron in the sediment. In a sediment rich in terrigenous material, the iron level is high enough to cause iron sulfide precipitation at or close to the decaying organic material. C. Pyrite accumulates at the source of the organic decay, and the fossil is covered with pyrite at this nucleation site. A–C from Speyer and Brett (1991). Reproduced with permission. D. A trilobite buried in a rapid burial event but in an oxygenated sediment so that only a minor amount of pyrite formed. E. A coiled trilobite buried under anoxic, low-organic (except at the site of trilobite decay), and iron-rich conditions, resulting in a total covering of pyrite, PRI 49664.

More complete information on the processes of fossil preservation can be found in collective volumes edited by Briggs and Crowther (1990); Allison and Briggs (1991a, 1991b); Donovan (1991); Einsele, Ricken, and Seilacher (1991); and Martin (1999).

Trilobite Taphofacies

Various aspects of fossil preservation can be combined into the recognition and description of taphonomic facies or taphofacies (Speyer and Brett 1986, 1991). Together with lithofacies, biofacies, and ichnofacies, taphofacies tend to vary predictably with sedimentary environments, as shown by studies in modern marine settings (Parsons and Brett 1991). The modes of preservation of fossils can provide important insights into a number of features of mudrock deposition, including (1) the sedimentary environment (depth, temperature, salinity, oxygen level, substrate consistency); (2) the dynamics of sediment accumulation, average rates, as well as evidence for episodicity of sedimentation and erosion; (3) the temporal scope of individual mudrock units; and (4) the sediment geochemistry and early diagenetic environments.

The overall condition of trilobite skeletons can be assessed qualitatively or quantitatively but certainly should be noted in the field. Semiquantitative indices can be formulated by determining the proportion of skeletal parts in different, arbitrarily defined preservational states. In such cases, it is commonly useful to determine a set of standards with which particular shells can be compared and assigned to a category, in much the same way that grain shapes and roundness indices have long been assessed on the basis of standardized profiles by sedimentologists. Skeletal condition is particularly valuable for recognizing qualitatively the differing relative extents of sedimentary time-averaging and therefore, may be related to burial rates.

In their original formulation of the notion of taphofacies, Speyer and Brett (1986) used Middle Devonian trilobites and their modes of preservation to exemplify the general model. They also subsequently emphasized the fact that fossils in a particular taphofacies may show both background and event taphonomic signatures. That is, there may be distinct aspects of preservation of fossils under day-to-day conditions or under episodic catastrophic burial conditions in a given environment (Speyer and Brett 1988). The well-known Devonian trilobite species *E. rana* and *Greenops* species occur in many distinct associations representing different environments. Yet, these trilobites show very distinctive modes of common preservation that are unique to particular onshore-offshore positions and sedimentation conditions. For example, in high-energy shallow water, most trilobite remains are disarticulated, abraded fragments, although articulated, typically outstretched individuals may occur occasionally, owing to pulses of burial. In low-energy, fully oxic settings, mainly unbroken, though typically disarticulated trilobite material is preserved; under episodes of higher sedimentation, complete typically outstretched, inverted individuals may occur, even in clusters. Under lower sedimentation the event signature comprises primarily intact molt ensembles, with few if any fully articulated carcasses. Finally, under low-energy, dysoxic (background) conditions, intact molt parts and tagma are the rule; the event signature of this setting is distinctive in showing an abundance of enrolled, commonly pyritized, specimens of trilobites.

Each taphofacies records different types of information. The distinctive modes of preservation provide information on sedimentary environments that cannot be determined otherwise. For example, the enrollment of trilobites may reflect a response to toxic stimuli triggered by a stirring up of anoxic, sulfide-rich muds. Similarly, pyritization suggests burial in anoxic, low-organic muds.

In this way, the study of trilobite preservation has helped to provide a new tool in paleoenvironmental analysis. Development of predictive models relating preservation to depositional environments and positions in sedimentary sequences, in turn, may aid paleontologists in prospecting for new fossil bonanzas, including spectacular trilobite beds.

Trilobite Lagerstätten

Lagerstätten (derived from the German mining term translated loosely as "mother lodes") are extraordinary fossil assemblages. Trilobite Lagerstätten include obrution deposits, reflecting a rapid smothering of benthic faunas by sediment, yielding fully articulated remains and Konservat-Lagerstätten, in which even soft parts are preserved by a combination of rapid burial, anaerobic decay, and early diagenetic mineralization (Seilacher et al. 1985). In this section we describe examples of trilobite Lagerstätten from New York State to illustrate the general taphonomic concepts.

A famous trilobite site is Beecher's Trilobite Bed in the Upper Ordovician Frankfort Formation just north of Rome, New York. A 5-mm-thick, light-gray layer within a dark-gray mudstone contains an unusual collection of fossil material (Cisne 1973). The *Triarthrus eatoni* specimens within this layer have their appendages and some internal organs replaced by pyrite. Beecher (1893, 1894, 1895) reported the discovery by Valiant and the preparation of specimens. Cisne (1975, 1981) prepared very-high-resolution radiographs of Beecher's specimens and was able to report on internal structures never seen before. Whittington and Almond (1987) also examined specimens from the beds and suggested that certain structural elements of the appendages were utilized in food transport, among other things. Replacement of organic tissue in such high resolution by pyrite is very unusual. Briggs, Bottrell, and Raiswell (1991) examined these trilobites and concluded that this type of soft tissue replacement was due to bacterial decay in anoxic conditions resulting in sulfide formation. The rocks represent deep-water turbidites, and anything buried probably would have been subject to anoxic conditions.

The concentration of organic matter in the sediment was relatively low, and the concentration of iron in the pore-water relatively high. In this situation the sulfide produced at local decay sites was precipitated as an iron monosulfide (a precursor to pyrite) at the site where it was formed, resulting in the nearly perfect replacement of soft tissue by pyrite (Briggs and Edgecombe 1993). Beecher's Trilobite Bed remains the most productive source of preserved trilobite soft tissue known.

Trilobites found on limestone bedding planes typically are compressed for the same reasons that trilobites are compressed on shale partings. Their history is similar in that they were buried by a calcareous mudflow with little transport. Some of the best trilobites, however, from the physical preservation standpoint, are found within limestones. In some instances, these trilobites were apparently caught up in a calcareous sediment flow, killed, probably transported some distance, and entombed in the settled sediment. In such limestones the trilobites typically are randomly oriented, with the bodies flexed in unusual postures, and retain much of their original three-dimensional character. Large numbers of *Isotelus gigas* have been taken from the limestones of the Middle Ordovician Trenton Group in central New York. The specimens from bedding planes are flattened and commonly upside down, while those from within the limestone retain their three-dimensional character.

In the Walcott-Rust Quarry within the Trenton Group, several layers of the thinly bedded limestone have yielded excellent trilobites (Brett et al. 1997, 1999). One thin micritic limestone yielded large numbers of articulated *C. pleurexanthemus*, *Flexicalymene senaria*, and *Meadowtownella trentonensis* (but no *I. gigas*). More than 98% of the trilobites on the base and those within the layer were upside down while 60% of those on the upper surface, including many that were still partially within the limestone, were right side up. The thinner bed probably records a large population of trilobites caught in a current and transported far enough to be mixed within the small amount of sediment involved. Most of the trilobites were killed in the process. The larger more robust ones may have managed to struggle to the surface and die there or were covered and killed by a subsequent event. The observation that no *I. gigas* were involved suggests that they were large and strong enough to escape. Two other, thicker beds had no trilobites on the base or top but well-preserved specimens internal to the limestone. These internal trilobites, which included *I. gigas*, were randomly oriented.

Within the thin bed, coiled and semicoiled *C. pleurexanthemus* and *F. senaria* were discovered by C. D. Walcott (who also discovered the Burgess Shale), and there was evidence of appendages preserved as calcite infillings (Walcott 1876, 1877b). This unique mode of preservation is believed to result from anoxic bacterially induced calcite precipitation within the appendage, followed by calcite in-filling as the appendage material decayed (Brett et al. 1997, 1999). The *Ceraurus* bed is a thin, 15 to 50-mm bed that normally would be expected to be heavily bioturbated. A possible explanation is that the subsequent thicker beds were deposited a short time after this one, essentially sealing it and making any decay **anaerobic**. The low level of iron in the sediment precluded significant pyrite formation.

This form of appendage preservation is unique and was only found when Walcott made sections of the trilobites and observed them by transmitted light. The very fine calcite in-filling in a carbonate matrix is very difficult to see and evaluate by reflected light. This difficulty raises the question as to whether this mode of preservation is actually more common but generally unobserved.

The Lower Silurian Rochester Shale contains several horizons yielding numerous trilobites. Specimens of the large lichid genus *Arctinurus*, when articulated, are usually found right side up but flattened. *Dalmanites limulurus*, from several layers, is almost always inverted and often occurs in narrow, elongate "windrows" that may show some evidence of a preferred orientation. The living *Arctinurus* animals probably were buried in place by a heavy blanket of sediment, which resulted in their death. The *D. limulurus*, however, were transported somewhat before the burial process, aggregated, and current aligned in windrows. The fact that most are upside down is less easily explained. Convex surface down may be a preferred orientation during transport in a current, or the living animal may have been tumbled by the current and the upside-down trilobites may not have been able to right themselves. Again, alignment probably indicates slight transport of dead individuals by a current. Alternatively, it is possible that these trilobites swam upside down to escape the sediment and were buried in this position. A possible model for this latter behavior comes from the living horseshoe crab, *Limulus*, which normally crawls around the bottom but more rarely swims upside down.

The siltstones, mudstones, shales, and limestones of the Middle Devonian Hamilton Group have yielded enormous numbers of articulated specimens of trilobites. One well-studied bed, the Browns Creek Bed, a lime mudstone in the lower Centerfield Limestone, yields superbly preserved *Eldredgeops rana*, *Monodechenella macrocephala*, and *Pseudodechenella rowi*. This bed can be traced from Centerfield, Ontario County, to East Bethany, Genesee County, a distance of over 49 km (31 miles). The trilobites are found in random orientation, indicating they were swept up, tumbled, and transported some distance from their original position in the sudden event forming this bed.

Horizons within the lower Wanakah and the Windom members of the Hamilton Group are especially productive of *E. rana*. Especially notable are the "Grabau Trilobite Beds" (lower Wanakah Shale) along the Lake Erie shore near Eighteenmile Creek and the Smoke Creek Beds in the former Penn-Dixie Quarry at Hamburg, Erie County. These trilobite beds yield three-dimensional clusters of *E. rana*, which Speyer and Brett (1985) divided into molt and body clusters. The clusters are nearly intact, which suggests little transport. These beds are the result of very rapid burial and death, with very little turbulence during the burial event.

The *E. rana* specimens from the lower Wanakah Murder Creek Beds are found tightly coiled, fully outstretched, and in various semicoiled configurations in between these two extremes. Typically, in the semicoiled specimens, the pygidium appears to be missing. On close examination, however, the pygidium can always be found inside the body cavity of the trilobite. It is suggested that the pygidium was displaced when the tightly coiled trilobite, upon decay, partially opened and viscous mud filled the now-empty cavity.

The trilobites in the former Penn-Dixie Quarry are mostly in the Smoke Creek Beds of the Windom Member named for their outcrop on Smoke Creek, Erie County.[1] Well-preserved individual trilobites are common in these beds, and occasional clusters of 5 to 20 individuals are known. The general preservation is similar to that of the Murder Creek Beds.

In summary, taphonomic analysis is a major factor in understanding trilobite deposits, as well as fossil deposits in general. The understanding that comes from such analyses not only is intellectually rewarding but also provides a foundation for interpreting the geological history of fossil sites. There is also a predictive factor that cannot be ignored. For example, storm deposits over stable sea bottoms, below storm wave-base, often bury a living fauna, resulting in well-preserved articulated fossils such as trilobites and crinoids. Fossil preservation provides important clues in the interpretation of fossil deposits and ancient environments. Taphonomy and its concepts are powerful tools for general fossil collectors as well as professional paleontologists.

[1] The Penn-Dixie Quarry is a public fossil site operated by the Hamburg Natural History Society (P.O. Box 772, Hamburg, NY 14075; 716-627-4560).

4 The Paleozoic Geology of New York

The Paleozoic strata of New York State are a classic repository of fossils, including, at many levels, trilobites and a host of other invertebrate and even vertebrate and plant fossils. For details on the paleoecology and fossils, we recommend the publications by Linsley (1994), Isachsen et al. (1991), Landing (1988), Shaw (1968), and the references therein.

The purposes of this chapter are twofold: first, to give the reader both some general background regarding the tectonic, climatic, and paleoenvironmental history of life on earth (Archean through Pleistocene, Figure 4.1A), and more specifically of New York State and adjacent ancestral North America during the early to middle Paleozoic Era (Cambrian through Devonian, Figure 4.1B), the time during which trilobites lived and were deposited in New York's sedimentary rocks; and second, to provide some details on the stratigraphy, sedimentology, and paleoecology of the intervals from Early Cambrian to Middle Devonian that have yielded abundant trilobite remains. To these ends, the discussion of each Paleozoic time interval is subdivided into two portions: first, an overview of global and New York geological history, and second, details of the stratigraphy and sedimentary environments of the trilobite-bearing intervals, listed chronologically. Such general discussion of geology might seem out of place in a book devoted to trilobites; however, we believe that students of New York's trilobites should be well aware of these broader contexts in order to understand the environments and ecology of these remarkable ancient organisms.

Overview

The Paleozoic deposits of New York record a portion of the history of ancestral North America or **Laurentia** during some 200 million years of geologic time (Figure 4.1B), and as such, they cannot be understood in isolation from the region as a whole. During this interval, present-day New York lay generally in southern subtropical to warm temperate latitudes, with the equator running approximately centrally through Laurentia. Climates ranged from hot and arid to warm and humid during this time.

Prelude to the Paleozoic: Late Proterozoic Collisions and the Grenville Orogeny

The oldest rocks in New York, exposed in the Adirondack Mountains and the Hudson Highlands, are somewhat more than 1 billion years old. These are crystalline rocks that were metamorphosed or altered from older rocks by enormous heat and pressure (Figure 4.2). Some were originally igneous rocks, formed from the cooling and crystallization of magmas, and others are sedimentary deposits, such as quartz sandstones, limestones, and shales. These were transformed (metamorphosed) by recrystallization and under intense heat and pressure; for example, sandstones were altered to tough metaquartzites, limestones to marbles, and shales to mica-rich schists. By looking at the types of minerals formed within these rocks by metamorphism, geologists can be certain that the rocks now exposed in the Adirondacks were once buried up to 25 km within Earth during a great **orogenic** or mountain-building episode—about a billion years ago. This event, the Grenville Orogeny, apparently resulted when the (present) eastern edge of ancestral North America was overridden by another continent, perhaps the northwestern side of present-day South America. This enormous collision helped to weld together a supercontinent known as **Protopangea** or **Rodinia** (Figures 4.3 and 4.4). For nearly half a billion years, this supercontinent held together, and the massive

A

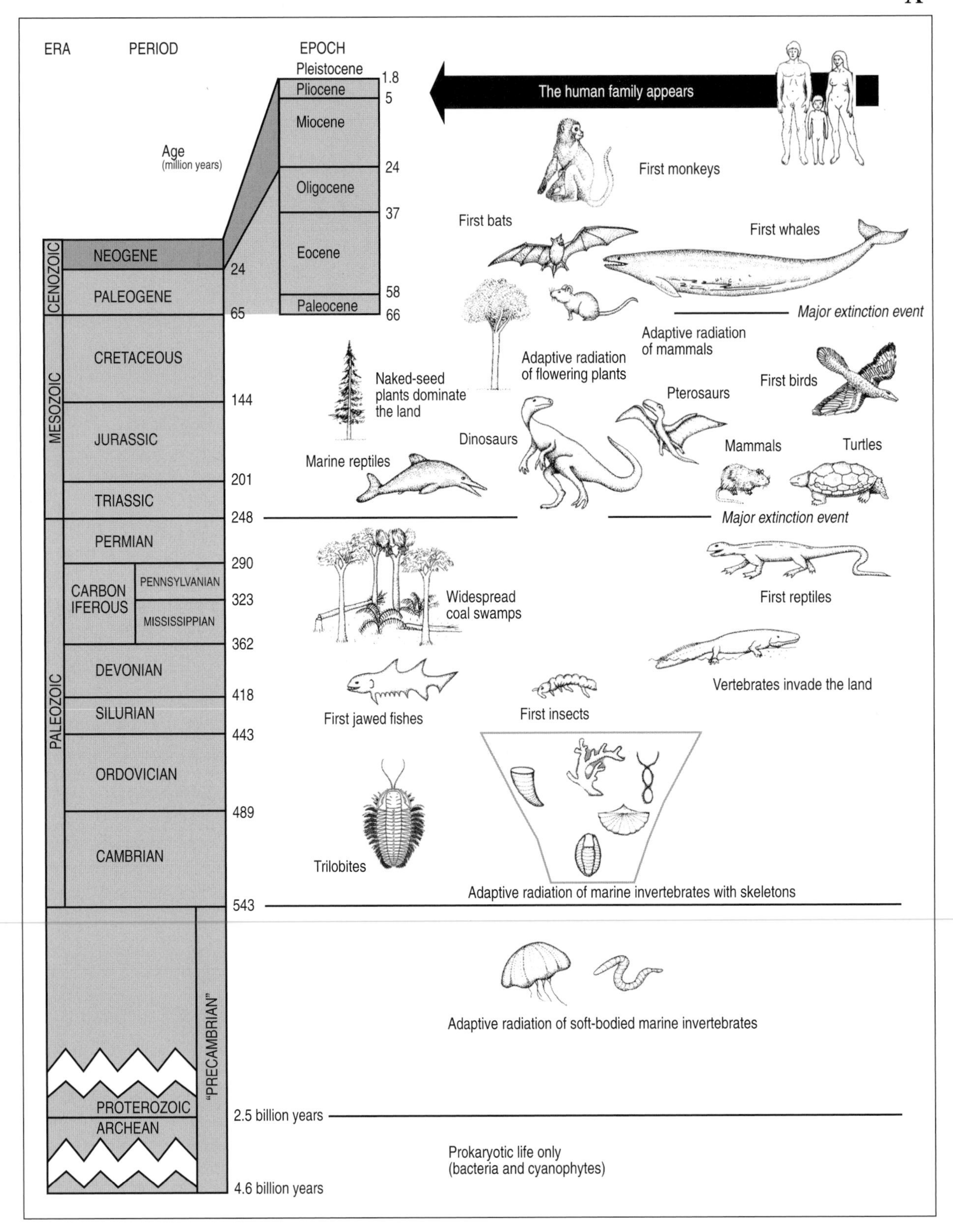

FIGURE 4.1. A. Time scale for Earth history and for life on Earth. The New York Paleozoic is from the early Cambrian to the end of the Devonian. Adapted from S. M. Stanley, *Exploring Earth and Life through Time*. New York: W. H. Freeman, 1989. B. Generalized geology and detailed scale for the Paleozoic rocks in New York. The vertical scale is linear and proportional to time; note that the dates of beginnings and ends of periods are given in millions of years before present. The right columns list the names of major unconformities and the names of Sloss supersequences. Most of the information is from a number of sources. The dates for the Cambrian are from Davidek et al. (2000) and Landing et al. (1998a,b). The T in the Cambrian time period indicates where trilobites first appear in the fossil record, 519 million years before present. The dates in the Epochs are the most current. The dates in the Stages have yet to be reconciled in the literature.

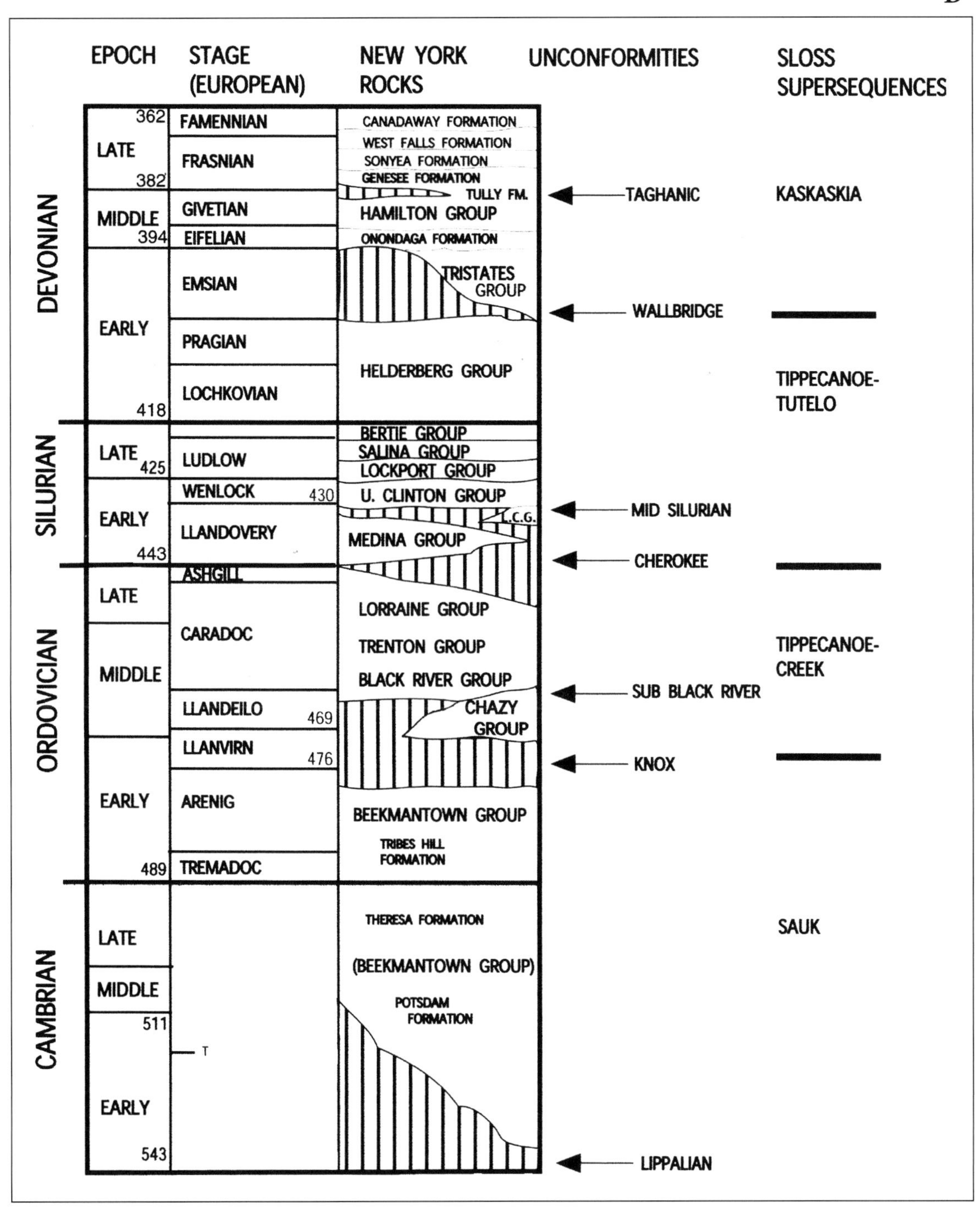

FIGURE 4.1. *Continued.*

Grenville mountain belt, fully formed by a billion years ago, was exposed in a life-less continental interior to the forces of weathering and erosion. We know that by about 550 million years ago, an entire thickness of continental crust had been removed and erosion had exposed the roots of the ancient Grenville Mountains.

Laurentia is the term geologists apply to the ancestral Paleozoic core of North America, lacking certain areas such as the present eastern seaboard region (eastern Massachusetts, eastern Nova Scotia, eastern Newfoundland), Florida, California, and other marginal areas that were added later. Laurentia was isolated as a separate continent during rifting and opening of new ocean basins that began approximately 700 to 600 million years ago during the interval of time referred to as late **Proterozoic** (or Neoproterozoic). Notably, on the present east side of Laurentia,

FIGURE 4.2. Precambrian Grenville (1.0 BP) metamorphic/igneous rocks. Note the granitic pegmatite (a) and smaller folded dikes (b) cutting gneiss. Rte. 12, near Alexandria Bay, Jefferson County.

rifting or fracturing of the crust began over 600 million years ago—some 400 million years after the Grenville Orogeny. The fractures and faults ultimately tore the supercontinent of Rodinia apart. Evidently, ancestral South America, which had collided with proto-Laurentia to form the Grenville Mountains, now pulled back away. A new ocean, the **Iapetus**, or Proto-atlantic, began to open along a line that would pass through present-day central New England and southward to the east-central Carolinas, a bit east of the Blue Ridge Mountains in the Appalachian chain (these mountains formed later).

At the beginning of the Paleozoic Era, about 545 million years ago, a narrow, but widening Iapetus Ocean lay slightly to the east of present-day New York State (Figure 4.5A). East of the present Berkshire (Massachusetts) and Green (Vermont) Mountains lay the edge of Laurentia, which, much like the present eastern edge of modern North America, formed a continental shelf bordered eastward (southward at that time) by a relatively steep drop-off along a continental slope into the deep water of the Iapetus Ocean.

Seawater was displaced upward out of the Iapetus Ocean, partly because of the expanding midocean ridge (or spreading center, an area of hot upwelling magma). This seawater spread out onto the craton of Laurentia and caused a major rise of the shoreline (transgression) up onto the old weathered remnants of the Grenville rocks.

Initially, a large volume of sediment that eroded from the ancient Grenville terrane was shed off the old weathered craton and into the narrow Iapetus basin. However, as the sea level continued to rise during the latest Proterozoic and Early Cambrian, the old land was finally flooded and the source of sediments cut off (Figure 4.5A). The entire eastern border of Laurentia had become a passive continental edge; a sea with a very broad continental shelf ultimately extended from the area of central New England and the mid Atlantic states region westward to the present Mississippi Valley during the Cambrian to Early Ordovician (Figure 4.5A, B). As spreading ensued in the Iapetus midocean rift, the ocean basin grew wider, at least up to the Early Ordovician time, about 480 million years ago, before the process began to reverse and the Iapetus basin began to shrink and ultimately close.

Cambrian Period

The Cambrian Period, as it is now dated, spans approximately 54 million years from about 543 to 489 million years before present. Yet this was one of the most significant times in the history of life, for it was during this time that nearly all phyla or major groups of animals appeared in the rock record.

The onset of the Cambrian was marked by a time of low-diversity fossil assemblages typified by "small shelly" skeletons

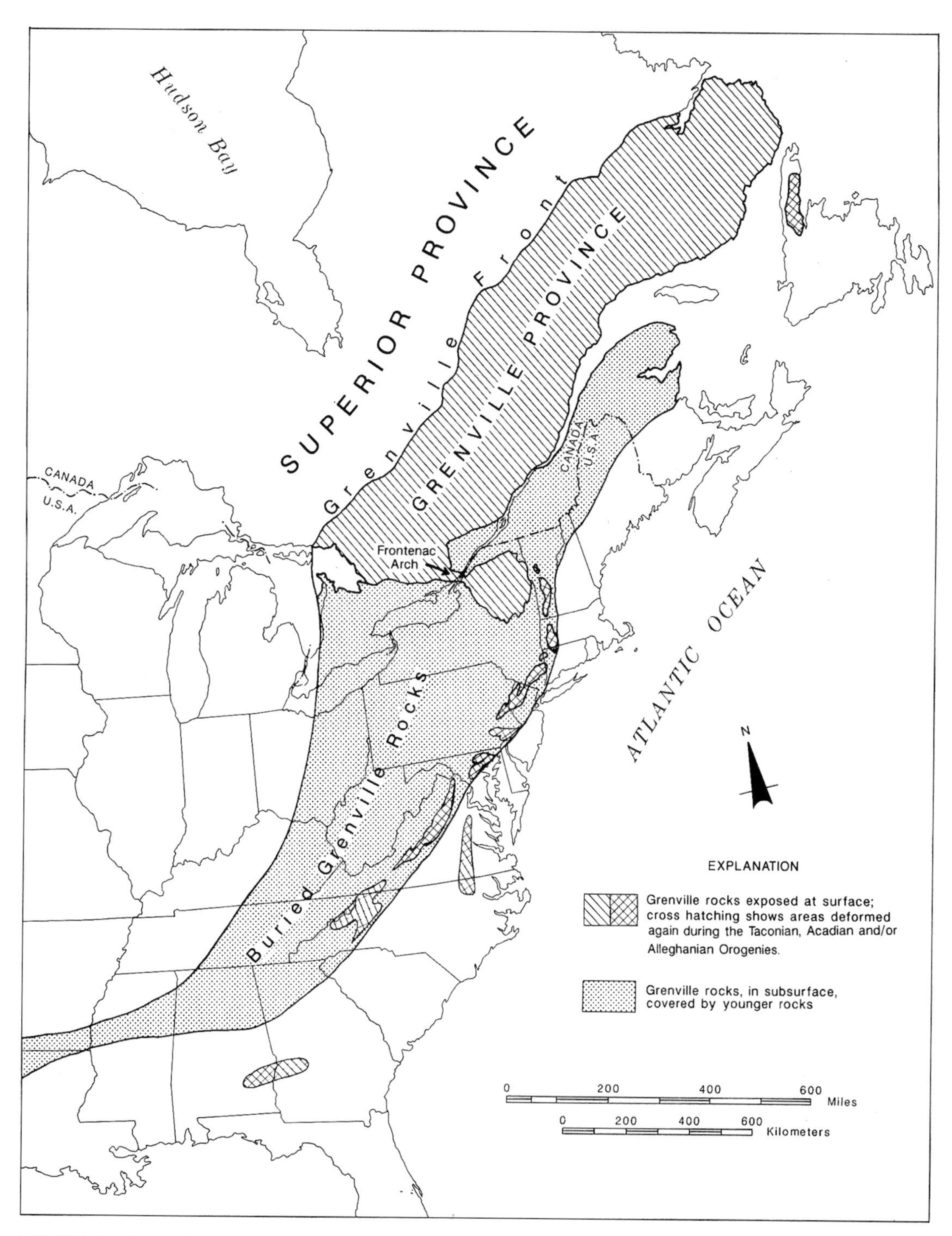

FIGURE 4.3. Map showing the extent of the Grenville belt in eastern North America. Rocks in this area were deformed and metamorphosed about 1 billion years ago. The dotted pattern shows regions where Grenville rocks are buried beneath younger strata; the slanted lines indicate areas where Grenville rocks are exposed; and cross-hatching shows areas where the Grenville rocks are deformed by later orogenies. Modern Adirondack Mountains lie just to the east of the Frontenac Arch (labeled). From Isachson et al. (1991). Printed with permission of the New York State Museum, Albany, N.Y.

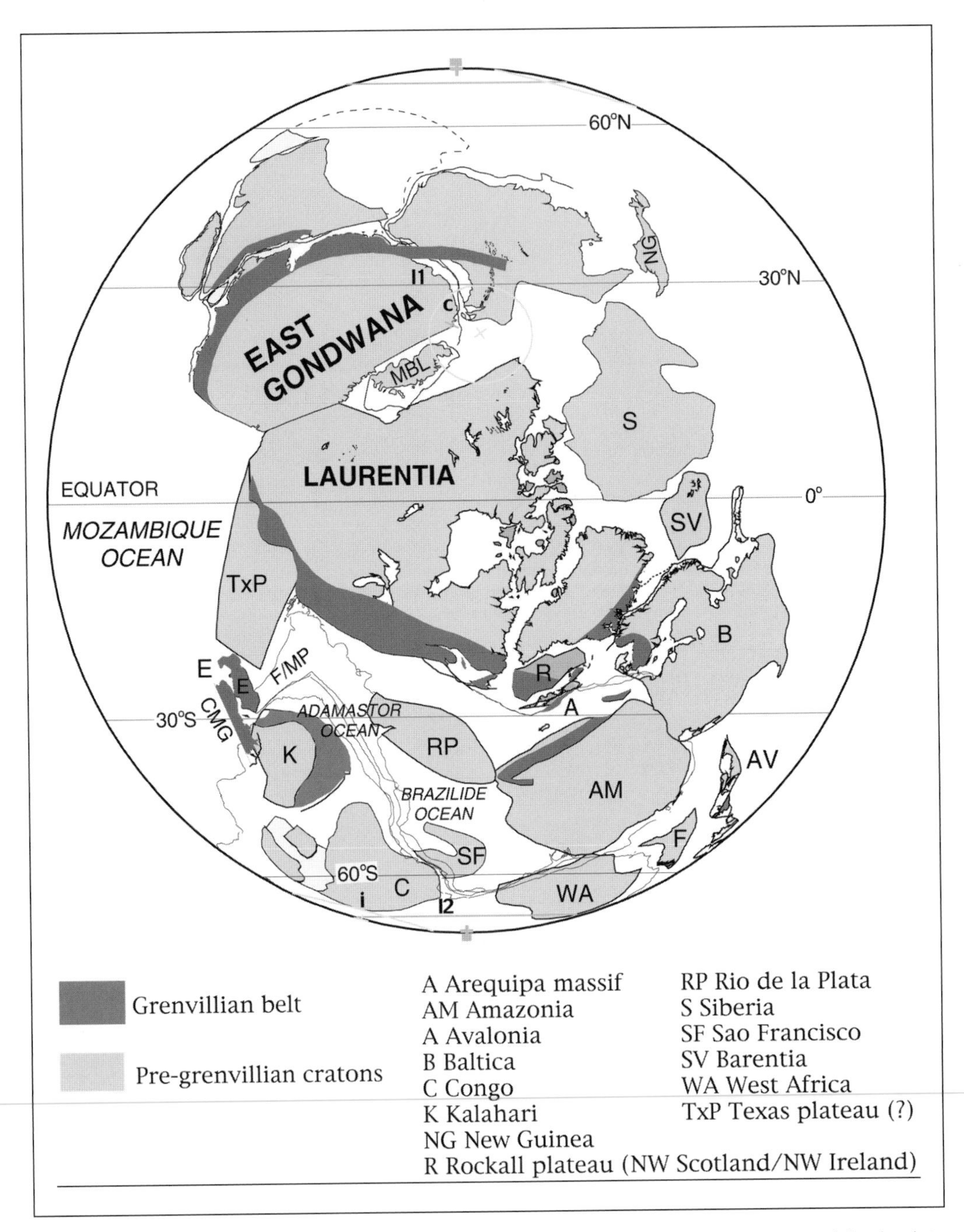

FIGURE 4.4. Paleomagnetic reconstruction of the supercontinent Rodinia as it existed in the late Proterozoic, about 700 million years ago. Note the position of Laurentia, near the center. The dark belt shows the position of the Grenville Orogenic belt. After Dalziel (1997), reproduced with permission.

including a variety of calcareous phosphatic tubes, rods, and plates. Only in later Early Cambrian time (~519 millon years ago) did trilobites first appear as body fossils, although fossil trackways (traces of walking on the bottom) and resting pits (*Rusophycus*) suggest that unpreserved soft-bodied trilobites, or similar arthropods, occurred earlier.

The Cambrian rocks of New York State are generally subdivided into two great packages of strata (Figure 4.7). The **autochthonous** (nontransported) rocks, which are in the site of original deposition, a shallow-shelf environment, and the Taconic **allochthonous** (displaced) rocks, mostly deep-water shales (now often slates) that were originally deposited to the east of New York in oceanic environments. The Taconic rocks were transported westward 80 km during the Middle Ordovician

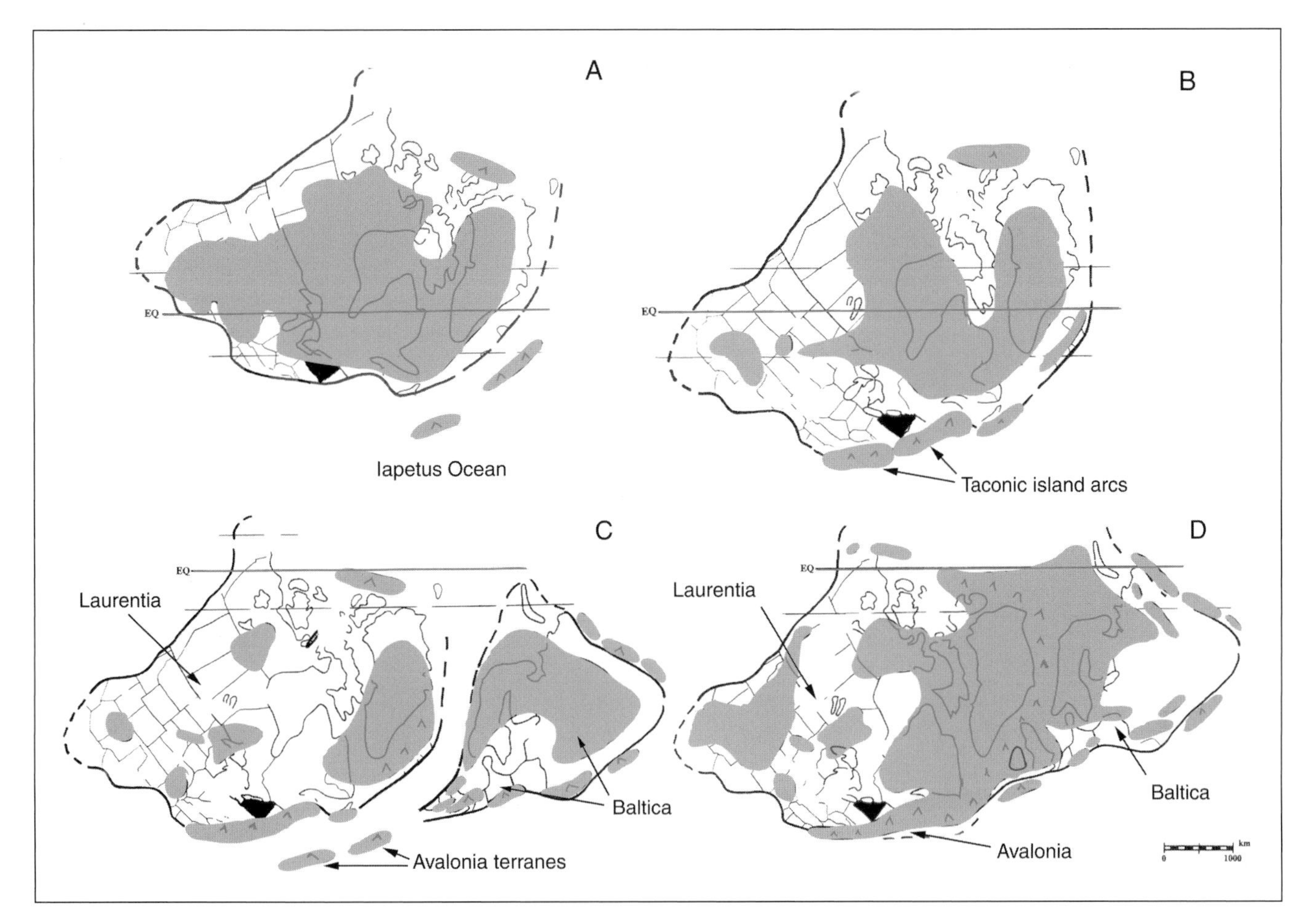

FIGURE 4.5. Position of the Laurentian plate and neighboring Baltica and Avalonia terranes from the Cambrian to the Devonian. Areas of exposed land are dark, areas covered by water are white, and New York is highlighted in black. A. Middle Cambrian as most of what is the United States was beginning to be flooded by shallow seas resulting in carbonate depositions in the Late Cambrian (note Taconic island arcs). B. Middle Ordovician with almost complete flooding of Laurentia and the beginning of the Taconic Orogeny with collision in the southeast. C. Silurian, with Baltica colliding with Laurentia from the east. Also note the smaller "islands of Avalonia." D. Middle Devonian, with the Acadian Orogeny fully developed due to collision of the Avalon terranes; note heavy sedimentation from the eastern mountains. Modified from Witzke (1990). Reproduced with permission.

Taconic Orogeny (a major mountain-building event) along major thrust faults. This deformation apparently was related to the convergence of an island arc complex with eastern Laurentia (or ancestral North America).

Cambrian Autochthon

The autochthonous rocks of the central Appalachians, which include basal sandstone and limestone or dolostone, represent extensive shallow, subtropical platform seas, sometimes referred to as the Great American Tidal Flat (Figures 4.5, 4.7). In southeastern New York and southern Pennsylvania these beds may be up to 5 km thick and were mainly deposited in very shallow, tide-influenced environments. The immense thickness is accounted for by active subsidence, or downward sinking, of the Cambrian continental shelf. Following the interval of **rifting** that produced the Protoatlantic or Iapetus Ocean, the continental shelf, which is represented by the region from New England westward into central New York, underwent rapid subsidence due to cooling. Lime mud and silt probably were produced by organisms such as algae, the growth of which was evidently able to keep up with the rate of subsidence so that these shelf environments remained in very shallow water, above the normal wave-base. Cambrian to Early Ordovician rocks form a package of sandstones and carbonates—the Sauk Sequence—bounded above and below by major unconformities.

During Cambrian and Early Ordovician time, eastern North America lay in a subtropical position, perhaps 25° south of the paleoequator (Figure 4.5). Today this zone is known as the "subtropical desert belt" because it is here that some of the driest

FIGURE 4.6. Cambrian rocks in New York. A. Upper Cambrian Potsdam Sandstone at Ausable Chasm near Plattsburg, Clinton County. B. Tilted Precambrian-Cambrian (Lippalian) nonconformity. Approximately 500 million years of nondeposition and erosion separate dark amphibolites (a) (metamorphic rocks) from the light gray, Upper Cambrian Little Falls Formation (b). Cut along U.S. Rte. 5S near Fonda, Montgomery County.

conditions on Earth develop. There is some evidence that ancestral North America in the Cambrian was relatively arid in the (present) eastern regions. There is also evidence for the buildup of slight **hypersalinity** in the Cambrian waters. Cambrian dolostones contain vugs and sometimes molds of evaporite crystals such as gypsum, anhydrite, and halite. The occurrence of **ooids** (small spherical, concentrically ringed calcium carbonate grains) and **stromatolites** (moundlike structures formed of sediment trapped by cyanobacteria or blue-green algae) also is typical of hypersaline (elevated-salinity) waters in the subtropics today. Whatever the case, much of the Cambrian seafloor in the New York area was relatively low in shelly organisms.

The broad continental shelf of the Cambrian sea was bordered to the east, in what are today central Vermont, western Massachusetts, and Connecticut, by an abrupt slope into deeper water. In the region flanking the continental shelf of North America, sediments accumulated very gradually (Figures 4.7 and 4.9).

Cambrian Trilobite-Bearing Autochthonous Rocks

The lowest and shallowest-water deposit of the Cambrian autochthonous rocks belong to the Potsdam Sandstone or equivalent sandy Little Falls Formation (Figures 4.6 and 4.8). The Potsdam Formation consists of up to 140 m of clean quartz **arenites** or quartzose sandstones that display **cross-stratification**, ripple marks, and other features indicative of deposition in shallow-wave and sometimes tide-dominated environments (Figure 4.8). The Potsdam sediments represent sand eroded from deeply weathered areas of the North American craton. The presence of herringbone **cross-stratification** (inclined bedding formed by alternate migration of ripples in opposite directions) is an indication of the oscillatory currents associated with tidal action. Some portions display large-scale trough cross-bedding and may represent wind-formed sand dunes in coastal areas.

The Potsdam is generally sparse in body fossils, although trace fossils (burrows, tracks, and trails) of a variety of forms are present, including the bizarre and huge (by Cambrian standards) *Climactichnites*. This elongate trail up to 30 cm (or more than a foot) wide closely resembles marks made by a tractor tire in soft sand. This trace is found in flat-bedded sands that may represent the upper foreshore or beach. Just what large organism in the Cambrian was able to come out into very shallow water is quite unclear. Traces may have been preserved by the sun drying initially wet sands of coastal areas. Yochelson and Fedonkin (1993) argued that it might have been a large gastropod-like mollusk. In slightly more offshore Potsdam facies, trace fossils such as vertical shafts (*Skolithos*) and U-shaped burrows (*Diplocraterion*) are quite abundant. Lingulid brachiopods and small fragments of trilobites also have been obtained in a few levels, but in general body fossils are rare.

The higher beds of the Upper Cambrian and those straddling the Ordovician boundary are carbonates. The Little Falls Dolostone of the Mohawk Valley, the Hoyt Limestone of the Saratoga area, the Whitehall Formation of Lake George, and the Theresa Dolostone of the Saint Lawrence region are carbonates: limestones, and dolostones, formed very late in the Cambrian or in earliest Ordovician time (Figures 4.7 and 4.9). They display an abundance of stromatolites. Some of the most famous stromatolites in the northeastern part of North America occur in the Petrified Gardens within the Hoyt Limestones near Saratoga Springs, New York (Figure 4.9).

The Little Falls Formation stromatolitic dolostones (up to 30 m thick) are only very sparsely fossiliferous. Only a single free cheek of the trilobite *Elvinia* has been found from the Cambrian Little Falls Formation, but this is sufficient to bracket its age within the second to last or Franconian Stage of the Cambrian. (The Little Falls is noted for its characteristic vugs or cavities that contain beautiful, doubly terminated quartz crystals referred to as Herkimer Diamonds; such crystals formed much later during deep burial of the Little Falls sediments.) Just why most Upper Cambrian carbonates are so poor in body fossils, including trilobites, but rich in stromatolites is as yet unclear. It may be that the seas were somewhat hypersaline. However, the Hoyt Limestone contains not only the famous stromatolites but also beds of oolitic limestone and some fossiliferous limestone in which **sclerites** (skeletal pieces) of trilobites belonging to a number of Late Cambrian species are found; Ludvigsen and Westrop (1983) recently described these species from the Hoyt and Galway limestones.

These trilobites are associated with other fossils, including brachiopods, molluscan fragments, and even plates of the world's oldest chitons. This diverse assemblage indicates relatively favorable, normal-salinity conditions in the area of Saratoga Springs during this period. However, relative high energy due to wave action and slow deposition prevented the easily disarticulated trilobites from being preserved whole. The trilobites of the Galway and Hoyt are:

GALWAY FORMATION

Calocephalites cf. *C. minimus*
Dellea saratogensis
Drabia cf. *D. menusa*
Cameraspis convexa
Drabia cf. *D. curtoccipita*
Elvinia granulata

HOYT LIMESTONE

Dellea? landingi
Keithiella depressa
Plethopeltis saratogensis
Prosaukia tribulis
Hoytaspis speciosa
Plethopeltis granulosa
Prosaukia hartti
Saratogia (Saratogia) calcifera

Cambrian of the Taconic Allochthon

Initially, a substantial amount of silt, sands, and muds was swept from the Grenville basement of Laurentia, which had been exposed to weathering and erosion for over 400 million years.

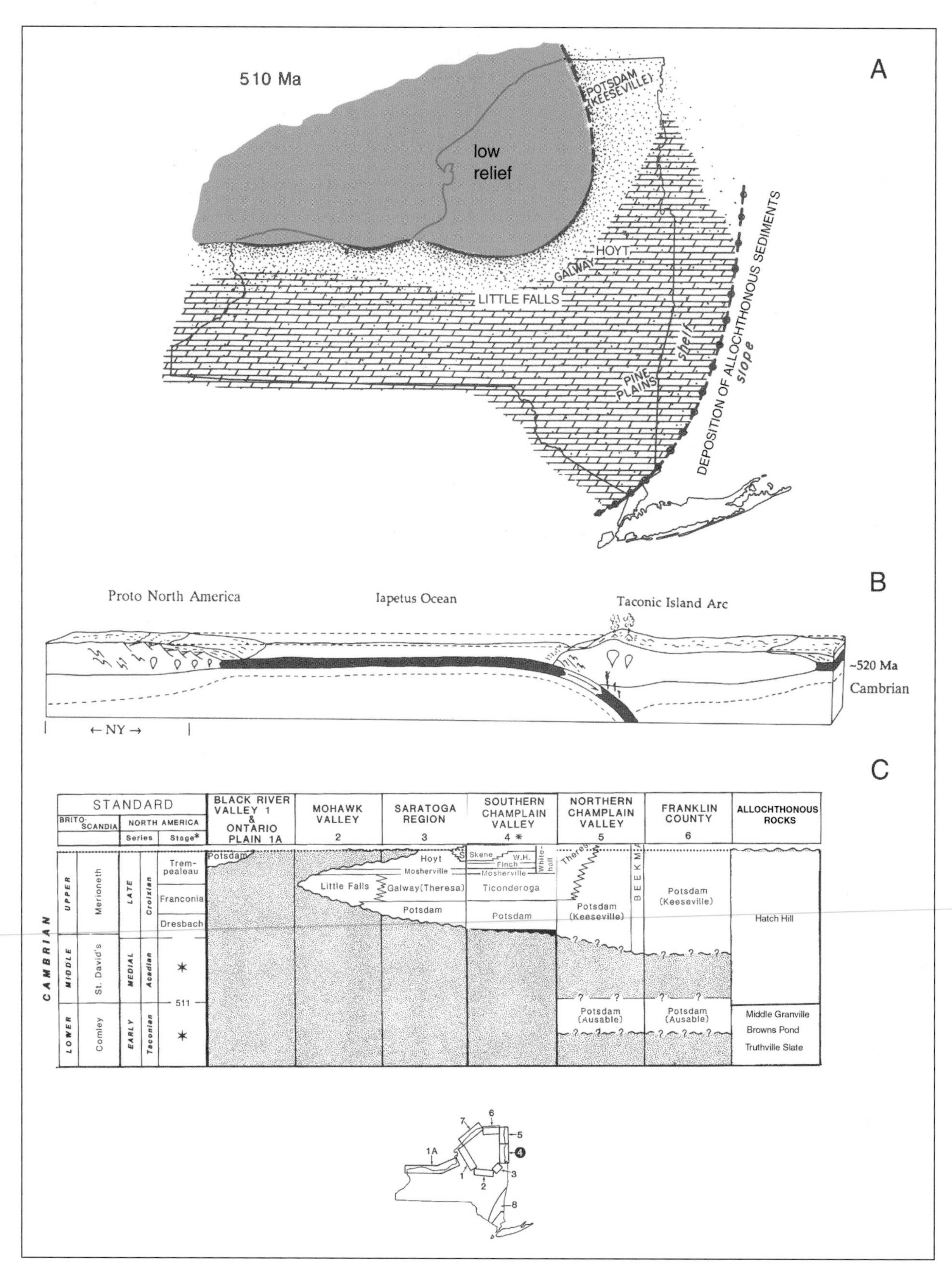

FIGURE 4.7. New York in the Cambrian. A. New York in the Upper Cambrian, showing the carbonate bank over much of the state. B. Cross section of the plate movement during the Upper Cambrian. C. Stratigraphic chart of the Cambrian exposures in New York. From Isachsen et al. (1991). Printed with permission of the New York State Museum, Albany, N.Y.

FIGURE 4.8. Close-up of Upper Cambrian Potsdam Sandstone showing sets of cross-bedded quartz-rich sandstone. Cut along Street, Whitehall, Washington County.

The coarser sands derived from this erosion accumulated in nearshore areas to form the Potsdam Sandstone facies (Figures 4.8 to 4.10). However, substantially larger amounts of fine-grained sediment (clay minerals) were either carried in dilute suspensions to offshore areas, where they settled out as **hemipelagic** "rain" of mud, or perhaps were blown offshore during dust storms, eventually to settle out and accumulate as deep-water deposits.

The rocks that were deposited in the continental slope and rise belt are no longer found in their area of original accumulation. Rather, they are found as a series of thrust sheets that lie east of the Hudson River Valley in present eastern New York, referred to as the Taconic Mountains (see under Ordovician). The earliest, pretrilobite, portion of the Cambrian is poorly recorded in New York. However, some of the thick, muddy sandstones and siltstones of the high Taconic Mountains may represent this time interval.

The low or western portions of the Taconics display well-preserved successions of Cambrian–Early Ordovician strata that comprise a series of alternating green to purple or black slaty shales (Truthville, Browns Pond, Middle Granville). The alternation between purple, green, and black mudrocks indicates differing redox (oxidation states) conditions on the seafloor. At certain intervals, the bottom water seems to have been better oxygenated, leading to the development of reddish or green slates with little or no accumulated organic matter (Figure 4.10).

Cambrian Trilobite-Bearing Allochthonous Rocks

The green and purple Early Cambrian shales or slates of the Taconic Mountains (up to 600 m thick) are commonly quarried as roofing slates in eastern New York and Vermont, but these quarries are not major fossil localities. In general, it appears that relatively few organisms inhabited the deep sea during the Cambrian time, and most slates are barren, even as life was just bursting forth in shallow-shelf settings. Very few fossils have been found within these beds, although an unusual branching trace fossil, *Oldhamia*, is abundant in some of the purple shales low in the Taconic succession. These very small trace fossils may represent some of the earliest deep-water grazing animals. A few localities have yielded trilobites. The Lower Cambrian Middle Granville or Nassau Formation, in the vicinity of Troy, Rensselaer County, has yielded articulated remains of *Elliptocephala asaphoides* as well as a fairly complete growth series of this trilobite. The Lower Cambrian trilobites found are:

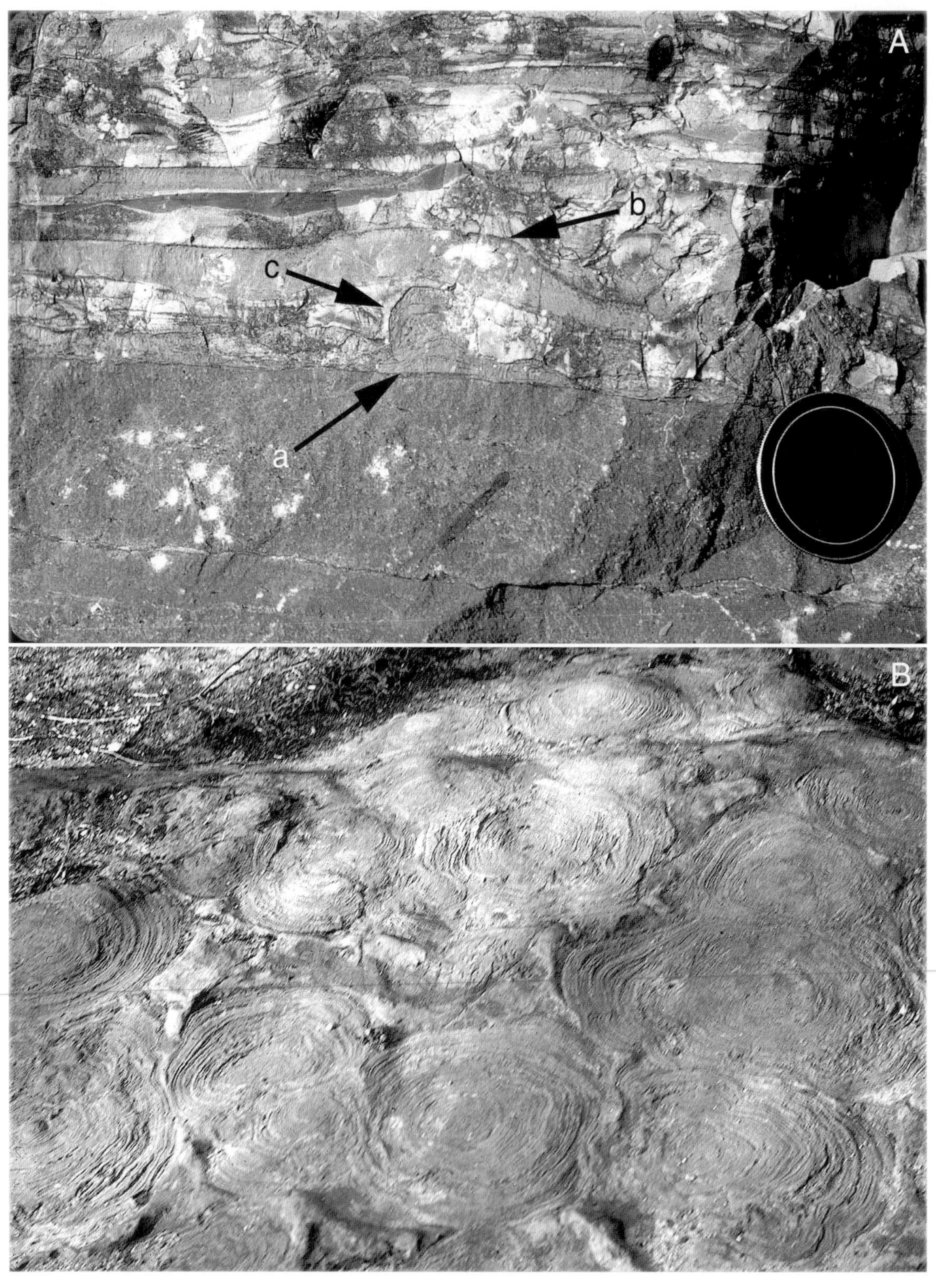

FIGURE 4.9. Upper Cambrian limestones. A. Upper Cambrian limestone, Whitehall Formation. Note the darker gray oolitic limestone (a) in sharp contact with fine-grained "ribbon limestone" (b). Also note the small stromatolite (c) attached to the Upper contact of the oolitic limestone. Warner Hill Quarry, Whitehall, Washington County. B. Domal stromatolites ("cryptozoan") in upper Cambrian Hoyt Formation. Cryptozoan ledge, Petrified Gardens Road, Lester Park, Saratoga County.

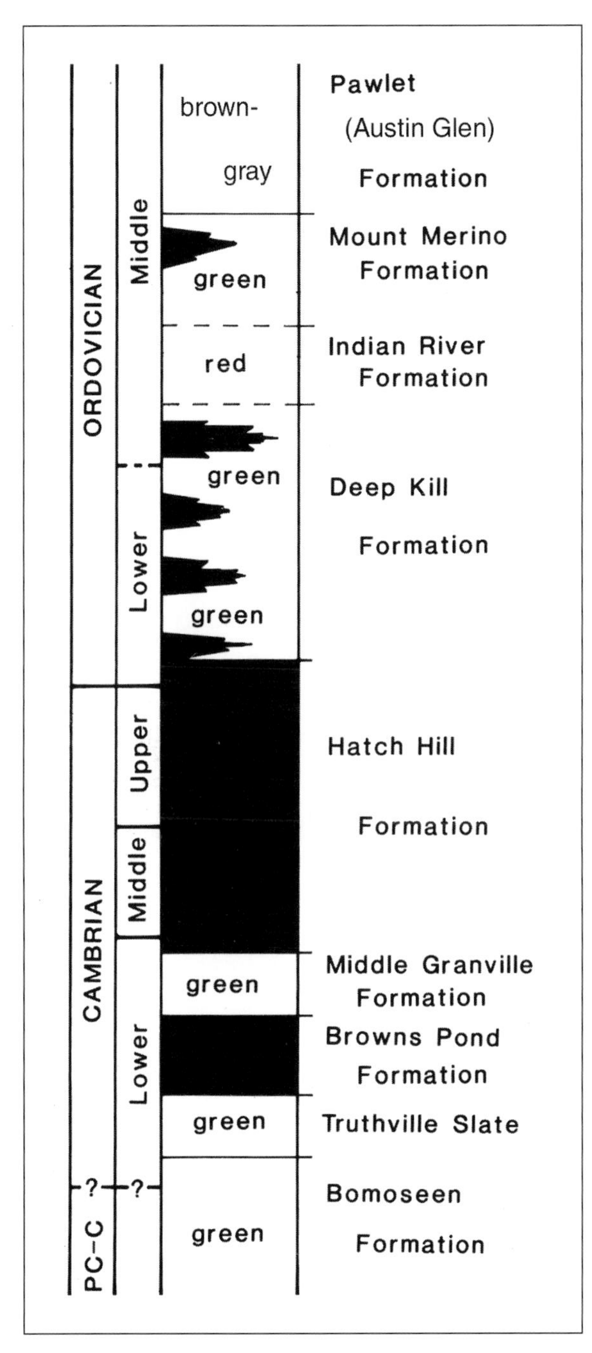

FIGURE 4.10. Simplified stratigraphy of the Cambrian-Ordovician rocks in the Taconic allochthon. Adapted from Landing (1988), with permission.

MIDDLE GRANVILLE FORMATION

Calodiscus lobotus | *Elliptocephala asaphoides*
Fordaspis nana | *Kootenia fordi*
Serrodiscus speciosus

The dark shaly intervals with more numerous interbeds represent a period of time either when the bottom was more stagnant or when productivity in the surface waters was increased, allowing the accumulation of organic matter. These intervals also contain beds of limestone breccia or angular conglomerates that represent debris flows broken from the edge of the continental shelf that avalanched down into deeper water. Actually, a rather diverse fauna of trilobites has been obtained from some of the limestone breccias or conglomerates.

The dark facies include the Lower Cambrian Browns Pond Formation and the Lower to Upper Cambrian Hatch Hill Formation (a 150-m interval of dark slaty shales) (Figures 4.6C and 4.11). Both units contain numerous interbeds of several types, including ripple cross-stratified sandstones that probably record turbidites. **Turbidity currents** (masses of suspended sediment that flow downslope under the influence of gravity) swept fine-grained siliciclastic silt and sand off of shelf regions into the deeper water. Another common type of bed consists of light-gray weathering bands of very fine-grained limestone. Careful examination of some of these beds reveals that they contain very fine laminations or even cross-laminations and display sharp bases. Therefore, they have been inferred to have been deposited as relatively dilute turbidites of suspended carbonate silt and mud that were exported into the deeper water from the carbonate bank near the top of the slope. Such deposits rarely contain fossils, though trilobite fragments are known from some turbidites. Another type of accumulation within the Taconic dark shale consists of **brecciated** (fragmented) limestones. Perhaps the best known of these is the Lower Cambrian "Schodack Limestone" found near Castleton Cutoff in the Hudson Valley. Relatively few organisms could actually live in the low-oxygen deeper waters. But the remains of shallow-water organisms were abruptly transported into some of these environments as debris flows from the shallow-shelf regions above where these animals lived. The fossiliferous breccia beds are a key to the stratigraphy of the lower part of the Taconic rocks. These thin limestone-clast conglomerates contain a sandy matrix that yields abundant fossils of a variety of trilobites, including agnostids, brachiopods, and some of the world's oldest bivalves. These fossils, particularly the trilobites, are invaluable for dating the succession.

A few Lower Cambrian allochthonous rocks are rich with fragmentary, small trilobites, particularly agnostids and eodiscids. Rasetti (1946, 1952, 1966a, b, 1967) with Theokritoff (1967) reported the following:

LOWER CAMBRIAN

Acidiscus birdi | *Acidiscus hexacanthus*
Acimetopus bilobatus | *Analox bipunctata*
Analox obtusa | *Atops trilineatus*
Bathydiscus dolichometopus | *Bolboparia elongata*
Bolboparia superba | *Calodiscus agnostoides*
Calodiscus fissifrons | *Calodiscus lobatus*
Calodiscus meeki | *Calodiscus occipitalis*

FIGURE 4.11. Lower Cambrian allochthonous beds of the low Taconic Mountains. Thin-striped white beds in the lower view are lime turbidites. A thin bed of broken or brecciated limestone (a) occurs (to the left of the hammer). Upper beds are black shaly Hatch Hill Formation (b). Rte. 9 south of Hudson, Columbia County.

Calodiscus reticulatus
Calodiscus theokritoffi
Chelediscus chathamensis
Eoagnostus acrorhachis
Hyolithellus micans
Kootenia fordi
Litometopus longispinus
Oodiscus binodosus
Pagetia bigranulosa
Pagetides amplifrons
Pagetides leiopygus
Pagetides rupestris
Peronopsis cf. *P. primigenea*
Rimouskia typica
Serrodiscus speciosus
Serrodiscus subclovatus
Stigmadiscus stenometopus

Calodiscus schucherti
Calodiscus walcotti
Elliptocephala asaphoides
Fordaspis nana
"Kochiella" fitchi
Leptochilodiscus punctulatus
Neopagetina taconica
Oodiscus subgranulatus
Pagetia connexa
Pagetides elegans
Pagetides minutus
Peronopsis evansi
Prozacanthoides eatoni
Serrodiscus griswoldi
Serrodiscus spinulosus
Stigmadiscus gibbosus
Weymouthia nobilis

MIDDLE CAMBRIAN

Baltagnostus angustilobus
Bathyuriscus eboracensis
Corynexochides? expansus

Baltagnostus stockportensis
Bolaspidella fisheri
Hypagnostus parvifrons

Olenoides stockportensis
Pagetia erratica
Ptychagnostus punctuosus

Pagetia clytioides
Ptychagnostus gibbus

UPPER CAMBRIAN

Prosaukia briarcliffensis

Plethometopus knopfi

Ordovician Period

Unlike the Cambrian, the Ordovician Period was a relatively long interval, spanning about from 489 to 438 million years ago. During this long span, North America continued to straddle the paleoequator, and New York lay in the southern subtropics (Figure 4.5). Early Ordovician saw a continuation of the passive Great American Tidal Flat environment. But during the Middle Ordovician time the eastern edge of Laurentia began to encounter an offshore volcanic island arc (chain) (Figure 4.12). Ultimately this collision pushed (thrust) a great mass of deep-sea sediments up onto the present eastern edge of Laurentia, creating the Taconic Mountains and causing the continental edge to collapse into a **foreland basin**.

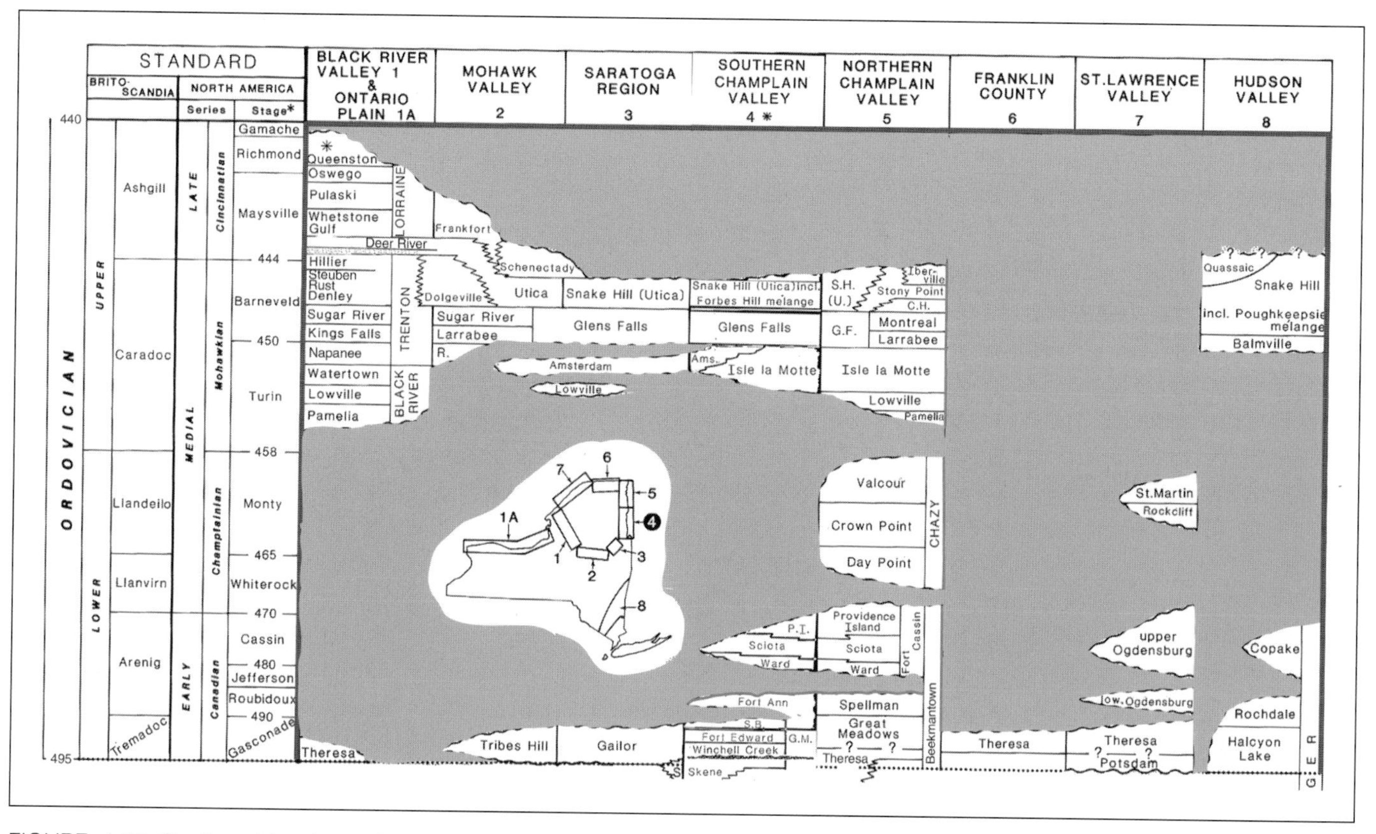

FIGURE 4.12. Stratigraphic chart of the Ordovician exposures in New York. Modified from Isachson et al. (1991). Printed with permission of the New York State Museum, Albany, N.Y.

Early Ordovician

The Lower Ordovician interval in eastern North America is locally referred to as the *Canadian Series* (Figures 4.12 and 4.13). This interval, some 20 million years in duration, is set off from the higher Ordovician rocks in New York State and in most parts of North America by the major Knox Unconformity, a 25- to 30 million-year gap in the geological record. Moreover, the type of sediments and the style of stratigraphy change markedly between the Lower Ordovician and the late Middle Ordovician rocks that overlie the unconformity. Lower Ordovician rocks contain a sparse and rather poorly documented fossil fauna, dominated by certain small mollusks, such as various gastropods and nautiloid cephalopods and rare trilobites. This restricted Early Ordovician fauna suggests that somewhat unusual, perhaps slightly hypersaline, conditions continued in the North American interior seas during this interval of geologic time.

The Lower Ordovician strata in New York generally are assigned to the upper part of the Beekmantown Group that takes its name from an area near Lake Champlain. In the outcrop belt the Lower Ordovician rocks are best exposed and most complete in areas around Lake Champlain and southward to about Lake George. In this vicinity at least four distinct formations, each bounded by a minor unconformity, are represented in the Canadian Series. South of the Adirondacks, in the central Mohawk River Valley, the Lower Ordovician rocks are relatively thin (about 35 to 50 m) and are assigned to a single formation, the Tribes Hill (Figure 4.12).

The Lower Ordovician (Canadian) Series rocks crop out in a roughly concentric belt around the Adirondack region. To a large extent, this outcrop belt is controlled by the rather recent (late Cenozoic) uplift of the Adirondacks. However, it should be noted that Canadian or Lower Ordovician rocks are thin to absent in a belt running northward from Utica, New York, to an area north of Watertown. This suggests that although most of North America was covered by shallow seas during the Early Ordovician, a low peninsular area of land extended eastward off the Canadian Shield roughly in the area of the Thousand Islands and Adirondacks of the present day. This peninsula probably has nothing to do with the present expression of the Adirondacks. It represents an ancient arch that was present in the continent and probably developed during the time of rifting of North America from another continent in the late Proterozoic (late Precambrian). This region is referred to as the *Frontenac Arch*. The Ordovician carbonates thickened regularly to the south away from this area. **Isopach maps** (maps showing variations in thickness of a particular rock unit) for the Lower Ordovician reveal a pattern comparable to that seen in the Cambrian in which the

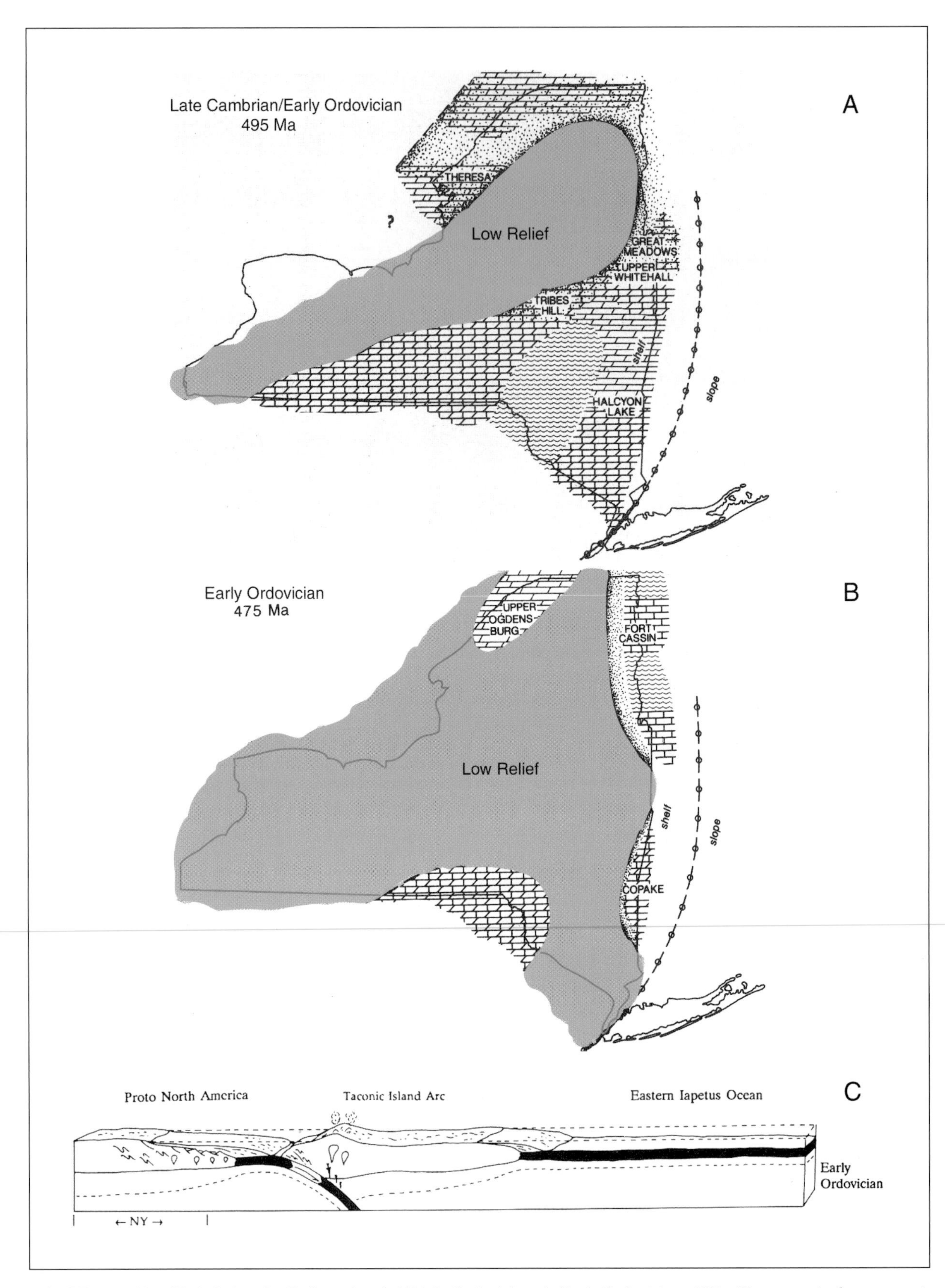

FIGURE 4.13. New York during the Early and early Middle Ordovician. A. Early Ordovician, 495 million years before present. B. Early Ordovician, 475 million years before present C. Cross section of plate movement during the Early Ordovician showing the Taconic Orogeny. D. Chazyan time. E. Black River times. F. Cross section of the plates during the Middle Ordovician. From Isachson et al. (1991). Printed with permission of the New York State Museum, Albany, N.Y.

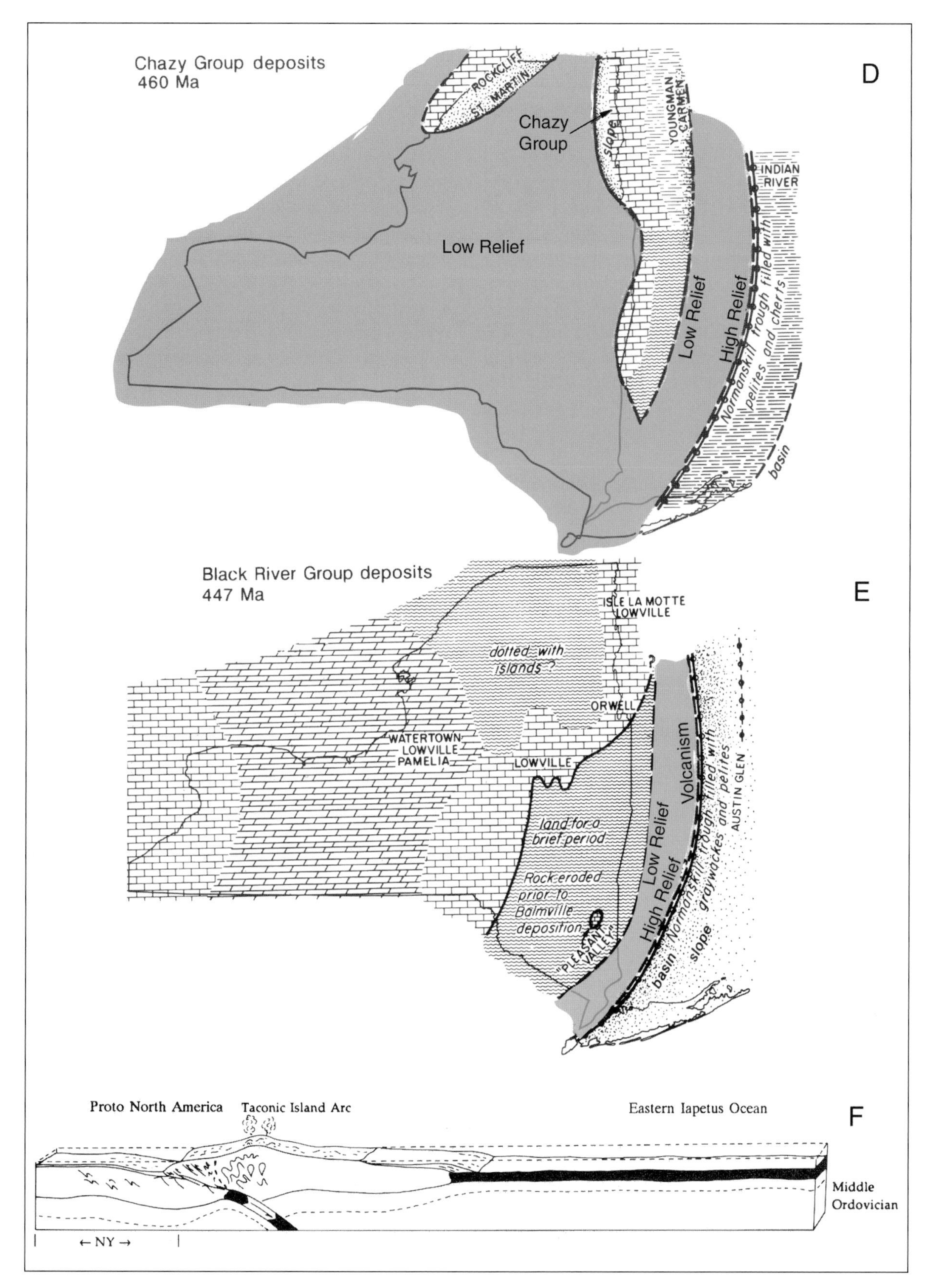

FIGURE 4.13. *Continued.*

rocks thicken dramatically southward from the approximate zero line (pinch-out of the Lower Ordovician strata) that coincides approximately with the present position of the south Lake Ontario shoreline. Ordovician or Canadian age strata attain a thickness of nearly 1500 m in the subsurface near the New York State–Pennsylvania border and thicken still more into central Pennsylvania, where they may exceed 3000 m in thickness. The rocks also thicken eastward from the area of the eastern Adirondacks into New England where they are represented as metamorphosed carbonates (marble), including the Stockbridge Marble, about 1000 m thick. North of the Frontenac Arch, relatively thin sandy dolostones of the uppermost Theresa and Ogdensburg formations represent the Lower Ordovician carbonates. The high sand content in these carbonates and in some thin intervals within the Lower Ordovician of the Mohawk Valley suggests that some **terrigenous** sediment continued to be swept from the now deeply weathered and eroding highlands of the Frontenac Arch into adjacent shallow seas.

The upper units of the Lower Ordovician, particularly the Fort Cassin Formation (Figures 4.12 and 4.13), are restricted to areas east of the Adirondack Mountains and the Champlain Valley. No trace of these units is found in the Mohawk Valley farther west. This suggests that toward the end of the Early Ordovician, the area of deposition was restricted to a relatively narrow, north-south trending basin lying close to the present eastern New York State line and into New England. This major change from widespread shallow seas over much of eastern North America to a narrow eastern basin is not fully understood. In part, it may reflect major **regression** (shallowing or lowering of sea level) associated with the end of the Sauk **Supersequence** (Figure 4.13B). Another factor may be tectonic disturbance of the eastern margin of North America. Such a disturbance may have produced a **subsiding** (deepening due to crust deformation) trough in the region east of New York and New England, while at the same time the former shelf area to the west was uplifted in a broadly upwarped archlike feature. It is notable that the Fort Cassin Formation bears a number of faults that are truncated by the overlying unconformity. These faults must have occurred following the deposition of the Fort Cassin Formation but before the deposition of the overlying Middle Ordovician strata. Evidently, the eastern edge of the continent was undergoing some stresses, perhaps associated with an initial encounter of eastern North America with a trench and the development of an offshore volcanic island arc complex. Some geologists have argued that minor volcanic ash input was already coming into the basin during the time of the late Canadian Epoch.

Early Ordovician deposition was terminated throughout eastern North America by a major fall in sea level, and an erosional unconformity, commonly referred to as the *Knox Unconformity*, was initiated (Figure 4.13A–D). During the middle portion of the Ordovician, in places such as Mohawk Valley, the older carbonates that had been deposited during the Early Ordovician were exposed to the atmosphere for some 25 to 30 million years and became deeply eroded. Because they had a relatively low siliciclastic content when exposed to rainwater, they mainly underwent solution, although some thin residues of siliciclastic mud may have been developed on the **karstic** or solution surface. Major solution features, such as sink holes and collapse **breccias**, developed at this time owing to karstification or dissolution and cave formation and collapse of the older Ordovician and Cambrian carbonates. In places, the unconformable surface has a relief of up to tens of meters.

Lower Ordovician Strata and Trilobites

The oldest of the Lower Ordovician Canadian Series of rocks are represented by the Whitehall Formation (Figures 4.12 and 4.13) that is well exposed in the region north of Lake George near the town of Whitehall. This formation technically spans the Cambrian-Ordovician boundary, but an upper unit within the Whitehall appears to be set off by a minor unconformity that occurs close to that boundary. An erosion surface and overlying sandstone and siltstone unit, the Winchell Creek Member, appears to be a signature of a drop and initial rise in sea level. This may represent a widespread regression that occurred near the end of the Cambrian but still within the overall Sauk Supersequence. The Winchell Creek Siltstone is overlain by somewhat fossiliferous limestone that has yielded occasional fragments of trilobites as well as other marine fauna, suggesting partially normal marine conditions.

The next higher and somewhat better-known interval, the Tribes Hill Formation (0 to 30 m), in the Mohawk Valley, again commences with a sandy or silty carbonate in New York State, the Palatine Bridge Member. Again, this silt and sandstone unit overlies an unconformity that may represent an interval of minor sea-level drop. The Palatine Bridge is overlain by burrow-mottled and somewhat fossiliferous Wolf Hollow Member that represents shallow marine shelf deposition. The most fossiliferous unit within the Tribes Hill is the Fonda Member. The Fonda is a fossil-rich limestone. Some beds display reddish to greenish color due to the presence of iron mineralization, especially the clay mineral glauconite. This latter mineral is believed to form in open marine environments during times of relative sediment starvation that enables mineral precipitate to become concentrated around decaying organic matter, especially fecal pellets. The glauconite granules of the Fonda Member are also associated with hashy, finely broken down, skeletal remains of many types of organisms. Particularly prevalent are several species of small gastropods, cephalopods, and a bivalve-like organism referred to as a *ribeiroid rostroconch*. The rostroconchs represent a nonhinged bivalve mollusk that possibly was ancestral to the bivalves. The Fonda Member also contains rare, disarticulated fragments of trilobites.

Finally the upper portion of the Tribes Hill is represented again by stromatolitic dolostones assigned to the Chuctununda Creek Member. This unit contains large, but very poorly preserved stromatolites referred to as "hippo backs" where they crop out, especially along Canajoharie Creek, Montgomery County.

The two higher packages of predominantly dolomitic but mollusk-containing carbonates, the Rochdale and the Fort Cassin Formation, make up the remainder of the Beekmantown Group of Lower Ordovician in New York (Figure 4.12). The Fort Cassin shows a repeat of the same pattern observed in the lower units, particularly the Tribes Hill; that is, it commences with a widespread silt and sandstone unit, the Ward Siltstone Member, and this in turn is overlain by a fossiliferous condensed limestone that may record maximum marine flooding of the craton on another higher cycle. The upper part of the Fort Cassin, the Bridport Member, consists mostly of thin-bedded to massive stromatolitic dolostones and records the final regression in the late part of the Canadian Series. Brett and Westrop (1996) recently reviewed the trilobites from the Fort Cassin Formation.

Trilobites reported, in total, from the Lower Ordovician are as follows:

Acidiphorus whittingtoni	*Bathyurus levis (?)*
Bathyurus? perkinsi	*Bathyurellus platypus*
Bellefontia gyracanthus	*Bellefontia* sp.
Benthamaspis striata	*Bolbocephalus seelyi*
Clelandia parabola	*Eoharpes cassinensis*
Strigigenalis caudatus	*Grinnellaspis* cf. *G. marginiata*
Hystricurus conicus	*Hystricurus crotalifrons*
Hystricurus cf. *H. oculilunatus*	*Hystricurus ellipticus*
Isoteloides canalis	*Isoteloides peri*
Isoteloides whitfieldi	*Paraplethopeltis* sp.
Robergiella cf. *R. brevilingua*	*Scotoharpes cassinensis*
Shumardia pusilla	*Strigigenalis cassinensis*
Symphysurina sp.	*Symphysurina* cf. *S. woosteri*
Symphysurus convexus	

Allochthonous Rocks of Early Ordovician Age

Lower Ordovician allochthonous rocks exposed in the western part of the Taconic Mountains closely resemble those of the Upper Cambrian. Lowest Ordovician strata are represented by beds of the upper Hatch Hill Formation, as indicated by the presence of distinctive index fossils including **conodonts. Graptolites** first become common in the strata of the Taconic Mountains in the earliest Ordovician, where they are represented by **dendroid** graptolites such as the genus *Dictyonema.* Hatch Hill Shales, now metamorphosed in places to slates, continue upward from the Cambrian to the Ordovician. Thin carbonates, sandstones, and carbonate breccias, occurring at or near this level as well, represent sediments that were washed off the shallow platform of North America at the end of the Cambrian and earliest Ordovician. The Hatch Hill dark shales give way upward in the stratigraphic succession to olive-greenish, slaty shales with some thin limestones but no debris flow breccias. These strata have been assigned to the Poulteney or Deep Kill Formation (Figure 4.10). Although predominantly green, some thin dark shale partings occur within these strata, particularly in the vicinity of Deep Kill, a small creek near Melrose, Rensselaer County. These Deep Kill black beds have yielded a highly diverse and well-preserved assemblage of graptoloid graptolites, typically preserved as silvery carbonized remnants on the dark slaty bedding planes. These graptolites have been used to correlate the Deep Kill rocks with portions of the Lower Ordovician in other parts of the world. However, the bulk of the predominantly green Poulteney or Deep Kill Formation is sparsely fossiliferous. In the Taconic allochthon succession as on the craton, the Sauk Unconformity or the Knox Unconformity appears to be present as a gap or break in sedimentation. Just why this should be so in deeper water is unclear. In fact, one might anticipate the occurrence of an increase in the influx of sediments, at least if shales were uplifted and exposed to erosion during the long span of the Knox Unconformity. However, the continental slope and rise area of New England appear to have been relatively sediment starved during this interval of time. Siliceous layers within the Poulteney represent probable volcanic ash beds that were being implaced within the Iapetus Ocean. Also, some cherty beds represent some of the world's oldest **radiolarian**-based, deep-sea silica ooze deposits. The Poulteney is overlain, probably with an unconformity, by the Indian River red slates. Trilobites are not present in these deposits.

In the Early to Middle Ordovician, the Iapetus Ocean no longer continued to widen, and a portion of seafloor attached to the east edge of Laurentia began underthrusting the rest of the seafloor (Figures 4.5B and 4.13C). This action produced a deep trench on the Iapetus Ocean floor and a so-called **subduction zone**, wherein the western plate or slab was forced downward beneath the eastern or overriding slab. This interaction produced an offshore island arc—the Taconic or Ammonoosuc Arc, a chain of volcanoes that formed on the overriding plate of proto-Atlantic seafloor. The **magmas** (melted rock) were generated by frictional heating and partial melting of the downgoing slab and broke through to the surface, forming the volcanic chain.

As subduction of the proto-Atlantic seafloor continued, the eastern edge of Laurentia itself eventually was brought close to the subduction zone. Ultimately, the sediments that bordered Laurentia along the continental slope and deep ocean floor were scraped off from the downgoing slab, forming a somewhat chaotically deformed series of slabs of strata—an **accretionary wedge**. This mass would become the rocks of the Taconic Allochthon (a mass of rock ultimately displaced some 80 km west of its site of origin into the area of present-day eastern New York). The Taconic allochthon was thrust or pushed up onto the continental shelf of Laurentia, or one might say that the east edge of the continental shelf was subducted beneath this mass of deformed sediments. The latter area also was being compressed and thrust westward by collision of the Ammonoosuc Arc with the accretionary wedge and Laurentia.

Middle Ordovician

The Knox Unconformity is manifested as the sharp upper contact of the Beekmantown Group or the underlying Cambrian units. The Knox Unconformity, with a relief of up to several meters, due to karstification, is one of North America's major stratal breaks and forms the subdivision between two huge packages of strata, referred to as *Sloss supersequences*: the Sauk Supersequence below and the base of the Creek phase of the Tippecanoe Supersequence above (Sloss 1963; Figure 4.1).

Middle Ordovician Chazy Group

The oldest Middle Ordovician strata to accumulate above the Knox Unconformity in New York are the sandstones and carbonates of the Lake Champlain area, assigned to the Chazy Group (Figure 4.13D). These strata are of middle Middle Ordovician or Chazyan age (the Llandeilo Series of the British terminology) (Figures 4.13D, 4.14, 4.15). The Chazy Group strata evidently accumulated in a relatively narrow, restricted trough or foreland basin that existed in present-day northeastern New York State and once extended farther to the north and east into central Quebec and Vermont. Laterally equivalent sediments of the Youngman and Carmen formations of western Vermont consist of thin-bedded, ribbon-like limestones and interbedded dark shales that represent more basinal accumulations (Figure 4.13D). The localized nature of the Chazy basin probably reflects additional subsidence of the outer portion of the continental margin of Laurentia, which, during the Middle Ordovician, was beginning to encounter the trench or subduction zone associated with the collision of a volcanic island arc.

Chazy strata typically commence with cross-bedded sandstones containing sparse fossils, but including lingulid brachiopods, referred to as the *Day Point Formation.* These siliciclastic sands apparently were recycled from older sandstones that had been eroded during the long span of the Knox Unconformity. They were deposited in shallow subtidal to intertidal environments. The overlying beds of the Crown Point and Valcour in the Chazy Group are carbonates that display generally deepening upward trends. Coarse-grained limestones representing cross-bedded shoals of crinoidal and other skeletal debris occur low in the Crown Point Formation but are interbedded with and succeeded by nodular to wavy-bedded fine-grained limestones containing abundant fossil fragments. Locally, within the Crown Point, small **bioherms** or reeflets were developed on skeletal shoals (Figure 4.14). These were composed of sponges and bryozoans, with some primitive rugose and tabulate corals. A small patch reef or bioherm in a cow pasture on Isle Lamotte, Vermont, is often said to be the world's oldest coral reef, although most of these bioherms were not composed of corals.

Trilobites as disarticulated elements are rather common and highly diverse in some of the Chazy nodular limestone beds. The trilobite fauna of the Chazy (Shaw 1968) includes over 60 species of trilobites:

Acanthoparypha? sp.	*Amphilichas minganensis*
Apianurus narrawayi	*Basilicus (Basiliella) whittingtoni*
Bumastoides aplatus	*Bumastoides comes*
Bumastoides gardenensis	*Bumastoides globosus*
Calyptaulax annulata	*Carrikia setoni*
Ceratocephala triacantheis	*Ceraurinella latipyga*
Ceraurus granulosa	*Cheirurus (Nieszkowskia) mars*
Cybeloides prima	*Cyrtometopinid* sp.
Dimeropyge clintonensis	*Dolichoharpes* sp.
Eobronteus sp.	*Eoharpes antiquatus*
Gabriceraurus hudsoni	*Glaphurina lamottensis*
Glaphurus pustulosus	*Heliomeroides akacephala*
Hemiarges turneri amiculus	*Hibbertia* sp.
Hibbertia valcourensis	*Hyboaspis depressa*
Illaenus crassicauda	*Isotelus angusticaudum*
Isotelus beta	*Isotelus canalis*
Isotelus giganteus	*Isotelus harrisi*
Kawina? chazyensis	*Kawina?* sp.
Kawina vulcanus	*Lonchodomas chaziensis*
Lonchodomas halli	*Nanillaenus? punctatus*
Nanillaenus? raymondi	*Nieszkowskia billingsi*
Nieszkowskia? satyrus	*Nileoides perkinsi*
Otarion spinicaudatum	*Paraceraurus ruedemanni*
Physemataspis insularis	*Platillaenus erastusi*
Platillaenus limbatus	*Pliomerops canadensis*
Proetus clelandi	*Pseudosphaerexochus? approximus*
Pseudosphaerexochus vulcanus	*Remopleurides canadensis*
Sphaerexochus parvus	*Sphaerocoryphe goodnovi*
Thaleops arctura	*Thaleops longispina*
Uromystrum brevispinum	*Uromystrum minor*
Vogdesia bearsi	*Vogdesia? obtusus*

Middle Ordovician Allochthonous Rocks

As previously mentioned, the Ammonoosuc island arc, lay outboard of North America in the proto-Atlantic in the region that today would be occupied by central New England. The old cratonic edge of Laurentia may have been uplifted to the east of the Chazy trough. This relatively narrow trough area lay to the east, between the old continental margin (present-day central Vermont and Massachusetts) and an accretionary prism, consisting of the sedimentary rocks and some ocean-floor volcanics, which were being **obducted** off of the downgoing oceanic-floor plates (Figure 4.13B, part E, F). The uppermost or youngest sediments of the Taconic allochthon succession accumulated in this trench during the Middle Ordovician and are approximately the

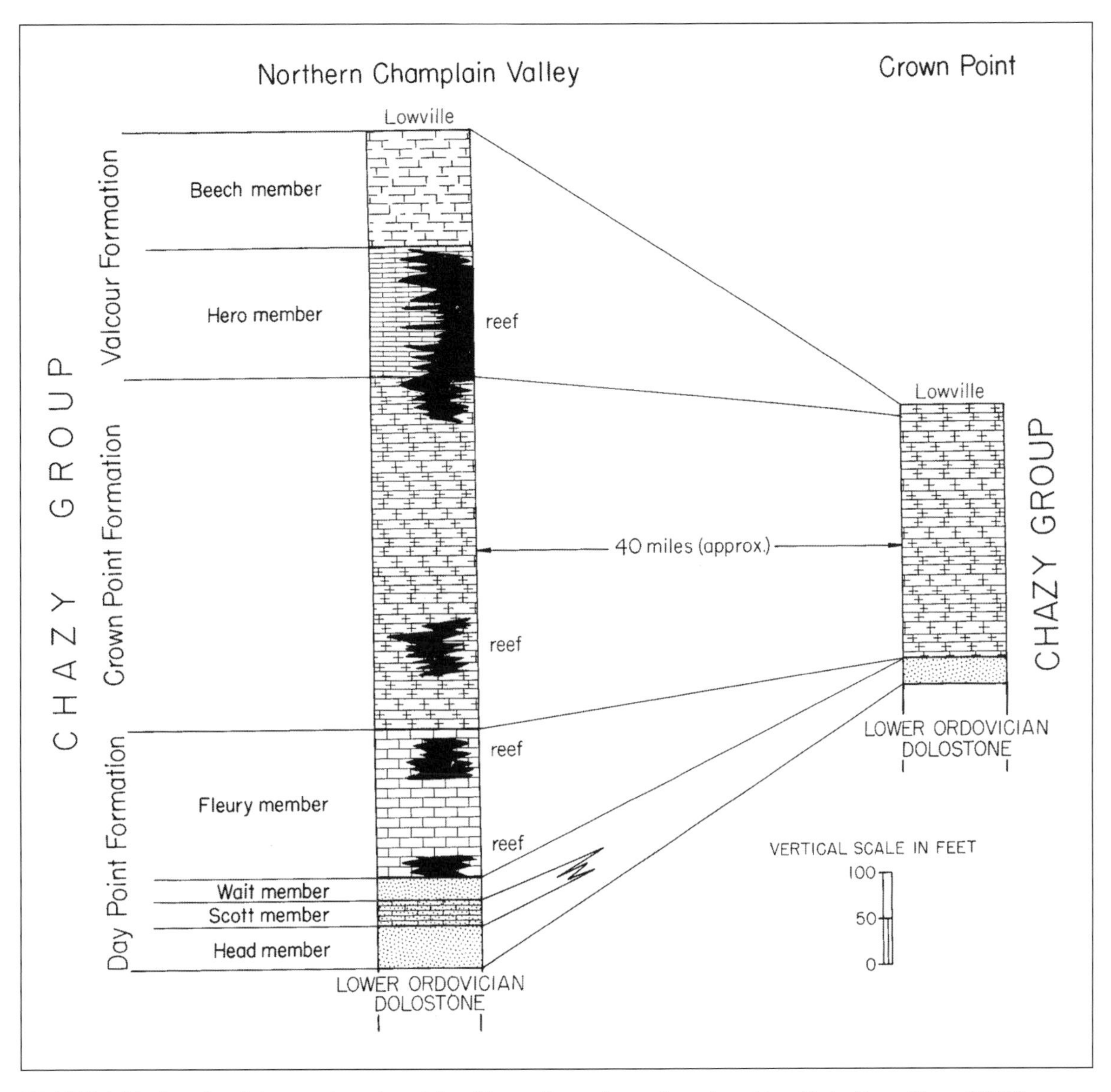

FIGURE 4.14. Details of the stratigraphy of the Chazy Group in northeastern New York. From Shaw (1968), reproduced with permission.

same age as the Chazy carbonates of New York State. These strata belong to the Normanskill Group and have been subdivided into three major formations (Figure 4.10). The lowest is a very distinctive brick-red shale or slate referred to as the *Indian River Formation.* This slate is exposed in many places in the low Taconics and has been quarried extensively for roofing slates in the region of the New York–Vermont border. These red slates lack body fossils but contain small trace fossils apparently made by deep-sea burrowing organisms. The red coloration of the Indian River Slate is unusual and makes it particularly attractive as building stones. It reflects oxidized iron concentration within the muddy sediments. However, the source of these iron enrichments is still poorly understood. It might reflect highly weathered soil that developed during the long-term exposure of the craton during development of the Knox Unconformity. However, it also reflects a high degree of oxygenation of the bottom waters during this time. This might be associated with the elevation of the seafloor due to buckling and arching as the continent edge of Laurentia was driven into a trench.

Overlying sediments of the Mount Merino Formation are mostly dark gray or greenish gray siliceous shale. They are also noted for thin, ribbon-like beds of distinctive light green cherts. These cherts perhaps represent accumulations of radiolarian (siliceous microfossil) skeletal oozes on the proto-Atlantic seafloor. Body fossils are, again, extremely rare within the Mount Merino Formation. No trilobites are known.

The highest and thickest portion of the Normanskill Group is composed of the Austin Glen Formation (commonly referred to as **graywacke** or mud-rich sandstone). In places, the Austin Glen is enriched in small fragments of metamorphic rocks, including slates that probably represent uplifted and metamorphosed chunks of the original proto-Atlantic seafloor sediments. The Austin Glen is noted for its thick sandstone deposits that are interpreted as **turbidites**, the product of deposition from basinward-flowing masses of suspended sediment and water that moved due to gravity (Figure 4.15). It is evident from distinctive **flute** and **groove-casts** on the undersides of many of these thick turbidite beds that the sediment source now was from the east, probably off the erosion of the rising accretionary prism and volcanic island arc. These rocks comprise the highest portion of the Taconic strata, and they were thrust westward some 80 km to their present resting position, which approximates the position of the modern Hudson River Valley. The Austin Glen does contain occasional fossils in places, including some marine benthic fauna of brachiopods, bryozoan scraps, and very rare trilobites.

Black River Group

The rocks of the Chazy Group in the Champlain region are conformably overlain by another series of typically pale gray, rather pure limestones referred to as a part of the Black River Group (Figures 4.13E, 4.16, and 4.17A). The Black River Group, named for exposures in the Black River Valley near Watertown, Jefferson County, is a very widespread and distinctive package of shallow marine limestones that ranges up to 70 m thick (Figures 4.13E, 4.16, and 4.17). Unlike the restricted Chazy, however, the Black River limestones extend over most of central and western New York State and into the midcontinent. In places, the Black River carbonates rest directly on the Knox Unconformity and comprise the basal portion of the Creek phase of the Tippecanoe Supersequence. In other areas to the northwest, the Black River Group rests directly on Cambrian or even the Grenville (billion year old) basement rocks. In these regions, the basal unit, referred to as *Pamelia Formation* in New York or *Shadow Lake* Formation in Ontario, consists of greenish gray to reddish mudstones and sandstones, typically as poorly sorted thin layers. Where it rests on Precambrian basement, the Shadow Lake may incorporate cobbles of quartzite or other metamorphic rocks. This unit is normally unfossiliferous and is interpreted as extremely shallow marine or nonmarine fluvial sediment. However, a few very scrappy remains of fossils, including trilobites, have been discovered from carbonate beds in the Shadow Lake Formation in Ontario.

The carbonates of the Black River Group were deposited at a time when shallow, rather monotonous conditions existed over very broad tracts of seafloor. Judging from the persistence of features, such as beds of mudcracks over hundreds of square miles (Figure 4.16A), it appears that vast areas were exposed and wetted periodically by particularly high tides, at least during shallow-water intervals of the Black River deposition. There are no really good modern analogs of such vast tidal flats with which to compare the Black River depositional environments. Probably the closest would be some of the extensive platform of the Bahama Banks of today. Presumably, the Black River sediments accumulated in subtropical environments, approximately 25° south of the equator (Figure 4.7). However, some workers have argued that the overlying Trenton sediments may actually have developed under cooler temperate rather than warm subtropical conditions. Perhaps a climatic fluctuation did characterize the transition from the Black River to the Trenton.

The broad, flat-shelf conditions of the Black River deposition indicate tectonic quiescence. However, the presence of thick **bentonites** (volcanic ash beds) within the Black River strata indicates that volcanism was occurring, perhaps in the Ammonoosuc Arc at this time.

The main body of the Black River Group consists of the Lowville Formation, also called Gull River Formation in Ontario. This is distinct, very pale gray weathering, massive micritic limestone. In past times it was commonly referred to as the "bird's eye limestone" because of dark calcite spar-filled **vugs**. This term originally referred to burrows of vertically excavating worms, *Phytopsis* (Figure 4.16). However, the term "bird's eye structure" has come to refer to smaller vug- or pit-filling areas of dark calcite also common in the Lowville or Gull River Formation of the Black River Group. These are thought to represent originally gas bubble holes that developed from desiccation of carbonate sediments in a tidal flat or perhaps from the decay of algal or bacterial filaments within the sediments. Black River limestones contain evidence of extremely shallow-water deposition. For example, desiccation crack polygons are beautifully displayed on many bedding planes (Figure 4.16A). Stromatolites also may be common locally, as are flat pebble conglomerates. The unit tends to be rather sparsely fossiliferous, although in places it contains spectacular, large nautiloid and endocerid cephalopods. Scattered trilobite fragments have been found within these rocks, but they are not particularly common or well preserved. Nonetheless, the laterally equivalent Gull River Formation in Ontario has yielded beautifully articulated remains of the trilobite genus *Bathyurus* and the large asaphid *Isotelus* sp. These trilobites evidently ranged into relatively shallow water, as their remains are associated with "bird's eye" structures and other features commonly taken to indicate the upper subtidal to intertidal zone (Figure 4.16B). Perhaps these represent carcasses that were stranded within the inner lagoon or tidal mud flat environment and buried intact.

Slightly darker gray limestones within the Black River Group, particularly near the top in what has been referred to as the *House Creek and Watertown formations*, may contain abundant thickets of the distinctive coral *Tetradium*, so called because of the fourfold symmetry of its corallites and septa. These strata reflect shallow-shelf lagoonal environments.

FIGURE 4.15. Flute casts on Austin Glen graywacke ("dinosaur leather"). Basal surface of a vertically dipping bed of Middle Ordovician sandstone turbidites shows sole marks including groove casts (a), ripple-like scours, and smaller flute casts (b). These features clearly indicate deposition from a turbidity current. Allochthonous beds of low Taconics Austin Glen Formation. Cut in Rte. 9W near Coxackie, Greene County.

In a few locations, trilobites are found in the Lowville but mostly in the upper or Watertown Formation (Young 1943a, 1943b; DeMott 1987). The trilobites reported from the Black River are as follows[1]:

Basilicus ulrichi
Basiliella barrandei
Bathyurus johnsoni (Lowville)
**Bumastoides milleri* (Lowville)
Eoharpes pustulosus
Illaenus latiaxiatus
Isotelus simplex
Raymondites longispinus
Thaleops ovata
Basilicus? vetustus
Bathyurus extans (Lowville)
Bumastoides billingsi
Ceraurinella scofieldi
Failleana indeterminata
Isotelus homalonotoides
Pterygometopus schmidti
**Raymondites spiniger*

[1] The base of the Trenton and the top of the Black River were not well distinguished by early authors. The trilobites with an asterisk may be early Trenton.

FIGURE 4.16. Black River limestones. A. Desiccation crack polygons on the upper surface of a bed of fine-grained limestone. Lowville Formation, Black River Group, East Canada Creek near Ingham Mills in Fulton County. B. Beds of laminated limestone (intertidal environment) with vertical burrows (a) (*Phytopsis*). Note the sharp contact (b) with the dark gray limestone (deeper subtidal). East Canada Creek at Ingham Mills in Fulton County.

FIGURE 4.17. Middle Ordovician Black River Group. A. Overview of Black River Group (Middle Ordovician) at East Canada Creek, near Ingham Mills in Fulton County. Note the channel in the upper left and the sharp contact (arrow) of light gray vertically burrowed limestone over dark gray fossiliferous limestone. B. Sharp contact (arrow) between the Middle Ordovician Watertown Limestone (Black River Group) and thin-bedded Napanee Limestone (Trenton Group), Sugar River, Lewis County (north of Booneville, Oneida County).

Trenton Group (General)

Among the most fossiliferous carbonate rocks within the New York Ordovician section are those of the 30- to 130-m-thick Trenton Group (Figures 4.12 and 4.18 to 4.23). The Trenton rocks record a major change from shallow carbonate shelf into a deep foreland basin due to the thrusting of the Taconic allocthon. This interval, which is classically exposed at Trenton Falls on West Canada Creek (Figures 4.21 to 4.23), is a very widespread, highly fossiliferous and complex carbonate succession.

The lower portion of the Trenton rests sharply on the underlying Black River Group with an erosion surface that represents a regional unconformity. Hence, the base of the Trenton represents a lesser sequence boundary within the Tippecanoe Creek Supersequence (Figure 4.1). Presumably, at the termination of Black River Group deposition, seas were briefly withdrawn from much of eastern North America. However, the overlying Trenton strata appear to represent relatively deeper-water facies. All of the Trenton Group beds are highly fossiliferous.

Apparently, the **transgression** or deepening that began the Trenton sedimentation was a strong or rapid one, of perhaps a few tens of thousands of years, such that offshore lime muds and silts were the first deposits that accumulated above the top Black River erosion surface in many localities, as the Napanee or Amsterdam Formation (Figure 4.12).

Lower Trenton Group Shelf Facies and Trilobites

The Napanee Formation (Figure 4.18) (0 to 10 m) consists of thin-bedded, brownish carbonate mudstones or calcisiltites interbedded with dark brownish gray shales that reflect a significant increase in the amount of siliciclastic mud input into the depositional system relative to the underlying clean carbonates of the Black River Group. The Napanee carbonates are highly fossiliferous and are characterized particularly by the brachiopod *Triplesia*, but also a high diversity of other brachiopod, bryozoan, mollusk, and trilobite fossils. These beds are some of the lowest that contain abundant remains of *Flexicalymene* trilobites. The depositional environment of the Napanee has been a source of debate. Some have argued that the muddy carbonates record the deposition of a shallow lagoon system inboard of and protected by a carbonate shoal. However, the widespread nature of these beds and the resemblance of their lithology and fauna to those of some of the indisputably deeper-water upper Trenton beds suggest a very different interpretation.

FIGURE 4.18. Close-up of sharp Black River/Trenton (Watertown-Napanee) contact. Note the massive limestone of the Watertown (a) and the thinly bedded brown limestone and shale of the Napanee (b). East Canada Creek near Ingham Mills in Fulton County.

The Napanee Limestone is abruptly overlain by a very thin interval (0.5 m) of beds that display a shallow-water Black River–like lithology with the return of *Phytopsis* and *Tetradium* rugose corals. These beds give way upward to the distinctive Kings Falls Limestone, which contains a mixture of skeletal **grainstones** that represent high-energy shoal environments and are composed primarily of the fragmentary remains of brachiopods and crinoids (Figure 4.18). The grainstones are interbedded with thinner-bedded limestones and dark gray shales that may represent shallow, subtidal deposits that accumulated outboard from the shoals. In turn, the Kings Falls carbonate beds pass gradationally upward into the thicker Sugar River Formation (15 to 25 m), which comprises, for the most part, wavy-bedded, rather massive, slightly shaly wackestones and **packstones**.

The Sugar River and Kings Falls units thin eastward to the vicinity of Canajoharie, Montgomery County, where they are locally pinched out near the crest of an apparent arch feature on the Ordovician seafloor referred to as the *Canajoharie Arch* (Figures 4.12 and 4.20). A sharp discontinuity in this area separates the Sugar River, or Glen Falls Limestone, the eastern New York equivalent of the Sugar River, from the overlying dark gray calcareous shales.

These beds are particularly noted for their diverse assemblages of bryozoans, brachiopods, and trilobites. In the Kings Falls, one finds *Erratencrinurus vigilans* and *Hemiarges paulianus*, two trilobites rarely found, if at all, in the higher units of the Trenton. *Calyptaulax callicephalus* occasionally found in the higher Trenton is common in the Kings Falls. The trilobites notably abundant are specimens of *Flexicalymene senaria* and in the Sugar River specimens of the trinucleids *Cryptolithus tessellatus* and *Cryptolithus lorettensis;* the latter two fossils are considered to be an index of the Sugar River (Shorehamian age). They disappear from the stratigraphic sections above the Sugar River in New York, but a related species reappears considerably higher, in the Lorraine Group.

In the west, the upper Sugar River Formation includes a bundle of thin-bedded micrites or **lutites** (fine-grained, light gray weathering limestones), or Rathbun Member, that alternate with dark gray shales capped by more skeletal grainstones. These seem to represent distal storm deposits or lime mud turbidites that were accumulated from shallower-shelf regions from the north and west. The Rathbun appears to pass eastward into dark brownish gray shales in the vicinity of Little Falls, Herkimer County. They are typically rich in small, ramose bryozoans and also may carry abundant trilobites of a number of species, but particularly dominated by *Flexicalymene* sp. *Cryptolithus tessellatus* is mostly replaced in the Rathbun by a variant called *C. lorettensis*.

FIGURE 4.19. Close-up of storm beds in the Middle Ordovician Kings Falls Limestone (Trenton Group). Note the channeled base of bed (arrow) midway up the hammer handle. East Canada Creek near Ingham Mills in Fulton County.

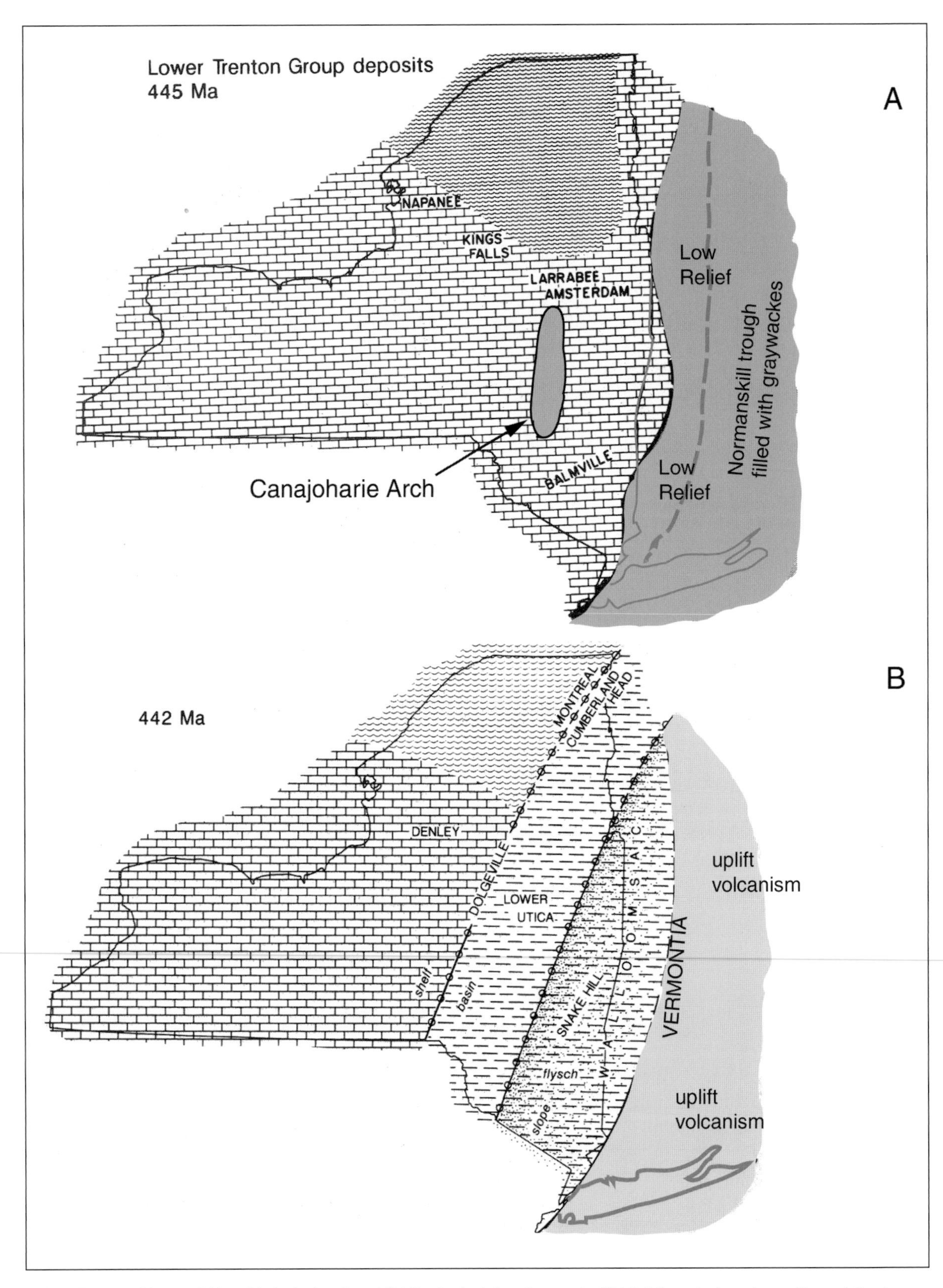

FIGURE 4.20. Maps of New York during late Middle Ordovician times. A. Kirkfieldian during deposition of the lower Trenton. B. Shermanian during the deposition of the middle Trenton. From Isachson et al. (1991). Printed with permission of the New York State Museum, Albany, N.Y.

Trilobites of the lower Trenton, up through the Sugar River are as follows:

Amphilichas trentonensis	*Bathyuropsis schucherti*
Bumastoides porrectus	*Bumastoides trentonensis*
Calyptaulax callicephalus	*Calyptaulax eboraceous*
Ceraurus pleurexanthemus	*Cryptolithus lorettensis*
Cryptolithus tessellatus	*Cyphoproetus* cf. *C. wilsonae*
Encrinuroides cybeleformis	*Encrinuroides trentonensis*
Eomonorachus convexus	*Erratencrinurus vigilans*
Flexicalymene senaria	*Gabriceraurus dentatus*
Gravicalymene magnotuberculata	*Hemiarges paulianus*
Illaenus cf. *I. conradi*	*Illaenus latidorsatus*
Isotelus gigas	*Isotelus jacobus*
Isotelus latus	*Platylichas inconsuetus*
Raymondites ingalli	*Sceptaspis bebryx*
Triarthrus beckii	

Middle and Upper Trenton Group

The middle portion of the Trenton has been referred to as the *Denley* or *Denmark Formation* (Figures 4.21 and 4.22). Kay (1937, 1943, 1968) subdivided the middle Trenton into two members: the Poland and the Russia, both defined in the Trenton Falls area. More recently, the middle Trenton units of the Poland and Russia Members were grouped into the Denley Limestone. These units represent a transition from deep-shelf to more shallow-shelf encrinal limestones. They pass eastward in dark gray and black shales, the Flat Creek Formation, that signify collapse of the eastern Laurentia shelf area into a foreland basin as the Taconic Orogeny set in.

In the area around Middleville, Herkimer County (in particular, outcrops on City Brook), the thin beds of the Rathbun Member are capped by a distinctive nodular, fossiliferous limestone about 1 m thick that marks the base of the Denley Formation. This interval, the "City Brook bed" of the Poland Member, is exceptionally rich in fossil nautiloid and endocerid-cephalopods. The City Brook bed carries the distinctive coiled nautiloid *Trocholites* and was previously referred to as the *Trocholites* bed, although the latter fossil is not particularly common. This interval is also rich in the remains of *Flexicalymene*, mostly disarticulated, but some of which are preserved as enrolled individuals. This cephalopod- and trilobite-rich bed is considered to represent a condensed horizon, that is, a bed that records a considerable amount of time in a relatively thin interval. Such cephalopod beds are typical of times of relatively rapid deepening when siliciclastic sediments appear to become trapped in nearshore areas and carbonates are not produced at a high rate.

The remainder of the Poland Member consists of thin- to medium-bedded, fine-grained limestones with chocolate-brown shale partings (Figure 4.22). Many of the limestones are rather fossiliferous and composed mainly of carbonate silt (calcisiltite). Commonly the beds display a basal surface that may be overlain by a thin (millimeters thick) layer of shell and trilobite hash. Upper parts of the beds may be laminated but typically show disruptions due to burrowing. Contacts with the overlying chocolate-brown shales may also display well-preserved fossils, such as articulated trilobites. The trilobite *Flexicalymene senaria* is particularly common in beds of the Poland near Trenton Falls. Large specimens of *Isotelus gigas* are also locally abundant, and a bed near the top of the Poland at Trenton Falls apparently yielded the majority of the well-preserved, black, articulated *I. gigas* specimens that are found in museums around the world. This is a relatively coarse bed of skeletal debris, approximately 30 cm thick, which fine upward into carbonate silt and trilobites were preserved, mostly inverted, on the base of this bed. These evidently represent organisms that were caught up and buried in a graded debris bed deposited by storm waves.

Various features of the Poland beds strongly suggest that they represent storm-deposited carbonates transported to offshore, normally quiet water environments. A relatively diverse fauna occurs in these beds. In addition to trilobites, varied brachiopods and bryozoans, especially small specimens of *Prasopora*, are abundant within some layers in the Poland. For a listing of the trilobites of the Denley Formation, see the publications by Titus (1982, 1986) and Delo (1934) for discussions of the trilobites and their communities.

Toward the top of the Poland interval occurs a series of thin bentonite beds. Weathering of these beds typically forms notches in outcrops, such as those seen near the top of the classic Sherman Falls section at Trenton Falls, Herkimer County (Figure 4.22). The top of the Poland is bounded by a distinctive widespread interval including two key bentonites. These two beds, separated by 0.5- to 1-m nodular limestone, form the base of the Russia Member (Figure 4.23). The Russia, overall about 23 m thick, is divisible into a series of small-scale coarsening upward cycles. Each cycle commences with dark shale and thin tabular, very-fine-grained micritic limestones similar to those seen in parts of the Poland. The tops of these cycles are composed of more thickly bedded, nodular, fossil-rich limestone. A particularly distinctive package of fine, evenly bedded, light gray weathering limestones occurs near the top of the Russia in areas around Trenton Falls (Figure 4.23). Like the Poland, the Russia Member yields abundant and commonly well-preserved fossils. Certain bedding planes again display completely articulated trilobites and crinoids, indicating very rapid pulses of burial. As with the Poland, it is thought that many of these thin limestones represent **distal** washoff of carbonate silt sands and muds following times when storms disrupted the shallow shelf. Storms resuspended finer carbonate silts and muds and incorporated them into basin-flowing **gradient currents**. These currents carried the sediments down a gently sloping shelf or ramp to the southeast. Nodular beds within the Russia represent somewhat shallower water environments that were disturbed both by **winnowing**

FIGURE 4.21. Middle Trenton at Trenton Falls. A. Contact of the Poland (a) and Russia (b) Members of the Denley Formation (Middle Ordovician Trenton Group), marked by a deep re-entrant (arrow at base of shadow) of Kayahoora bentonite beds. West Canada Creek above Sherman Falls, Trenton Falls, Oneida County. B. Overview of the Middle Ordovician Trenton Group, showing the Russia Member (a) of the Denley Formation and the Rust Formation (b). Upper High Falls, Trenton Falls on West Canada Creek, Oneida County.

FIGURE 4.22. Middle Ordovician Trenton Group limestone (mainly Poland Member of the Denley Formation) at Sherman Falls, West Canada Creek, Trenton Falls, Oneida County. Note the deeply recessed notches (arrow) marking the positions of bentonites (volcanic ash beds) near the top of the falls.

currents and by burrowing organisms that flourished in the somewhat better-oxygenated shelf environments. The sediment shutoff associated with the tops of the cycles allowed development of some thin skeletal hash limestone deposits that capped the cycles. Abrupt deepening above these caps is recorded by a shift back to dark shales and then to very-fine-grained limestones.

Overlying the Denley Formation, near Trenton Falls, is the Rust Formation, approximately 30 m of rather nodular, fossiliferous, poorly bedded, thin-bedded limestone (Figure 4.23). The Rust begins abruptly with a very distinctive bentonite bed exposed in the upper High Falls of the Trenton Gorge on West Canada Creek in Herkimer County, referred to as the *High Falls Ash* (Figure 4.23A). This critical marker bed occurs at the base of a thin shaly interval at the bottom of the Rust Formation.

Approximately 17 m above the base of the Rust is an interval up to 3 m thick in which the beds are heavily contorted and deformed (Figure 4.23B). In places, individual beds appear to have been overturned upon one another and doubled along recumbent folds. In other areas such as near the spillway of the power dam within Trenton Gorge, these deformed intervals appear to be confined to lenticular channel-like features. These beds record an interval of slumping of sediment on the Trenton seafloor that may have been the result of tectonic steepening of the carbonate ramp. This deformation together with the

FIGURE 4.23. Trenton at Trenton Falls. A. View of Upper High Falls showing contact of the thin-bedded, light gray weathering Russia Member, Denley Formation (a), overlain by the darker Rust Formation (b). Contact is marked by a re-entrant (arrow) at the position of the High Falls bentonite. Trenton Falls on West Canada Creek, Oneida County. B. Contorted beds (a) in the middle portion of the Rust Formation. Thin beds near the base (b) are the Walcott-Rust Quarry beds. Spillway at the power dam within the Trenton Falls Gorge, Herkimer County.

abundant thin bentonite (or volcanic ash) layers in the Denley-Rust interval signals the onset of major tectonic activity in the Taconic belt east of the present Hudson River Valley. These channel-like features are still poorly understood but might represent minor bypass channels eroded into the underlying carbonate muds and then infilled with nodular shaly limestones, which in turn were slumped during a time of tectonic disturbance. There is no doubt that this slumping deformation occurred in relatively soft, wet sediments on the seafloor. The upturned edges of the folded beds in the middle part of the Rust are truncated by an erosion surface and overlapped by a bed of crinoidal grainstones that contains a breccia of pebbles derived from the underlying deformed beds. The highest beds of the Rust Formation are thin-bedded, coarser-grained crinoidal limestones that appear to become somewhat thicker and coarser upward, perhaps in transition to the overlying massive crinoidal grainstones of the Steuben Formation.

The Rust limestones contain very abundant trilobite and crinoid debris, bryozoans, brachiopods, and particularly nautiloid cephalopods. The beds tend to be rather poorly defined and commonly somewhat **amalgamated** (bundled together) in the lower portion of the Rust.

A package of thin-bedded limestones referred to as the Walcott-Rust Quarry beds, occurring about 12 m above the base of the Rust Formations, has yielded exquisitely preserved fossils (Figure 4.23B). This thin-bedded and rather fine-grained interval in the Rust Formation sharply overlies burrowed fossiliferous limestone that caps a shallowing cycle within the lower Rust. The presence of more distal, deeper-water facies suggests that this interval represents a time of minor deepening within the basin due to either tectonics or sea-level rise. The Walcott-Rust Quarry beds interval exposed near the former Rust farm estate was discovered and initially excavated by Charles Walcott. Collections from this site were sold to the Museum of Comparative Zoology at Harvard in the 1870s and were listed in a paper by Delo (1934). The Walcott-Rust Quarry beds consist of sharply based and, in some cases, graded layers of fossil debris and carbonate silt or mud. Fossils occur both on bedding planes and within some of the very-fine-grained beds. Remains of these organisms were caught up within the turbid flows that carried the carbonate mud. Carcasses or living organisms may have been transported a short distance locally before being deposited and very rapidly covered by the settling carbonate muds. However, in some instances, organism remains were buried exactly in situ as evidenced by upright cystoids and bryozoans encased within a limestone bed. Most of the carbonates represent individual carbonate event beds, either gradient current deposits or turbidites that followed turbulent scouring and resuspension in upslope areas. The original quarry site was re-excavated and the layers examined for trilobites in some recent work (Brett et al. 1999). In particular, a bed at the base of the quarried interval yielded a series of partially enrolled *F. senaria* and *C. pleurexanthemus*, in which the fine structure of appendages was well preserved by very early calcite infilling. These fossils were the first trilobites for which appendage structure was carefully analyzed and documented (Walcott 1875a,b,c,d). This same bed yielded abundant specimens of *C. pleurexanthemus*, most (at least 95%) in an inverted position, along the lower surface and in contact with the brownish gray interbedded shale. This parting also has produced nearly complete crinoids (without the holdfasts) and several other species of whole trilobites. The quarry limestone beds appear to represent distal deposits of sediments that were resuspended, probably during storms in shallower water areas, and imported onto a southeastward-sloping carbonate ramp environment. Typical conditions in this area were low-energy, soft, muddy substrate settings that seemed to favor a variety of delicate bryozoans, crinoids, cystoids, and at least 21 reported species of trilobites, including the following:

Achatella achates
Amphilichas cornutus
Bumastoides decemsegmentus
Bumastoides porrectus
Calyptaulax eboraceous
Diacanthaspis parvula
Gabriceraurus dentatus
Hypodicranotus striatulus
Isotelus walcotti
Nanillaenus americanus
Sphaerocoryphe robusta
Amphilichas conifrons
Amphilichas inaequalis
Bumastoides holei
Calyptaulax callicephalus
Ceraurus pleurexanthemus
Flexicalymene senaria
Gerasaphus ulrichiana
Isotelus gigas
Meadowtownella trentonensis
Proetid sp.

The strata above the quarry beds interval including the upper disturbed zone comprise thin-bedded, very nodular, shaly fossili-menid ferous wackestones and packstones that are particularly rich in strophomenid brachiopods such as *Rafinesquina* and *Platystrophia*. Nautiloids are also common, but trilobites are less abundant than below and are typically fragmentary. These beds record heavily burrowed sediments deposited in shallow, storm-wave influenced environments.

The Rust grades up into the Steuben Limestone, a heavy bedded crinoidal grainstone. The Steuben is composed of **winnowed** crinoid remains and indicates a period of shallow water above wave-base. Although trilobites have been reported from the compressed shale between the beds of the Steuben, they are uncommon.

In the Poland, Herkimer County, area the entire upper Trenton also appears to undergo a rapid thinning and passage into gray shales and calcilutites to the southeast. The Rust beds and the underlying Russia are both very condensed and thin in the vicinity of Middleville, Herkimer County, where the entire Rust Member, over 27 m at Trenton Falls, is thinned to just under 2 m.

North and west of the Mohawk Valley, the Hillier Limestone overlies the Steuben Limestone in a generally deepening upward progression. The Hillier consists of shaly, nodular limestone with abundant cephalopods and a few *Flexicalymene specimens.* **Lag** deposits between the Hillier and the overlying black shales (Deer River Formation) with bored phosphate pebbles contain the trilobites *Ceraurus* sp., *Flexicalymene* sp., and *Pseudogygites latimarginatus.* The latter, a common trilobite from the equivalent beds in Ontario, is only known from the Hillier and the base of the Deer River in New York.

Trilobites are common and diverse in the upper Trenton strata from Middleville northwest into Ontario. Within the Trenton above the Sugar River are found the following:

Achatella achates	*Amphilichas conifrons*
Amphilichas cornutus	*Amphilichas inaequalis*
Bumastoides decemsegmentus	*Bumastoides holei*
Bumastoides porrectus	*Calyptaulax callicephalus*
Calyptaulax eboraceous	*Ceraurinus marginatus*
Ceraurus pleurexanthemus	*Diacanthaspis parvula*
Erratenerinurus vigilans	*Flexicalymene senaria*
Gabriceraurus dentatus	*Gerasaphus ulrichiana*
Gravicalymene magnotuberculata	*Hypodicranotus striatulus*
Isotelus gigas	*Isotelus walcotti*
Kawina (Pseudosphaerexochus) trentonensis	*Lonchodomas hastatus*
Meadowtownella trentonensis	*Nanillaenus americanus*
Proetid sp.	*Pseudogygites latimarginatus*
Sphaerocoryphe robusta	*Triarthrus beckii*
Triarthrus eatoni	

Eastern Trenton Equivalents and Utica Shale

The eastward fate of the upper portions of the Trenton Group remains somewhat uncertain, pending correlation of bentonite beds within the section. The Poland and Russia Members appear to undergo a transition into predominantly dark gray to black platy shales in the vicinity of Caroga Creek, Fulton County, and St. Johnsville, Montgomery County. However, portions of the Rust appear to extend eastward as a calcareous zone, "the Wintergreen Flats beds" that carry *Prasopora* and calymenid trilobite material at least as far east as Flat Creek near Sprakers, Montgomery County. Dark Flat Creek (Canajoharie) Shales, particularly those overlying the Wintergreen Flats *Prasopora* bed, contain an abundance of the trilobite *Triarthrus beckii.* The shales also become quite rich in graptolite fossils that are dated as near the top of the *Corynoides americanus* graptolite zone. A dark gray or black shale bed overlies the Wintergreen beds in the vicinity of Canajoharie and appears to represent a major deepening within the section. This interval may coincide with the Walcott-Rust Quarry beds in the Trenton Falls area. It is the zone that carries the most abundant *Triarthrus beckii* in the central and eastern Mohawk Valley region. This dark gray shale, newly termed the "Valley Brook Shale" (Brett and Baird, in press), appears to pass upward with a series of thin ash beds into the overlying Dolgeville Formation.

To the east, in the Hudson Valley area, the black Flat Creek or Canajoharie shales grade into a thick succession (over 500 m) of dark gray shales and turbidite sandstones referred to as the *Snake Hill Formation* (Figure 4.20B). These sediments accumulated rapidly in a trough that lay just west (modern directions) of the rising Taconic thrust mass. This is evident not only because the thrust sheets lie on, or in, the Snake Hill Shale, but also because the Snake Hill includes pebbles and even huge blocks of rocks derived from the old shelf edge to the east. These include limestone blocks that contain disarticulated trilobite parts, evidently "bulldozed" by the incoming Taconic thrust sheets and then collapsed as debris flows into the Snake Hill Shales.

The eastern equivalent of the highest Trenton beds is a very distinctive rock unit, well exposed along the New York State Thruway, near the Little Falls exit, called the Dolgeville Formation. Dolgeville consists of thin (5 to 20 mm) platy beds of very-fine-grained limestone alternating with black shale (Figure 4.24A). The black and light bed striping are a very striking motif in weathered outcrops. The thin limestones are believed to be turbidites of lime mud washed off the shallow platform (Rust to Steuben depositional area) to the west. The shaly intervals of the Dolgeville contain *Triarthrus beckii.*

The top of the Dolgeville beds shows deformation presumably due to slumping on the seafloor prior to deposition of the overlying black shale. All of this suggests strong tectonic influence of the Taconic Orogeny in the foreland basin at this time (Figures 4.20B and 4.25). The Dolgeville Formation and Steuben Limestone are abruptly overlain by black shales of the Indian Castle (Utica) Formation (Figure 4.24B). This implies abrupt deepening over much of New York State due to an episode of tectonic Subsidence.

On the basis of the metabentonites and graptolites, it can be shown that the Indian Castle Shale is a westwardly thinning wedge of rock with progressively higher units onlapping the Trenton Unconformity in a westwardly direction. These black shale beds appear conformable with the underlying Dolgeville in eastern Mohawk Valley. Finally, near Middleville, Herkimer County, black shales of the upper Utica beds rest with sharp disconformable contact on a corrosion surface in the upper Steuben Limestone.

The dark Flat Creek and Snake Hill shales that are laterally equivalent to the upper Trenton are sparsely fossiliferous and represent deeper-water environments with high rates of sediment accumulation. Graptolites and *Triarthrus* specimens are found sparingly.

In the Hudson Valley the Snake Hill Shale also contains debris flows and blocks of exotic material that were thrust in from the eastern area. One such mass is the Rysedorph Conglomerate.

Two miles southeast of Rensselaer, New York, Rysedorph Hill is underlain by an unusual conglomerate, the Rysedorph Con-

FIGURE 4.24. Middle Ordovician Dolgeville and Utica rocks. A. Middle Ordovician Dolgeville Formation sharply overlain by Indian Castle (Utica Group) black shale (upper white arrow at the contact). Note the thin "ribbon limestones" and black shale, and the small folds (lower white arrow) in the Dolgeville beds. New York State Thruway, 1.6 km (1 mile) west of the Little Falls exit, Herkimer County. B. Black Indian Castle (Utica Group) shale, with light weathering limestone beds. East Canada Creek, Dolgeville, Herkimer County.

glomerate. Various pebbles within this unit contain fossils from Cambrian to Upper Ordovician. Some of the Middle Ordovician trilobites obtained by Ruedemann (1901) are not found elsewhere in New York, but some are known in Virginia. Most of these remains are fragmentary, and the actual species assignments may not be correct in all cases. The trilobites include the following:

Achatella achates
Calyptaulax callicephalus
Ceraurus pleurexanthemus
Cybele sp.
Gerasaphus ulrichiana
Lonchodomas hastatus
Otarion? matutinum
Remopleurides tumidus
Thaleops ovata
Tretaspis diademata
Calyptaulax eboraceous
Ceraurus sp.
Eobronteus lunatus
Flexicalymene senaria
Isotelus maximus
Nanillaenus americanus
Remopleurides linguatus
Sphaerocoryphe major
Tretaspis reticulata

Late Ordovician

Utica Shale–Schenectady Formation

During later Middle Ordovician, carbonate deposition ceased and was abruptly followed by deposition of widespread, black Utica Shale facies (Figure 4.25). The eastern portion of the basin was beginning to receive increased amounts of coarse siliciclastic sediments, primarily siltstone and sandstone turbidite beds, Schenectady Formation, which occur closer to the eroding Taconic Mountains during the same time interval (Figure 4.25B).

Throughout much of the Mohawk Valley, alternating carbonates and shales of the Dolgeville Formation (probable equivalents of the uppermost Rust or Steuben limestones) are abruptly overlain by a black, rusty weathering laminated shale, Utica (Indian Castle), with abundant thin (1 to 2 cm thick) clay layers that represent metabentonites.

These black shales are referred to herein as the Indian Castle Formation of the Utica Shale Group. The trilobite *Triarthrus spinosus,* normally seen in the Collingwood Formation of Ontario, is found in the upper Indian Castle Shale immediately above the Steuben Limestone near Holland Patent, Oneida County.

Throughout the Mohawk Valley region, black shales of varying age ranging up to 300 m thick constitute a relatively monotonous facies composed largely of black laminated graptolite-rich shales. Only certain beds of the Indian Castle contain other fossils, such as highly flattened orthoconic nautiloids and the trilobite *Triarthrus eatoni.* Again, these trilobites appear to have been particularly adapted to the low-energy, low-oxygen environments represented by the Utica Shale deposition. Waters were probably relatively deep and stagnant in the center of the Utica trough, which extended eastward toward New England.

Lorraine Group

Strata of the late portion of the Ordovician Period in New York State reflect the filling and overfilling of the Taconic foreland basin. In eastern and east-central New York State these strata above the Schenectady Formation have been removed by a subsequent period of uplift and erosion prior to onlapping of Silurian and Early Devonian deposits. However, in the western Mohawk Valley region and northwestern New York State, the filling phases of the Taconic Basin are recorded in the Lorraine Group with its Frankfort, Whetstone Gulf (western equivalent of the Frankfort), and Pulaski members, and the Oswego and Queenston formations (Figures 4.12 and 4.25). The Frankfort-Pulaski succession commences with dark gray shales and shows a general coarsening-upward trend through siltstones and sandstones. Parallel with this is a change from anoxic to fully oxygenated seafloor conditions reflected in increasingly diverse fossil assemblages. Overall, the seafloor was shallower due to fill-in or **progradation** of the sediments from the east.

Lorraine Group Environments and Trilobites

The Frankfort Formation, of the Lorraine Group, consists of a series of dark gray to black shales with interbedded, thin, silty turbidites and one significant package (Hasenclever) of fine-grained, muddy sandstones. This package appears to represent a minor sea-level **lowstand** or a drop in the relative sea level that caused progradation or westward migration of the coarser siliciclastic silt and sand facies over much of central New York. This bundle, however, is overlain by a return to dark gray shales, referred to as *Moyer Creek* Member in the Utica region, and Deer River dark shales to the northwest. These upper Frankfort shales are notable in the area in Rome, New York, for yielding extraordinarily preserved fossils. These are displayed in thin beds of silty shale referred to as *Beecher's Trilobite Bed* that is exposed along Six Mile Creek north of Rome. These beds have yielded spectacularly pyritized *Triarthrus eatoni* specimens with preserved appendages, in significant numbers in one thin band. These trilobites were buried very rapidly by a thin, silty turbidite, probably coming out of the then-developed deltaic regions to the east. Very early infilling of the appendages by fine-grained pyrite preserved these fossils exceptionally well. They have been the subject of numerous anatomical and taphonomic studies. The similarly pyritized trilobites *Cryptolithus bellulus* and *Cornuproetus? beecheri* are present but relatively rare.

Exceptionally well-pyritized cephalopod remains and *Triarthrus* specimens are also known from the Frankfort (Deer River) shales in the vicinity of Constableville, Lewis County, and Whetstone Gulf, Lewis County, in northwestern New York State. These beds show increasingly thick, silty and sandy beds that represent distal or more proximal turbidites or storm deposits; these

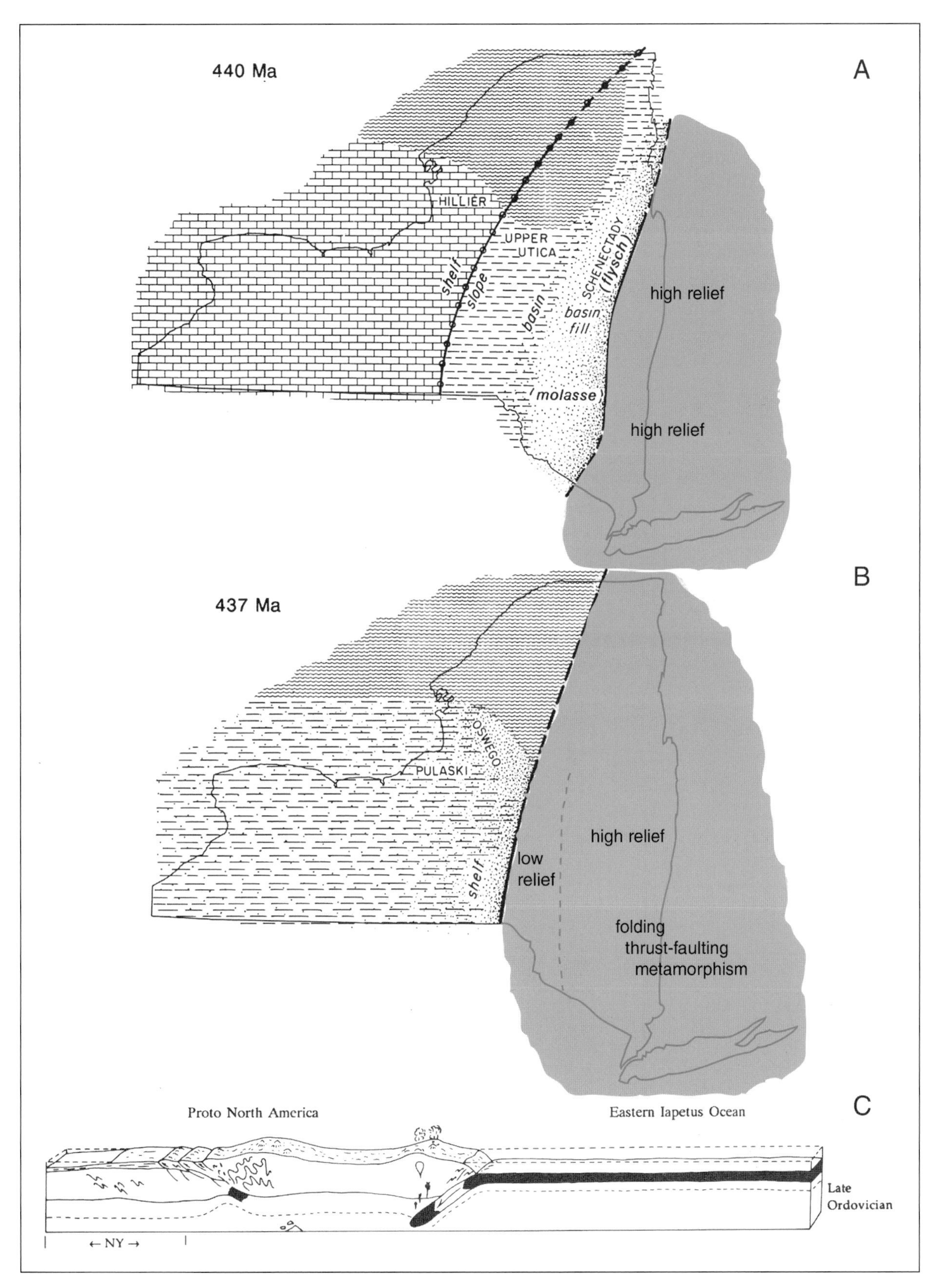

FIGURE 4.25. New York maps during the late Middle Ordovician and the Upper Ordovician. A. Late Middle Ordovician. B. Late Ordovician. C. Cross section of the plate movement during Late Ordovician. From Isachson et al. (1991). Printed with permission of the New York State Museum, Albany, N.Y.

beds are interbedded with medium to brownish gray shales that may represent background conditions. The storm beds typically display a hash or lag of shelly material that includes brachiopods, bivalves, and crinoid stems on their bases. These coarser skeletal debris layers grade upward into fine grainstones and siltstones that commonly display hummocky cross-lamination; these low-mounded laminations, typically convex-upward, are thought to be formed by the interference of storm waves and currents in relatively shallow waters. Tops of sandstone beds also display features such as interference and wave-formed ripple marks that further indicate a relatively shallow-water origin for these beds. The Pulaski beds are dominated by brachiopods, clams, and a few species of crinoids. However, trilobites are known at certain levels and include *Cryptolithus bellulus*, as well as some *Flexicalymene* species and *Isotelus* species, which may occur in thin, skeletal hash beds. These beds represent some of the highest occurrences of these trilobites. In New York the overlying beds of the Oswego and Queenston formations represent nearshore or nonmarine facies inappropriate for the preservation of trilobites, although these genera persisted, with abundant *Flexicalymene* and *Isotelus* remains common in the Richmond Group in southwestern Ohio. *Cryptolithus* and *Isotelus* and the entire trinucleid and asaphid trilobite groups to which they belong appear to have become extinct in the Late Ordovician crisis, which affected more than 20% of marine families.

The reported trilobites of the Upper Ordovician are as follows:

Calymene? conradi
Ceraurinus marginatus
Cornuproetus? beecheri
Flexicalymene granulosa
Homotelus stegops
Odontopleura ceralepta
Proetus spurlocki
Triarthrus eatoni
Triarthrus spinosus
Calyptaulax cf. *C. callicephalus*
Ceraurus sp.
Cryptolithus bellulus
Flexicalymene meeki
Isotelus pulaskiensis
Otarion? hudsonica
Pseudogygites latimarginatus
Triarthrus glaber

Upper Ordovician Oswego and Queenston Formations

The record of benthic life in the highest part of the Ordovician deposits in New York is sparse. The Pulaski gives way upward to coarser gray-and-maroon-mottled sandstones of the Oswego Formation. Here, abundant wave marks, desiccation cracks, and other features toward the top indicate shallowing of the Ordovician seafloor effectively to sea level. This shallowing was produced by the rapid in-filling of clastic sediments that were being shed out of the now eroding Taconic Highlands to the east. The effect of this was the outward-building or progradation of the so-called Queenston delta complex into the Appalachian foreland basin.

The overlying Queenston Formation, well exposed along the south shore of Lake Ontario from Oswego County westward to the Niagara Gorge, consists of maroon-red blocky mudstone with occasional streaks of greenish gray siltstone and sandstone (Figure 4.28). Locally, bundles of fine-grained sandstone, some of them containing animal burrows, are also present within the Queenston, for example, at Rochester. However, the presence of small, limey nodules, referred to as *calcretes*, in certain horizons suggests the development of soil-forming features within the Queenston Formation. The presence of calcretes implies long periods of aerial exposure during the deposition of these muds. We visualize the Queenston as representing a broad alluvial floodplain of possibly meandering streams that originated in the Taconic Highlands and spread sheetlike masses of sediment over a wide tract of the Appalachian Basin. Red coloration, together with the calcretes, indicates **subaerial** exposure for much of the unit in New York State, at least. However, to the west, the Queenston interfingers with the gray calcareous mudstone and even brachiopod- and bryozoan-rich limestones northwest of Hamilton, Ontario. Still farther west, the Queenston appears to be equivalent to the Richmond Group of the Cincinnati, Ohio, area, part of the section noted for exceptionally well-preserved fossils, including abundant trilobites of the genera *Flexicalymene*, *Isotelus*, and others. Hence, these organisms lived on in the later part of the Ordovician in the midcontinent, while New York State developed into a delta plain and tidal flat complex.

The latest portion of the Ordovician Period (Gamachian Stage) is not recorded in New York, or in most of the midcontinent of North America. The Queenston is terminated by a major erosional unconformity (referred to as the *Cherokee Unconformity*; Figures 4.1 and 4.26). This erosion surface increases in magnitude to the east along the Queenston clastic wedge. Ultimately, it cuts down through the entire thickness of the Queenston (over 400 m) and through the underlying Pulaski, Frankfort, and upper parts of the Utica or Schenectady formations (Figures 4.12 and 4.26). This suggests that several things were happening in the Late Ordovician. First, there was a period of uplift in the eastern portions of the old Queenston delta. This might reflect renewed orogenic or mountain-building activity at the very end of the Ordovician or Early Silurian, or possibly **isostatic** forces (buoyancy forces) that caused the eastern areas, formerly part of the mountain belt, to bob upward in much the same way that land rebounds after removal of the weight of a glacier (Figure 4.25B). This "pop-up" phenomenon or isostatic rebound would be the result of erosional redistribution of sediments from the old thrust belt of the Taconic Mountains and out into a wider tract of the interior of North America. A second event that may have contributed to the development of a major widespread unconformity or erosion surface at the end of the Ordovician was the development of extensive continental glaciers in the region that today is Saharan Africa. These Late Ordovician ice sheets would have locked up a considerable amount of water as glacial ice and thereby caused a drop in sea level from the Late Ordovician on the order of perhaps 100 m. This is reflected in successions of rock worldwide as a major erosion surface associated with regression of seas. This time also corresponds to one of the great mass

FIGURE 4.26. Ordovician/Silurian unconformities. A. Ordovician/Silurian unconformity. Upper Ordovician, black Frankfort Shale (a) is sharply overlain by Lower Silurian, light gray Oneida conglomerate and sandstone (b), which grades upward into gray Sauquoit Shale. Cut along Rte. 171, Frankfort Gorge of Moyer Creek, Frankfort, Herkimer County. B. Angular Cherokee unconformity (arrow) between Ordovician Austin Glen Formation (b) (nearly vertical beds on right side) and gently dipping uppermost Silurian Rondout Formation (a). Note that the unconformity, produced by erosion of the folded Taconic terrane, was later rotated to 45 degrees during the Devonian Acadian Orogeny. Exit from Rte. 23 to 23B, Catskill, Greene County.

extinctions in Earth's history and one that decimated many trilobites, such as the genux *Isotelus*.

Silurian Period

The Silurian Period was a relatively short span (438 to 408 million years ago) (Figures 4.1 and 4.27) that commenced with recovery from major Late Ordovician extinction and ended with minor extinction before the Devonian. There is really not a major Silurian-Devonian break in most areas. The Silurian marks a return to greenhouse conditions following the Late Ordovician glaciation, but there appears to have been some lingering glaciation in the paleo-Andes Mountains of South America through the Early Silurian. Middle and Late Silurian climates appear to have

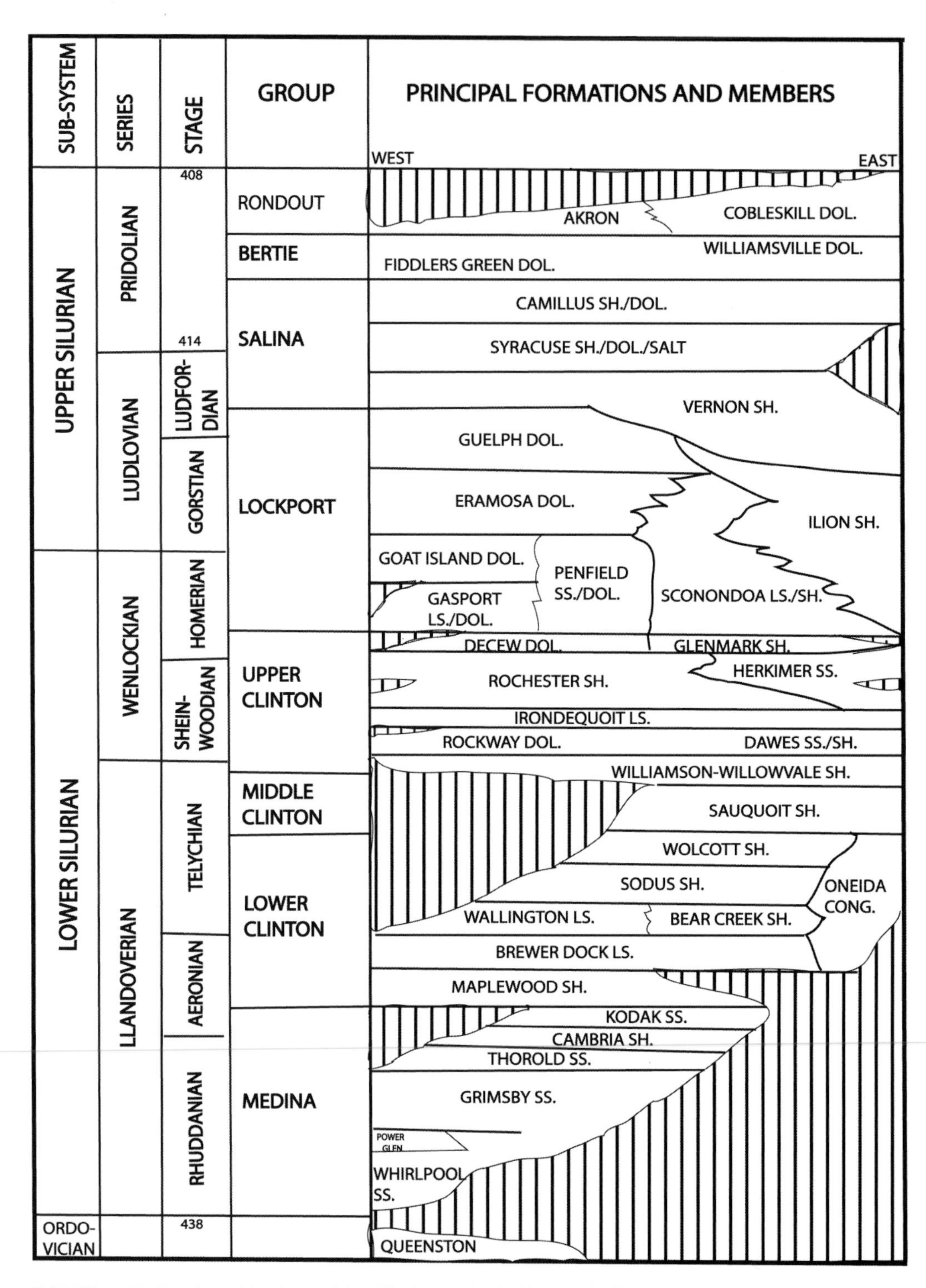

FIGURE 4.27. Stratigraphic chart of the Silurian rocks in New York. The numbers are millions of years before present. Modified from Brett et al. (1985).

been quite equable worldwide, with lots of shallow carbonate deposition and some return to stagnation and black shale development in deeper basins.

The eastern side of Laurentia, now about 25 to 30° south latitude, experienced the final rumbles of the Taconic Orogeny (Figures 4.5 and 4.26). Ettensohn and Brett (1998) recognized a third and final tectophase of Taconic Orogeny in the Early Silurian; this mountain-building activity produced a new clastic wedge, the Tuscarora-Medina formations, and pulse of westward subsidence.

Baltica (Ancestral Europe) had moved northward through the Ordovician and now collided with northern Laurentia to form the widespread Caledonian Orogeny in Europe and created Euramerica (or "Old Red Continent") straddling the paleoequator (Figure 4.5). At the same time the microcontinent of Avalonia was converging on (present-day) eastern North America, making initial contact in the northern Maritime and New England region in the Late Silurian or Early Devonian. There appears to have been some **orogenesis** during the Middle Silurian in eastern North America (Laurentia), termed the Salinic Orogeny, as evidenced by a renewed pulse of subsidence and westward migration of the foreland basin.

Sea level rose in the Early Silurian, following the lowest sea level at the end of a regression that produced the Cherokee Unconformity in the latest Ordovician. The Silurian–Early Devonian bundle of strata forms the Tutelo (second) Phase of Sloss's Tippecanoe Supersequence. Sea level was quite high throughout the mid Silurian but was beginning to fall in the later Silurian. Smaller-scale (third and fourth order) sequences were developed owing to shorter-duration rises and falls of sea level during the Silurian in eastern North America.

Although the Silurian is a relatively shorter interval of time than the Ordovician or the Devonian, Silurian rocks in New York State are highly varied, indicating a broad range of depositional environments. The Silurian can be subdivided into five major divisions that are generally construed as groups (Figure 4.27). In ascending order these are the (1) Medina Group: predominately shales and sandstones of early Silurian age; (2) Clinton Group: a heterogeneous group composed of shales, minor sandstones, and shell-rich carbonates with minor hematite beds; (3) Lockport Group: predominantly dolomitic carbonates in western New York, grading eastward into dark gray or greenish shales and thin stromatolitic limestones; (4) Salina Group: Upper Silurian shales, carbonates, and evaporites; and (5) Bertie and Rondout Groups: dolomitic limestones and dolostones with some shales and minor evaporites.

Early Silurian

Medina Group

The Early Silurian Medina Sandstone of New York and its equivalents, the Tuscarora sandstones and conglomerates of the central Appalachian region, represent a renewed influx of siliciclastic sediments following the major end-Ordovician unconformity or period of erosion. The Medina Group is confined to western and west-central New York, pinching out eastward in the vicinity of Oneida, Oneida County (Figure 4.28). The Medina is characterized by quartz-rich sandstones and shales, many of which are red.

In western New York and Ontario the basal unit is a quartzose sandstone with **trough cross-bedding**, the Whirlpool Sandstone, which may represent, in part, nonmarine stream deposits (Middleton et al. 1987). These sediments appear to represent a reworking of older Ordovician siliciclastics, as the Whirlpool locally contains very minor thin seams of greenish gray shale that not only contain early Silurian **acritarchs**, small microfossils representing probable algal resting cysts, but also acritarchs reworked from the Ordovician.

Trilobites evidently were present in some of these environments, but their body fossils are rare. The upper portion of the Whirlpool Sandstone is distinctly marine and locally shows hummocky cross-laminated sandstone beds with minor fossil debris. Remnants of starfish and crinoids have been found in the upper layer of the Whirlpool in Ontario, and farther west the Whirlpool grades laterally into or is overlain by the Manitoulin Dolostone, a rather massive fine-grained dolomitic carbonate. This unit locally contains abundant brachiopods and minor disarticulated trilobite material. The probably equivalent massive Tuscarora Sandstone of Pennsylvania is generally devoid of body fossils but contains abundant traces including *Rusophycus*, the probable resting traces of trilobites.

In Niagara County and westward into Ontario the Whirlpool Sandstone is overlain by dark gray to medium greenish gray shales and with some thin quartz-rich sandstones. In western New York this Power Glen Shale tends to be sparsely fossiliferous. Only minor trace fossils, including *Rusophycus*, are found on the base of sandstones at Lockport. At Niagara the unit is somewhat more fossiliferous, yielding a small fauna including snails, small twiglike bryozoans, nautiloid cephalopods, and a single species of small crinoid. Westward in Ontario the unit becomes even more fossiliferous, and near Hamilton, Ontario, it contains abundant remains of several species of bryozoans, brachiopods, crinoids, and starfish. Trilobites tend to be uncommon in these beds, but a few remains of calymenids and dalmanitids have been collected.

The Power Glen Shale and laterally equivalent Cabot Head Shales appear to have accumulated in shallow offshore shelf or pro-deltaic settings of early Medina. Dark gray and greenish coloration, as well as an abundance of pyrite at some levels, suggests accumulation under low-oxygen, quiet water conditions. Perhaps the inner-shelf muds were deposited in a sheltered, muddy lagoon. However, somewhat more offshore sediments in the vicinity of Hamilton, Ontario, do show abundant storm deposits, indicating deposition by storm waves and currents.

The higher beds of the Medina Group in western New York, assigned to the Grimsby Formation, are primarily reddish and white-mottled, quartz-rich, fine-grained sandstones and siltstones interbedded with green and red mudstones. Generally, these sediments are sparsely fossiliferous, particularly east of Lockport, New York. But in the west they may contain abundant shells of lingulid brachiopods, a few species of clams, and even bryozoans at Niagara Gorge. The presence of abundant burrowing with many primary sedimentary structures, such as wave and current ripple marks, minor trough and hummocky

FIGURE 4.28. A. Ordovician-Silurian (Cherokee) unconformity (arrow). Upper Ordovician, dark maroon Queenston Shale (a) is sharply overlain by light gray Whirlpool Sandstone (Lower Silurian) (b). In turn, the Whirlpool grades abruptly upward into gray Power Glen Shale (Medina Group) (c). Cut on West Jackson Street, Lockport, Niagara County. B. Ordovician-Lower Silurian succession. Unconformity (arrow) at the top of the Upper Ordovician Queenston Shale (a) is overlain by Lower Silurian Medina Group (b). The white band at the top of the Medina is the Kodak Sandstone (c). This unit is overlain, in turn, by green Maplewood Shale (d) of the Clinton Group. Genesee River Gorge, Rochester, Monroe County.

cross-stratification, rare desiccation cracks, evidence of channeling and soft sediment deformation in the form of load casts and **ball-and-pillow-structures** all point to deposition in very shallow waters associated with outer- to inner-tidal flats. Within the Medina, one may recognize small-scale, coarsening, upward cycles in western New York that range from maroon mudstones upward to hummocky cross-laminated sandstone beds that may be capped by thin lag beds of bryozoans and *Lingula* brachiopod shell hash representing minor deepening events. In Rochester, New York, somewhat different, fining upward cycles seem to have sharp-based, channel-fill sandstones at the bottoms, which are often extensively burrowed by large wormlike organisms that produced structures called *Arthrophycus* or *Daedalus*, apparently as feeding burrows in muddy sands. The upper reddish mudstones in higher parts of the cycles apparently reflect upper tidal-flat conditions, where low-energy conditions prevailed, allowing fine-grained muds to drop out of suspension. All of these sediments apparently accumulated from continued erosion and deposition off mountainous highlands to the south-southeast. At present, it appears that a late phase of the Taconic Orogeny occurring within the early part of the Silurian may have uplifted or re-uplifted these source terrains.

Trilobites are extremely rare body fossils in the upper sandstones of the Medina Group. Two persistent, often light gray or pinkish sandstones, the Thorold and the Kodak, contain abundant trace fossils but very few body fossils. Near Rochester the Kodak bears remains of beautifully preserved traces of trilobite activity, although as yet no body fossils have been found. These include the coffee bean–shaped *Rusophycus* as well as delicately detailed scratch marks produced by trilobites walking on the tips of their claws. The fact that slabs of sandstone from Glen Edith near Irondequoit Bay, Monroe County, display trilobite trackways superimposed on mud cracks suggests that these trilobites may have lived in a very shallow water area that was subject to periodic exposure and desiccation.

Toward the end of Medina deposition, uplift of a broad arch-like feature, the Algonquin axis or Findley Arch, to the west of New York State elevated parts of the Medina seafloor above sea level and caused erosion to bevel down through the highest beds of the Medina, producing a widespread unconformity that separates this unit from the overlying Clinton Group (Figure 4.28B).

The Tuscarora-Shawangunk-Medina clastic wedge consists of quartz-rich sandstones overlying the angular Taconic Unconformity in the Appalachians, which seem to record renewed uplift and sedimentation. In the Niagara region, these show a transgressive succession from nonmarine and nearshore sandstones, to deeper marine gray shales, followed by a shallowing (regression) into tidal flat and nonmarine red beds.

Middle Silurian

Lower Clinton Group

Middle Silurian (late Llandovery series) mixed shales, carbonates, and ironstones of the lower Clinton Group record the wearing down of Taconic mountainous source terranes to the east and a shift back to open marine sediments. One distinctive feature is the widespread "Clinton iron ores"; these famous fossil-rich and **oolitic** hematites have been mined extensively from near Birmingham, Alabama, to central New York for making steel and red paint oxides. These beds probably were associated with enrichment of iron in sediments due to deep weathering of Taconic uplifts. Hematite coatings on grains precipitated under conditions of fluctuating oxidizing-reducing conditions in muddy sediments, during times of sediment starvation. Hence, ironstones are commonly associated with phosphatic nodules and **flooding surfaces** (surfaces with very low sedimentation rates during deepening episodes).

The lowest part of the Clinton Group comprises greenish gray shales, the Neahga or Maplewood of western New York, overlain by thin limestones. The Neahga is a highly fissile, very soft, chippy shale that like the Power Glen seems to represent nearshore but dysaerobic muddy bottom conditions. The Maplewood contains very abundant microfossils of acritarchs, algal resting cysts, and some early plant spores but is almost devoid of macrofossils. A very meager fauna, including one species of calymenid, one odontopleurid trilobite, and a few brachiopods, has been obtained from a basal phosphate-rich layer that immediately overlies the unconformity at the bottom of the Maplewood. However, in general, the conditions appear to have been unfavorable for benthic or bottom-dwelling life during deposition of the Maplewood muds.

The Maplewood grades eastward into a sandy, hematite-rich conglomerate that is generally lacking in fossils. However, it grades upward and laterally to the west into thin-bedded limestones and fossil-rich hematite beds. These limestones, the Reynales Formation, are rich in brachiopods, particularly the large robust *Pentamerus*, which appears to have formed shell banks in shallow shoal areas on the seafloor. To the west, these brachiopod-rich shoals and crinoidal grainstones pass laterally into more nodular, shaly, and probably more offshore limestones. These facies contain smaller brachiopods, abundant and diverse bryozoans, crinoids, and a few trilobites. Particularly notable in these beds are abundant cephala and pygidia of the trilobite *Encrinurus*.

Reported from the Reynales are the following species of trilobites:

Bumastus sp.
Calymene sp.
Encrinurus cf. *E. raybesti*
Eophacops trisulcatus
Scutellum niagarensis

Green and purplish shales of the overlying Sodus Formation in west-central New York represent a return to shallow water muddy conditions similar to but somewhat different from those of the Maplewood. These muddy seafloors supported a modest diversity of brachiopods dominated by the small atrypid *Eocoelia*, small bryozoans, a few crinoid **ossicles**, the conical fossil *Tentaculites*, and scattered specimens of small tabulate corals. A few trilobites make up the remainder of the fauna. Both *Diacalymene rostrata* and *Eophacops trisulcatus* have been extracted from these beds. Here conditions alternated between dysoxic and somewhat better oxygenated, as indicated by the green and purple banding typical of the unit, and the seafloor was at least temporarily hospitable to benthic assemblages of a very shallow-water restricted type. These muds apparently represent accumulations in a broad shallow-water lagoonal setting inboard of offshore, brachiopod-rich shoals, which at that time were as far west as Ohio.

The Wolcott Formation represents a return to Reynales-like conditions in which abundant pentamerid brachiopods once again flourished and formed shell banks in west-central New York. However, an unconformity has removed western portions of the Wolcott so it is not possible to trace these facies westward past Wayne County into offshore nodular sediments comparable to those seen in the western Reynales.

Thin hematite bands, usually only a few centimeters thick but traceable over distances of tens of kilometers, are found at several levels within the lower part of the Clinton Group. As noted, these appear to represent periods of marine sediment starvation in a shoal margin environment in which ferrous iron became concentrated in pore-waters and precipitated out as coatings and ooidlike grains on the seafloor. Reworking of these grains into windrows or piles concentrated the iron oxide components, making them lean iron ores. These ores are mostly "fossil ores," meaning that they contain abundant fragmentary fossils, especially crinoid ossicles, bits of bryozoans, and shell fragments. Again trilobites, especially calymenids, are present but only as highly **comminuted** debris.

The middle part of the Clinton is represented by the Sauquoit Shales and their eastern reddish sandstone equivalent the Otsquago Formation in the Utica area and eastward. These units may be in excess of 30 m thick, and expand dramatically to the southwest into a succession, up to 140 m thick, of purplish green shales referred to as *Rose Hill* in Pennsylvania and Maryland. The Sauquoit Shale from the middle part of the Clinton Group represents conditions quite similar to those of the Sodus and Maplewood muds, and like them it tends to carry a restricted fauna, although the trilobite *Calymene* is relatively common and the genus *Dalmanites* has also been reported from these shales.

Upper Clinton Group

A very widespread unconformity separates the middle Clinton and lower Clinton shales from the overlying upper part of the Clinton Group. During this medial part of Silurian time, more dramatic arching of the sea bottom in the vicinity of the Algonquin Arch (near Hamilton, Ontario) and farther east exposed major parts of the older Clinton sediments to erosion and produced a widespread beveling-surface unconformity that eroded off the western edges of the middle and lower Clinton units. Hence, a good deal of Middle Silurian time may be missing at this important unconformity.

The upper part of the Clinton Group is a varied succession of shales and carbonates that are readily grouped into two unconformity-bound sequences or packages. The lower consists of the Westmoreland Hematite (locally), Williamson-Willowvale shales, and the Rockway Formation and its lateral eastern equivalent, the Dawes Sandstone. The basal hematite bed, or Westmoreland, rests unconformably on the beveled edges of the underlying middle and to the west of the lower Clinton units. It is again a particularly oolitic-rich ironstone that contains abundant, small (1 to 2 mm) spheroidal bodies of hematite that appear to have formed by accretion of iron-rich coatings on grains that were at least intermittently rolled on the sea bottom. This unit probably represents a deposit accumulated during sea-level deepening in which the fine-grained portion was winnowed away, associated with an episode of a very widespread rise in sea level that occurred globally during mid Silurian times. It is well dated by ostracods and sawblade-like graptolites that occur in the interbedded shales.

Beds of black-green shale of the mid Silurian Williamson Formation and its eastern equivalent Willowvale Shale carry an abundant fossil assemblage. Fossils typically are poorly preserved in the shales but they may include over a hundred species of brachiopods, bryozoans, the small "button" coral *Palaeocyclus*, crinoids, and trilobites. Both *Dalmanites* and *Calymene* species are relatively abundant in the Willowvale of central New York. To the west, the medium gray shales give way laterally to green satiny-smooth clay shales, alternating with black-laminated, sometimes sandy shales that yield a great abundance locally of the sawblade monograptid graptolite *Monograptus clintonensis*, which enables correlation of the Willowvale and Williamson Shales with other deposits globally that represent this mid Silurian highstand or sea-level rise event. The Williamson Shales in the Rochester area generally contain few other benthic fossils, although some beds carry small brachiopods and rare specimens of the trilobite *Calymene* have been found. Overall the Williamson-Willowvale Shales probably record the deepest-water interval during deposition of the Silurian in New York State. Bottom conditions range from anoxic or having very low oxygen, near the basin center, to fully oxygenated conditions favorable to life in central New York Willowvale sections.

The overlying Rockway Dolostone and equivalent Dawes Sandstone represent somewhat shallower water conditions in which rhythmic limestones or calcareous sandstones accumulated, alternating with shales. Again, fossils tend to be sparse because of poor preservation in these beds. But a few species of brachiopods, including the large form *Costistricklandia*, and graptolites, mostly dendroids (or brushy graptolites), and rare trilobites, such as *Dalmanites*, have been found in these beds.

In western New York the Rockway consists of pale to medium gray, argillaceous dolostones with interbedded greenish gray shales. Overall, the Rockway appears to represent an offshore muddy carbonate unit deposited well below wave-base and under somewhat restricted conditions. It grades eastward and southeastward into the Dawes Sandstone of central New York, an interbedded, hummocky cross-laminated, fine-grained sandstone and shale unit with the beautiful trilobite trace fossil, *Rusophycus*.

The Rockway and its equivalents are sharply overlain by a thin, widespread blanket deposit of crinoidal limestone. This unit,

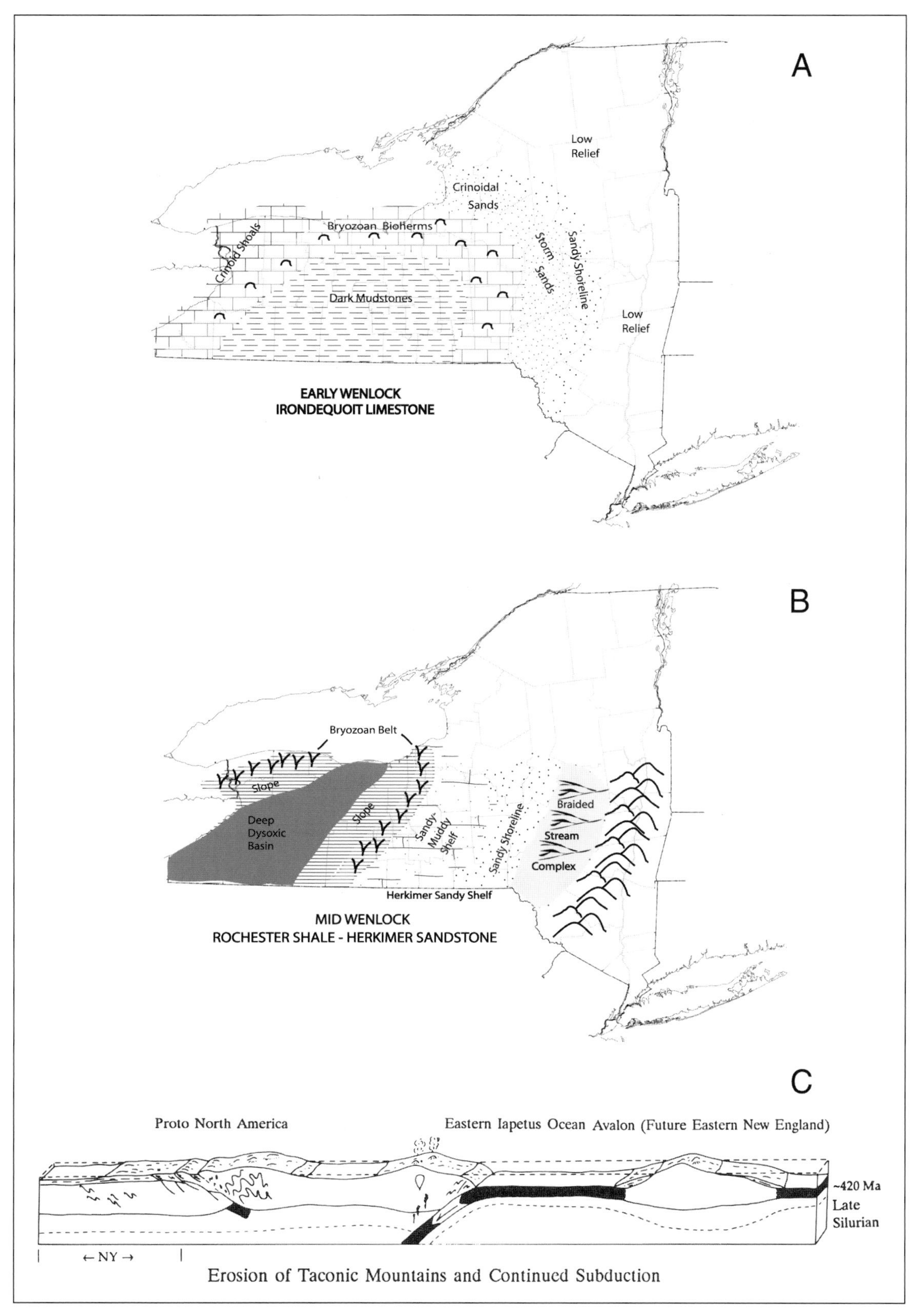

FIGURE 4.29. Maps of New York. A. Early Wenlock times during the deposition of the Irondequoit Limestone. B. Mid Wenlock during the deposition of the Rochester Shale. C. Cross section showing the approach of Avalon with proto North America. C is from Isachson et al. (1991). Printed with permission of the New York State Museum, Albany, N.Y.

FIGURE 4.30. Irondequoit-Rochester bioherm at the upper part of the Middle Silurian Clinton Group. The sharp contact at the knee level of William Goodman is the unconformable contact of Rockway Dolostone (a) and massive Irondequoit Limestone (b). The lens-shaped body at the top of the Irondequoit is a small bioherm ("reefs") (arrow) of bryozoans and algae. It occurs at the contact of the dark gray Rochester Shale (c). Niagara Gorge, Lewiston, Niagara County.

referred to as the *Irondequoit Limestone* (Figure 4.29), consists almost entirely of the disarticulated and commonly abraded skeletal remains of crinoids and cystoids. However, there are some thin, interbedded shaly units, and in the Rochester area, the Irondequoit is somewhat finer-grained and carries a lower diversity of fauna typified by a few species of brachiopods, of which *Whitfieldella* is a common element. Also in this area, as well as in Niagara County, small (up to 5 m across and 2 m high), irregular, lenticular mounds, or bioherms, occur within the Irondequoit (Figure 4.30). In Niagara County these occur mainly at the upper contact of the crinoidal limestone and extend upward, sometimes as much as 1 to 2 m (4 to 5 feet), into the overlying Rochester Shale. These are rather amorphous masses of very-fine-grained (micritic) limestone with some leaflike bryozoans that apparently helped support the mounds. The bioherms represent very small buildups or "reeflets" that developed on the seafloor during a time of deepening. These mounds, however, have yielded some intriguing fossil deposits that are not found elsewhere within the Irondequoit Limestone or the overlying Rochester Shale. For example, pockets of greenish shale within the cream-colored, fine-grained limestone of the mounds in Niagara County yield extremely abundant remains of the trilobite genera *Bumastus* and *Illaenoides*. These relatively large fossils, up to 3 cm across, appear to represent the remains of reef- or bioherm-dwelling trilobites that accumulated or collected in the small pockets and cavities within the bioherm framework. Also found in these associations are a *Cheirurus* sp., *Scutellum rochesterense*, and a new species of *Diacalymene* that is also rather large and represented by nearly complete articulated remains from the Niagara Gorge.

The Irondequoit and its bioherms represent a shallow, highly agitated shelf environment. Water depths were close to normal wave-base, ranging from perhaps 10 to 20 m in western New York and no more than 20 to 30 m in the slightly deeper waters around Rochester. The skeletal remains of crinoids and cystoids were reworked and accumulated to thicknesses of up 3 or 4 m. Considering that it takes about 15,000 average-sized crinoids to make 1 m^3 of limestone, the total number of echinoderms represented in these beds is stunningly high (10^{10} or even 10^{12} individuals). These organisms flourished in shallow, relatively clean waters associated with a major lowering of relative sea level in New York State. To the southeast, the Irondequoit skeletal limestone grades first into hematitic limestone (the Kirkland Formation) and beyond into quartzose, cross-bedded sandstones of the upper Herkimer Formation. A few trilobites are reported from the Irondequoit, including the following:

Bumastus sp.	*Diacalymene* sp.
Calymene sp.	*Cheirurus* sp.
Liocalymene clintoni	*Scutellum rochesterense*

Irondequoit and its lateral equivalents, the Kirkland and Keefer formations, are overlain sharply, but apparently conformably, by gray mudstones of the Rochester Formation. Perhaps no Silurian unit in North America is more noted for its exquisite fossils than the Rochester. Although much of the unit is relatively sparsely fossiliferous, a total fauna of over 200 species of invertebrate fossils is known from the Rochester Shale, including over 20 species of trilobites. Thin lenses of brachiopod- and bryozoan-rich mudstones and argillaceous limestone yield prolific faunas, with over 80 species of bryozoans alone known from the middle beds of the Rochester. In western New York the unit is divisible into two members: a fossil-rich lower half, bounded on its top by a bundle of limestone bryozoa beds and referred to as the *Lewiston Member*, and an upper sparsely fossiliferous, dolomitic shale unit, the Burleigh Hill Member.

The Lewiston Member has yielded most of the diverse trilobite and echinoderm faunas for which the Rochester is justifiably famous. Fossils are exceptionally well preserved, especially in beds near the transition from a lower, highly shell-rich interval into a middle barren portion of the Lewiston and in the upper transition from this sparsely fossiliferous interval to the overlying bryozoan-rich beds. These beds contain a number of trilobites, some of them in an extraordinary state of preservation. The well-known examples of the large lichid *Arctinurus boltoni*, as well as abundant *Dalmanites limulurus*, *Calymene niagarensis*, and *Trimerus delphinocephalus*, occur in these levels. The exquisite preservation of some of the fossils suggests that they were buried instantly by sediment plumes or turbidity currents.

The Burleigh Hill Member consists of medium to dark gray shales with thin calcisiltites, especially toward the top, that grade upward into the overlying DeCew Formation. This upper portion of the Rochester, like the middle portion of the Lewiston Member, is characterized by dark gray, nearly barren shales in much of western New York. Nonetheless, some bedding planes contain abundant remains, mostly disarticulated, as well as articulated specimens of the trilobite genera *Dalmanites* and *Trimerus*. These dark, sparsely fossiliferous shale facies, representing deeper portions of the Rochester sea, were deposited during relative rises in the sea level. Here, high rates of sedimentation combined with low-oxygen conditions near the substrate made the bottom conditions inhospitable for most invertebrate species. Nonetheless, some trilobites, as scavengers, appear to have thrived in these environments. Shallower portions of the Rochester are represented by banks of bryozoans and echinoderms (crinoids and the cystoid *Caryocrinites*). Trilobites were less dominant here but are represented by a greater diversity of species.

Herkimer Sandstone, the eastern equivalent of the Rochester Shale, represents sandy shallow shelf to nearshore environments. It yields *Trimerus* and *Dalmanites* specimens and excellently preserved trilobite trace fossils, especially the large resting trace *Rusophycus*.

The paleogeographic distribution of the trilobites in the Lewiston Member is of interest. The trilobites east of Rochester are different from those west of this area. Differences are at the species as well as genus level. Certain genera such as *Arctinurus*, *Decoroproetus*, *Deiphon*, *Dicranopeltis*, *Illaenoides*, *and Radnoria* have been found only in the west, while *Cheirurus*, *Maurotarion*, and *Staurocephalus* have been found to the east. The genera *Dalmanites* and *Calymene* occur both east and west of Rochester, but they are represented by different species. *Bumastus* and *Trimerus* are found in both the east and the west. *Trimerus delphinocephalus* is also found in equivalent age beds in England.

Rochester Shale trilobites reported from western New York are as follows:

Acanthopyge sp.	*Arctinurus boltoni*
Bumastus ioxus	*Calymene niagarensis*
Calymene sp.	*Cheirurus* sp.
Dalmanites limulurus	*Dalmanites* sp.
Decoroproetus corycoeus	*Deiphon pisum*
Diacalymene sp.	*Dicranopeltis nereus*
Illaenoides cf. *I. trilobita*	*Maurotarion* sp.
Odontopleurid	Proetid
Radnoria sp.	*Staurocephalus* sp.
Trimerus delphinocephalus	*Trochurus halli*

The Rochester Shale seems to reflect siliciclastic muds that were derived from erosion of newly developed highlands to the east or southeast. Initiation of this new input is represented through much of the Appalachian Basin by the occurrence of rather thick sands known as Keefer Sandstone in Pennsylvania and the Herkimer Sandstone in central New York State (Figure 4.29). Other circumstantial evidence suggests that a new pulse of tectonic activity occurred about the middle part of the Silurian. This comes in the form of the shifting of the depositional basin itself. In the time period represented by the Williamson Shale to the upper part of the Rochester Shale, the area of thickest sediment accumulation, dominantly of shale, shifted westward approximately 100 km from the region of Oneida Lake to the area of Wayne County. This migratory pattern together with the increased input of coarser sediments suggests that renewed thrust faulting may have taken place in the former Taconic landmass. Possibly this is the result of new ocean-floor subduction underneath the eastern margin of North America (Figure 4.29C). The Rochester Shale represents a relatively shallow but muddy-bottom sea that ranged upward to over a hundred meters of water depth. The Rochester Shale facies grades eastward to the sandy shoreline facies of the Herkimer Sandstone in central New York and to the northwest into carbonate banks.

Throughout much of western New York, the Rochester Shale is overlain by a rather unusual unit in the DeCew Dolostone. This unit is a thin (3 to 5 m) and uniform interval of buff-colored carbonate silt with hummocky cross-stratification. The result of storm deposition, the DeCew is also highly convoluted and contorted and in some places displays overturned folds. We suggest that this represents a period of seismic shock (earthquake) activity on the floor of the foreland basin or interior continental sea. This effect was widespread; we have found evidence for a similar deformation in DeCew age strata in central Pennsylvania and southern Ohio. Because of the rapid deposition of the carbonate silts, this is not a highly fossiliferous unit, although occasional remains of some fossil crinoids and the trilobite *Trimerus* have been found within the unit. The source of the DeCew carbonate silts and fine sands is somewhat unclear, but it probably was represented by carbonate shoals in the form of crinoid gardens that existed to the northwest of the main New York depositional area. These shoal facies encroached upon New York during the deposition of the succeeding Lockport Group.

An eastern shaly facies equivalent to the DeCew, the Glenmark Shale, yields a diverse fauna of small rugose corals, brachiopods, and trilobite genera, including *Dalmanites* sp., *Trimerus*, *Encrinurus* sp., calymenid, *Maurotarion* sp., *Cheirurus* sp., and proetids.

Lockport Group

The Lockport Group is an interval over 65 m thick of predominately dolomitic carbonate rocks. However, to the east of the Rochester area, these dolostones and dolomitic limestones interfinger with dark gray shales of the Ilion Formation. In the central Appalachians the Lockport Group is represented by a dominantly dark, shaly interval with thin ribbon-like limestones termed the *Mackenzie Formation*. The Lockport represents a depositional sequence. It is bounded at its base by a major erosion surface that locally cuts into the DeCew Dolostone and removes it in areas near Hamilton, Ontario, where the unconformity continues to cut downward into the Rochester Shale until the base of the Lockport Group comes to rest on the underlying crinoidal grainstone of the Irondequoit.

During the Late Silurian a vast carbonate bank, somewhat like that which existed in the Cambro-Ordovician, developed once again in Laurentia. Very widespread shoals were developed from sand- and gravel-sized skeletal pieces (mainly columnals) of crinoids and cystoids. Under similar, high-energy environments, tabulate corals and **stromatoporoids** established reefs in the Great Lakes area. A barrier reef complex developed around a circular, subsiding area—the Michigan Basin—and patch reefs developed on the tops of crinoidal skeletal shoals from Indiana, to Ohio, and into New York. Many such reefs or bioherms show a good succession of pioneer thicket formers to climax communities dominated by head corals and stromatoporoids. Trilobites were a minor but significant component of the reefal faunas.

The basal Lockport unit, Gasport Limestone, consists of dolomitic limestones and some argillaceous dolostone. Much of it can be described as an echinoderm packstone or grainstone, like the underlying Irondequoit, that consists of disarticulated and commonly abraded fragments of crinoid and cystoid plates and columnals. Typically these crinoidal sands and gravel deposits were moved by submarine currents. Consequently the unit shows medium- to small-scale cross-stratification (Figure 4.32A). Some portions of the Gasport also show bimodal cross-stratification, two opposite orientations of cross-bedding that might reflect the influence of oscillating tidal currents. The Gasport sediments evidently accumulated during a time of initial transgression following a major regression of seas from western New York that produced the unconformity beneath the unit. During a time of rising sea level, but still in environments of shallow, agitated waters, perhaps no more than a few meters deep, crinoid banks developed extensively over much of western and west-central New York. In the Rochester area, these banks apparently interfingered with quartz sand bars represented by the Penfield Formation. These cross-stratified dolomitic sandstone units seem to reflect the input of quartzose sediment from a northern or northeasterly source terrain. They form an elongate shoal or barlike feature that extended southward from Rochester to the subsurface of New York State. To the east of this area, in Wayne County, the sands give way to muddy dolostones and dolomitic limestones, many of them containing remains of ostracods, some stromatolites, and the trilobite genera *Trimerus*, *Calymene*, *Encrinurus*, and *Maurotarion*. Still farther to the east, these interfinger with dark gray shale of the Ilion Formation. Within the upper part of the Gasport Formation, deepening produced a change in sedimentation patterns, in western New York, to thinner-bedded, more argillaceous dolostones. At the same time, small patch reefs, which had begun developing on the now-stabilized surface of the crinoidal shoals, built upward, forming the famed Gasport reefs or bioherms (Figure 4.32B). These reefs were built primarily of favositid corals and stromatoporoids. The Gasport is not particularly noted for trilobites, although some

FIGURE 4.31. Mid Silurian successions. A. Lowest light limestone ledge, Irondequoit Limestone (a) is overlain by about 20 m of dark gray Rochester Shale (b). Thick layers of the Lockport Group (c) protrude at the top of the cliff. Niagara River Gorge, south of Lewiston-Queenston Bridge, Niagara County. B. Lowest rhythmic thin beds are Rockway Formation (a). These are sharply overlain by Irondequoit Limestone (b) bearing small bioherms ("reefs") (arrow) that protrude slightly into the overlying Rochester Shale (c). Genesee River Gorge, Rochester, Monroe County.

FIGURE 4.32. Silurian carbonates. A. Cross-bedded crinoidal limestone, Gasport Formation, cut along Robert Moses Parkway, Lewiston, Niagara County. B. Bioherm or reef mound of stromatoporoids and tabulate corals, Goat Island Formation, overlies thin-bedded upper Gasport Formation on the right end of the cut. Road cut on Rte. 429, Niagara escarpment, Pekin, Niagara County.

specimens of calymenids are found. An area of interfingering (shaly) dolostones with typical Gasport crinoid grainstones occurs in the Sweden-Walker Quarry at Brockport, Monroe County. This transitional facies has yielded the remains of *Dalmanites*, *Trimerus*, and another rare homalonotid trilobite. Also, some of the biohermal faunas have yielded fragmentary remains of the calymenid, *Diacalymene*, and other trilobites.

The overlying Goat Island Formation is separated in most places from the Gasport by a minor unconformity. It represents the second cycle of shallowing, followed by deepening and the development of crinoidal shoals and in places such as at Brockport, the development of small patch reefs of stromatoporoids and corals (Figure 4.31B). These reefs have many features in common with the underlying Gasport. However, in many places, the Goat Island has been affected severely by late diagenetic dolomitization, which has altered its texture and made it difficult to decipher the original fossil content and other features.

An argillaceous dolostone bed near the base of the Goat Island, however, has yielded an extraordinary biota of soft-bodied organisms, particularly algae, some worms, and graptolites. This deposit, which LoDuca (1995) studied extensively, appears to represent an accumulation of muds in shallow waters between patch reefs in the Goat Island Formation. Unfortunately, to date, it has yielded few, if any, trilobite remains.

The upper Goat Island in western New York, and particularly in Ontario, is highly argillaceous and frequently has been misidentified as the Eramosa Formation. It is presently termed the *Vinemount Shale Member*. Vinemount Shale beds have yielded several fragmentary fossil remains, including trilobites, small rugose corals, and brachiopods, although the extensive dolomitization of these argillaceous carbonates has obscured most of the details.

Near Ancaster, Ontario, and locally in Niagara County, New York, an upper portion of the Goat Island yields white chert nodules that contain the remains of numerous species of fossils in a rather good state of preservation. Particularly notable are small siliceous sponges that may hint at the source for the silica of the light-colored cherts. Also, a variety of brachiopods and several trilobite species, including calymenids and dalmanitids, have been obtained as fragments from these chert nodules. Evidently, early replacement by silica protected fossils from dolomitization, thereby leaving a reasonable record of these fossils.

Howell and Sanford (1947) described trilobites from the Goat Island Formation (formerly misidentified as Oak Orchard Member) in the Rochester area:

Calymene singularis	*Encrinurus* sp.
Proetus artiaxis	*Scutellum wardi*

The Eramosa Formation is represented over most of central and western New York by vuggy, massive dolostones that have commonly been termed *Oak Orchard Formation* in the past. These dolostones are characterized by small thickets or biostromes of corals, especially cladoporids. They represent thickets of small tabulates and other corals that grew in shallow waters. Stromatoporoids are also common as isolated heads within these beds, typically up to a foot across but highly altered by **diagenesis**. These beds, like the underlying Goat Island, are typically vuggy and contain abundant mineralized pockets that are well known to local mineralogists. In Ontario, Canada, however, the basal beds of the Eramosa that overlie Eramosa Shales and appear almost gradational into the latter, are platy argillaceous dolostones that seem to represent somewhat more offshore carbonate accumulation. These basal Eramosa carbonates are noted for abundant, although mainly disarticulated, trilobites. Quarries in the vicinity of Dundas, Ontario, have yielded thousands of specimens of the otherwise rare trilobite *Encrinurus raybesti*, along with calymenids, and a new dalmanitid species. The Eramosa represents a return to conditions somewhat more like those of the Rochester Shale, although pulses of rapid burial were rare in the Eramosa; hence, most trilobites are disarticulated. In New York and throughout at least the eastern Niagara Peninsula of Ontario, the upper beds of the Eramosa reflect an abrupt shallowing. Layers of stromatolites or bacterial mat **boundstones** were common and widespread at this time. These stromatolites formed domal heads up to 2 or 3 m across and a meter tall (Figure 4.33). The stromatolites are generally associated with a sparse low diversity of favositid corals, ostracods, and a few species of brachiopods. The incoming of these beds appears to reflect the stress of increased hypersalinity in the Appalachian Basin. To the east, beds of the Scondondoa and upper Ilion reflect these same sorts of changes. Stromatolitic limestones give way abruptly eastward to dark shales containing few normal marine organisms but remains of lingulid brachiopods and eurypterids. These in turn pass upward to the red beds of the Vernon Formation.

Although not generally considered a trilobite-rich interval, the Lockport Group in total has yielded the following trilobites:

Calymene singularis	*Calymene* sp.
Dalmanites sp.	*Diacalymene* sp.
Encrinurus raybesti	*Maurotarion* sp.
Proetus artiaxis	*Proetus tenuisulcatus*
Scutellum wardi	*Trimerus delphinocephalus*

Late Silurian

Salina Group

During the Late Silurian, a more arid climate prevailed over eastern Laurentia, and a landlocking of seas occurred in two areas. The later part of the Silurian in eastern North America is represented by the Salina Group. As the name implies, these deposits represent a mixture of some shales and dolostones but are primarily noted for evaporites. In the Michigan Basin, the

FIGURE 4.33. Domal stromatolites (arrow). Guelph Dolostone, Robert Moses Power Plant access road, Lewiston, Niagara County.

barrier reef complex restricted circulation, while in the eastern Appalachian region, a new pulse of sediments eroded off coastal mountains. The Vernon-Bloomsburg delta—was derived from Salinic Orogenic uplifts southeast (present directions) of New York State. The outward building (progradation) of the Vernon-Bloomsburg delta into south-central Pennsylvania formed a partial barrier to oceanic circulation in the northern Appalachian Basin of New York. These restrictive barriers prevented open circulation to the sea and enabled salinity to build up to the point of precipitating evaporites. A cyclic process of seawater influx then evaporation enabled evaporites, such as anhydrite, gypsum, and halite, to accumulate to considerable thickness. The salt layer under Detroit, Michigan, is over 1 km thick.

The Upper Silurian Vernon Formation consists of probably nonmarine or very shallow tidal-flat deposits of a red, rather massive mudstone. These beds contain few fossils, although an occasional interbedded dolostone yields remains of fossil jawless, armored fish. Small clams and brachiopods also may be present in these beds. To the west the Vernon gives way from red mudstones to greenish gray dolomitic shales and dolostones, many of which contain evaporite crystal molds. The Vernon in western New York is probably most noted as the source of key salt beds that have been mined for many decades at the Retsof Mines in the area of the Genesee Valley. The appearance of these evaporite beds within the Vernon reflects restriction of the Appalachian Basin. The Vernon clastic sediments appear to represent a final pulse of clastics derived from erosion of the newly uplifted mountainous source terrain to the southeast of New York State. In Pennsylvania, extremely thick red mudstones of the Bloomsburg Shale take the place of the Vernon and seem to point to a nearby source of muds and silts. North America at this time appears to have lain within the subtropical desert belt, probably in paleolatitudes between 20 and 30° south of the equator (Figure 4.5). Under such circumstances, the development of a large wedge of deltaic mudstones and sandstones in the Bloomsburg/Vernon Formation produced a restriction in the circulation in the northern Appalachian Basin region. This restricted circulation in combination with the arid climate led to the development of a hypersaline seaway, probably similar in some ways to the modern Dead Sea. As seawater evaporated away, salt and anhydrite or gypsum deposits began to accumulate in the lower sags of the basin, while on the margins some, at least seasonal, input of siliciclastic muds continued.

Although salt deposition occurred within the western New York/Genesee Valley region during deposition of the Vernon Shales at the base of the Salina Group, the locus of salt deposition shifted progressively eastward during the deposition of the Salina Group sediments. This eastward shifting also may have been associated with the minor uplift of the old Vernon/Bloomsburg deltaic land above sea level, which produced an erosional beveling of Late to Middle Silurian sediments in areas east of Utica, Oneida County. A large salt depositional basin existed in the central region of the Finger Lakes and eventually in the classic Syracuse area. These deposits, the Syracuse Formation, take their name from that city. The formation consists of thin-bedded dolostones and shales and in places thick accumulations of anhydrite or salt. Although the remains of normal marine organisms are uncommon within these beds, there are stray reports of some brachiopods and even of one occurrence of a trilobite, a calymenid, within the Syracuse Formation in the Syracuse area. Otherwise, conditions generally were far too harsh, owing to the elevated salinity, to support normal marine communities, and most of the beds of the Salina Group are barren of fossils. More diverse marine fossils, including calymenid trilobites, are known from the laterally equivalent Tonoloway Formation in Pennsylvania, Maryland, and West Virginia.

Salina deposition was terminated by the accumulation of the Camillus Formation, a relatively thick interval of mottled gray to slightly reddish barren shales with salt crystal molds and some dolostones, commonly with molds of small gypsum or anhydrite lathlike crystals. Not surprisingly, the Camillus is nearly barren of fossils. It too has served as a source for economically important gypsum deposits in western New York.

Bertie Dolostone and Rondout Group

During the very late phases of Silurian deposition in New York State, the Bertie Group sediments accumulated. The Bertie is an unusual rock unit that consists of argillaceous, buff-colored dolostone, referred to in the past as *waterlimes* because of their property as natural cement rocks, yielding cements, which hardened underwater. The Bertie, as with the underlying Salina Group, contains much evidence for deposition at or near sea level. Within the formation are beds of low domal stromatolites, extensive layers of fine, mud-cracked dolostone, gypsum crystal molds, very shallow water ripples, and other evidence for deposition in a restricted tidal flat to shallow lagoonal setting. The Bertie is most noted for its extraordinary eurypterids at certain horizons. The environment of these interesting **chelicerate** arthropods is still debatable. Although they are often preserved in rocks that contain evidence for hypersalinity, such as salt crystal casts, the eurypterids probably did not live under these highly saline conditions. Rather, the local heavy accumulations of carcasses of these animals probably represent dead remains that were washed out from estuaries that fed into the more saline Bertie sea; carcasses were rapidly buried in briney sediments. The appearance of small land plants, some of the oldest known in the world, along with the eurypterids suggests that these sediments accumulated in very close proximity to low-lying lands of the upper tidal flats and probably in small estuaries that emanated off the exposed land that were, at least periodically, brackish water in their composition (Figure 4.36A). Only a few species of normal marine fossils are found within parts of the Bertie. They include lingulid brachiopods, a few other species of articulate brachiopods, and a few species of mollusks (snails, clams, and orthoconic cephalopods). Evidently, salinity remained high and variable enough that normal marine communities still were not established in the offshore areas.

A new, unusual trilobite, apparently a lichid, was recently collected from the eurypterid-bearing waterlimes in Fort Erie, Ontario, near Buffalo. On the whole though, these beds reflect unusual hypersaline conditions, and trilobites are very rare.

However, finally within the latest part of the Silurian, normal marine conditions returned over New York, Pennsylvania, and much of the Appalachian Basin. The Keyser Formation of Pennsylvania and the laterally equivalent Rondout Group and Decker Formation in New York State contain a far more diverse assemblage of marine fossils than do the underlying Bertie and Salina beds. At least during times of the deposition of the Rondout, shallow marine conditions rather akin to those that developed during the Lockport deposition existed over portions of eastern and central New York State. At times, units such as the Cobleskill Limestone accumulated with abundant crinoid, coral, stromatoporoid, and other remains. Trilobites are scarce but are represented by a fair diversity of species such as the *Calymene camerata*, *Hedstroemia pachydermata*, *Richterarges ptyonurus*, and *Dalmanites aspinosus*. In Pennsylvania the laterally equivalent Keyser Formation contains small reef buildups of corals and stromatoporoids and other shaly beds, rather reminiscent of the much older Rochester Shale, that contain diverse bryozoan and brachiopod faunas and abundant, well-preserved cystoids. Again, some trilobites are reported from these Keyser beds. Preservation within the Keyser ranges from disarticulated to exquisite articulated remains. Thus, the last chapter of Silurian marine deposition records a breakdown of barriers to normal marine circulation, as well as possible climatic changes from the more arid conditions that characterized deposition of the Salina and Bertie Groups. As normal marine seas returned to the Appalachian Basin, so too did normal marine representatives, which must have immigrated in from outside the basin.

Devonian Period

The Devonian Period (415 to 360 million years ago) was relatively long, with many important events in Earth and life history (Figure 4.1). Major orogenies took place in Europe (end of Caledonian), eastern North America (Acadian), and for the first time in the Phanerozoic, western North America (Antler Orogeny) (Figure 4.34). Much of the Devonian was characterized by warm, "greenhouse"-type climates and a strong tendency for stagnation in deeper sea environments, leading to the formation of very widespread **anoxic** (oxygen-deprived) black shale deposits (Figure 4.35). However, toward the end of the period

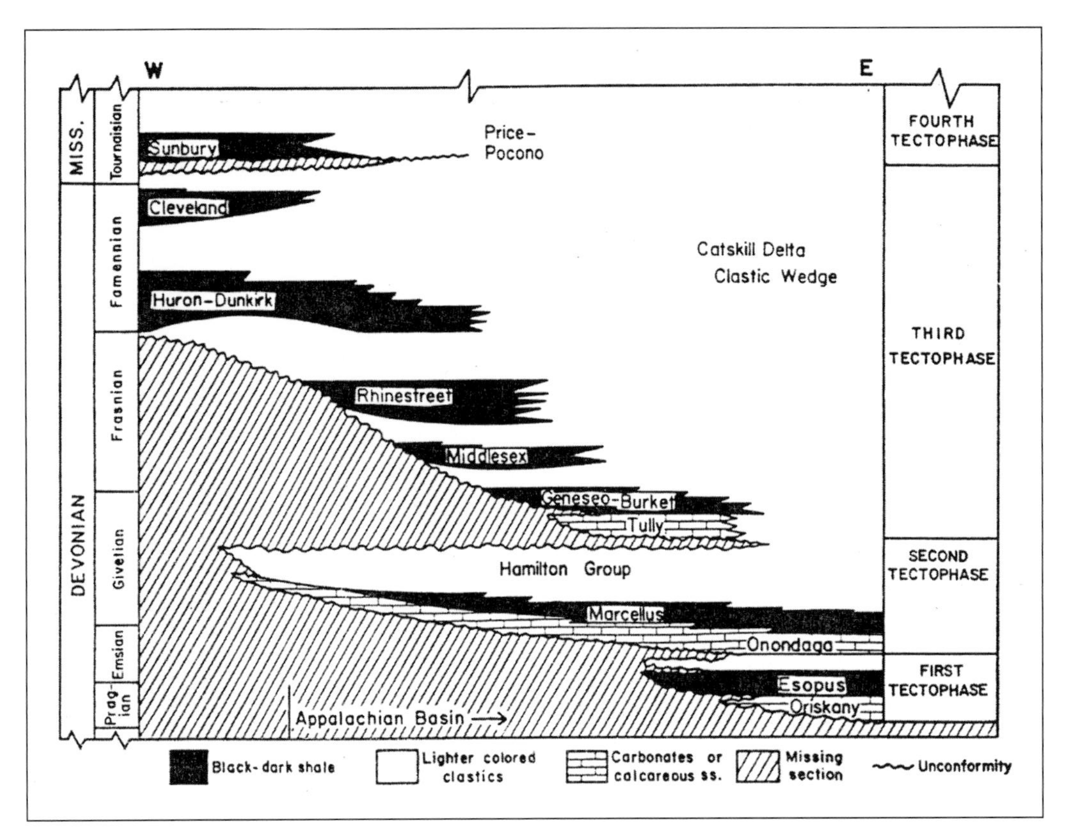

FIGURE 4.34. Composite stratigraphic chart for the northern and central parts of the Appalachian Basin, showing the depositional sequences and their relationship to tectophases. From Ettensohn (1987),

there is evidence for climatic stress associated with renewed glaciation in Gondwana (especially South America).

Early Devonian

The Lower Devonian rocks of New York State are represented primarily by the Helderberg and Tristates Groups. The final, regressive phases of Sloss's Tippecanoe (Tutelo phase) Supersequence are recorded in the Helderberg carbonates of eastern New York and Pennsylvania (Figure 4.35). The Helderberg Group is a series of limestones and minor shales that crop out in central to eastern New York State. Outstanding exposures of these rocks are cuts along Rte. I-88 in the Schoharie Valley, and a number of quarries and road cuts from the area of Albany, at the Helderberg Escarpment, southward to the state line at Port Jervis, Orange County.

It should be noted that during Helderberg deposition, the **axis** (deepest part) of the Appalachian Basin was at a position substantially farther east than during Silurian or later Devonian time. The basin axis appears to have shifted eastward during a time of tectonic quiescence beginning in the late part of the Silurian, and to have been in a position east of the Hudson Valley extending northeastward into New England at this point in Early Devonian time. For this reason, most of the Helderberg and overlying Tristates groups, including the more offshore facies, are confined to the Hudson River Valley and probably once extended northeastward into New England and Quebec.

Late Early Devonian

Helderberg Facies and Trilobites

The lowest unit of the Helderberg Group is the Thacher Member of the Manlius Formation (Figure 4.35C). However, the Manlius facies are distinctly diachronous, being younger in the area near Syracuse, New York, than in the Hudson Valley and equivalent in age to the Coeymans or even Kalkberg Formations in the Hudson Valley. The Manlius is considered to be close to the Silurian-Devonian boundary (Rickard 1975, 1981; Klapper 1981).

The Manlius Formation bears many resemblances to the Ordovician Black River Group. Both intervals contain a series of peritidal to very shallow subtidal facies. The Manlius represents the belt of low-energy but very shallow lagoonal to tidal-flat environments that were sheltered to shoreward by offshore shoals and bars. This suite of environments has been called the "Z" zone (Irwin 1965). These facies typically are arranged in 1- to 3-m scale, shallowing-upward cycles referred to as punctuated aggradational cycles (*PACs*) (Goodwin and Anderson 1985). Such cyclic facies are typical of the older Thacher Member of the

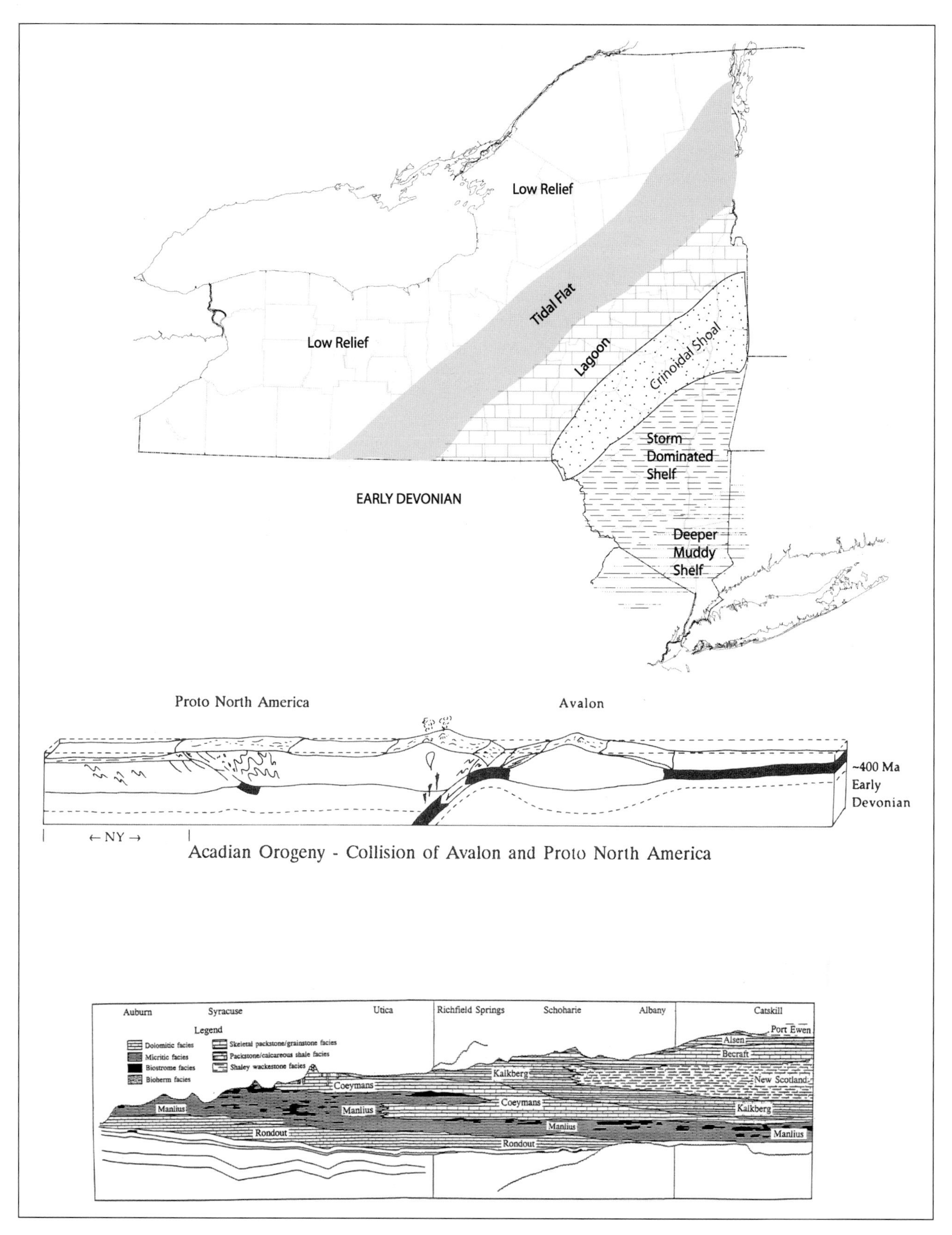

FIGURE 4.35. New York during the Early Devonian. A. Map of the state during this time. B. Cross section of the Acadian Orogeny. C. Stratigraphic chart of the Early Devonian. B, C from Isachson et al. (1991). Printed with permission from the New York State Museum, Albany, N.Y.

Manlius in the Hudson Valley. The typical Manlius cycle begins with somewhat nodular, medium gray, bioturbated limestone, commonly with abundant stromatoporoids and some tabulate corals. Fragments of a variety of other fossils, including gastropods, nautiloids, some brachiopods, and ostracodes, may be present. Such beds may be transitional into thin platy limestones that often have been referred to as "ribbon limestones" and may contain rather well-preserved fossils on some bedding planes. For example, famous outcrops near Litchfield, New York, have yielded bedding planes with abundant well-preserved crinoids and some cystoids (Goldring 1923). Trilobites, other than *Odontochile micrurus,* tend to be very rare in these facies. Shallower water portions of the Manlius consist of planar to wavy laminated micritic limestones that display a regular, even slightly wavy lamination. A regular alternation of thicker coarser and finer thinner laminae suggests that they have accumulated under the influence of tidal currents. The waviness of lamination within these facies suggests that thin bacterial mats may be responsible for some of the trapping of sediment deposited from ebbing tidal currents. Hence, these laminations are sometime referred to as "cryptalgal" or tidal laminites. This facies is generally sparsely fossiliferous. However, occasional thin layers or bedding planes, some representing hard grounds (early cemented seafloors), do display small fossils such as ostracodes, *Tentaculites,* minor bryozoans, small bivalves, and edrioasteroids. Again trilobites tend to be very rare in these facies.

Finally, the shallowest facies of the Manlius comprise thinly bedded, sometimes shaly dolomitic limestones, often with major **desiccation cracks** that may extend downward into the rock, forming polygonal prismatic columns as much as 50 cm into the section (Figure 4.36A). These mudcracked facies represent **supratidal** environments (accumulation above mean high-tide level).

The Manlius Formation extends the farthest west of any of the Helderberg units, being found in considerable thickness in the vicinity of Syracuse and extending as a thin feather edge westward to Cayuga Lake. However, portions of the Manlius are probably age equivalent to some of the middle portion of the Helderberg Group farther east. The upper Manlius Olney, Jamesville, and Pools Brook members in the Syracuse area are the shallow-water equivalents of the middle Helderberg beds and represent a marginal marine, carbonate mud-flat to lagoonal environment that existed on the northwest side of the Appalachian Basin.

In the Hudson Valley and Schoharie regions, the Manlius is sharply overlain by the Coeymans (Ravenna) Member. To the west, Coeymans appears to intertongue and grade into Manlius facies in the vicinity of Herkimer, Herkimer County, or Utica. The Ravenna Member is best developed in the Hudson Valley and consists of rather coarse-grained skeletal limestone composed mainly of ossicles of crinoids and the unusual cystoid *Lepocrinites.* It also contains the remains of robust shelled pentamerid brachiopods (*Gypidula*), but few trilobites. In the Jerusalem Hill area, Herkimer County, just east of Utica, *Dalmanites litchfieldensis* and *Odontochile micrurus* are found. Most material is disarticulated, although rare articulated specimens are known. The Ravenna Member represents a high-energy shoal environment, comparable in many ways to the older Silurian Gasport Formation, Irondequoit Limestone, or other similar units. In the traditional model of X, Y, and Z zones (a low-, high-, low-energy profile) associated with an offshore shoal, the Coeymans represents Y-zone (high-energy) facies. As such, it is thought that the Coeymans shoal-like environments sheltered the low-energy belt in which Manlius fine-grained carbonates accumulated in a series of low-energy lagoons and tidal flats.

The Coeymans grainstone grades upward into the Kalkberg and New Scotland shaly limestones. These intervals are typically composed of finer-grained carbonate muds and silts, with some thin shaly intervals. Fossils are moderately to highly abundant, although typically not making up the major component of these rocks, in contrast to the Coeymans skeletal limestone. The lower Kalkberg (Hannacroix) Member is typified by a series of black chert beds, each about 10 cm thick, which have been traced laterally in the Hudson Valley. The unit is notable for a fauna of brachiopods, bryozoans, and pelmatozoan fragments. The upper Kalkberg and portions of the New Scotland are typically somewhat darker gray, shaly limestones that display a banded appearance of alternating 20- to 30-cm-thick, buff and gray bands. The buff bands are somewhat dolomitic limestones, typically rich in brachiopod shells, whereas the darker gray bands are composed of shale, often with abundant fossil fragments and some minor phosphatic pebbles. Beds of winnowed skeletal hash in the Kalkberg, however, suggest the influence of storm-wave deposition at times.

The New Scotland shaly limestone appears to represent a culmination of a deepening cycle. The New Scotland may be compared in some ways to the older Rochester Shale facies in the Silurian. Like the Rochester it represents an area close to the lower effects of storm waves (commonly referred to in the Irwin Model of facies as the "X zone"). The New Scotland is a calcareous mudstone and argillaceous limestone. It is highly fossiliferous in some beds and more sparsely fossiliferous elsewhere. The New Scotland is rich in brachiopods, bryozoans, and mollusks and carries a fairly diversified trilobite fauna. In common with the Silurian Rochester Shale, these deeper-water facies in the Devonian appear to be dominated by dalmanitid trilobites such as *Dalmanites pleuroptyx, Neoprobilium nasutus,* and *N. tridens.* A more complete listing of the Kalkberg–New Scotland trilobites is as follows:

Acanthopyge (Lobopyge) consanquinea	*Cordania cyclurus*
Dalmanites pleuroptyx	*Dicranurus hamatus*
Gerastos protuberans	*Kettneraspis tuberculata*
Kettneraspis sp.	*Neoprobilium nasutus*

FIGURE 4.36. Upper Silurian/Lower Devonian rocks. A. Mudcrack polygons in the Upper Silurian Salina Group (Wills Creek Formation), near Cumberland, Maryland. B. Lower Devonian, upper Helderberg Group. Lowest white beds are Becraft crinoidal limestone (a). These are overlain by medium gray Alsen Formation (b) and the latter, by dark to light gray Port Ewen Formation (c). Note the anticline (fold) formed during the Acadian Orogeny. Rte. 199 north of Kingston, Ulster County.

Neoprobilium tridens
Oinochoe pustulosus
Scutellum pompilius
Oinochoe bigsbyi
Paciphacops logani

The top of the New Scotland is marked by the abrupt return to shallower-water facies. The unit is overlain by pinkish gray crinoidal grainstone referred to as the *Becraft Limestone*. This unit bears some resemblance to the older Coeymans Formation, although it is a more thoroughly crystalline (well-washed crinoid grainstone), pinkish gray limestone, typically with green shale partings near the base. The unit also is composed mainly of crinoidal fragments including the strange, disk-shaped shields, referred to as *Aspidocrinus*, which are thought to be a type of holdfast or attachment structure of crinoids. The brachiopod *Gypidula* occurs in this unit, as it does in the Coeymans, although is represented by a different species. The Becraft represents the return of shallow, high-energy shoal conditions. These environments do not appear to have been highly favorable to the growth or preservation of trilobites; very few have been reported from this unit.

The Becraft Limestone is followed by a succession of rocks that closely mimics that seen in the Coeymans–Kalkberg–New Scotland succession of the lower portion of the Helderberg Group. Thus, the Becraft is directly overlain by a somewhat cherty, fine-grained, buff- and gray-banded limestone, the Alsen Formation that strongly resembles the older Kalkberg. This layer is followed, in turn, by an analog of the New Scotland, referred to as Port Ewen Formation. The Port Ewen is brownish gray weathering with distinctive white, podlike, concretionary limestone layers. These layers somewhat resemble the bands of the New Scotland but are more distinctly lenticular in form. They appear to represent burrowed limestone that has undergone some alteration during compaction and cementing of the sediment. The Port Ewen and Alsen units also contain much the same fauna as the Kalkberg–New Scotland succession, including familiar bryozoans, brachiopods, and the dalmanitid trilobites. However, the latter are less common in the Port Ewen than in the New Scotland. Overall, the Port Ewen contains a substantial amount of sparsely fossiliferous, although heavily burrowed, muddy limestone and calcareous mudstone.

The highest unit of the Helderberg Group, the Port Jervis Formation, is only exposed in a narrow region of southeastern New York near the state line at Port Jervis, and in a small isolated outcrop belt referred to as the *Skunnemunk Outlier*. Port Jervis is somewhat argillaceous limestone, similar to the New Scotland, but it is noted particularly for an abundance of trilobites. Indeed, a landmark in the Port Jervis area is called "Trilobite Mountain," named for the abundance of the large dalmanitid trilobites *Phalangocephalus dentatus* found within this unit.

Acadian Orogeny

The major Wallbridge Unconformity truncates the Helderberg strata westward from the Hudson Valley and separates them from overlying Kaskaskia Supersequence. The higher Lower Devonian strata are assigned to the Tristates Group and comprise quartzose sandstones, dark gray mudstones and interbedded calcareous mudstones, and silty argillaceous siltstones (Rickard 1962, 1975). The early phase of sea-level rise resulted in deposition of basal transgressive quartz sandstones (Oriskany) analogous to Potsdam (Sauk Supersequence; see Figure 4.1)—the Oriskany sandstones and equivalent limestones, an orthoquartzite-carbonate suite—but these give way to mixed shales and carbonates.

Some uncertainties remain about the positions of continents. Gondwanaland was apparently getting close to Euramerica. Laurentia still sat on the paleoequator, though it had rotated slightly from its earlier position. The present east side was south of the equator, the northwest in the doldrums, while Alaska and Arctic Canada developed reefs and evaporites in the northern subtropics.

In Europe the Caledonian Orogeny continued from the Silurian and caused major folding and thrusting in Scandinavia and parts of northern Britain. A major theme for eastern North America involved the convergence of the subcontinent of Avalonia with Laurentia, to produce the Acadian Orogeny in a series of three or four **tectophases** (tectonic episodes); each follows a similar pattern (Figures 4.33 and 4.5D).

Some of the first activity occurs in the Northeast, even during the Helderberg times. Thin ash beds have been reported in the Kalkberg, and one in particular (the Bald Hill bentonite) in Cherry Valley, Otsego County, has been dated at approximately 417 million years before present. These ash beds indicate that volcanism was ongoing during the deposition of the Helderberg Group. Indeed, in northern Maine and the Maritime Provinces, Helderberg-age (Lower Devonian) volcanics are rather well developed. However, the first major effect in the New York–Pennsylvania area is reflected by the development of the Tristates Group.

Avalonia attached to the outer side of Laurentia through a suture or line of contact, in central England (southern England is part of Avalonia, Scotland is part of Laurentia), producing the Acadian Orogeny (Figure 4.5). Eastern Newfoundland (Avalon Peninsula) was welded to western Newfoundland, and the eastern Massachusetts area was added to Laurentia. Effects of the Acadian Orogeny are evident throughout New England and the Maritime Provinces; subduction of the Avalonian plate westward under the old Taconic land of New England created a new magmatic arc (Figure 4.35B). Many of the granite and gneiss domes of New Hampshire (the "Granite State"), as well as those of the Acadia National Park area in Maine (from whence the term *Acadian*), were formed in the Devonian. Sediments of basins associated with the magmatic arc in New England were heavily folded, thrusted, and metamorphosed at high grade.

The Acadian fold and thrust belt propagated from east to west, owing to the collisional compression, reaching as far west as the

Hudson Valley region of eastern New York. The Hudson Valley fold and thrust belt is evident along the thruway between Albany and Kingston, Ulster County. Farther west, a foreland basin developed and then migrated westward in response to episodes of thrust loading back to the east.

As noted, the Acadian Orogeny involved a series of pulses or tectophases; each shows a similar pattern: (1) quartz sandstone and limestone to (2) dark shales to gray mudstones and (3) sandstones including red beds. The sandstone-limestone succession reflects tectonic quiescence. The sudden switch to black shale (flysch) records an abrupt pulsed subsidence in the foreland basin, and the sandstone and red beds (molasse) represent the filling of the basin with clastic sediments.

Tectophase I of Early Devonian age is seen primarily in the north. The Tristates Group shows an upward change from orthoquartizes and limestone (Oriskany-Glenerie) through black shales with K-bentonites (Esopus Shale), shallowing upward to siltstones and sandstones and then a shift to carbonate-shale (Upper Esopus Shale–Schoharie Formation).

Tristates Group

The Oriskany Sandstone appears to represent nearshore sand that was reworked from older sandstones and redeposited during a major rise in sea level that followed a post-Helderberg regression of seas. Erosion that preceded the Oriskany beveled the Helderberg units successively to the west, such that moving westward from the vicinity of Catskill, Greene County, the Post Ewen, Alsen, Becraft, New Scotland, Kalkberg, and finally the Coeymans and Manlius formations underlie the Wallbridge Unconformity and are overlain by Oriskany Sandstone. The sands probably accumulated in a very shallow, high-energy beach and offshore sandbar type of environment. This environment was conducive to the growth of abundant large brachiopods and even some corals but does not appear to have been favorable to trilobites. During Oriskany time, waves must have broken near shore and produced pile-ups of clean, rather well-rounded quartz sand that ultimately were cemented to form the Oriskany Sandstone. The "XYZ" (low-to-high-to-low-energy) pattern typical of the Helderberg Group appears to have been broken down at this time, such that no low-energy, nearshore tidal-flat deposits are known. Offshore the sandbars appear to have graded into subtidal, below-wave-base accumulations of finer-grained sandy/silty muds and lime muds that form the Glenerie Limestone.

The Oriskany forms the base of the Tristates Group and in New York State is a generally thin, clean quartz sandstone, rarely exceeding 2 m in thickness. In many areas of central New York, the only remnant of the Tristates Group consists of thin stringers of Oriskany quartz sand, some of which extend downward as much as 50 cm as pipes or cavity fillings into the underlying eroded carbonates.

The Oriskany Sandstone thickens tremendously in southern Pennsylvania and Maryland, where it commonly is mined as a glass-making sand because of the purity of its quartz content. In these areas, the unit frequently exceeds 30 m in thickness and may hold up ridges in the Valley and Ridge sector of the Appalachians because of its resistance to erosion. The Oriskany itself generally lacks trilobites but contains large robust shelled brachiopods, such as *Costispirifer arenosus*, the large circular orthid *Hipparionyx*, and the terabratulid *Rensselaeria*. A thin remnant of the Oriskany Sandstone near Cayuga Lake also has yielded abundant specimens of favositid corals and some large *Naticonema* gastropods but few, if any, trilobites. To the southeast, in the Hudson Valley region, the Oriskany grades into a sandy to silty limestone referred to as the *Glenerie Limestone*.

In contrast to the Oriskany, the Glenerie was deposited in environments similar perhaps to those of the Kalkberg or Alsen Formation that were extremely favorable to a diversity of species; the unit even displays chert beds similar to those in the latter units. Evidence for storm waves is abundant in the Glenerie in the form of abundant storm hash layers or coquinas of shells and fragments of trilobites. The Glenerie is highly fossiliferous, containing an abundant, diverse fauna of brachiopods, snails, and trilobites. Again, large dalmanitid trilobites appear to be among the more common elements within this unit.

It should be noted that the trilobite fauna, although analogous in some ways to that found in the Kalkberg and Alsen, is quite distinct at a generic level from Helderberg faunas. A major faunal event separates the Helderberg faunas from those of the overlying Oriskany. This faunal event very likely is associated with the major sea-level drop that created the major post-Helderberg erosion surface or sequence boundary that underlies the Oriskany and Glenerie Formations. Glenerie trilobites include the following:

Cordania becraftensis
Homalonotus major
Paciphacops clarkei
Synphoria sopita
Synphoria stemmata stemmata
Dalmanites bisigmatus
Odontochile phacoptyx
Phacopina? correlator
Synphoria stemmata compacta
Trimerus vanuxemi

The Glenerie and Oriskany are abruptly overlain by a dark shale and mudstone unit up to 100 m thick in portions of the Hudson Valley and in the Skunnemunk Outlier. This unit, the Esopus Shale, represents a major influx of siliciclastic muds and silts that followed the rather clean sands and limestones of the Oriskany and Glenerie times. The occurrence of thin but widespread bentonite beds in the base of the unit suggests that the Esopus is associated with a pulse of tectonism that has been referred to as the first major tectophase of the Acadian Orogeny.

The Esopus, for all its thickness in the Hudson Valley (>100 m), appears to thin abruptly westward to a feather edge west of Cherry Valley. It is apparent that a deep but rather localized basin in the eastern New York and eastern Pennsylvania regions served as a sediment trap for the fine-grained muds and silts that were being deposited off eroding mountains to the southeast or east. This basin was created by the crustal bending forces produced in turn by the loading of thrust sheets as a result of mountain-building activity.

The Esopus Formation represents deep-water dysoxic environments. At times the bottom became sufficiently well oxygenated that it was inhabited by a great abundance of wormlike animals that produced the distinctive trace fossil *Zoophycos.* Portions of the Esopus Formation are so very heavily churned by this swirly trace that in the early days of New York geology, the unit was sometimes referred to as the *Cauda-galli* or "*roostertail*" *grits.* Relatively low oxygen, high turbidity, and possible sediment instability produced by these trace-making/sediment-feeding worms apparently made the Esopus a fairly inhospitable environment for most shelly organisms. As such, the unit is only very sparsely fossiliferous. A few layers do contain assemblages of relatively small, probably deep-water brachiopods such as the genus *Pacificocoelia.* Some mollusks, occasionally as pyritized specimens, are also known from the Esopus and its equivalent Beaver Dam Shale of Pennsylvania. However, as a whole the unit is very sparse and has yielded little insight into the ecology of the time, particularly with regard to trilobites.

Esopus shales are overlain by a somewhat more silty to sandy unit, the Carlisle Center Sandstone, that commonly contains some large quartz granules and the green clay mineral glauconite. This unit locally contains large phosphatic concretions near its base and displays a distinctive, sharp, sequence-bounding unconformity at the top of the Esopus. The Carlisle Center seems to represent a continuation and accentuation of coarser siliciclastic sediment input into the Appalachian Basin. The basal contact of this sandstone unit displays a remarkable assemblage of trace fossils, which is beautifully shown in cuts along U.S. Rte. 20 near Cherry Valley, Otsego County (Miller and Rehmer 1979). Among other traces are distinctive ribbon-like trails referred to as *Cruziana.* These display V-shaped chevron-like scratch marks and are probably the work of trilobites furrowing into muds of the underlying older and partially eroded Esopus Formation. Such trilobite traces probably represent feeding activity. These traces were produced in firm muds, and therefore even delicate scratch marks were preserved. However, despite these traces, there are few, if any, body fossils of trilobites found in the Carlisle Center Formation. To the west, the Carlisle Center becomes very thin, very rich in glauconite and phosphate, and appears to overstep the thin Esopus shale. In the Hudson Valley region, the unit is overlain and grades laterally into the highest formation of the Tristates Group, the Schoharie Formation, previously called "Schoharie Grit."

The Schoharie Formation is dated as close to the end of the Early Devonian (the Emsian Stage). Schoharie beds range from about 10 to over 30 cm (1 foot) in thickness and display a relatively regular cyclic alternation between more and less carbonate-rich, silty to sandy mudstone. Cream-colored beds are more calcareous and sometimes contain fairly abundant fossils. However, the fossiliferous nature of the Schoharie is masked in many outcrops by prominent slaty **cleavage** that was probably developed during the later phases of the Acadian (Devonian) mountain-building episodes. Cleavage crosscuts bedding, making collecting from the unit very difficult.

A curious feature of the Schoharie is its absence in the central portion of New York and the presence of deposits of this latest Early Devonian age both east and west of the region centered on Cayuga Lake. Recent work suggested that this may have been the result of arching of the seafloor that took place during late Early Devonian time, perhaps in response to the thrust loading that occurred in the Tristates tectophase of the Acadian Orogeny (Ver Straeten and Brett 2000). In areas west of Syracuse, the entire Tristates Group, except for scattered thin stringers of Oriskany Sandstone, has been removed at the Wallbridge Unconformity. Where Lower Devonian strata are absent, the Onondaga Formation of Middle Devonian age (see below) rests directly on eroded beds of the Upper Silurian. The relief on this unconformity then may represent at least two major episodes of erosion, one before and one after deposition of the Tristates Group. This composite unconformity shows local sinkhole fillings, cavities that were dissolved out of the top of the Silurian beds prior to deposition of the overlying Lower or Middle Devonian strata (Figures 4.37B and 4.38A).

The uppermost portion of the Schoharie, particularly near Albany and in the Schoharie Valley region itself, has been termed the *Rickard Hill Member.* This fossiliferous skeletal limestone is particularly noted for its exquisitely preserved fossils that are commonly weathered out. A rich fauna of brachiopods, corals, cephalopods, and over 15 species of trilobites, primarily dalmanitids, has been reported from the Schoharie beds near Albany (Goldring 1943). This unit represents environments again very similar to the older Glenerie and New Scotland Formations. That is, the unit was deposited in low-energy offshore environments that were subject to minor wave agitation during storms but were well below the normal wave-base. Thin hashes of shelly material are common in the Schoharie.

The Schoharie Formation was removed by erosion in areas west of Syracuse. A thin, hematitic, phosphate-rich sandy bed near Syracuse may represent some of the last vestiges of the unit in central New York. However, less silty argillaceous limestone, referred to as *Bois Blanc Formation,* is present intermittently in the area from near Phelps, Ontario County (north of Seneca Lake), westward to Buffalo and into Canada. The Bois Blanc becomes a prominent unit in Ontario and westward into the Michigan Basin. This unit is rich in brachiopods and corals that have species identical to those found in the Schoharie of eastern New York. A few trilobites, *Burtonops cristatus, Anchiopella anchiops, Maurotarion minuscula,* and *Crassiproetus* species, have also been obtained from these rocks.

The complete Schoharie–Bois Blanc trilobite listing is as follows:

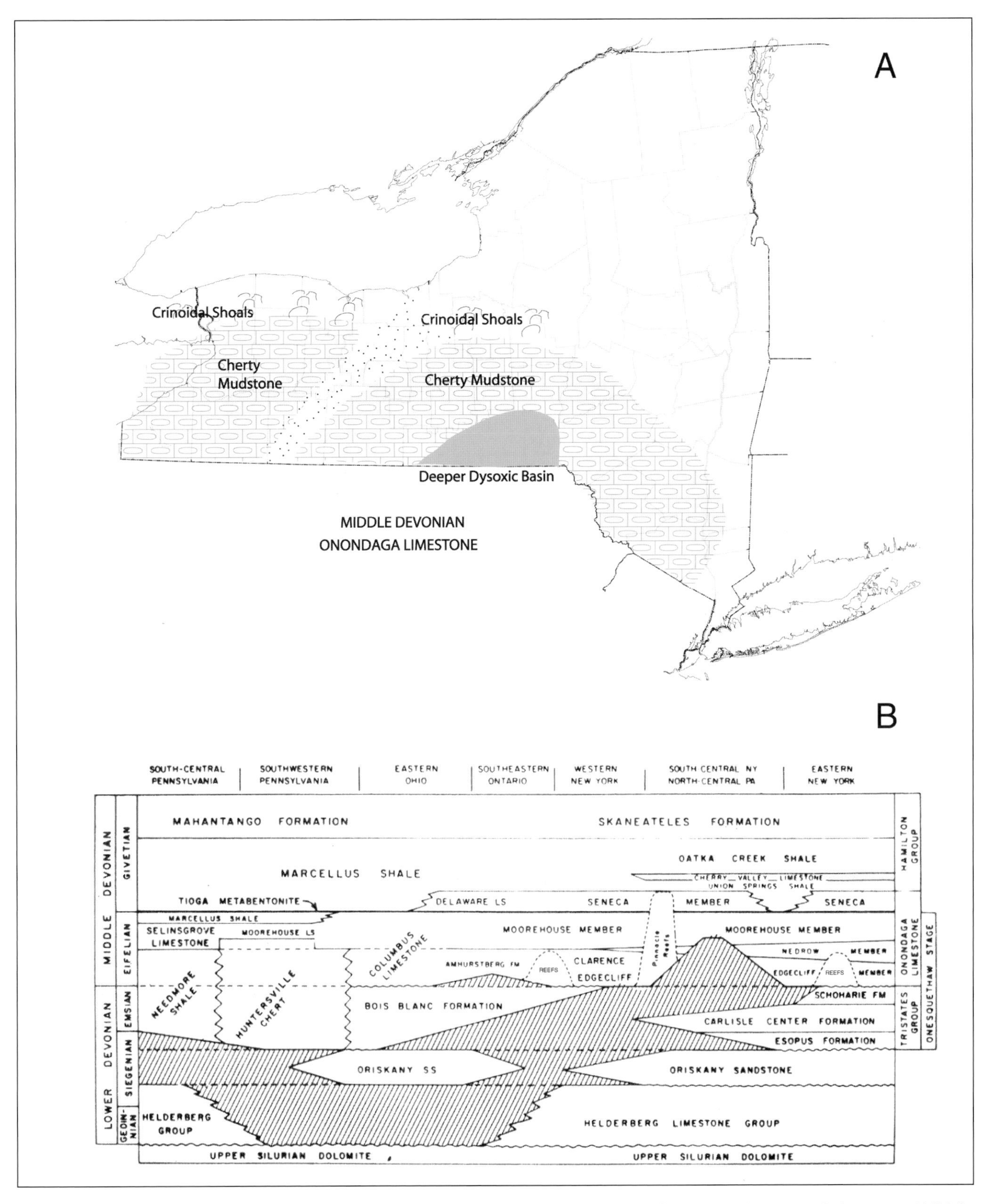

FIGURE 4.37. New York during the early Middle Devonian. A. During deposition of the Onondaga Limestone. B. Lower and Middle Devonian stratigraphic chart for the Northeast. From Cassa and Kissling (1982), New York State Geological Association. Reproduced with permission.

FIGURE 4.38. Upper Silurian/Onondaga Limestone. A. Silurian-Devonian Wallbridge Unconformity (arrow). Darker gray dolostone (Akron Formation (a), Upper Silurian) at the base of the picture is overlain along an irregular unconformity by light gray Onondaga Limestone (b) (Middle Devonian). Oaks Corners Quarry, Oaks Corners, Ontario County. B. Middle Devonian Onondaga Limestone, Nedrow Member. Note the cyclic banding of dark chert-rich limestone and lighter gray fossiliferous, noncherty limestone. Quarry off Rte. 5, Leroy, Genesee County.

Anchiopella anchiops	*Anchiopella anchiops sobrinus*
Burtonops cristatus	*Calymene platys*
Coniproetus angustifrons	*Coniproetus conradi*
Corycephalus regalis	*Crassiproetus schohariensis*
Echinolichas hispidus	*Kettneraspis callicera*
Maurotarion minuscula	*Mystrocephala arenicolus*
Phacops? clarksoni	*Pseudodechenella hesionea*
Schoharia emarginata	*Synphoria? concinnus*
Terataspis grandis	*Trypaulites erinus*

Middle Devonian

Tectophase II of the Acadian Orogeny occurred during Middle Devonian. A basal minor sandstone and thick limestone (Onondaga Group) were deposited in quiescent conditions (Figure 4.37A). A sudden shift to black shale (Marcellus Formation), then shallowing upward through siltstones and sandstones into red beds, records first tectonic deepening and then filling of the foreland basin.

Onondaga Limestone

The Middle Devonian (Eifelian Stage) boundary occurs at or near the base of the Onondaga Formation in New York State and its equivalents in Pennsylvania. The Onondaga is an extremely widespread unit that correlates with other limestones in the Midwest, such as the Jeffersonville Limestone of Indiana and Kentucky and the Columbus Limestone of Ohio. The early Middle Devonian appears to have been a time of very widespread, rather uniform topography. This initial transgression of Middle Devonian seas occurred during a time of tectonic quiescence. Hence, there was little tendency for the development of deep basins or major archlike swells at this time. The Onondaga therefore spread as a sheetlike unit that was deposited at relatively uniform depths of a few tens of meters over a vast tract of eastern North America. The unit does grade laterally into shallow-water sands in regions of southern Pennsylvania and the southern Appalachians. However, from the Hudson Valley to Buffalo, the Onondaga displays considerable uniformity, varying primarily only in the relative abundance of **chert** and skeletal-rich beds (Figure 4.38B).

The basal Edgecliff Member is a coral-rich crinoidal grainstone that passes upward laterally into very chert-rich, fine-grained limestone. Shaly beds of the Nedrow Member probably record maximum transgression. Upper portions of the Moorehouse and the overlying Seneca Member become increasingly cherty and **micritic** in composition. They also contain some of the most widespread bentonite beds known in the Appalachian Basin. Particularly notable in this cluster, which is sometimes termed the *Tioga Bentonite cluster*, is the ash referred to as Onondaga Indian Nations Ash (Figure 4.39A). This unit is locally up to 30 cm (1 foot) thick and contains a mixture of shale and actual volcanic ash beds. Extracted zircon crystals from this ash recently have been dated using uranium-lead dating, to yield a very precise date of 390 ± 0.5 million years ago. The Onondaga ash beds are a harbinger of renewed tectonic activity in the Acadian mountain terrain and in the Appalachian foreland basin.

The base of the Middle Devonian Onondaga Formation is marked by a locally sandy, pinkish gray crinoidal limestone unit in many areas, a portion of the Edgecliff Member. The Edgecliff contains some of the largest crinoid columnals known in any unit in the Paleozoic. Trilobites, as well as diverse corals, are abundant and represented by *Viaphacops bombifrons*, *Calymene platys*, and *Crassiproetus crassimarginatus.*

The depositional environment of the grainstone facies is thought to represent offshore (Y-zone) shallow-water shoal environments, similar to older units such as the Lower Devonian Becraft, Coeymans, and portions of the still-older Silurian Lockport Group. However, in places, where the water was somewhat deeper, at least the upper Edgecliff is composed of relatively thick micritic or fine-grained limestone with extremely abundant bluish gray to chalky white chert. The source of this siliceous material is poorly known. However, the chert-rich beds may have been derived from finely particulate organic silica, such as sponge spicules, that must have developed in enormous quantity in the somewhat quieter water areas of the Edgecliff deposition. Chert is not common in the high-energy shoal facies, perhaps because the fine-grained silica required for its formation was winnowed from these areas and removed to the slightly deeper regions of the basin.

Fossils are generally sparse in the Onondaga cherty facies but include moderately abundant fragments and articulated specimens of trilobites. The trilobites found are *Viaphacops bombifrons, Kettneraspis callicera*, and *Odontocephalus aegeria.* A large slab of articulated *Odontocephalus humboltensis* was excavated from the cherty facies during the construction of the Humbolt Parkway in Buffalo.

The Edgecliff Member also is noted for the occurrence of small to medium-scale bioherms or reefs. In areas of eastern New York near Albany and Syracuse, and in westernmost New York, near Buffalo and Amherst, Erie County, these bioherms were developed as low mounds, up to perhaps 1 km in diameter and generally a few meters to about 10 m in height. Bioherms were composed primarily of rugose and tabulate corals. Thickets of colonial rugose corals near their bases pass upward to lime-mudstone mounds with abundant favositid and some other colonial and solitary rugose corals. Trilobites and even brachiopods are scarce in these facies, which seem to have been fully dominated by corals and crinoids. The source of the lime mud is a mystery, but it may represent algal deposition. However, shaly beds on the margins of some of the bioherms evidently represent a type of protected quieter water environment, perhaps in the lee of the bioherms. While trilobites are not common in the reef framework of the Edgecliff bioherms,

FIGURE 4.39. Middle Devonian rocks. A. Sharp contact between the light gray Onondaga Limestone (a) (Middle Devonian) and black Bakoven (Union Springs) Shale (b) of basal Hamilton Group. Partings with plants mark the positions of bentonites (arrows). Kaaterskill Creek, below Rte. 23A, Catskill, Greene County. B. Middle Devonian Mount Marion Formation shows coarsening and thickening upward from lower shaly beds to upper sandstones. High Falls, Kaaterskill Creek, near Saugerties, Ulster County.

interfingering shaly beds have yielded some remarkable finds, including the enormous spiny trilobite *Terataspis grandis*, one of the largest known of all Devonian trilobites. Fragments of this unusual lichid were formerly rather abundant in the Vogelsanger Quarry at East Amherst, Erie County. *T. grandis* has also been found in the Onondaga in Ontario, and fragments have been found in central New York.

The Edgecliff Member is sharply overlain in many localities by a much more argillaceous, fine-grained, and noncherty limestone and calcareous shale, referred to as the *Nedrow Member*. In places the Nedrow contains very dark gray or almost black shale beds representing dysoxic muddy-bottom conditions. These beds, not surprisingly, are very sparse in fossils. However, some of the greenish gray shaly limestones of the Nedrow are extremely rich not only in the solitary rugose and small tabulate corals but also in a diversity of brachiopods and trilobites, including the proetids, *Pseudodechenella clara* and *Coniproetus folliceps*, the small odontopleurid, *Kettneraspis callicera*, and the phacopid *Viaphacops bombifrons*.

The upper two members of the Onondaga record a shallowing and deepening once again. The Moorehouse Member records conditions rather similar to portions of the Edgecliff. It varies from crinoidal packstone with abundant rugose corals, albeit with no development of reefs, to shaly and chert-rich limestone that contains very abundant brachiopods and trilobites. The trilobites *Viaphacops* and *Odontocephalus selenurus* as well as the large synphoriid *Coronura aspectans* occur in abundance in some of the shaly beds of the Moorehouse.

The fine-grained limestone beds of the upper Moorehouse and Seneca (and their equivalent in the Selinsgrove Formation of Pennsylvania) tend to be sparsely fossiliferous, but certain bedding planes do contain small brachiopods and, most notably, an abundance of *O. selenurus* trilobites. This synphoriid appears to have been among the most tolerant of organisms of the low-oxygen, lime-mud sea bottom. The facies of the upper Onondaga resembles dalmanitid-rich portions of the Helderberg, such as the New Scotland and Port Ewen, and even the sparsely fossiliferous calcareous mudstones of the Rochester Formation in the Silurian. These facies throughout the Silurian to Middle Devonian time appear to have harbored an abundance of dalmanitid trilobites.

The highest beds of the Onondaga in New York, belonging to the Seneca Member, overlie a very widespread ash bed, the Onondaga Indian Nations Bentonite. The Seneca is generally sparsely fossiliferous, dark micritic limestone in the Finger Lakes area that grades to more fossiliferous beds in western New York. The Seneca typically contains brachiopods, small burrows (*Chondrites*), and a few scattered fragments of *Odontocephalus* trilobites.

At the close of Onondaga deposition, many families of trilobites disappeared from New York. Lichiids, odontopleurids, calymenids, dalmanitids, and synphoriids have never been reported from beds above the Onondaga. However, proetids show little change, and Phacopidae and Homalonotidae continue but in somewhat modified forms.

A complete listing of the reported Onondaga trilobites follows:

Acanthopyge (Lobopyge) contusa	*Asaphus? acantholeurus*
Australosutura gemmaea	*Bellacartwrightia pleione*
Calymene platys	*Ceratolichas dracon*
Ceratolichas gryps	*Coniproetus folliceps*
Coronura aspectans	*Coronura helena*
Coronura myrmecophorus	*Corycephalus pygmaeus*
Crassiproetus brevispinosus	*Crassiproetus crassimarginatus*
Crassiproetus neoturgitus	*Crassiproetus stummi*
Echinolichas eriopsis	*Echinolichas hispidus*
Harpidella stephanophora	*Harpidella* sp.
Kettneraspis callicera	*Mannopyge halli*
Mystrocephala varicella	*Odontocephalus aegeria*
Odontocephalus bifidus	*Odontocephalus coronatus*
Odontocephalus humboltensis	*Odontocephalus selenurus*
Odontocephalus sp.	*Otarion? diadema*
Otarion? hybrida	*Otarion? minuscula*
Paciphacops logani	*Proetus delphinulus*
Proetus microgemma	*Proetus ovifrons*
Proetus stenopyge	*Proetus tumidus*
Pseudodechenella canaliculata	*Pseudodechenella clara*
Synphoria concinnus	*Terataspis grandis*
Trypaulites erinus	*Trypaulites macrops*
Viaphacops bombifrons	*Viaphacops pipa*

Certain beds near the top of the Seneca Member are distinctive in containing an abundance of fossil fish remains. In the area between Syracuse and the Hudson Valley, the erosional top of the Onondaga Limestone is marked by a very distinct bone bed with abundant fish teeth (known as *Onychodus*) and bones. These bone beds probably accumulated during a period of very slow sedimentation on the seafloor. They are abruptly overlain by the sooty black shales of the Union Springs Formation, which ushered in a new tectophase with abrupt deepening of the basin and input of a large amount of dark siliciclastic mud that evidently was being imported from uprising mountains of the second Acadian tectophase (Figure 4.21). With the Union Springs Formation begins the Middle Devonian Hamilton Group.

Hamilton Group—General

The Middle Devonian Hamilton Group (Figure 4.40) is one of the most richly fossiliferous of the New York Devonian. Its fossil faunas have been well studied since the times of James Hall. Generally, the fossil assemblages are arranged in a series of associations, or **biofacies**, that represent distinct environments from deeper-water black shales with low-diversity brachiopod assemblages to shallow-water coral beds, and nearshore, sandy, clam-dominated assemblages.

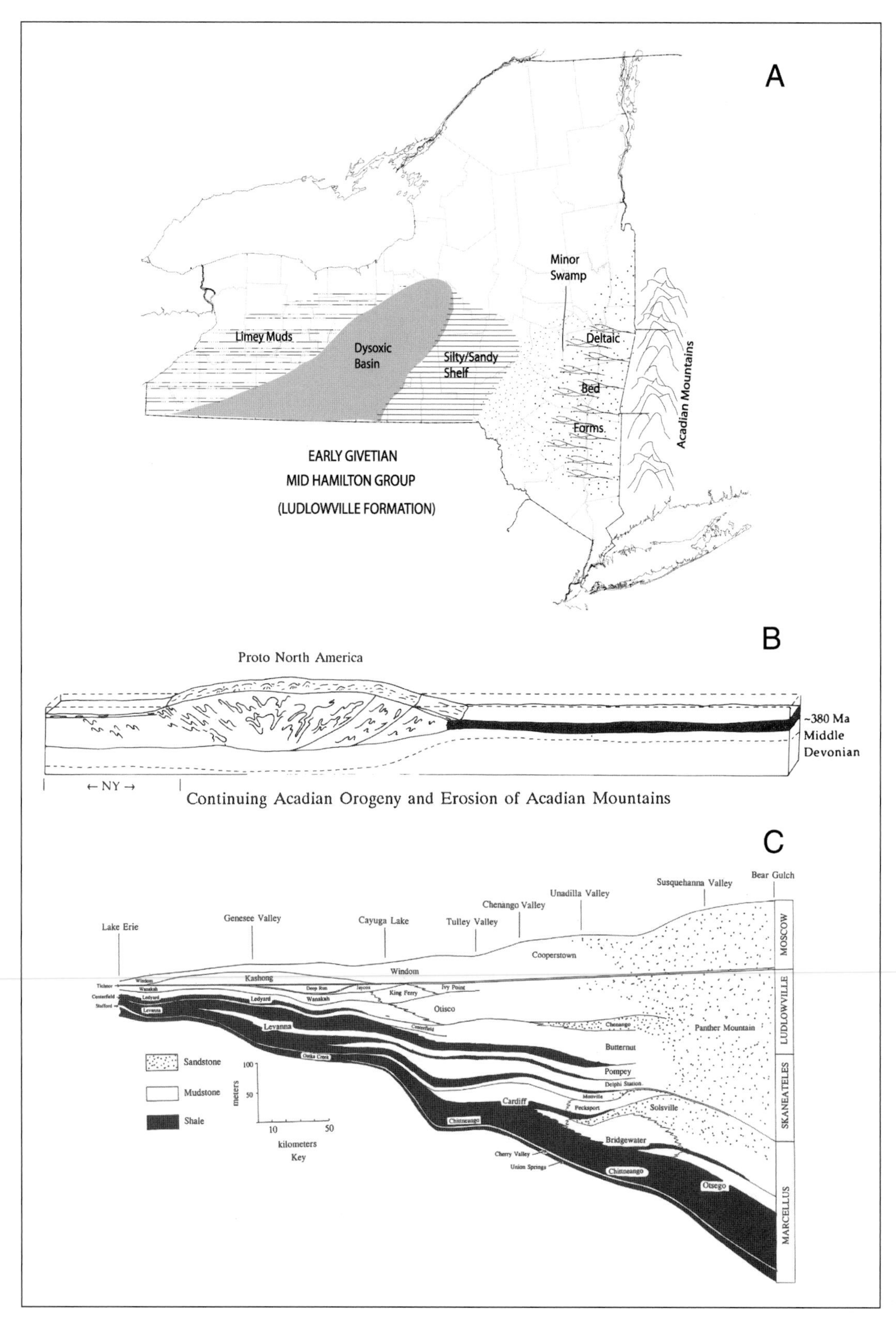

FIGURE 4.40. New York during the Middle Devonian Hamilton depositions. A. Map during the middle Hamilton time. B. Cross section showing the eroding Acadian Mountains. From Isachson et al. (1991). Printed with permission from the New York State Museum, Albany, N.Y. C. Stratigraphic cross section of the Hamilton formation deposits. From Linsley (1994). Reproduced with permission.

Lower Hamilton Marcellus Formation

The lower Marcellus or Union Springs beds that overlie the Seneca Member are primarily black shales. During deposition of these sediments, oxygen levels were very low in the deeper parts of the seafloor (late Eifelian Stage). However, during deposition of the upper portion of the Union Springs, there was evidently a change. Higher beds of the Union Springs in the Hudson Valley, referred to as the Stony Hollow Member, are characterized by calcareous siltstones, which contain a relatively sparse and low-diversity fauna of auloporid corals, some brachiopods, and notably, the trilobite *Dechenella haldemani*. Common pygidia of this trilobite and rare cephala occur in a few beds of a portion of the Stony Hollow. This trilobite has been traced widely in rocks that date from the late Eifelian Age. For example, it is found in a thin (20 cm), pale gray limestone (the Chestnut Street bed) that occurs just below the Cherry Valley Limestone throughout central and western New York. Very similar proetids also are found in a comparable position in the southern and central Appalachians as far south as Maryland. A related species is found in the latest Eifelian of the Michigan Basin region. *Dechenella haldemani* is just one of a suite of unusual species of fossils found in this assemblage. The brachiopods, the proetid trilobites, crinoids, and other invertebrate fossils display few similarities with the older Onondaga fauna and even fewer with those of the overlying Hamilton beds. The Stony Hollow fauna seems to represent the brief incursion of an assemblage of organisms that normally lived in warmer tropical areas to the northwest, represented in the present-day by the Canadian Arctic.

The Cherry Valley Limestone, a thin (0.5 to 3.0 m) unit, overlies the Union Springs Formation throughout New York, Pennsylvania, Maryland, and West Virginia. Cherry Valley is an unusually good example of a condensed bed that formed during a period of sea-level rise. Following the shallowing Stony Hollow conditions, there must have been a period of minor seafloor erosion, as the base of the Cherry Valley is typically sharp. The Cherry Valley itself is composed of the remains of small conical pelagic organisms, referred to as *styliolinids*, and with cephalopods, for which it is particularly noted. Some of the earliest abundant goniatitic ammonoids (*Agoniatites vanuxemi*) occur within the Cherry Valley Limestone. However, no trilobites are known from the Cherry Valley proper.

The Cherry Valley Member is overlain abruptly by Oatka Creek (Chittenago) black shale that resembles the underlying Union Springs Formation. Most of this unit represents deep-water accumulation of anoxic muds. However, there are hints of a profound faunal change within these rocks. Fossils are rare in the Oatka Creek black shales. But a few gray layers contain forms typical of the higher Hamilton Group and not the underlying Union Springs or Onondaga. Most notable is a fossiliferous bed that occurs variably from just a few centimeters in western New York up to about 50 m (160 feet) in the east, above the Cherry Valley Limestone. This thin bed (10 to 50 cm) of medium gray shale and thin limestones contains a rich, diverse fauna across the state. In the Hudson Valley region, this unit is a coral-rich bed that has been referred to as the Halihan Hill bed (Griffing and Ver Straeten 1991). Farther westward the corals become less common and are absent around Schoharie, New York, but a diverse fauna of brachiopods, bryozoans, bivalves, and other fossils occurs from here westward. Trilobites are scarce, but the first representatives of the *Eldredgeops rana* lineage, known in New York, occur within the Halihan Hill bed and its lateral equivalent, the LeRoy bed in the western part of New York. A *Harpidella* sp. has also been found. The fauna of this horizon marks the incursion of many of the typical Hamilton species. The species that occur within this bed are forms that persist throughout the entire remainder of the Hamilton Group. Moreover, individual assemblages within the Halihan Hill/Leroy bed recur at multiple levels of the higher Hamilton succession. The fauna of this bed shows virtually nothing in common with similarly shallow-water, coral- and brachiopod-rich faunas of the Union Springs/Stony Hollow members. A few species are recurrent Onondaga taxa that seem to have come back into the Appalachian Basin following a period of outage associated with the unusual Stony Hollow incursion.

In the Hudson Valley area, the Hallihan Hill bed is overlain by a thick (up to 500 m) succession of silty shales, siltstones, and sandstones of the Mount Marion Formation. The Mount Marion Formation displays a series of small-scale cycles, a few tens of meters in thickness, superimposed on a generally coarsening- and shallowing-upward trend. The gray mudstones are sparsely fossiliferous, but the caps of the small cycles are formed by shell beds with a number of species of brachiopods and clams. The trilobite *E. rana* and asteropyge species have been reported from these levels. *Dipleura dekayi* may occur as well in some of the sandy beds, but it is not common.

Westward, the various units display a gradual transition from silty mudstones into medium and dark gray shales. The lower portions of the Mount Marion above the Halihan Hill bed grade into the Chittenango Shale Member of the Marcellus Formation in central New York. This black, rusty shale with concretions contains very few fossils other than scraps of wood, some cephalopods, and styliolinid remains. However, it passes upward into medium gray shales of the Bridgewater Member and the overlying Solsville Sandstone and Pecksport Shale. These units are noted for the occurrence of exceptionally well-preserved brachiopod and molluscan remains. A very few specimens of the trilobite *E. rana* showing primitive patterns in terms of eye structure (18 vertical files of lenses) have been recorded in these beds from the region of the Hamilton-type area in the Chenango Valley (Eldredge 1972). To the west only the upper two-thirds to one-fourth of the Oatka Creek (Marcellus) Formation represent medium gray mudstones (Cardiff Formation); these are typically only very sparsely fossiliferous.

The thick Mount Marion Formation (Figure 4.39B) represents muds, silts, and sands shed from Acadian uplifts into a rapidly subsiding foreland basin. This succession thins abruptly westward from more than 500 m to less than 20 m in western New York, where thin muds of the Oatka Creek Shale accumulated deep anoxic waters. Rapid accumulation of sediments eventually outstripped subsidence near the Hudson Valley, where the Mount Marion shallows upward from deep basinal conditions to shallow sandy shelf. The upper beds of the Mount Marion Formation are sandstones with some conglomerates. The latter contain quartz and chert pebbles, some of which may be reworked from older Devonian units uplifted during the Acadian Orogeny. These pass, in turn, upward into gray flaggy siltstone and sandstone beds, assigned to the Ashokan Formation, containing primarily plant material and a few horizons of low-diversity brachiopod assemblages. These beds are interpreted as tidal-flat sands to nonmarine beds. They are overlain gradationally by maroon or red strata of the Plattekill and Manorkill Formations. These nonmarine, meandering river deposits typically show channeled sandstones that fine upward through a few meters into red mudstones containing plant-root marks and mud cracks. Such evidence indicates that the relatively deep basins that occupied the Hudson Valley region during Middle Devonian times were completely infilled with terrigenous detritus derived from the second tectophase of the Acadian Orogeny (Figure 4.34).

Skaneateles Formation

In central to western New York, the Pecksport or Cardiff Shale (upper Marcellus) is gradationally to sharply overlain by the second major Hamilton limestone or calcareous siltstone, that is, the Stafford and Mottville Members at the base of the Skaneateles Formation. At this level a diverse Hamilton fauna reappears for the first time above the Halihan Hill coral bed, and it displays minor modifications, most notably the loss of a few Onondaga holdover species and a greatly increased abundance of trilobites. Both *Eldredgeops* species and various forms of astropygids, *Kennacrypheus harrisae* and a *Greenops* species (assigned in the past to *G. boothi*), are common for the first time in the Mottville beds of central New York. Joining them is the homalonotid trilobite *D. dekayi*. This species occurs in considerable abundance in some of the silty beds of the Mottville, and especially in a siltstone capping a small cycle approximately 10.5 to 12.0 m above the top of the Mottville Member. This latter unit, the Cole Hill tongue of the Delphi Station Shale, is widely known as a source of these large trilobites, particularly in the vicinity of Sangerfield, Oneida County, where large numbers of specimens have been obtained in beds rich in large bivalves and certain brachiopod species.

Three higher, major shallowing-upward cycles in the Skaneateles correspond to three members of central New York: the Delphi Station (40 m), Pompey (25 m), and Butternut Members (up to 60 m). The top of the Pompey contains limestone concretions that have been traced from western New York eastward into the Chenango Valley region. The upper Butternut Member is typically the darkest shale of the Skaneateles Formation throughout most of western and central New York and is only sparsely fossiliferous. The Delphi Station and Pompey are relatively rich in small brachiopods and trilobites.

As with the Marcellus shales, the Skaneateles Formation tends to pass westward into poorly differentiated, dark gray to black shales that generally contain few fossils but are characterized in many areas by leiorhynchid brachiopod fauna or small bivalves. These deeper-water facies of the Levanna Shale yield relatively abundant *E. rana* and some asteropyge specimens, but *D. dekayi* is very rare. Somewhat more diversified brachiopods, auloporid coral, and trilobite assemblages are found at levels that seem to correspond with the caps of the Delphi Station and Pompey Members.

Ludlowville Formation

The Centerfield Member at the base of the overlying Ludlowville Formation displays an abrupt shallowing trend. In western New York, the Centerfield Member is a classic example of a shallowing to deepening shale to carbonate to shale cycle within the Hamilton Group. A thin basal succession of gray shales passes upward into mudstones and finally skeletal limestones. A nearly symmetrical transition back through calcareous mudstones to medium or dark gray shales of the Ledyard Member occurs in the upper half of the Centerfield. In places, a phosphatic pebble bed marks the top of the unit. The calcareous mudstone facies of the Centerfield pass eastward in central New York into calcareous siltstones and cross-bedded sandstones, marking out a major shallowing-up cycle that commences with dark gray shales.

A complete spectrum of Hamilton biofacies is observable within the Centerfield and adjacent units. It overlies and underlies dark gray shales with leiorhynchid brachiopod fauna that probably represent deep offshore muds, well below the storm wave-base. These shales are transitional into gray shales characterized by a ambocoeliid brachiopod-rich fauna and very abundant trilobites, such as *Bellacartwrightia jennyi*, *Eldredgeops rana*, *Harpidella craspedota*, *Monodechenella macrocephala*, and *Mystrocephala baccata*, and finally into the highly diversified coral and brachiopod assemblages that typify the calcareous mudstones and limestones of the middle Centerfield. These beds are particularly noted in western New York as a source of rather abundant proetid trilobites. In particular, one thin (10 cm) concretionary limestone about 30 cm (1 foot) below the coral-rich limestones of the middle Centerfield contains unusually preserved *Pseudodechenella rowi* in contorted orientations, which suggests these organisms were caught up in an unusually rapid storm burial event. This bed has been traced for over 80 km from the Buffalo region westward to the type section at Centerfield, Ontario County, near Canandaigua Lake. The Chenango Member siltstone and sandstones facies also display *Dipleura dekayi*, as do similar facies in the underlying Skaneateles Formation.

The bulk of the higher Ludlowville Formation is composed of black and dark to medium gray shales and mudstones of the

Ledyard and Wanakah Members in western New York and their equivalents in central New York, the silty mudstones of the Otisco Shale and the siltstone of the overlying Ivy Point Member.

The Ledyard Shale in western New York is medium to dark gray shale and mudstone. It is particularly noted for an occurrence of an interval about 6 m (20 feet) above the Centerfield that is very rich in small spheroidal nodules of pyrite. These beds, termed the Alden Pyrite Beds, yield a moderately diverse fauna dominated by brachiopods but also containing very abundant remains of trilobites *E. rana* and *Greenops grabaui,* typically as enrolled individuals and often as the nuclei of pyrite concretions. Babcock and Speyer (1987) reasoned that these enrolled trilobite beds reflect unusual conditions on the seafloor under which storm disturbance stirred up hydrogen sulfide–rich muds. Trilobites responded to the toxicity of the waters and perhaps their turbidity with the typical escape reaction of enrollment but were later buried by clouds of mud suspended during the same storm event. The Alden beds cannot be traced into the Finger Lakes, where the Ledyard Shale is predominately fissile or platy black shale and represents low-oxygen environments. Trilobites, however, are again common in the gray silty mudstone facies of the equivalent Otisco Member that replaced the typical Ledyard facies east of Owasco Lake. The Otisco is also noted for the occurrence of two coral-rich submembers, the Staghorn Point and the Joshua coral beds. These are formed of thickets or biostromes of solitary rugose corals with some tabulates. Other fossils tend to be rare within the coral thickets, but on the periphery of the Joshua biostrome, clusters of *E. rana* have been collected.

The Wanakah Member gray shales are exceptionally rich in fossils, at certain levels, including corals, brachiopods, bryozoans, crinoids, trilobites, and others. Over 200 fossil species have been reported from the Wanakah Shale in western New York. Among these are about six or seven species of trilobites. These species are most abundant in a series of thin argillaceous limestones that occur low in the Wanakah from Lake Erie at least to Canandaigua Lake. These are the famed "trilobite beds" of Amadeus Grabau (1898–1899). Along with large numbers of *E. rana* are found *P. rowi*, *Bellacartwrightia whiteleyi*, and *G. grabaui.* Such trilobite bed facies, typically associated with small solitary (*Stereolasma*) corals, represent a distinctive biota that occupied an offshore mud bottom during times of low net sediment input. These thin limestones have much in common with the Browns Creek Bed of the Centerfield. Like the latter bed, they show evidence for both rapid entombment of trilobites and a longer-term signature of a cyclical increase and decrease in carbonate within the sediment. Trilobites frequently occur in clusters of up to 100 individuals. Small-scale cyclicity, apparent in the lower Centerfield and in the lower Wanakah trilobite beds, suggests possible climatic oscillations that controlled the carbonate content of the offshore muds.

The highest unit of the Ludlowville Formation in western New York is the Jaycox Member, a richly fossiliferous mudstone. These beds are sources of high-diversity fossil assemblages very similar to those seen in the older Centerfield Member. Again, well over 100 species of brachiopods, bryozoans, crinoids, mollusks, and trilobites, including relatively large astropyges, *Bellacartwrightia* and *Greenops*, species, and additionally *E. rana* and *M. macrocephala* have been obtained from the Jaycox beds. They also represent the first appearance of the trilobite *Australosutura gemmaea.*

Moscow Formation

The highest formation of the Hamilton Group, the Moscow, is bounded at its base by an unusually widespread skeletal limestone composed primarily of ossicles of crinoids and some fragmentary or complete colonies of favositid and rugosan corals. This unit, the Tichenor Member, has been traced from Lake Erie eastward to the Schoharie Valley region. It appears to represent a transgressive lag deposit formed in shallow, relatively sediment-starved waters. It overlies an erosional disconformity formed during a major regression at which beds of the Jaycox and subjacent units of the Ludlowville were locally truncated. Some of the Tichenor Limestone may actually be composed of fossils reworked from the older Ludlowville units. The Tichenor passes upward into calcareous, sparsely fossiliferous mudstone of the Deep Run Member that is the only occurrence of the trilobite *Cyphaspis*. The Deep Run is overlain by a thin (30 to 50 cm) silty Menteth Limestone, which is most noted as a source of silicified fossils (Beecher 1893a) that includes very early growth stages of many species, such as protaspides of trilobites. The lowest part of the Moscow Formation (Tichenor, Deep Run, Menteth) is also noted as the source of some of the largest specimens of trilobites. The largest known *E. rana*, *Bellacartwrightia phyllocaudata*, and *M. macrocephala* are obtained from the Deep Run, but similarly large individuals are found in the Menteth Limestone and Kashong Shale. These muddy-bottomed, shallow-water environments were highly favorable to the growth of these trilobites. However, an enigma associated with these beds is the relative paucity of small individuals. Because trilobites molt, one should see a record of these various growth stages observed as disarticulated parts. The absence of these early stages may suggest that trilobites were rather more mobile creatures and that the shallow-water muddy environments represented by the Deep Run and Kashong facies were inhabited primarily by mature individuals. Some modern crustaceans are known to undergo widespread migrations of this sort. Overall, the Tichenor–Deep Run succession represents a gradual deepening from near wave-base to offshore subtidal mud-bottom environments influenced strongly by intermittent storms.

The blue-gray mudstones of the overlying Kashong Member generally are rather sparsely fossiliferous but contain an abundance of trace fossils. However, local patches within these mudstones display exceptionally well-preserved fossils, including brachiopods of the *Tropidoleptus* fauna, various species of brachiopod bivalves, crinoids, bryozoans, and trilobites. Probably the most famous portion of the Kashong is the "Retsof beds" found in the excavations for a railroad near the former Retsof salt mines in Livingston County. These beds have been a prolific source of crinoids and blastoids and, again, relatively large specimens of the trilobites *Eldredgeops*, *Greenops*, and *Monodechenella*.

The highest member of the Hamilton Group, the Windom Shale, has many features similar to the mid Ludlowville. It has a number of small-scale cycles capped by beds rich in fossils of the typical diverse Hamilton biofacies. Again, trilobites, *Greenops barberi, E. rana, Bellacartwrightia* sp. and *P. rowi*, occur in a number of levels and are exceptionally abundant in cyclically bedded, calcareous mudstones and argillaceous limestones that form another series similar to the "trilobite beds" of the lower Wanakah. Other portions of the Windom Shale display the low-diversity, leiorhynchid brachiopod-rich, dark shale facies, and *Ambocoelia* biofacies in medium gray mudstones that are often highly enriched in phacopid trilobites, and coral-rich beds that display a diversity of forms including proetids, the unusual trilobite *Australosutera gemmaea*, and the occurrence of *Phacops? iowensis.*

The trilobite distribution in the Hamilton shows both stasis, little change over a long period of time, and significant change in the case of the Asteropyginae. *Greenops* and *Bellacartwrightia genera* show species change, often after a transgression (Liebermann and Kloc 1997). One *Kennacrypheus* species makes only a brief appearance, and there are now still undescribed species to be accommodated.

The complete listing of Hamilton Group trilobites follows:

Australosutura gemmaea	*Bellacartwrightia calderonae*
Bellacartwrightia jennyi	*Bellacartwrightia phyllocaudata*
Bellacartwrightia whiteleyi	*Bellacartwrightia* sp.
Cyphaspis sp.	*Dechenella haldemani*
Dipleura dekayi	*Eldredgeops crassituberculatus*
Eldredgeops norwoodensis	*Eldredgeops rana*
Greenops barberi	*Greenops grabaui*
Harpidella craspedota	*Kennacryphaeus harrisae*
Monodechenella macrocephala	*Mystrocephala ornata baccata*
Phacops? iowensis	*Proetus jejunus*
Pseudodechenella arkonensis	*Pseudodechenella rowi*

Tully Limestone

The shales of the Moscow Formation are unconformably overlain throughout central New York by the unusual fine-grained (micritic) limestone of the Tully Formation (Figures 4.41, 4.42A and 4.43). The Tully is a rather enigmatic limestone, as it formed at a time of general substantial siliciclastic input to the Appalachian Basin. It is suggested that the development of minor folds on the sea bottom in central New York (near Chenango Valley, Chenango County) served to entrap most siliciclastic sediment that was still being shed from the second tectophase of the Acadian Orogeny and to prevent these sediments from moving westward.

The Tully was preceded by an interval of substantial erosion of the older seafloor sediments. Evidence from the unconformity surface itself suggests that the seafloor was buckled into a series of low folded areas that became erosionally truncated prior to and during Tully deposition.

The lower portion of the Tully Limestone, in western New York, is unusually clean, fine-grained limestone that apparently accumulated in very-shallow-water conditions. It contains an unusual assemblage of brachiopods, most species of which are not common to the Hamilton Group below. In addition, the typical Hamilton trilobites and a few other species are found rarely in the lower part of the Tully. An undescribed member of Asteropyginae has been collected from this lower portion of the Tully. It is not a *Greenops* species, as previously reported. However, it is very similar to a species found in the Cedar Valley Limestone of Iowa. A minor unconformity separates these lower limestone beds from the upper Tully. At Bellona, Yates County, apparently truncated algal stromatolites have been found at this surface, suggesting a substantial period of shallowing, followed by erosion.

The upper portion of the Tully, a relatively clean limestone resembling the much older Onondaga, displays a fauna of recurrent Hamilton forms. Most of the common species within the upper Tully are forms also present to abundant within the Hamilton Group shales below. The return of trilobite bed facies is quite evident. The trilobites *P. rowi*, *Bellacartwrightia* sp., *Eldredgeops norwoodensis*, and *Greenops?* sp., as well as small rugose corals that typify the Hamilton calcareous "trilobite beds," occur in several beds of the upper Tully. The unusual trilobite *Scutellum tullius* appears to be associated with a small patch reef facies found near Skaneateles at Borodino, Onondaga County, but may occur rarely in other beds in the upper part of the Tully. The upper Tully seems to represent the last stand for most common Hamilton genera and for their particular associations of biofacies. The trilobites of the Tully, in total, are as follows:

New Asteropyginae referred to as *Greenops*

Bellacartwrightia sp.	*Eldredgeops norwoodensis*
Harpidella spinafrons	*Monodechenella macrocephala*
Pseudodechenella rowi	*Scutellum tullius*

In western New York west of Canandaigua Lake, the Tully Limestone has been removed, probably by deep seafloor erosion and corrosion of the carbonates in undersaturated water. In these areas, the black Geneseo Shale rests unconformably on the upper beds of the Hamilton Group. However, in most localities localized small lenses, up to 10 or 20 cm thick, of reworked fossil bone and pyrite material are observed intermittently at this contact. These beds of the Leicester Pyrite do contain occasional enrolled trilobites. What is most unusual about these trilobites is not simply that they are pyritized, but that they are probably clasts removed by erosion of the underlying shale and reworked into the lag deposits represented by the Leicester Pyrite.

FIGURE 4.41. Devonian successions. A. Devonian succession at Lake Erie. Lowest beds are Wanakah Shale (a) (Middle Devonian Ludlowville Formation). A distinctive limestone ledge, Tichenor Member (b), marks the base of the Moscow Formation, which is overlain by gray Windom Shale (c). A second thin limestone, Genunduwa (d), marks the approximate position of the Middle-Upper Devonian boundary dark gray to black shale (e) at about a third of the cliff height. Higher beds show an alternation of black West River-Middlesex-medium gray, concretion-bearing Cashaqua Shale (f), and finally black Rhinestreet Shale (g) at the top of the cliff. Lake Erie Shore south of Eighteen Mile Creek, Evans, Erie County. B. Middle Devonian Ludlowville and Moscow Formations (Hamilton Group). Main cliff below the falls is the Wanakah and Jaycox shales. Lower limestone of the falls cap is the Tichenor Member (a) (marking the base of the Moscow Formation), which is overlain by 2 m of Deep Run Shale (b). The falls are capped by the Menteth Limestone (c). The bank above the falls is gray Kashong Shale (d). Wheeler Falls, Jaycox Creek, north of Geneseo, Livingston County.

FIGURE 4.42. Middle Devonian/Upper Devonian successions. A. Middle Devonian succession. Windom Shale (a) (Moscow Formation, Hamilton Group), below ledges, is unconformably overlain by thick-bedded Tully Limestone (b). Taughannock Creek, Trumansburg, Tompkins County. B. Middle-Upper Devonian succession. The base of the falls is slightly above the Tully limestone. The main falls face is Geneseo Shale (a). The falls are capped by the Lodi beds (b) of the Sherburne Formation, near the Middle-Upper Devonian boundary. Higher cliffs are in the Penn Yan (c) (Sherburne) and Renwick Shale of the Upper Devonian. Taughannock Falls, Trumansburg, Tompkins County.

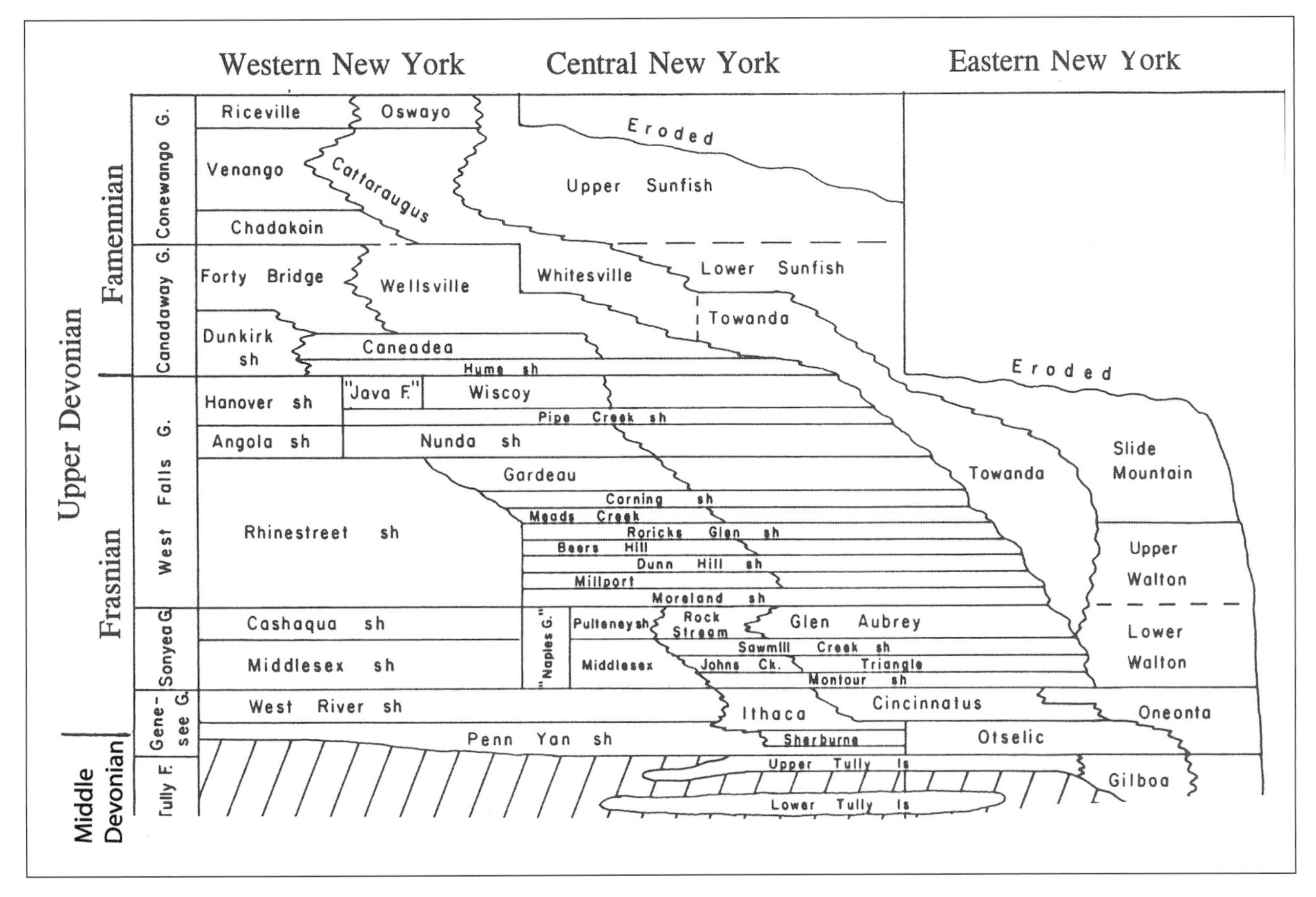

FIGURE 4.43. Stratigraphy of the Upper Devonian and the upper Middle Devonian of New York. From Sevon and Woodrow (1985), after Rickard (1975).

Late Devonian Strata

During tectophase III in the late Middle–early Late Devonian, the basal carbonate (Tully Limestone) gave way to black shales (Genesee facies). This was followed by a general progradation of siltstone, sandstone (prodelta, Portage facies; deltaic platform; Chemung facies), and red beds (subaerial delta platform; Catskill facies) (Figures 4.34 and 4.43).

Upper Devonian strata occupy most of the southern tier of New York State and are up to 2 km thick. To the east they are represented by red siliciclastic muds, silts, and coarse conglomerates that cap some of the high peaks in the Catskill Escarpment. These muds and conglomerates represent both meandering and braided stream facies that accumulated rapidly in a subsiding basin in front of a rising set of Acadian Mountains. The abrupt increase in siliciclastics near the base of the Upper Devonian, in association with the abrupt deepening of the basin to the west and the occurrence of very widespread black shales, signals the onset of the third and largest tectophase of the Acadian Orogeny. This major mountain-building pulse was the source of the vast Catskill deltaic complex, an enormous set of clastic wedges that prograded or built seaward into deepened basins of central New York and elsewhere. The deep basin again implies a period of tectonic thrust loading that caused flexure of the crust throughout much of western and central New York. That basin center migrated abruptly westward from its position in late Middle Devonian (Tully) time, when it lay close to the Chenango Valley region. The black shales of this tectophase are actually still of Middle Devonian (Givetian) age in western and central New York. This black mud deposition represents the greatest deepening during the entire Devonian an event referred to as the *Taughannock onlap* or *transgression*. The basal Geneseo Shale is characterized by the black shales but displays intervals of gray, marly, calcareous mudstones and thin limestones. Fossils tend to be quite scarce in these beds, presumably because of the strongly **dysoxic** (reduced oxygen level) to anoxic conditions that prevailed on the muddy seafloor through most of this time. Driftwood fragments, goniatites, nautiloids, and conodonts are common in some beds. Trilobites are extremely rare in the Genesee and higher formations of the Upper Devonian.

The base of the Frasnian Stage of the Upper Devonian occurs high within the Genesee Formation. The basal beds of the Frasnian are typically siltstones alternating with medium gray mudstones. These beds, like the Genesee below, pass eastward into much coarser-grained siltstone and sandstone facies within the Finger Lakes that, in turn, pass into massive marine sandstones with **coquina** (shell accumulations). The Sonyea and West Falls Groups make up the remainder of the Frasnian Stage. Some beds within the silty facies contain abundant coquinas of brachiopods and clams. However, corals, bryozoans, and trilobites are only rarely seen within these beds.

Most environmental interpretations of the facies of the Upper Devonian follow the model of a delta. In the deepest most offshore setting, fine-grained siliciclastic muds and a fairly large amount of organic detritus settled out in quiet waters. Black shale facies record these settings. Eastward of this, the seafloor sloped upward from deep water to near storm wave-base (a few tens of meters of water), and this area is sometimes termed the *prodelta region*. This sloped area experienced generally quiet water conditions with settle-out of fine muds, but episodically storm or seismic shock phenomena generated sediment-filled turbidity currents that laid down a series of flaggy and fine-grained sandstones with flute casts and drag marks on their lower surfaces. Such marks indicate that current directions were mainly out of the southeast.

Still farther east, the equivalent sandstones with shell coquinas suggest deposition near or slightly above storm wave-base and approaching normal wave-base. These facies have been termed the "*Chemung* facies" and are interpreted as representing a submarine delta platform. Obviously the red beds with their channel sands ("Catskill facies"), which are the final facies to the east, represent a nonmarine, subaerially exposed alluvial plain, perhaps a tidal-flat setting, in part.

The westward migration of these facies through the Upper Devonian is quite dramatic, with the red beds eventually coming to be deposited as far westward as the Genesee Valley region. The fact that these individual facies transcend or cut time lines established within the deltaic complex indicates that the delta was prograding or building forward into the seaway. As it did so, each successively more shoreward facies built "piggyback" over the top of the next most-seaward facies, yielding a generally shallowing-up pattern. However, this simplistic model of a prograding delta must be modified in some ways. For example, we know there were intervals, probably associated with major events causing a rise in sea level, when dark shales were spread widely both over the prodelta slope facies and over the Chemung-shelf environments.

The Appalachian Basin shifted southwestward as progressively younger black shales (Cleveland–New Albany–Chattanooga) spread into the Midwest. This pattern was recognized in the mid 1900s as the pattern of basin filling-overfilling by deltaic progradation of the great Catskill Delta, the flysch-molasse of the Acadian Orogeny. Eventually it was sorted out that these facies were **diachronous**, older to the east.

Why did discrete tectophases occur and why did the locus of basins shift southwestward? The Laurentia-Avalonia collision was mainly completed by Middle to early Late Devonian, with rotational scissor-like closure and different portions of the Avalon **terranes** hitting different promontories of the Laurentian margin.

Only a few fragments of phacopids, possibly *E. rana*, have been found in the Geneseo. These black shales pass eastward into gray mudstone and siltstone, with a somewhat more diverse brachiopod fauna. Trilobites were never again common within the Devonian of New York.

There are only two trilobites unique to the Upper Devonian in New York:

Scutellum senescens *Otarion? laevis*

Reports of *Eldredgeops* and *Greenops* species are questionable because the beds in which they were found are now possibly part of the upper Middle Devonian.

Chemung-shelf environments appear very similar to silty and sandy portions of the older Hamilton Group. Therefore, it is somewhat surprising that many characteristic Hamilton species including the trilobite *D. dekayi* are completely absent from the Upper Devonian portions of the successions. Indeed, trilobites are rare in almost all facies of the Upper Devonian. An extinction near the end of the Middle Devonian eliminated many of the typical Hamilton species and permitted an abrupt overturn to the typical Genesee faunas of the latest Middle Devonian to earliest Late Devonian age. Among the organisms that were decimated locally by the Middle Devonian extinctions were the trilobites and many, if not most, of the associated rugose corals. Hence, the gray silty and sandy sandstone and mudstone facies that occupy so much of the southern tier, although a rich source of brachiopods, bivalves, and other mollusks, are very poor for trilobite material. Trilobites were clearly on the wane by the end of the Devonian, and that last part of their history is obscure in New York State.

Marine ecosystems show intervals of relative stability punctuated by abrupt turnovers at several apparently global events, each associated with widespread anoxia and possible climatic changes in Earth's history. During the Late Devonian (Frasnian-Famennian stage boundary), one such mass extinction wiped out about 25% of marine families; decimated reef-building corals and stromatoporoids; and caused extinction of cystoids and pentamerid and atrypid (orders) brachiopods, most ammonoids, and most remaining trilobites. The terrestrial system was not so drastically affected. The causes of the events remain unclear. Like other mass extinction episodes, the Late Devonian crisis seems to have involved sea-level fluctuation, widespread hypoxia, and climate stress.

5 The Trilobites

In this chapter, all the trilobites of New York State known to us are listed by family. The listing is first by order, with the orders appearing in the same sequence used in the classification chapter of the *Treatise* (revised): Agnostida, Redlichiida, Corynexochida, Lichida, Phacopida, Proetida, Asaphida, and Ptychopariida (see also Fortey, 2001). The families are listed alphabetically within their respective order. Some New York trilobites mentioned in the literature did not have the author or publication listed, and a limited search for this information was unsuccessful. These trilobites are listed, however, for completeness.

The location information includes the county, and only New York counties are mentioned. We did not attempt to make this location information all-inclusive; thus, for most species, it is only representative. Many New York trilobites can be found wherever the appropriate rock unit is exposed.

We identified the trilobites by surveying the literature and by going through the collections of a few major museums. The referenced museums are the American Museum of Natural History (AMNH); Canadian Geological Survey (Ottawa) (GSC); Carnegie Museum in Pittsburgh (CM); Museum of Comparative Zoology at Harvard (MCZ); National Museum of Natural History (USNM), sometimes called just "the Smithsonian"; Natural History Museum of London (formerly the British Museum (Natural History)) (BM); New York State Museum (NYSM); Paleontological Research Institution (PRI); Peabody Museum at Yale (YPM); Rochester Museum and Science Center (RMSC); Royal Ontario Museum (ROM); Buffalo Museum of Science (BMS); Ohio State University (OSU); San Diego Natural History Museum (SDMNH); and University of Michigan Museum of Paleontology (UMMP). When known, the location of the holotype is given, but in most instances only the location of representative specimens is given. Any designations other than holotype, such as cotype, appear as they are on the label. Whenever geographic data such as the presence of the species outside New York are known to us, they are given.

When the term *type species* is given at the beginning of the descriptive paragraph, it indicates that this species was used to define the genus. In those instances where there may be difficulties in identifying similar trilobites, the diagnostic characteristics are presented in tabular form. An asterisk before the name indicates that the name is no longer valid. These names were included because they are represented in museums or often listed in the literature.

5.1 ORDER AGNOSTIDA

The agnostids are small trilobites, usually less than 12 mm (0.5 inch) long, characterized by having cephala and pygidia of about the same size and never more than two or three thoracic segments. There are two suborders, Agnostina and Eodiscina. In the past, some workers considered the Agnostida as a different arthropod class than trilobites, but current thinking is that they are indeed part of the class Trilobita. Agnostids are found from the Cambrian through the Ordovician, but in New York they have only been reported from the Cambrian rocks, primarily the allochthonous rocks on the western edge of the Taconics. The order is significantly revised in *Treatise* (revised).

Suborder Agnostina

The agnostins have no eyes, two thoracic segments, and no facial sutures. The pygidium has three or fewer axial rings. As is so often the case with agnostids, the thoracic segments are unknown. Table 5.1 lists the families and species within this suborder.

Table 5.1. Trilobites of the suborder Agnostina

FAMILY	NAME	SPECIMENS	LOCATION
Diplagnostidae	*Baltagnostus angustilobus* (Rasetti, 1967)	Holotype USNM 156565	Middle Cambrian, Nutten Hook, Columbia County; see Bird and Rasetti (1968, p. 32).
	Baltagnostus stockportensis (Rasetti, 1967)	Holotype USNM 156567	Middle Cambrian, Stockport Station, Columbia County; see Bird and Rasetti (1968, p. 26).
Peronopsidae	*Peronopsis evansi* (Rasetti and Theokritoff, 1967)	Holotype MCZ 8546	Lower Cambrian, Washington County, New York.
	Peronopsis primigenea (Kobayashi, 1939)	Lectotype USNM 18328	Lower Cambrian, Washington County, New York.
Ptychagnostidae	*Ptychagnostus gibbus* (Linnarsson, 1869)	Plesiotypes USNM 156552, 156553	Middle Cambrian, Columbia County; see Bird and Rasetti (1968, p. 11).
	Ptychagnostus punctuosus (Angelin, 1851)	Plesiotypes USNM 156551	Type species. Middle Cambrian, Columbia County; see Bird and Rasetti (1968, p. 28).
Spinagnostidae	*Eoagnostus acrorachis* (Rasetti and Theokritoff, 1967)	Holotype MCZ 8544	Lower Cambrian, West Castleton Formation, Washington County, New York.
	Eoagnostus primigeneus (Kobayashi, 1939)	Cotype USNM 18328	Lower Cambrian.
	Hypagnostus parvifrons (Linnarsson, 1869)	Plesiotypes USNM 156561 to 156563	Type species. Middle Cambrian, Columbia County; see Bird and Rasetti (1968, p. 68). The holotype is from Sweden.

Suborder Eodiscina Plate 1

The eodiscins have two or three thoracic segments and four or more axial segments on the pygidium, and may be without eyes or facial sutures. In the *Treatise* (revised), the suborder has been significantly taxonomically revised, and new family structures assigned. Table 5.2 lists the families and species within this suborder.

5.2 ORDER REDLICHIIDA

The redlichiids are the most primitive and earliest known of the trilobites. They generally have a relatively large, semicircular cephalon with long genal spines, numerous thoracic segments, and a very small pygidium. The only well-known member of this order in New York is *Elliptocephala asaphoides*, a Lower Cambrian trilobite from the suborder Olenellina, a suborder representing the earliest of the known trilobites. The holaspids of the olenellins do not show facial sutures.

Family Holmiidae

***Elliptocephala asaphoides* Emmons, 1846 Plate 2**

Type NYSM 4954

Type species. This trilobite is from the Lower Cambrian, Nassau Formation, Troy, Rensselaer County. It also is reported from numerous Lower Cambrian limestone sites in Columbia County (see Bird and Rasetti 1968). This is an apparently deep-water trilobite found in Lower Cambrian allochthonous dark shales. The specimen in Plate 2 is rust colored on a black matrix.

5.3 ORDER CORYNEXOCHIDA

Family Dolichometopidae

***Athabaskiella* cf. *A. proba* (Walcott)**

Middle Cambrian, Stockport Station, Columbia County. See Bird and Rasetti (1968, p. 28).

***Bathyuriscidella* cf. *B. socialis* (Rasetti)**

Middle Cambrian, Stockport Station, Columbia County. See Bird and Rasetti (1968, p. 28).

***Bathyuriscus eboracensis* (Rasetti, 1967)**

Holotype USNM 156654

Middle Cambrian, Stockport Station, Columbia County. See Bird and Rasetti (1968, p. 26).

***Bathyuriscus* cf. *B. fibriatus* (Robison)**

Middle Cambrian, Stockport Station, Columbia County. See Bird and Rasetti (1968, p. 28).

***Corynexochides? expansus* (Rasetti, 1967)**

Holotype USNM 156651

Middle Cambrian, Stockport Station, Columbia County. See Bird and Rasetti (1968, p. 26).

Family Dorypygidae

***Fordaspis nana* (Ford, 1878)**

Hypotypes NYSM 11047, 11048

Type species. Lower Cambrian, Nassau Formation. See Goldring (1943).

Table 5.2. Trilobites of the suborder Eodiscina

FAMILY	*NAME*	*SPECIMEN*	*LOCATION*
Calodiscidae	*Calodiscus agnostoides* (Kobayashi, 1943)	Holotype USNM 116356	Lower Cambrian, Schodack Formation, Salem, Washington County.
	Calodiscus fissifrons (Rasetti, 1966)	Holotype USNM 146004	Lower Cambrian, Columbia County.
	Calodiscus lobatus (Hall, 1847)	Syntype AMNH 210	Type species. Lower Cambrian, Nassau Formation, found at Troy, Rensselaer County, and in Washington County.
	Calodiscus meeki (Ford, 1876)	Holotype NYSM 4587	Lower Cambrian, Schodack Formation, Troy, Rensselaer County.
	Calodiscus occipitalus (Rasetti, 1966)	Holotype USNM 146003	Lower Cambrian, Columbia County.
	Calodiscus reticulatus (Rasetti, 1966)	Holotype USNM 146006	Lower Cambrian, Columbia County.
	Calodiscus schucherti (Matthew, 1896)	Syntype ROM 138	Lower Cambrian, Schodack Formation, Troy, Rensselaer County.
	Calodiscus theokritoffi (Rasetti, 1967)	Holotype GSC 105	Lower Cambrian, Malden Bridge roadcut, Columbia County; see Bird and Rasetti (1968, p. 11).
	Calodiscus walcotti (Rasetti, 1952)	Holotype USNM 26710	Lower Cambrian, Schodack Formation, near Greenwich, Washington County.
	Chelediscus chathamsis (Rasetti, 1967)	Holotype USNM 156584	Lower Cambrian, Malden Bridge roadcut, Columbia County; see Bird and Rasetti (1968, p. 11).
Eodiscidae	*Pagetia bigranulosa* (Rasetti, 1967)	Holotype USNM 1566	Lower Cambrian, Griswold Farm, Columbia County; see Bird and Rasetti (1968, p. 5).
	Pagetia connexa (Walcott, 1891)	Holotype USNM	Lower Cambrian, Washington County.
	Pagetia clytioides (Rasetti, 1967)	Holotype USNM 156615	Middle Cambrian, Griswold Farm, Columbia County; see Bird and Rasetti (1968, p. 5).
	Pagetia erratica (Rasetti, 1967)	Holotype USNM 156613	Middle Cambrian, Griswold Farm, Columbia County; see Bird and Rasetti (1968, p. 5).
	Pagetia laevis (Rasetti, 1967)	Holotype USNM 156610	Lower Cambrian, Riders Mills and Griswold Farm, Columbia County; see Bird and Rasetti (1968, p. 5).
	Pagetides amplifrons (Rasetti, 1945)	Plesiotypes USNM 156632, 156633	Lower Cambrian, Griswold Farm, Columbia County; see Bird and Rasetti (1968, p. 5).
	Pagetides elegans (Rasetti, 1945)	Plesiotypes USNM 156618	Type species. Lower Cambrian, numerous sites in Columbia County; see Rasetti (1967).
	Pagetides leiopygus (Rasetti,1945)	Plesiotypes USNM 156626 to 156628	Lower Cambrian, Griswold Farm, Columbia County; see Bird and Rasetti (1968, p. 5).
	Pagetides minutus (Rasetti, 1945)	Plesiotypes USNM 156621 to 156625	Lower Cambrian, numerous sites in Columbia County; see Rasetti (1967).
	Pagetides rupestris (Rasetti, 1948)	Plesiotypes USNM 156629 to 156631	Lower Cambrian, Griswold Farm, Columbia County; see Bird and Rasetti (1968, p. 5).
Hebediscidae	*Hebediscus marginatus* (Rasetti, 1967)	Holotype USNM 156637	Lower Cambrian, 1.6 km (1 mile) east of Salem, Washington County.
	Neopagetina taconica (Rasetti, 1967)	Holotype USNM 156634	Lower Cambrian, Griswold Farm, Columbia County; see Bird and Rasetti (1968, p. 5).
Weymouthiidae	*Acidiscus birdi* (Rasetti, 1966)	Holotype USNM 145987	Type species. Lower Cambrian, Griswold Farm, Columbia County; see Bird and Rasetti (1968, p. 5).
	Acidiscus hexacanthus (Rasetti, 1966)	Holotype USNM 145989	Lower Cambrian, Columbia County.
	Acimetopus bilobatus (Rasetti, 1966)	Holotype USNM 145991	Type species. Lower Cambrian, Columbia County.
	Analox bipunctata (Rasetti, 1966)	Holotype USNM 145093	Type species. Lower Cambrian, Columbia County.
	Analox obtusa (Rasetti, 1967)	Holotype USNM 156575	Lower Cambrian, Griswold Farm, Columbia County; see Bird and Rasetti (1968, p. 5).
	Bathydiscus dolichometopus (Rasetti, 1966)	Holotype USNM 145955	Type species. Lower Cambrian, Columbia County.
	Bolboparia elongata (Rasetti, 1966)	Holotype USNM 146001	Lower Cambrian, Columbia County.

Table 5.2. *Continued*

FAMILY	NAME	SPECIMEN	LOCATION
	Bolboparia superba (Rasetti, 1966)	Holotype USNM 145998	Type species. Lower Cambrian, Columbia County.
	Leptochilodiscus punctulatus (Rasetti, 1966)	Holotype USNM 146009	Lower Cambrian, Griswold Farm, Columbia County; see Bird and Rasetti (1968, p. 5); also found in Quebec.
	Litometopus longispinus (Rasetti, 1966)	Holotype USNM 146012	Type species. Lower Cambrian, Columbia County.
	Mallagnostus desideratus (Walcott, 1891)	Holotype USNM 18327	Washington County, New York. See Rasetti and Theokritoff (1967) for a discussion of this species.
	Microdiscus? (Eodiscus) connexus (Walcott, 1890)		Lower Cambrian, Nassau Formation; the genus *Microdiscus* is no longer used so this species is probably *Eodiscus*.
	Oodiscus binodosus (Rasetti, 1966)	Holotype USNM 146016	Lower Cambrian, Columbia County.
	Oodiscus longifrons (Rasetti, 1966)	Holotype USNM 146018	Lower Cambrian, Columbia County.
	Oodiscus subgranulatus (Rasetti, 1966)	Holotype USNM 146014	Lower Cambrian, Columbia County.
	Serrodiscus griswoldi (Rasetti, 1967)	Holotype USNM 156596	Lower Cambrian, Griswold Farm, Columbia County; see Bird and Rasetti (1968, p. 11).
	Serrodiscus latus (Rasetti, 1966)	Holotype USNM 146024	Lower Cambrian, Griswold Farm, Columbia County; see Bird and Rasetti (1968, p. 5).
	Serrodiscus speciosus (Ford, 1873)	Lectotype NYSM 4588 (Rasetti, 1952)	Type species. Lower Cambrian, Schodack Formation, Troy, Rensselaer and Washington Counties.
	Serrodiscus spinulosus (Rasetti, 1966)	Holotype USNM 14602	Lower Cambrian, Malden Bridge roadcut, Columbia County; see Bird and Rasetti (1968, p. 11).
	Serrodiscus subclovatus (Rasetti, 1966)	Holotype USNM 146022	Lower Cambrian, Columbia County.
	Stigmadiscus gibbosus (Rasetti, 1966)	Holotype USNM 146031	Lower Cambrian, Griswold Farm and Malden Bridge, Columbia County; see Bird and Rasetti (1968, p. 5, 11).
	Stigmadiscus stenometopus (Rasetti, 1966)	Holotype USNM 146029	Lower Cambrian, Griswold Farm and Malden Bridge, Columbia County; see Bird and Rasetti (1968, p. 5, 11).
	Weymouthia nobilis (Ford, 1873)	Plesiotype MCZ 105039	Lower Cambrian, Nassau Formation, Troy, Rensselaer County. The type is from Massachusetts.

Kootenia fordi (Walcott)
Lower Cambrian, Nassau Formation.

Olenoides stissingensis (Dwight)
Specimen NYSM 17011
Stissing Limestone, Stissing Mountain, Dutchess County, New York. The specimen is a sandstone slab with many disarticulated trilobite parts.

Olenoides stockportensis (Rasetti, 1967)
Holotype USNM 156668
Middle Cambrian, Stockport Station, Columbia County. See Bird and Rasetti (1968, p. 26).

Family Illaenidae

Illaenids are first found in the Lower Ordovician and disappear by the end of the Silurian. In New York they are first found in the lower Middle Ordovician Chazy Group, and the last representative is in the Lower Silurian Rochester Shale. The family is characterized by a smooth, vaulted cephalon with no glabellar furrows. The posterior of the glabella is often outlined by weakly depressed furrows. These furrows rarely extend to the anterior portion of the cranidium. The facial sutures are opisthoparian, and the eyes are broadly set near the cephalic margins. There are eight to ten unfurrowed thoracic segments. The pygidium is similarly vaulted and smooth. The pygidium, when found separately, often resembles a dark thumbnail. Lane and Thomas (1983) moved a number of former illaenids into the family Styginidae. A large number of species have been described from New York, but this may well be oversplitting, particularly in the Trenton species, which need redescription. Shaw (1968) redescribed most of the Chazy material.

Bumastoides aplatus (Raymond, 1925)
Holotype MCZ 101150
This illaenid is from Middle Ordovician Chazy. For location information, see Shaw (1968). The cranidium is semicircular in

dorsal view and about half as long as wide. The axial furrows are faint. The thorax is unknown. The pygidium is smooth, vaulted, and without furrows. The smooth cephalon and pygidium differ from all other Chazy illaenids. Raymond also listed it as from Vermont and Tennessee.

***Bumastoides? bellevillensis* (Raymond and Narraway, 1908)**
Type CM 1900; specimens MCZ 720, 719

Both MCZ specimens are from the Middle Ordovician. Number 720 is from the Trenton Limestone at Sugar River, Lewis County, and 719 is from the Black River Limestones at Buck's Quarry near Poland, Herkimer County. Both are cranidia in C. D. Walcott's collections.

***Bumastoides billingsi* (Raymond and Narraway, 1908)**
Type CM 5472, hypotype GSC 331

The type of this species is from Hull, Quebec. The species differentiation from other similar Trenton specimens is based primarily on size. The length, 60 and 83 mm, is much greater than that of *B. trentonensis* and *B. milleri*; the dorsal furrows on the cephalon are also stronger and those on the thorax are wider apart.

***Bumastoides gardenensis* (Shaw, 1968)**
Holotype NYSM 12456

This small illaenid from the Middle Ordovician Chazy is very similar to *B. aplatus*. For location information, see Shaw (1968).

***Bumastoides globosus* (Billings, 1859) Plate 3**
Lectotype GSC 1090B (Shaw, 1968)

This Middle Ordovician Chazy species is defined as having a wide axis, well-developed cephalic axial furrows, and a broad smooth pygidium. For location information, see Shaw (1968). The specimen in Plate 3 is a rare, articulated specimen from the Chazy Group. Articulated trilobites from the Chazy are rare, probably because the Chazy commonly represents a high-energy environment not conducive to the preservation of articulated specimens.

***Bumastoides holei* (Foerste, 1920) Plates 4 and 5**
Hypotype MCZ 101148

The type of this species is from the Middle Ordovician Kimmswick Limestone of Missouri. It was redescribed by Raymond (1925) with material from the Walcott-Rust Quarry, Herkimer County. The specimens from New York have a cephalon much larger than the pygidium. The cephalon is evenly convex, with a slight longitudinal median depression. The eyes are small, wide apart, and close to the posterior margin. There is a small median tubercle just in front of the posterior cephalic margin. The axial lobe of the thorax is very wide. The thorax has ten thoracic segments. The pygidium has no trace of an axial lobe. The entire exoskeleton is granulose with corresponding pits on the internal cast. These pits have not been seen on specimens from locations other than the Walcott-Rust Quarry. The specimen in Plate 4 is partially exfoliated and shows, on the left side, the lunette and its position relative to the eye. Plate 5 is a side view of another specimen from the same collection, also showing the lunette.

***Bumastoides milleri* (Billings, 1859) Plate 6**
Holotype GSC 1319b; specimens USNM 72265; MCZ 721, 722; PRI 6455

The holotype is from Ontario. The specimens in the USNM are from the Middle Ordovician Lowville Member of the Black River Group. They are listed from Watertown, Lewis County. Other specimens are from quarries near Newport, Herkimer County. Specimens in the MCZ are from Walcott's collections and are labeled from the Black River Limestones at Buck's Quarry near Poland, Herkimer County, and from the lower Trenton at Rawlin's Mills (near Saratoga, Saratoga County). DeMott (1987) listed specimens from Illinois and Wisconsin. The specimen in Plate 6 is from Watertown and was listed on the label from the Trenton Group.

***Bumastoides porrectus* (Raymond, 1925) Plate 7**
Holotype MCZ 101147

The type for this Middle Ordovician trilobite is from the Walcott-Rust Quarry, Herkimer County, in the Rust Member of the Trenton Group. The species was described to replace *B. trentonensis* because, in the opinion of Raymond, the latter has a poor definition. The cephalon has faint dorsal furrows, and the eyes are situated far back and wide apart. The thorax has a wide axial lobe and ten thoracic segments. The pygidium is short and without a trace of an axial lobe. The entire exoskeleton is smooth. The holotype is figured in Plate 7. DeMott (1987) also listed *B. porrectus* from Wisconsin and Iowa. Bolton (1966) listed specimens from Ontario and Quebec.

****Bumastoides trentonensis* (Emmons, 1842)**
Plastotype AMNH 847; hypotypes NYSM 4158, 10757

NYSM 4158 is labeled as the "Type of Geology of New York, Report of the 2nd District, 1842, p. 390" and "Geological Survey of Minnesota, 1896, p. 720, Figure 32." There is also a label with the specimen for *Bumastus porrectus* Raymond. NYSM 10757 is listed from the Rockland (Napanee) Member of the Trenton at Wells, Hamilton County (Fisher). This Middle Ordovician Trenton species was abandoned by Raymond (1925) because of its poor definition. Given the existence of the hypotypes, there is reason to question Raymond's decision.

***Illaenus arcturus* (Hall, 1847)**
Holotype NYSM 4499

The type of this species, labeled from the Middle Ordovician Chazy Group, was studied by Shaw (1968), who concluded that the specimen was in too poor condition for comparison with other illaenids of the Chazy. He did believe it had definite nanillaenid character and the matrix was unlike any of those he was familiar with in the Chazy. Chazy specimens assigned by Raymond (1910b) as *Thaleops arctura* became the basis for *Thaleops longispina* (Shaw, 1968).

***Illaenus consimilis* (Billings, 1865)**
Specimens USNM 72319, 72317

USNM 72319 is from the Middle Ordovician Trenton Walcott-Rust Quarry, Herkimer County, and USNM 72317 is from the Trenton at Ellisburg, Jefferson County. The types are from Table Head, Newfoundland (Bolton, 1966).

***Nanillaenus americanus* (Billings, 1859) Plate 8**
Specimens NYSM 17013, 17014

This Middle Ordovician trilobite is well represented in NYSM, USNM, and MCZ collections from Walcott-Rust Quarry, Herkimer County, in the Middle Trenton Rust Limestone. The cephalon on this small trilobite is considerably larger than the pygidium. Axial furrows on the cephalon are only developed on the posterior region. The eyes are set well back and far apart. The free cheeks have a prominent blunt angle. The furrows on the thorax are well defined. There are eight thoracic segments. The pygidium has a similarly well-defined axis over half the length of the pygidium. The specimen in Plate 8 is from the MCZ and illustrates the difference between these trilobites and *Bumastoides* specimens found in the same rocks.

***Nanillaenus* cf. *N. conradi* (Billings, 1859)**
Syntypes GSC 1320, 1320a; hypotype NYSM 10749; specimen NYSM 17012

Type species. This trilobite is listed from the Middle Ordovician Upper Black River Chaumont Limestone in Wells, Hamilton County, by Fisher (1957). The syntypes are from the Middle Ordovician Leray beds of Quebec (Bolton 1966).

***Nanillaenus latiaxiatus* (Raymond and Narraway, 1908) Plate 9**
Type CM 5471

The type is from the Middle Ordovician Black River Limestones at Mechanicsville, near Ottawa, Canada. Raymond and Narraway suggested that many of the *Nanillaenus americanus* specimens reported from the Black River are this species. The species is similar to the latter trilobite except in the thorax and the pygidium. The thorax has ten segments, versus eight in *N. americanus*. The pygidium is about as long as the thorax, somewhat rectangular in outline, three-fifths long as wide. The sides of the pygidium are abruptly truncated to right angles with the anterior margin. The axial lobe is strongly convex and outlined by deep furrows on the sides. The axial lobe is about half the length of the pygidium. This trilobite is reported in the Black River rocks at Pattersonville, Schenectady County, and Newport, Herkimer County, and in the Wells Outlier, Hamilton County, by Fisher (1957). The specimen in Plate 9 is labeled from the Bowmanville Quarry in Ontario. This would place it high in Trenton age rocks.

***Nanillaenus? punctatus* (Raymond, 1905)**
Holotype CM 1278, specimen NYSM 17015

This rare middle Chazy species is one of the most easily characterized, as the entire surface of the cephalon is covered with small pits or puncta, and the anterior slope of the cephalon is covered with terrace lines so distinct that Raymond referred to them as concentric wrinkles. For location information, see Shaw (1968).

***Nanillaenus? raymondi* (Shaw, 1968) Plate 10A**
Holotype NYSM 12491

Nanillaenus? raymondi from the Middle Ordovician Chazy is characterized as being very similar to *N.? punctatus*, with the exception that the cephalic outline in dorsal view is subtriangular rather than semicircular and the former possesses a spine on the free cheek. For location information, see Shaw (1968).

***Thaleops longispina* (Shaw, 1968) Plate 10B**
Holotype NYSM 12922

This trilobite is from the Middle Ordovician Chazy limestones. For location information, see Shaw (1968). *Thaleops longispina* is a short, wide trilobite with a broadly rounded cephalon and pygidium. The glabella is outlined by deep furrows, which are parallel for half their length and then turn outward and down on the front of the cephalon. The eyes are raised on tapered stalks, which reach up and out at about a 45-degree angle. There are long, narrow genal spines. The thorax has ten segments. The pygidial axis is outlined by deep grooves and extends about half the length of the pygidium. Both the cranidium and pygidium are covered with small pits.

***Thaleops ovata* (Conrad, 1843)**
Holotype AMNH 1011/3

Type species. The type is from Wisconsin, in rocks with fossils that are clearly Middle Ordovician Trenton age (Hall, 1847). Reports of *T. ovata* from the Chazy of New York are wrong (see Shaw 1968). Ruedemann (1901) found two pygidia in a gray crystalline limestone within the Rysedorph Conglomerate that agreed with the descriptions of *T. ovata*. He also stated that specimens like those of *T. ovata* from other areas outside New York are found in strata corresponding to the Black River Lowville Limestone. DeMott (1987) listed *T. ovata* from Illinois and Wisconsin.

Family Styginidae

Lane and Thomas (1983) revised the family Styginidae to include the family Thysanopeltidae and some illaenids. The genus *Bumastus* is now considered a styginid. Holloway and Lane

(1998), however, consider that of the New York trilobites assigned to *Bumastus*, only *Bumastus ioxus* is assigned with confidence. Styginids are rarely encountered in New York. In general, the glabella expands strongly forward, the genal angle is acute, the pleurae have narrow spines beyond the edge of the thorax, and the pygidium is large and the back of the anterior edge is oval. They are found from the Middle Ordovician to the Upper Devonian.

Bumastus ioxus **(Hall, 1852) Plate 11**
Holotype AMNH 1821/1

The holotype for *B. ioxus* is from Wisconsin, and the specimens from the Lower Silurian Rochester Shale in western New York have long been assigned to this species. Small whole specimens are often found, sometimes in groups, in the lower parts of the Lewiston Member in Niagara County (Tetreault 1994). This trilobite is smooth and oval. The cephalon has widely spaced eyes. The axial furrows are faint and fade toward the front. The pygidium has a semicircular shape. The specimen in Plate 11 is much larger than those usually found. The large specimens and the smaller ones are found in different horizons within the Rochester Shale. Compression due to the weight of the overlying sediments has flattened the specimen, causing the edges of the cephalon and pygidium to crack.

Eobronteus lunatus **(Billings, 1857) Plate 12**
Holotype GSC 1781, hypotype NYSM 4150, specimen NYSM 17016

Ruedemann (1901) reported this species from Middle Ordovician Trenton pebbles in the Rysedorph Conglomerate, Rensselaer County. It is geographically widely distributed, being reported from Canada, Minnesota, and New Jersey. The holotype is from the Middle Ordovician of Ontario. The specimen in Plate 12 is from Snake Hill Shales in Saratoga County.

Failleana indeterminata **(Walcott, 1877) Plate 13**
Holotype MCZ 104928

The holotype is from the Middle Ordovician Black River Group limestone at Buck's Quarry, Poland, Herkimer County. The relatively large type specimen was figured for the first time by Raymond (1916) and is shown in Plate 13. The specimen is from the Walcott collection at MCZ. Wilson (1947) reported the trilobite from the lower Trenton Rockland beds of Canada.

Illaenoides* cf. *I. trilobita **(Weller, 1907) Plates 14 and 15**
Hall and Clarke (1888, plate 66, Figures 1–5)

The presence of a second illaenid in the Lower Silurian Rochester Shale, other than *Bumastus ioxus*, was long unrecognized. *Illaenoides trilobita* is easily differentiated by the comparatively larger cephalon and the distinctive border on the pygidium, and the axial furrows on both the cephalon and the pygidium are well defined. Weller (1907) first recognized the new genus *Illaenoides* and the trilobites he describes from Wisconsin do not differ from those found in the Rochester Shale.

Platillaenus erastusi **(Raymond, 1905)**
Lectotype YPM 7410A (Shaw, 1968)

This trilobite and the following *P. limbatus* are from the Middle Ordovician Chazy Group. See Shaw (1968) for detailed descriptions and locations.

Platillaenus limbatus **(Raymond, 1910)**
Lectotype YPM 23301 (Shaw, 1968)

Middle Ordovician Chazy Group. For location information, see Shaw (1968).

Scutellum barrandi **(Hall, 1859)**
Type NYSM 4149

An undisclosed location in the Lower Devonian Coeymans Limestone.

Scutellum niagarensis **(Hall, 1852) Plate 16**
Holotype AMNH 1830

Scutellum niagarensis was originally found in a float boulder at Niagara Falls, Niagara County. Only the pygidium was originally described. The pygidium is very broad and semicircular. Plate 16 also shows the previously undescribed cephalon. It is also reported from the Middle Silurian Reynales Limestone in Wayne County (Gillette 1947).

Scutellum pompilius **(Billings, 1863)**

Goldring (1943) listed this species as from the Lower Devonian Kalkberg in Columbia County.

Scutellum rochesterense **(Howell and Sanford, 1946) Plate 26B**
Holotype Princeton University 57696a (now USNM)

Lower Silurian Irondequoit Limestone of the Rochester, Monroe County, area.

Scutellum senescens **(Clarke, 1889) Plate 17**
Holotype NYSM 4151; hypotypes NYSM 4152, 4153

Upper Devonian Chemung of western New York. The hypotype specimens were found while recovering the glass sponge *Hydnoceras* from a Chemung outcrop near Avoca, Steuben County. Trilobites are very rare in the Upper Devonian of New York, and the one illustrated in Plate 17 is an excellent example of a very uncommon species.

Scutellum tullius **(Hall and Clarke, 1888) Plate 18**
Syntypes NYSM 4154, 4155

Middle Devonian Tully Limestone. NYSM 4154 is from Kingsley's Hill near Otisco, Onondaga County. Specimens are reported from a quarry near Spafford, Onondaga County. Cooper and Williams (1935) described the trilobite and listed a number of locations. The illustrated specimens were found by Professor Wells of Cornell University and the reconstruction is an unpublished drawing of his.

Scutellum wardi **(Howell and Sanford, 1947)**
Holotype USNM 488132

The type is a single pygidium from the Upper Silurian Oak Orchard Member of the Lockport Group in the Rochester, Monroe County, area. The axis is very short, and the entire surface is smooth.

Family Zacanthoididae

Prozacanthoides eatoni **(Walcott, 1891)**
Holotype USNM

Lower Cambrian, Washington County, New York.

5.4 ORDER LICHIDA

Family Lichidae

Trilobites of the family Lichidae are first found in the Lower Ordovician and disappear at the end of the Devonian. In New York they range from the Middle Ordovician to the Middle Devonian. Lichids are very distinctive in their glabellar outline with two prominent lateral lobes, which because they are so distinctive are given their own name, bullae. Figure 5.1 shows lichid cephala with the distinctive bulla. Lichids are common nowhere, but in certain horizons their characteristic exuviae can be found by carefully searching. The lichids range in size from the quite small *Hemiarges paulianus* (10 to 15 mm) to the large and heavily ornamented *Terataspis grandis* (> 30 cm). Whole specimens of the lichids are rare and very prized by both museums and collectors. Thomas and Holloway (1988) extensively reviewed the classification and phylogeny, but the New York species need revision.

Acanthopyge (Lobopyge) consanquinea **(Clarke, 1894) Figure 5.1D**
Holotype NYSM 4530

A lichid from the Lower Devonian New Scotland Limestone in Schoharie County.

Acanthopyge (Lobopyge) contusa **(Hall and Clarke, 1888) Figure 5.1C**
Types NYSM 4531, 4532

Hall and Clarke reported this trilobite from the Upper Helderberg (Lower Devonian) of eastern New York, Albany and Otsego Counties, and from Onondaga (Middle Devonian) float boulders in west-central New York, Ontario County. The illustrated specimen is from Hall and Clarke (1888).

Amphilichas conifrons **(Ruedemann, 1916)**
Holotype NYSM 9607; paratype NYSM 9608; specimens MCZ 1560, MCZ 1561

Amphilichas conifrons was described from a specimen in the NYSM from the Middle Ordovician Trenton collections of William Rust. It is probable that the specimen is from the Walcott-Rust Quarry, Herkimer County. Ruedemann compared this specimen to *A. halli* (Upper Ordovician, Ohio) and found significant differences. He made no comparisons with any of the New York amphilichids. Ruedemann did report evidence that the holotype cranidium was covered with thin spines, an observation not reported in any of the other New York amphilichids nor does it appear to be supported by our own examination of the holotype and other specimens. The glabella has a low rounded shape quite different from that of *A. cornutus.* The illustrated pygidium from Ruedemann (NYSM 9608) differs from *A. cornutus* in that the pygidal lappets closest to the axis are drawn as forked on their ends. (Close examination of the paratype does not appear to support this conclusion that the pygidial lappets are forked.) The holotype is identical to specimens in the MCZ, from the Walcott-Rust Quarry, confirming that NYSM 9607 is a different species than *A. cornutus.*

Amphilichas cornutus **(Clarke, 1894) Plate 19**
Holotype NYSM 4533; specimens USNM 26341, 72629

The holotype, illustrated in Plate 19, is an articulated specimen that was badly damaged during the preparation. It does have a prominent, cone-shaped glabella. The pygidium is intact, however, and serves, along with the glabella shape, as a primary identifying feature. A cranidium with this name in the USNM (USNM 26341) is pustulose. The location for the holotype is given as Trenton Falls, which encompasses the entire middle Trenton; however, the location label indicates the specimen was purchased from William Rust. Since most of Rust's material came from the Middle Ordovician Trenton Walcott-Rust Quarry, Herkimer County, and other specimens have been found there, it is likely that the holotype is from that location. The USNM specimen 72629 has a "museum label" name. It is a prominent glabella from the Middle Ordovician Trenton Limestone near Trenton Falls. The prominent glabella indicates it is most likely *A. cornutus.* Labels of material like this from "near Trenton Falls" often mean the Walcott-Rust Quarry, Herkimer County.

Amphilichas minganensis **(Billings, 1865)**
Lectotype GSC 1332a (Shaw, 1968), hypotypes NYSM 12219-34

Amphilichas minganensis is reported from the Middle Ordovician Chazy limestones and is the earliest lichid reported from New York. For location information, see Shaw (1968). Only cranidia and pygidia are known from mature specimens. This is one of the few trilobites in New York from which significant information of growth stages is known from silicified material. Raymond (1925, p. 125) reported it to be also found in Quebec and Virginia.

Amphilichas trentonensis **(Conrad, 1842)**
Plesiotype MCZ 4411

The type is from Middle Ordovician Trenton age rocks in Pennsylvania. Specimens from New York that were referred by Hall to this species are from the lower Trenton at Middleville,

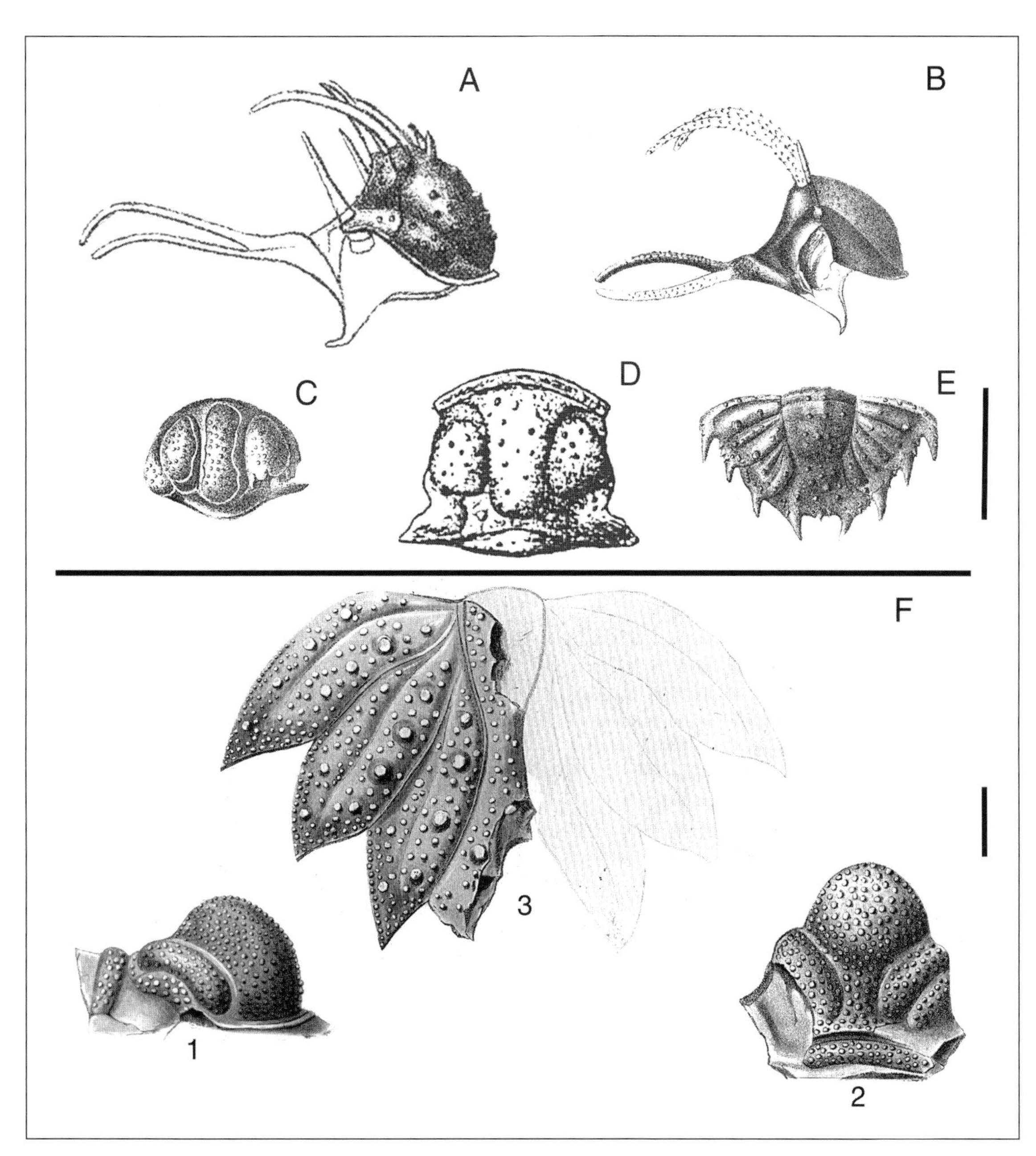

FIGURE 5.1. Lichids from the lower and lowest Middle Devonian of New York. A. *Ceratolichas dracon* from the Onondaga Limestone in Genesee County. The figured cephalon is a reconstruction from Hall and Clarke (1888). B. *Ceratolichas gryps* from the Onondaga Limestone at Schoharie, Schoharie County. The figured cephalon is a reconstruction from Hall and Clarke (1888). C. *Acanthopyge contusa* from the Onondaga Limestone in Ontario County. The figured glabella is from Hall and Clarke (1888). D. *Acanthopyge consanquinea* from the New Scotland Limestone in Schoharie County. The figured glabella is from Clarke (1894). E. *Echinolichas hispidus* from the "Schoharie Grit," Schoharie, Schoharie County. The figured pygidium is from Hall and Clarke (1888). F. *Oinochoe bigsbyi* from the shaly limestones of the lower Helderberg Group. Drawings 1 and 2 show side and dorsal views of the glabella. Drawing 3 shows a partially reconstructed pygidium. The figures are from Hall (1861b).

Herkimer County. The whole specimen figured by Hall (1847) is from Ohio and probably a different species. A specimen in the USNM (USNM 6303) is from Pennsylvania.

Arctinurus boltoni **(Bigsby, 1825) Plates 20 and 21**
Holotype BM It 15690; hypotype NYSM 9631; specimens USNM 499543, PRI 42095

Arctinurus boltoni is found in the lower half of the Lower Silurian Rochester Shale in western New York. *Arctinurus boltoni* is one of the premier trilobites from New York. This large, flat ovate trilobite has a broad axis. The cephalon has a distinctive anterior tongue-shaped prow and hook-shaped longitudinal furrows on the glabella. The pygidium is large and has three pairs of lobate spines. The large size (100 to 150+mm) and the pustulated exoskeleton make a striking display. These trilobites are very rare as whole, articulated specimens, and the types were only discovered during the digging of the Erie Canal in Lockport, Niagara County. Recently, large areas of the trilobite-rich layers of the Rochester Shale were uncovered in a shale quarry near Middleport in Orleans County, from which sizable numbers of whole specimens have been taken. The specimen in Plate 20, USNM 499453, from the Rochester Shale has epibionts adhered to the surface, suggesting that the trilobite was mature and had not molted since the brachiopod spat and other larvae lodged on the surface. There are two different brachiopods, bryozoans and *Cornulites*, species on the dorsal surface. The rhynchonellid brachiopod apparently shows four growth stages, indicating a time interval since the last molt. The trilobite in Plate 21, PRI 42095, shows healed damage to the left side, possibly as a result of predator attack. Three size classes of brachiopods are evident on the surface.

Autoloxolichas? inconsuetus **(Raymond, 1925)**
Holotype MCZ 101164

The holotype, and only specimen we are aware of, is a cranidium from Middle Ordovician Glens Falls Limestone (basal Trenton) at the (former) bridge at the falls of Flat Creek, 1.6 km (1 mile) south of Sprakers, Montgomery County. No comparisons can be made with other Trenton lichids.

Ceratolichas dracon **(Hall and Clarke, 1888) Figure 5.1A**
Types NYSM 4534 to 4536

The illustrated specimen is from the Onondaga Limestone in Hall and Clarke (1888). Hall and Clarke reported the trilobite from the upper Helderberg, Lower Devonian, at Schoharie, Schoharie County, and the Middle Devonian at LeRoy, Genesee County.

Ceratolichas gryps **(Hall and Clarke, 1888) Figure 5.1B**
Syntypes NYSM 4550 to 4552

The illustrated specimen is from the Onondaga Limestone and is in Hall and Clarke (1888). The trilobite is reported from the upper Helderberg of Clarksville, Albany County; Cherry Valley, Otsego county; and "decomposed chert boulders" at Canandaigua, Ontario County.

Dicranopeltis fragosa **(Phleger, 1937)**
Holotype MCZ 100865

Lower Silurian Rochester Shale. This trilobite is known only from a single cranidium and may be *Dicranopeltis nereus.*

Dicranopeltis nereus **(Hall, 1863) Plates 22 and 23**
Type AMNH 1825, specimen PRI 49623

Lower Silurian Rochester Shale. *Dicranopeltis nereus* is found in the lower Rochester Shale in its western exposures (Tetreault 1994) and at Middleport, Niagara County. The postaxial lobe of the pygidium has a rounded outline with a small median notch. The cranidium of *D. nereus* has never been described. However, the cranidium of *D. fragosa* described by Phleger (see above) could belong to *D. nereus.* The specimen in Plate 23 is from near Middleport and shows extensive healed damage to the left side, while the specimen in Plate 22 has healed damage to the right upper thorax.

Echinolichas eriopsis **(Hall, 1863) Plate 24**
Syntypes NYSM 4537 to 4539

The illustrated specimens are from the Onondaga Limestone and reported in Hall and Clarke (1888) from the upper Helderberg at Schoharie, Schoharie County, and "decomposed chert boulders" at Canandaigua, Ontario County.

Echinolichas hispidus **(Hall and Clarke, 1888)**
Holotype NYSM 4553

The trilobite is reported from the Schoharie Formation at Clarksville, Albany County, and Onondaga Limestone at LeRoy, Genesee County.

Hemiarges paulianus **(Clarke, 1894) Plate 25**
Lectotype USNM 4243a (Rudkin et al. 1994)

This small lichid is from the Middle Ordovician, early Trenton in New York. It has a wide geographic distribution west to Minnesota and north into Canada. In New York cranidia and pygidia are commonly found in the upper part of the Kings Falls Formation of the Trenton Group, particularly in exposures just east of Watertown in Lewis County. The trilobite in Plate 25 is from Trenton age quarries in Ontario.

Lichid sp. Plate 26A

Pygidium of an unidentified lichid from the Middle Devonian Onondaga Limestone in the Cheektawaga Quarry, Erie County. The specimen in Plate 26A differs from the known lichids in the Onondaga.

Oinochoe bigsbyi **(Hall, 1859) Figure 5.1F**
Type AMNH 2610/1

The illustrated specimen is from Hall and Clarke (1888) and is in the New Scotland Limestone.

Oinochoe pustulosus **(Hall, 1859) Plate 27**
Syntypes NYSM 4529, 4558 to 4564

Oinochoe pustulosus is found in the New Scotland Limestone of east-central New York. The cranidium shown, NYSM E-975, illustrates the source of the species name *pustulosus*.

Richterarges ptyonurus **(Hall and Clarke, 1888) Plate 28**
Types NYSM 4555 to 4557

Richterarges ptyonurus is reported by Hall and Clark from the "coralline limestone," now known as the Cobleskill Formation, in the area of Schoharie, Schoharie County. The median part of the glabella, particularly in front of the level of the lateral glabellar furrows, S1, is flattened; the palpebral lobes are short and the anterior branches of the facial sutures converge slightly forward. The pygidium has a long axis and broader postaxial ridge. *Richterarges* trilobites are almost entirely restricted to northern Laurentia. The only exception is *R. ptyonurus*.

Terataspis grandis **(Hall, 1861) Plate 29**
Holotype NYSM 4543

Type species. *Terataspis grandis* is found in both the lower Onondaga Limestone of western New York, particularly the former Fogelsanger Quarry in Williamsville, Erie County, and the upper Schoharie Formation of eastern New York. The large size and spiny nature of this trilobite make it one of the more spectacular ones. A number of models have been made and are featured in Devonian reef dioramas, such as the one at the Rochester Museum and Science Center, Rochester, New York. Only one specimen is known from New York; it is complete enough for a reasonable reconstruction. This specimen is upside down, and half, longitudinally, is missing. It is in the Buffalo Museum of Natural History, Buffalo, New York. Plate 29 illustrates a genal spine from a large *Terataspis* specimen. The An *Eldredgeops* specimen is included for scale. The drawing from the *Treatise* shows where the genal spine appears on the exoskeleton. Using the relative scale of the drawing and the size of the genal spine, we can estimate an axial length, excluding pygidial spines, of about 38 cm (14.8 inches).

Trochurus bulbosa **(Phleger, 1937)**
Holotype MCZ 1555

Lower Silurian Clinton Group, found in Clinton drift (i.e., displaced rocks of Clinton age probably moved by glacial action) near Trenton Falls, Oneida County.

Trochurus halli **(Foerste, 1917) Plates 30 and 31**
Holotype AMNH 1826

This trilobite from the Silurian Rochester Shale was originally designated as *Trochurus phlyctainoides* by Hall (1852). *Trochurus halli* species differs chiefly from *T. phlyctainoides* in the curvature of the median lobe of the glabella, from front to rear. This curvature is greater anteriorly, but it is not strongly accentuated anterodorsally. Viewed from the dorsal side, the anterior part of the median lobe is less prominent. The furrow separating the third lateral lobes from the compound anterior lateral ones diverges more to the front laterally. The specimen in Plate 30 is exceptional. The glabella of this specimen is missing, probably as a result of molting. Note the *Cornulites* worm tubes on the right pygidial pleural surface. An example of the cranidium, which is missing from Plate 30, is shown in Plate 31. This specimen is from Wayne County.

Trochurus phlyctainoides **(Green, 1837)**

Trochurus phlyctainoides is a small ornate lichid reported from the Rochester Shale. The species name means having the form of blisters or pustules, which is an apt description. The type is from Ohio, and its actual presence in New York is questionable. The reported material we are aware of is *T. halli* above.

Family Odontopleuridae

Odontopleurids are found in New York from the Middle Ordovician to the base of the Middle Devonian. With the exception of the genus *Dicranurus*, they are small trilobites, easily overlooked. The glabella tapers slightly forward, and the glabellar lobes are prominent. The genal spines are long, and short spines project down, rakelike, from the lateral edges of the free cheeks. Often there is one or more occipital spines, which can be curved. The thoracic pleura end in long spines. The pygidium has a long pair of border spines, with one or two pairs of shorter ones inside.

Apianurus narrawayi **(Raymond, 1910)**
Topotype MCZ 7697

Apianurus narrawayi is an uncommon Middle Ordovician Chazy species. The type is lost, and Shaw (1968) redescribed the species based on silicified material. Both *A. narrawyi* and *Ceratocephala triacanthaspis* from the Chazy have two diverging occipital spines. *Ceratocephala triacanthaspis* is distinguished in that its cranidium is covered with short spines. All the silicified material is from very small specimens. For location information, see Shaw (1968).

Ceratocara shawi **(Chatterton, Edgecombe, Vaccari, and Waisfeld, 1997)**
Holotype NYSM 15482

This species is from silicified material in the Crown Point Formation of Valcour Island, Clinton County. Ramsköld (1991) first noticed, in illustrations by Shaw (1968), that the remains of *Ceratocera shawi* were mixed in with those of *Ceratocephala triacantheis*. *Ceratocara shawi* differs from *C. triacantheis* in lacking sculpture on the glabella, other than the major pairs of spinose tubercles on the median lobe, and in lacking sculpture on the lateral glabellar lobes.

Ceratocephala triacantheis (Whittington and Evitt, 1954)

This species is a Middle Ordovician, middle Chazy trilobite that Shaw (1968) identified from silicified material. For location information and illustrated specimens, see Shaw (1968). The type is from Virginia and is probably in the USNM.

Diacanthaspis parvula (Walcott, 1877) Plate 32

Holotype MCZ 4372, paratype MCZ 4872

In New York *D. parvula* is primarily known from the Middle Ordovician, middle Trenton Walcott-Rust Quarry, Herkimer County. It is, however, fairly uncommon. *Diacanthaspis parvula* is found along with *Meadowtownella trentonensis*. Both are small trilobites with similar general shapes. The former can be readily distinguished from *M. trentonensis* because of its significantly more pustulose character and the presence of prominent pustules on the axis and pleural regions of the thoracic segments. It is widely distributed geographically, being reported from Quebec, Pennsylvania, and Minnesota.

Dicranurus hamatus (Conrad, 1841) Plate 33

Hypotypes NYSM 4190 to 4194

Lower Devonian New Scotland beds. This trilobite is only known from disarticulated parts in New York, but a close relative, *D. elegantus*, from Oklahoma is complete. Plate 33, from an Oklahoma specimen, shows how *D. hamata* must have looked. The Oklahoma species is similar enough to be considered a subspecies by Campbell (1977). The curved hornlike occipital spines are very distinctive. The spines are often encrusted by a variety of epibionts such as bryozoa, brachiopods, crinoids, corals, and worm tubes, as well as borings referred to as *endoliths*. There is evidence that these encrusters were on the trilobite while it was alive (Kloc 1992, 1993, 1997).

Kettneraspis callicera (Hall and Clarke, 1888) Plate 34

Hypotypes NYSM 4186 to 4189

Upper Lower/lower Middle Devonian, Schoharie Grit, Clarksville, Albany County, and Onondaga Limestone at Canandaigua, Ontario County. Note in Plate 34 how the genal spine ends with a “hook” shape.

Kettneraspis tuberculata (Conrad, 1840) Plate 35

Hypotypes NYSM 10704, 10705, 4196 to 4200

Lower Devonian New Scotland beds. Disarticulated parts of this species are common in some of the limestones and shales of the New Scotland Formation. Articulated specimens such as the one in Plate 35 are rare. Note the abundance of prominent tubercles on the exoskeleton, which prompted the name.

Kettneraspis sp. Plate 36

This specimen has been collected from the New Scotland Limestone. It is distinguished from known material by the very long cephalic border spines.

Meadowtownella trentonensis (Hall, 1847) Plates 37 and 38

Holotype AMNH 853/2, specimen MCZ 111717

The type is from the Middle Ordovician Trenton of Belleville, Ontario, and although it is not a good specimen, Ross (1979) believed it to be a different trilobite from the *Meadowtownella* specimen given this name in New York. The New York species has been taken in reasonable numbers from the middle Trenton Walcott-Rust Quarry, Herkimer County, and is well represented in many collections. Often the matrix with this trilobite contains bryozoan and cystoid colonies. The trilobite is ornate, small, and rarely over 2.5 cm (1 inch) long. It makes an attractive specimen. Plates 37 and 38 show the dorsal and ventral exoskeletal anatomy of this trilobite.

Odontopleura ceralepta (Anthony, 1838)

Hypotype NYSM 9803

The type for this species is from the Upper Ordovician of Ohio. It has been reported from the Upper Ordovician Whetstone Gulf Shales near Rome, Oneida County.

Odontopleura sp. (Fisher, 1953)

Specimen BMS E16735

Lower Silurian Maplewood Shale, Rochester, Monroe County. Fisher illustrated a small articulated specimen and referred to it as an *Odontopleura* species. Ludvigsen (1979b, p. 62) illustrated a trilobite referred to as *Leonaspis*, which is now known as *Exallaspis* (Ramsköld and Chatterton 1991), and reported from the Cataract Group of Canada. The Maplewood Shale is younger than the Cataract Group (Brett et al. 1995). This trilobite may be *Exallaspis*, but more material is needed to make identification possible.

Primaspis (Miraspis?) crosota (Locke, 1838) Plate 39

Hypotypes NYSM 9804 to 9806, USNM 34617

Primaspis crosota is an Ohio species from the Upper Ordovician. It has been reported from the Middle Ordovician Trenton age, Canajoharie and Dolgeville Shales, Montgomery County, to the Upper Ordovician Lorraine Group shales at Totman Gulf in Jefferson County. Plate 39 is USNM 23600 from the Upper Ordovician Frankfort Formation in Oneida County.

5.5 ORDER PHACOPIDA

This post-Cambrian order contains many of the more common trilobites from the Ordovician through the Devonian of New York. The suborder Phacopina contains the only trilobite families with schizochroal eyes.

Family Acastidae

The family Acastidae (Delo 1934) was resurrected by Edgecombe (1993) and includes *Greenops* and related species that are popular with trilobite collectors in New York. In New York the family is primarily restricted to the Middle Devonian and then mostly to the Hamilton Group. These trilobites have schizochroal eyes, which, because they are raised above the plane of the cephalon, are often damaged or lost during the collecting process. The subfamily Astropyginae, which includes *Greenops*, currently is under revision because of the observation that many more species apparently are represented in the specimens in museums and private collections than are currently recognized. A revision (Lieberman and Kloc 1997) divides the forms known in New York as *Greenops boothi* and *G. calliteles* into new genera and species. Tables 5.3 and 5.4 give descriptive information not included below.

Subfamily Asteropyginae Plate 40

Specimen in private collection (G. Kloc)

The specimen is from the Tully Limestone in Groves Creek off Cayuga Lake. It is characterized by broadening of the anterior border in front of the glabella, S2 meeting the glabellar furrow, prominently pustulose a cephalon, and rounded pleural segments on the pygidium.

Bellacartwrightia calderonae **(Lieberman and Kloc, 1997) Plate 41**

Holotype AMNH 45273

Windom Member, Hamilton Group, Ontario County. The cephalon has a broad flat anterior margin, axial nodes on the thorax, and pustules on the lappets.

Bellacartwrightia calliteles **(Green, 1837)**

Holotype lost

It is evident from Green's (1837a) description that *Cryphaeus calliteles* would be assigned to *Bellacartwrightia*, owing to the greater number of axial rings in the pygidium compared to *Greenops* species. Because the holotype is lost and no neotype from Pennsylvania has been designated, the presence of this trilobite is unknown in New York. Most specimens in New York previously assigned to *G. calliteles* have been reassigned to other species (Lieberman and Kloc 1997).

Bellacartwrightia jennyae **(Lieberman and Kloc, 1997)**
Plates 42 and 43

Holotype AMNH 45312, paratype AMNH 45310, specimen NYSM 4235

Type species. This trilobite is found in the Centerfield Limestone Member of the Hamilton Group in Livingston County. This type species is characterized by the smooth pygidial lappets and their triangular cross section. Plate 42 is the paratype. The cluster in Plate 43 of three (one hidden) *Harpidella craspedota* and two *Bellacartwrightia jennyae* specimens was first illustrated by Hall and Clarke (1888, Plate 24.15) and is from Centerfield in Ontario County. The smaller *B. jennyi* has small pustules on the thoracic pleural lobes and axis that are not found on the larger specimen.

Bellacartwrightia phyllocaudata **(Lieberman and Kloc, 1997) Plate 44**

Holotype AMNH 45230

Windom and Deep Run Members, Hamilton Group, Livingston County. The specimen in Plate 44 is the holotype. The pygidial lappets on this species are pustulose, flat, and sharply pointed with a thin sharp extremity, and the axial lappet is broader at the base than the other lappets.

Table 5.3. Defining features of the New York asteropygins

	Bellacartwrightia	*Greenops*	*Kennacryphaeus*
Cephalon	Moderate to broad cephalic border	Narrow cephalic border	Cranidial anterior border lengthened slightly medially, developed as a narrow lip
	Eye: maximum of 8 to 10 lenses in a dorsoventral file	Eye: maximum of 6 lenses in a dorsoventral file	Eye: maximum of 8 lenses in a dorsoventral file
	Posterior border furrow is deflected backward on the genal spine	Posterior border furrow is straight, ending halfway across the genal spines	Posterior border furrow is deflected backward on the genal spine
Thorax	Axial spines or nodes on thoracic segments	No axial thoracic spines, sometimes nodes	Thoracic axial spines
Pygidium	14 to 16 pygidial axial rings	11 pygidial axial rings	14 pygidial axial rings
	Tops of pygidial pleural segments rounded in anteroposterior view	Tops of pygidial segments flat in anteroposterior view	Tops of pygidial pleural segments rounded
	Posterior portion of pygidial axis prominently excavated	Posterior portion of pygidial axis faintly excavated	Pygidial pleural field flanking posterior portion of pygidial axis faintly excavated
	Posterior portions of pygidial pleural segments elevated above anterior portion	Interior and posterior portions of pygidial pleural segments of equal elevation	Interior and posterior portions of pygidial pleural segments of equal elevation

Table 5.4. Features of the genera *Bellacartwrightia* and *Greenops*

	B. jennae	*B. whiteleyi*	*B. calderonae*	*B. phyllocaudata*	*B. pleione*	*G. boothi*	*G. barberi*	*G. grabaui*
Horizon	Hamilton, Centerfield	Hamilton, Wanakah	Hamilton, Windom	Hamilton, Deep Run, Windom	Onondaga	Frame Group in Pennsylvania	Hamilton, U. Wanakah, Windom	Hamilton, Centerfield, Ledyard, Wanakah
Cephalon	Axial furrow narrow and shallow, nearly straight anterior of SO, diverging forward at about 25 degrees	Axial furrow narrow and shallow, nearly straight anterior of S1, diverging forward at about 35 degrees	Axial furrow narrow and shallow, nearly straight anterior of S1, diverging forward at about 35 degrees	Axial furrow narrow and shallow, diverging more strongly anterior of S1, diverging forward at about 45 degrees	Unknown	Prominent axial tubercle on LO	Faint or absent axial tubercle on LO	Faint or absent axial tubercle on LO
						Numerous and large circular perforations on glabellar and palpebral lobes		
Thorax	Has circular perforations	Has circular perforations	No circular perforations	No circular perforations	Unknown	Two transverse rows of circular perforations on anterior band of pleural segments	One transverse row of circular perforations on anterior band of pleural segments	One transverse row of circular perforations on anterior band of pleural segments
	Axial spines	Axial spines	Axial nodes	Axial spines		No axial nodes or spines	No axial nodes or spines	No axial Nodes or spines
Pygidium	Pygidial lappets are triangular in cross section	Pygidial lappets are oval in cross section	Pygidial lappets are oval in cross section	Pygidial lappets are flat	Pygidial lappets are oval in cross section	Tips of lappets developed as blunt triangles	Tips of lappets pointed	Tips of lappets developed as blunt triangles
	14 pygidial axial rings	14 pygidial axial rings	14 pygidial axial rings	15 to 16 pygidial axial rings	15 pygidial axial rings			
	Pygidial lappets lack, or have very fine, prosopon of small granules	Pygidial lappets have prosopon of small granules	Pygidial lappets have prosopon of small granules	Pygidial lappets have prosopon of small granules	Pygidial lappets have prosopon of small granules			
	Terminal pygidial lappet narrow	Terminal pygidial lappet narrow	Terminal pygidial lappet narrow	Terminal pygidial lappet broad	Terminal pygidial lappet narrow	Terminal pygidial lappet broad anteriorly, subrectangular, convex posteriorly	Terminal pygidial lappet developed as a narrow, sharp triangle	Terminal pygidial lappet subrectangular
	Circular perforations in the pleural area	Circular perforations in the pleural area	No circular perforations in the pleural area	No circular perforations in the pleural area	Circular perforations in the pleural area	Medial margins of lappets curved, lateral margins curved	Medial margins of lappets curved, lateral margins curved	Medial margins of lappets straight, lateral margins curved

Bellacartwrightia pleione **(Hall and Clarke, 1888)**
Holotype AMNH 4248

This species is found from the Middle Devonian Onondaga Limestone through the Tully Formation in numerous localities in western New York. The holotype is from Kentucky.

Bellacartwrightia whiteleyi **(Lieberman and Kloc, 1997)**
Plates 45 and 145
Holotype AMNH 45313, paratype AMNH 45314

Wanakah Member, Hamilton Group, Genesee County. Plate 45 illustrates the paratype. *Bellacartwrightia whiteleyi* has pygidial lappets, which are covered with small pustules and are semicircular in cross section. The axial lappet is narrow and sharply pointed.

Bellacartwrightia **sp. Plate 46**
USNM 89959

This undescribed species from the Tully Limestone is characterized by the very broad anterior border and the triangular pygidium.

Bellacartwrightia? **sp. Plates 47 and 48**

The specimen in Plate 47 cannot be identified by the internal exoskeletal anatomy. It was prepared to show the internal exoskeletal characteristics. The illustrated specimen in Plate 48 is undescribed but has some characteristics similar to *B. whiteleyi.* It differs in the broad shape of the pygidium and wider spacing between the lappets.

Asteropyginae aff. *Greenops* Plate 49

The specimen in Plate 49 is a small trilobite from the Pompey Member in central New York.

Greenops barberi **(Lieberman and Kloc, 1997) Plate 50**
Holotype PRI 41947, paratype AMNH 45277

Upper Wanakah and Windom Members, Hamilton Group, Erie County. The pygidial lappets on *G. barberi* are relatively long and narrow, and both the anterior and posterior edges are curved posteriorly. The axial lappet is narrow, relatively long, and triangular. Plate 50 illustrates the paratype.

Greenops boothi **(Green, 1837) Plate 51**
Holotype lost, neotype YPM 35807 Lieberman and Kloc, 1997

Type species. The long lost type for *G. boothi* is from the Middle Devonian Frame Member of the Hamilton Shales of central Pennsylvania. In describing the New York species, Hall accepted all asteropygins with short, rounded pleural lappets as *G. boothi* and all with long, spinelike lappets as *G. calliteles. Greenops "boothi"* is a common fossil of the Hamilton Shales, primarily as whole cephalon and pygidial exuviae. Whole specimens are infrequently encountered. Along with *Eldredgeops rana*, *G. "boothi"* is one of the trilobites found in the dark shales of the Hamilton, indicating their capability to live in poorly oxygenated waters. In the Ledyard pyrite beds, Genesee County, one finds whole, pyritized *G. "boothi"* specimens. *Greenops "boothi"* is reported from the Upper Devonian beds of south-central New York, Tompkins County, but little work has been done on these occurrences. In the NYSM there is an exfoliated pygidium from an acastid, possibly *Greenops*, from the Upper Devonian Rhinestreet or Cashagua Shale, Erie County. The Pennsylvania trilobite shown in Plate 51 (the neotype from the type locality) is included for comparison. Prior to the paper by Lieberman and Kloc (1997), all the New York acastids were lumped under *G. boothi* or *G. calliteles. Greenops boothi* has a double row of perforations on each of the thoracic pleural segments and a broad axial pygidial lappet with a triangular termination.

Greenops grabaui **(Lieberman and Kloc, 1997) Plates 52, 53, and 54**
Holotype BMS E25857

Centerfield, Ledyard, and Wanakah Members of the Hamilton Group in Genesee County. The pygidial lappets of *G. grabaui* are short and broad, with the anterior edge curved posteriorly and the posterior edge straight. The terminal lappet is short, broad, and subrectangular. Plate 53 illustrates a meraspid (degree 10) of this common *Greenops* species.

Greenops **sp. Plates 55 and 56**
Undetermined species from the upper Hamilton.

Kennacryphaeus harrisae **(Lieberman and Kloc, 1997)**
Holotype AMNH 45298

Mottville Member and Stafford Limestone, Hamilton Group, Onondaga County.

Family Calymenidae

The family Calymenidae arose in the Early Ordovician in Europe and migrated to North America during the early Middle Ordovician. In general, the cephalon is semicircular, usually with no genal spines. The glabella is prominently convex and narrows anteriorly, yielding an outline that can range from bell shaped to parabolic. There is usually a distinct preglabellar area, or "lip," the shape of which has some species significance. There are three or four distinctive lateral glabellar lobes. The posterior pair of lobes, L1, are the largest. The pygidium is somewhat triangular to semicircular with a smooth edge. The North American calymenids have 13 thoracic segments. The exoskeleton of the calymenids is robust, and clearly recognizable pieces are often found in fossil lag deposits. Because of this, and their presence in a wide variety of paleoenvironments, calymenids share with *Isotelus* species the distinction of being the most common Middle Ordovician trilobites encountered in the field. Edgecombe and Adrain (1995) reviewed the Silurian calymenids of the United States.

Calymene camerata (Conrad, 1842) Plate 57
Hypotypes NYSM 4163 to 4165

Calymene camerata is reported from the top of the Upper Silurian Cobleskill Limestone in the Schoharie Valley, Schoharie County. This trilobite is characterized by the deep furrow in front of the glabella. This feature is interpreted to indicate that the preglabellar area is steeply upturned, similar to the familiar *Flexicalymene meeki*.

Calymene? conradi (Emmons, 1855)
Specimen USNM 78402

This species was described by Emmons to replace the use of *Flexicalymene senaria* for the Upper Ordovician Lorraine Shale calymenids of New York. It was described as follows: "Small, wide across the cheeks, cheek angles obtuse or rounded; posterior lobes of the glabella comparatively large and globular; thoracic lobes very convex, with a row of tubercles in the furrow or between the axis and the lateral lobes." It was found in the Lorraine Shales from an indeterminate location in central New York.

Calymene niagarensis (Hall, 1843) Plates 58 and 59
Holotype AMNH 31063

This species is from the Lower Silurian Rochester Shale, although it has been reported from rocks above and below. *Calymene niagarensis* has a bell-shaped glabella. The anterior border to the cephalon is short and strongly convex. The preglabellar area is about 10% of the cranidial length. The pygidium has seven completely defined axial rings. The exoskeleton is covered with dense, fine granules. *Calymene niagarensis* is the best known of the Lower Silurian calymenids because the quarrying activities in the Rochester Shale near Middleport in Orleans County have uncovered large numbers. The original material was undoubtedly from the digging of the Erie Canal in Lockport, Erie County. There are at least three calymenids in the Rochester Shale, and *C. niagarensis* is easily the most common. Plate 58 shows an individual of the species. Plate 59 shows a cluster of *C. niagarensis* specimens from Middleport, once again demonstrating the gregarious nature of some trilobites. Clusters of this kind have been described as molting or breeding assemblages.

Calymene platys (Green, 1832) Plate 60
Plastotype NYSM 4168

Calymene platys represents the latest occurrence of calymenids in New York. It is found in the top of the Lower Devonian Schoharie Grit and the base of the Onondaga in Schoharie, Albany and Erie Counties. The cast of Green's type in the USNM is pustulose. The specimen shown in Plate 60, from the Onondaga of Ontario, has a smooth surface; however, this may be due to weathering.

Calymene singularis (Howell and Sanford, 1947) Plate 61
Holotype USNM 488139

The type for this calymenid is from the Upper Silurian "Oak Orchard Member" of the Lockport Group in Monroe County. It was originally described as *Cheirurus singularis*. Holloway (1980) recognized it as a calymenid.

Calymene sp. Plate 62

This unidentified *Calymene* species is from the eastern Rochester Shale in Wayne County. It is characterized by smaller eyes, a more tapered anterior portion of the glabella, and the ornamentation on the cephalon.

Calymenid sp. Plate 63

There are at least two calymenids of the Upper Ordovician found in New York. They have been referred to as *Flexicalymene meeki* and *F. granulosa*. More work needs to be done on these specimens

Calymenid sp.
Specimen OSU 22763, 4

Leutze (1961) illustrated specimens collected from the *Camarotoechia* zone in the Upper Silurian Syracuse Formation in Onondaga County. The specimens are too fragmentary for specific identification. In Maryland in the Will Creek and Toholoway Formations, there are calymenid species that may be the same (Edgecombe and Adrain 1995).

Diacalymene rostrata (Vogdes, 1879)
Lectotype SDSNH 868

The type is from Georgia. *Diacalymene rostrata* is reported from the Lower Silurian, lower Clinton in Oneida County (Gillette 1947). Vogdes wrote: "Distinct projecting process in front of the glabella. The facial lines cut the anterior border at the apex, giving the frontal limb a triangular form; at their juncture the marginal border is raised and forms a triangular process which supports the projection." There is no basis to distinguish New York material as different from that from Georgia. The anterior border of the cephalon is moderately long and comes to a point. The pygidial axis has eight or nine distinct interring furrows and a long terminal piece. See Edgecombe and Adrain (1995) for a recent description of the species.

Diacalymene vogdesi (Foerste, 1887)
Holotype USNM 84790a

This Silurian trilobite from Ohio has been reported from the Lower Silurian Rochester Shale of New York. This occurrence cannot be verified. See Edgecombe and Adrain (1995) for more information.

Diacalymene sp. Plates 64, 65

Two undescribed *Diacalymene* species are known from the Rochester Shale in Monroe and Wayne Counties. The one illustrated in Plate 64 has a triangular anterior cephalic margin. The other, in Plate 65, has a more rounded margin. These will be described as different species.

Flexicalymene granulosa **(Foerste, 1909)**
Hypotype NYSM 9649

Ruedemann (1926) identified pustulose calymenids found in the Upper Ordovician Whetstone Gulf Shale in Jefferson County as *Calymene granulosa*. Foerste (1924) described *C. granulosa* as follows: "size, relatively small. Anterior border of the cephalon less abruptly elevated than in *C. meeki*. Surface covered by numerous granules, larger and more conspicuous than those in the latter species." Hughes and Cooper (1999) illustrated *Flexicalymene* cf. *F. granulosa* from Kentucky.

Flexicalymene meeki **(Foerste, 1910) Plate 66**
Holotype USNM 78822 (Ohio), hypotype NYSM 9655, specimen USNM 23627

Foerste described *Flexicalymene meeki* as follows: "glabella relatively short, with a tendency towards truncation anteriorly. Anterior border of cephalon turned up abruptly and separated from the front of the glabella by a narrow groove. Genal angles with a short, acute, pointed extremity. Ribs of the pygidium with only a very faint trace of an impressed zone along their median line." The USNM specimen is from the Upper Ordovician "Pulaski Drift" from north of Utica, Oneida County. Similar material is in the NYSM and the MCZ. The specimen in Plate 66 is from Cincinnati, Ohio, the type area for the species.

Flexicalymene senaria **(Conrad, 1841) Plates 67, 68, 69, and 70**
Neotype AMNH 29474 (Ross 1967)

Flexicalymene senaria was one of the earlier New York trilobites recognized as being different from its European calymenid counterparts. It is found nearly everywhere that the Middle Ordovician limestones are exposed in New York. The neotype, Plate 67, is from Middleville, Herkimer County, and is probably from the lower Trenton. The illustrated specimen in Plate 68 is from the lower Poland Limestone from Trenton Falls, Herkimer County. The trilobite in Plate 69 is from the Walcott-Rust Quarry in the Rust Formation of the Trenton Group; Ross (1967) suggested this was a separate species owing to its unusually pustulose appearance. The specimen in Plate 70 is from the uppermost Trenton, the Hillier Member, and is less pustulose. It is interesting, however, that in spite of the number of trilobite workers splitting off new species from early New York discoveries, it was not until 1926 that Ruedemann recognized a "subspecies" of *F. senaria*. In 1967 Ross similarly pointed out that there are at least four different trilobites in New York called *F. senaria*. Revision of this group is presently in progress.

Flexicalymene **sp. (Ross, 1967)**
Specimen NYSM 17004

Ross identified the specimen from the Walcott-Rust Quarry in the Rust Limestone Member of the Middle Ordovician Trenton Group, Herkimer County. This trilobite, part of the revision mentioned above, has a very pustulose exoskeleton and short genal spine on the fixed cheeks. The facial sutures are proparian. This species has been observed throughout the Rust Limestone in the Trenton Falls area.

Gravicalymene magnotuberculata **(Ruedemann, 1926) Plates 71 and 72**
Holotype NYSM 9650

This Middle Ordovician trilobite, first figured by Ruedemann as *Flexicalymene senaria*, was given subspecies rank in 1926. Ross (1967) pointed out in some New York calymenids that not only was the glabella bell shaped in outline (vs. the glabella in *F. senaria* being parabolic) but also the anterior border was thickened and rolled compared to the standards for *Flexicalymene*. Ross did not, however, mention Ruedemann's subspecies, and the assignment as *Gravicalymene* is based on Ross's general observations for the species and our examination of specimens. The genal angle is very rounded. The facial suture is gonatoparian. Ruedemann's specimens came from the Trenton age, Canajoharie Shales, and we have observed the same trilobite in Dolgeville-like facies within the Sugar River Formation of the Trenton. Plates 71 and 72 illustrate the dorsal and ventral exoskeletal surfaces.

Liocalymene clintoni **(Vanuxem, 1842) Plate 73**
Hypotype MCZ 104689

The location of the original type is unknown. *Liocalymene clintoni* is listed from the Lower Silurian, lower and middle Clinton Group in Oneida County (Dale 1953). The trilobite has a bell-shaped glabella with very large first glabellar lobes (L1). The glabellar lobes are much less inflated than those in *Calymene* species. It is easily distinguished from other calymenids in that the pleural region of the pygidium is smooth, without pleural grooves. See Raymond (1916) and Whittington (1971a).

Liocalymene cresapensis **(Prouty, 1923)**
Lectotype USNM 164064 (Edgecombe and Adrain 1995)

Liocalymene cresapensis was originally described from the Silurian of Maryland. Dale (1953) reported the trilobite from the Lower Silurian, upper Clinton Group shale of Oneida County.

Family Cheiruridae

Cheirurids are common trilobites of the Middle Ordovician of New York. They are also more rarely found in the Silurian. Generally, the glabella expands forward and there are genal spines. The thoracic pleurae have short spines, although sometimes these are blunt. There is an oblique furrow on the thoracic pleura next to the axis. The pygidium is small and usually with marginal spines. Pribyl, Vanek, and Pek (1985) extensively reviewed and modified the cheirurids in a work that is not readily available in most libraries. The genera referred to in the following are from their paper.

Acanthoparypha trentonensis
Holotype NYSM 4766

Raymond (1925) initially identified *A. trentonensis* as *Kawina*, with two cranidia from the Middle Ordovician material in the Walcott collection (MCZ), probably from Herkimer County. Raymond (1925, p. 145) additionally listed the species from the Black River of New Jersey.

***Acanthoparypha?* sp. (Shaw, 1968)**
Specimens NYSM 12435 to 12438

This Middle Ordovician Chazy trilobite is only known from a few pieces. For location information, see Shaw (1968).

***Ceraurinella (Ceraurinella) latipyga* (Shaw, 1968) Plate 86C**
Holotype NYSM 12442

A Middle Ordovician Chazy trilobite that was first described by Shaw. For location information, see Shaw (1968).

***Ceraurinella (Arcticeraurinella) scofieldi* (Clarke, 1894) Plate 74**
Hypotype MCZ 100814

The MCZ specimen is from the Middle Ordovician Black River Limestone near Poland, Herkimer County, and is part of the Walcott collection. This trilobite is also reported from the Chaumont Limestone in the Watertown Town Quarry, Watertown, Lewis County. The original described material is from the Midwest; see DeMott (1987).

***Ceraurinus marginatus* (Barton, 1913) Plate 75**
Holotype MCZ 100813

Type species. The holotype is from the Hillier Member, Middle Ordovician Trenton Group, Jefferson County. This is a reasonably common trilobite from the Cobourg of southern Ontario but rarely reported in New York. There are unnumbered specimens in the USNM, three small pieces and a very large glabella, 2.3 cm long. They are from 3 m (10 feet) beneath the top of the Middle Ordovician Trenton on the North Fork of Sandy Creek, 7.2 km (4.5 miles) south of Adams, Jefferson County. The specimen in Plate 75 is from the Trenton age rocks of Ontario.

***Ceraurinus* sp.**
Specimen USNM 92536

The two cranidia from the Middle Ordovician Black River, Herkimer County, are part of the Hurlburt collection and are quite possibly from the same exposures, Buck's Quarry, from which Walcott got most of his Black River trilobites. An unnumbered specimen in the USNM is from the Middle Ordovician Lowville (Black River) or basal Trenton at a now-abandoned railroad cut 3.2 km (2 miles) below Poland, Herkimer County.

***Ceraurопeltis ruedemanni* (Raymond, 1916) Plate 84C–E**
Holotype NYSM 9689, hypotype MCZ 101171

Type species. *Ceraurопeltis ruedemanni* from the base of the Middle Ordovician, upper Chazy resembles *C. pleurexanthemus* with the following differences. A crease joins the proximal ends of the glabellar furrows, giving the glabella a trilobed appearance, and the second pleural spines on the pygidium are elongated. All of the known material is from a single location (Shaw 1968). The fixed cheek has a wrinkled look. Specimens are huge. One in the MCZ has a glabella 3.7 cm long.

***Ceraurus montyensis* (Evitt, 1953)**
Holotype USNM 116693a

This Middle Ordovician Trenton cheirurid from northeastern New York near the town of Chazy, Clinton County, is very similar to *C. pleurexanthemus*. The silicified material Evitt studied has a less elevated eye on the free cheek, and the pygidium appears to be more scalloped or dentated than that of *C. pleurexanthemus*.

***Ceraurus pleurexanthemus* (Green, 1832) Plates 76, 77, and 78**
Holotype NYSM 4203; specimens MCZ 111708, MCZ 111715

Type species. *Ceraurus pleurexanthemus* is a common and widely distributed trilobite of Middle Ordovician Trenton age rocks in central New York. The distinctive cephalon with its long divergent genal spines and the two long marginal pygidial spines make it easy to identify. Articulated specimens are uncommon in most exposures, as it was not as robust as *Isotelus gigas* or *Flexicalymene senaria* found in the same rocks. The exception is the Walcott-Rust Quarry, Herkimer County, where large numbers of whole, articulated specimens of all sizes are found in and on a single layer. This same layer yielded specimens with appendages, which Walcott used in his first description of trilobite appendages. Plate 76 provides an excellent example of the species. Plate 77 shows the ventral exoskeletal anatomy. The apodemes of Plate 77 are clublike at the ends. The specimen in Plate 78 shows a ferriginous trace of the gut from the midpoint of the thorax to the end of the pygidium.

***Ceraurus?* sp. Plates 79 and 80**

The specimen shown in Plate 79 is from a flooding-surface lag deposit over the Middle Ordovician Hillier Limestone Member of the Trenton Group and under the Upper Ordovician Utica Shale. The species has some distinct differences from *C. pleurexanthemus*. Plate 80 illustrates another unidentified *Ceraurus*? from a quarry in the middle Trenton, near Watertown, Jefferson County. These, and some of the museum materials designated as *Ceraurus*, need to be redescribed. The specimens differ from *C. pleurexanthemus* in the larger eye, the more prominent tubercles on the glabella, and inward-curving genal spines.

***Cheirurus* sp. (Hall, 1867) Plate 81**
Specimen NYSM 9693

Lower Silurian Rochester Shale. The cheirurid from the Rochester Shale was previously assigned to the cheirurid *Hadromeros niagarensis*, a Wisconsin species. Holloway (1980) noted that the New York specimens are of the genus *Cheirurus* and are different from the Wisconsin specimens. The trilobite has a long, inflated glabella that expands toward the front where it overhangs the anterior border. The glabellar furrows are straight, and the basal lobe is isolated. The pygidium has a single median spine and three pairs of long, curved, pointed lateral spines.

***Deiphon pisum* (Foerste, 1893) Plate 84B**
Specimen USNM 88648

The USNM specimen is from Lockport, Niagara County, and labeled *D. forbesi*, an English trilobite not known from New York. The glabella is greatly inflated and covered with pustules. The fixigenae are reduced to long genal spines. This genus is not benthic but lived in the water column (pelagic).

***Forteyopsl(?) approximus* (Raymond, 1905)**
Holotype YPM 23300

This trilobite is known from two cranidia from the Middle Ordovician Chazy Group. For location information, see Shaw (1968).

***Gabriceraurus dentatus* (Raymond and Barton, 1913) Plates 82 and 83**
Holotype GSC 1775 is lost; paratype MCZ 100802

The paratype is from Watertown, Jefferson County. *Gabriceraurus dentatus* is a large cheirurid from the Middle Ordovician Trenton. Specimens are found mostly in the lower Trenton but have been collected well up into the group. It is nowhere as common as *C. pleurexanthemus*. It is distinguished by its significantly larger size and the eyes, which are set farther back on the cranidium than they are in the more common *C. pleurexanthemus*. This trilobite is also found in the Bobcaygeon Formation of Ontario and in the upper Esbataottine Formation, District of Mackenzie, Canada. Specimens are also found in the Midwest; see DeMott (1987). The specimen in Plate 83 is from the upper Trenton Lindsay Formation in Ontario and is flattened, as one often finds specimens on bedding planes. Plate 82 illustrates a specimen from the lower Trenton of Fulton County with a more three-dimensional aspect.

***Gabriceraurus hudsoni* (Raymond, 1905) Plate 84A**
Holotype CM 1271

Shaw (1968) identified *G. hudsoni* as the only cheirurid of the genus *Ceraurus* in the Middle Ordovician Chazy and placed the other two species reported, *Ceraurus pompilius* Billings, Raymond, 1905, and *C. granulosus* Raymond and Barton, 1913, in synonymy. Later investigators placed this trilobite in the genus *Gabriceraurus* (Pribyl et al. 1985). *Gabriceraurus hudsoni* differs from *C. pleurexanthemus* in that the former has raised tubercles on the fixed cheeks, whereas the latter free cheek is pitted. For location information, see Shaw (1968). Raymond (1925, p. 139, 140) reported the species also from Virginia and Tennessee.

***Heliomera (Heliomeroides) akocephala* (Shaw, 1968)**
Holotype YPM 23326

Heliomera akocephala is an uncommon trilobite from the Middle Ordovician, uppermost lower and lowermost middle Chazy. Only cranidia are clearly identified. For location information, see Shaw (1968).

***Kawina? chazyensis* (Raymond, 1905)**
Holotype CM 1277

This trilobite is known only from cranidia in the Middle Ordovician Chazy Limestone. For location information, see Shaw (1968).

***Kawina vulcanus* (Billings, 1865)**
Lectotype GSC 669 (Whittington 1963)

Type species. *Kawina vulcanus* is identified as a New York trilobite from a single specimen, now lost, described by Raymond (1905). The type is from Newfoundland. Shaw (1968) questioned that this is a Middle Ordovician Chazy trilobite.

***Nieszkowskia? satyrus* (Billings, 1865)**
Holotype GSC 1087

This distinctive cheirurid of the Middle Ordovician Chazy Limestone has a raised, cone-shaped glabella similar to those on some lichids. For location information, see Shaw (1968).

***Sphaerexochus (Parvixochus) parvus* (Billings, 1863)**
Holotype GSC lost, hypotype GSC 1330

Type species. *Sphaerexochus parvus* is widely distributed in the Middle Ordovician, lower and middle Chazy Limestone of the Champlain Valley. The small trilobite has an inflated glabella and long genal spines. For location information, see Shaw (1968).

***Sphaerocoryphe goodnovi* (Raymond, 1905) Plate 86A**
Lectotype YPM 23304 (Shaw 1968)

Sphaerocoryphe goodnovi, from the Middle Ordovician Chazy, as with all trilobites of this genus, has a highly inflated glabella, which stands well above the plane of the body. The specimen figured by Shaw (1968) differs from others in the genus in that the glabella appears oval in a dorsal view, while other New York *Sphaerocoryphe* species have a spherical glabella. For location information, see Shaw (1968).

***Sphaerocoryphe major* (Ruedemann, 1901)**
Holotype NYSM 4813

This is one of the unusual trilobites found in Middle Ordovician Trenton pebbles within the Rysedorph Hill Conglomerate,

Rensselaer County. It is larger than *S. robusta* but resembles it in general.

Sphaerocoryphe robusta **(Walcott, 1875) Plate 85**
Holotype MCZ 111709, specimen MCZ 110901

Walcott (1875b) first described this small trilobite from his collecting in the Walcott-Rust Quarry, Herkimer County, located in the Middle Ordovician Rust Member, Trenton Group. The trilobite is unmistakable, with its spherical glabella raised well above the plane of the body. It is not common in the quarry, but the work by Walcott and later by Rust and Hurlburt resulted in it being represented with fine specimens at MCZ, NYSM, PRI, and USNM. In New York *S. robusta* is found on only one bedding plane in one location, the Walcott-Rust Quarry. This is probably due to its very small size and low concentrations, requiring a significant area to be searched to be successful. The distinctive glabella is also found in the Verulam (Trenton) of Ontario.

Staurocephalus **sp. Plate 91B**

Undescribed specimens have been found in the Rochester Shale. Presently they are only known from their free cheeks.

Family Dalmanitidae

Dalmanitids are characterized as follows: They have flattened bodies. The genal spines are often present. The abaxial end of the thoracic segments is well defined, often forming small, pointed lappets. The glabellar lobes are well defined. Dalmanitids have large cephala and pygidia and schizochroal eyes, which in most cases are large. In New York they are found in the Silurian through the Middle Devonian. Many trilobites listed in the *Treatise* as dalmanitids were assigned to the family Synphoriidae (Lespérance 1975). A defining difference between Dalmanitidae and Synphoriidae is the ratio of the distance from the cephalic apodemal pit SO and S1 and the distance between S1 and S2 (Figure 5.2).

Corycephalus? pygmaeus **(Hall and Clarke, 1888)**
Cotypes NYSM 4357 to 4358

This species is known from only two very small fragmentary cephala from Middle Devonian Onondaga float boulders near Canandaigua, Ontario County. The two have an elongated glabella, and crenulations can be seen on the margins. The fragments are probably juveniles and even possibly meraspids.

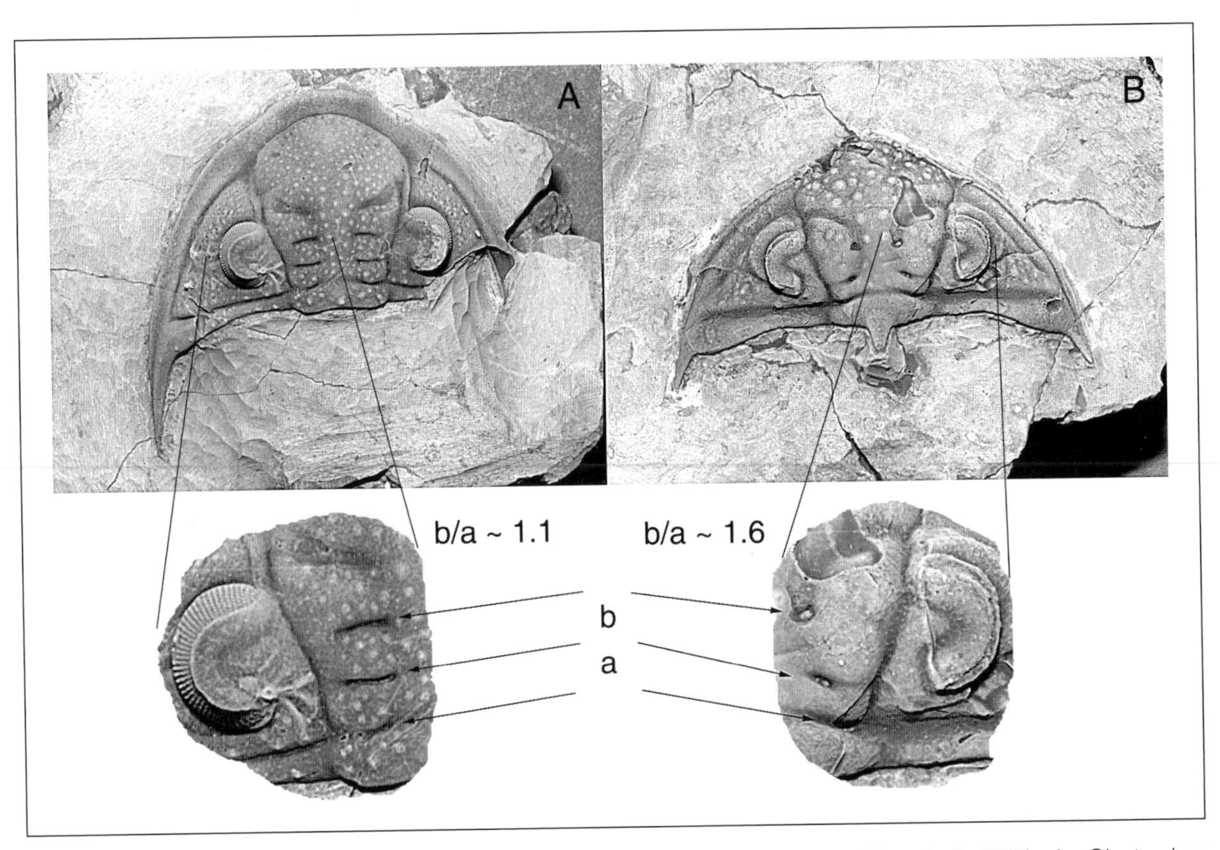

FIGURE 5.2. Key cephalic differences between Dalmanitidae and Synphoriidae (Campbell 1977). A. *Glyptambon amsdeni* from Tennessee, a dalmanitid, where the ratio in distance between the apodemal pits at S2-S1 and S1-SO is about 1.1. This trilobite was used because the apodemal pits show up well in the photograph. B. *Anchiopella anchiops*, a synphoriid, from Ontario, Canada, where the ratio in distance between the apodemal pits S2-S1 and S1-SO is about 1.6. This trilobite is also found in New York. Both specimens are in the Gerald Kloc collection.

Corycephalus regalis **(Hall, 1876) Figure 5.3A**
Holotype AMNH 2894, paratype AMNH 29247

This trilobite, of the Schoharie Grit, Albany County, is distinguished by the high convexity, large size, long genal spines, spatulate marginal denticles, and coarse tuberculation.

Dalmanites aspinosus **(Weller, 1903) Plate 87**
Holotype Walker Museum 10242

This Silurian trilobite is from the Decker Ferry Formation. It was found in the Nearpass Quarry 3.2 km (2 miles) south of Tristates, Orange County.

Dalmanites bisigmatus **(Clarke, 1900)**
Holotype NYSM 13367/1

Lower Devonian Oriskany/Glenerie Formation. Only the pygidium has been characterized, although there is a glabella with this label in the USNM (USNM 52699). The pygidium is triangular with 14 or 15 axial rings. There are prominent axial tubercles. There are 10 or 11 pairs of pleural ribs, which curve backward near the margins.

Dalmanites limulurus limulurus **(Green, 1832) Plates 88, 89, and 99**
Neotype NYSM 13381

Dalmanites limulurus limulurus is common in the Lower Silurian Rochester Shale of western New York in Niagara, Orleans, and Monroe Counties (Tetreault 1994). The cephalon has a distinct flattened border and long tapering genal spines. The pygidium is triangular; the ribs are almost equal in width and terminate in a long slender spine. This species is possibly the most common trilobite of the Rochester Shale. The exoskeleton is thin and often dissolved away. This subspecies is found in the western exposures while *D. l. lunatus*, with a shorter pygidial spine, is found in the more eastern exposures of the Rochester Shale. Both the specimens illustrated in Plates 88 and 89 are from the same quarry in Orleans County.

Dalmanites limulurus lunatus **(Lambert, 1904)**
Cotypes USNM 50459, 79120

This subspecies is similar to *D. l. limulurus*, but the frontal process is wider and longer, the cheeks are narrower, the terminal spine width and length are equal, and the posterior rib branches are larger. This trilobite is found farther east in the Rochester Shale, in Oneida County, than the more common *D. l. limulurus*.

Dalmanites (Synphoroides?) pleuroptyx **(Green, 1832) Plate 90**
Plastotype USNM 62486

This trilobite from the Lower Devonian Helderberg Group is listed from both the New Scotland Beds and the Kalkberg Limestone in Albany and Schoharie Counties. The cephalon has a broad crenulated frontal expansion and genal spines. The pygidium is triangular with 10 or 11 furrowed pleurae and 15 axial rings with axial nodes.

Dalmanites **sp. Plate 91A**

The specimen is from Lower Silurian Rochester Shale found in an unnamed creek in the Sodus area, Wayne County.

Forillonaria cf. F. russelli **(Lespérance, 1975)**
Specimen AMNH unnumbered

The referenced pygidium is from the Lower Devonian Central Valley Sandstone, in Orange County, and is the only specimen known from outside Quebec. There are 12 pleural ribs, which reach the margin. There are 20 axial rings and a small postaxial ridge. The surface is granular.

Neoprobilium nasutus **(Conrad, 1841) Figure 5.3B**
Hypotypes NYSM 4342 to 4347

Type species. This species is from the Lower Devonian New Scotland Formation. The cephalon has a long anterior, median process that terminates in two divergent rounded spines. The genal spines are long and slender. The pygidium is subtriangular with 13 or 14 axial rings and 10 or 11 pleural ribs, which bend posteriorly toward the margin. There is a terminal spine equal or greater in length than the pygidium. The spine is narrow, rounded in cross section, and comes to a sharp point. The surface of the pygidium is granulose.

Neoprobilium tridens **(Hall, 1859) Figure 5.3C**
Type AMNH 2608, hypotype YPM 6659

Only cephala are known of this species from the Lower Devonian New Scotland Formation. The cephala have a long anterior process that ends in two rounded divergent spines separated by a short, blunt, triangular process.

Odontochile litchfieldensis **(Delo, 1940)**
Cotype USNM 78089

Lower Devonian, Sallsburg Quarry, Litchfield, Herkimer County. Only two pygidia are known. They are as wide as long with a short acute termination. There are 17 complete axial rings and 2 or 3 incomplete ones. Each ring bears an irregular row of tubercles. There are 14 pleural furrows, each lobe with a row of tubercles along the crest. The border is granulose.

Odontochile micrurus **(Green, 1832) Plate 92**
Holotype NYSM 13385

This species from the Lower Devonian Coeymans Limestone in Herkimer County may be part of the genus *Huntonia* (Campbell 1977). The surface of the cephalon is ornamented with small tubercles. There are moderate-sized genal spines. The pygidial axis has 17 rings and there are 13 or 14 pleural ribs. The pygidium terminates in a short spine.

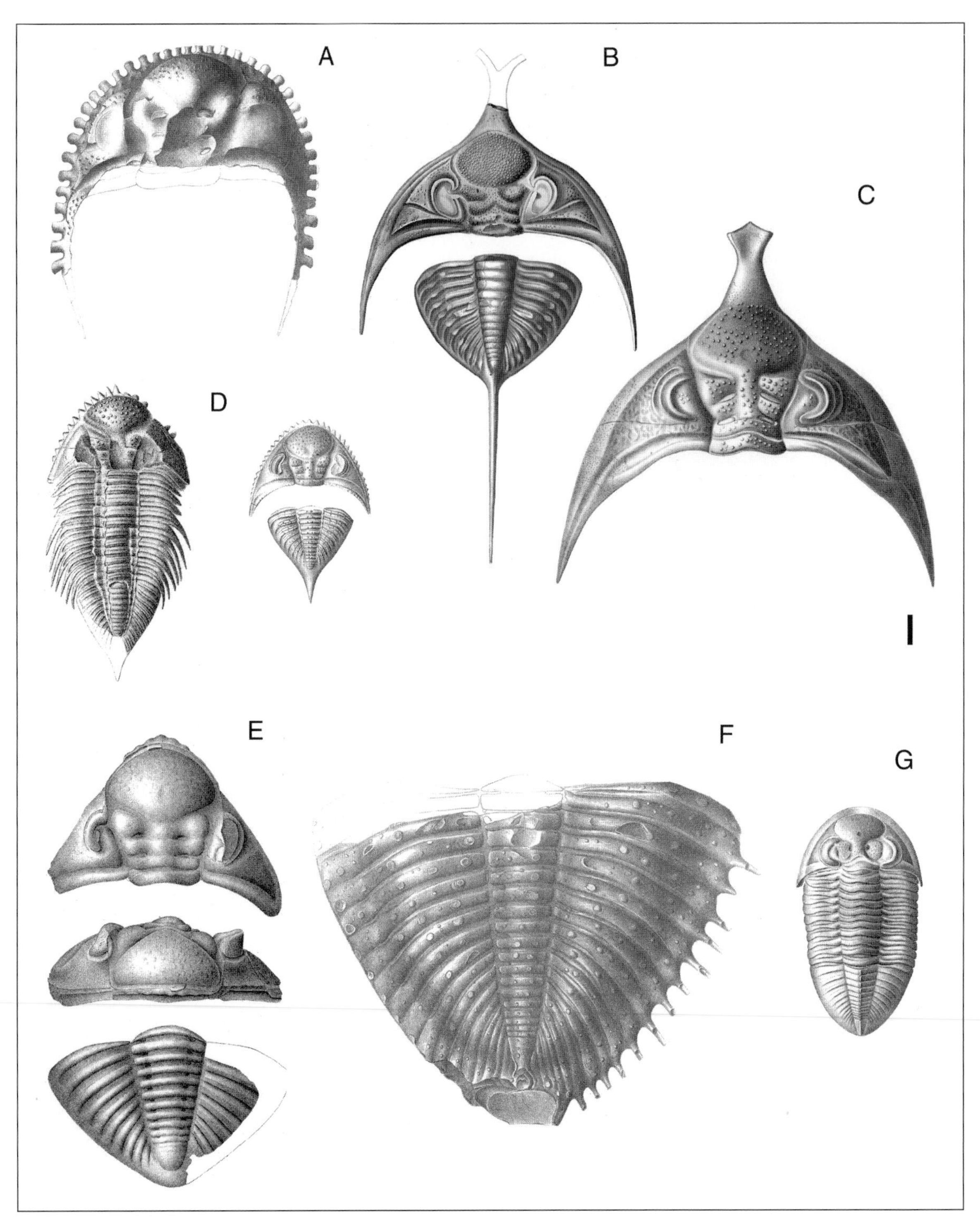

FIGURE 5.3. Trilobites of the families Dalmanitidae (A–D) and Synphoriidae (E–G) from the Lower and lower Middle Devonian. A. *Corycephalus regalis* from the "Schoharie Grit" of Albany County. The figured cephalon is from Hall and Clarke (1888). B. *Neoprobilium nasutus* from the New Scotland Formation. The figured cephalon and pygidium are from Hall (1861b). C. *Neoprobilium tridens* from the New Scotland Formation. The figured specimen is from Clarke (1908) and is YPM 6659. D. *Phalangocephalus dentatus* from the Port Jervis Limestone in Orange County. All three figures, a nearly complete trilobite and a cephalon and pygidium, are from Hall and Clarke (1888). E. *Synphoria stemmata stemmata* from the Oriskany/Glenerie Formations, Columbia County. The three specimens, a cephalon in dorsal and front view and a pygidium, are from Clarke (1900). F. *Coronura myrmecophorus* from the Onondaga Limestone at Kingston, Ulster County. The pygidium figured is from Hall and Clarke (1888). G. *Trypaulites calypso* from the Columbus Limestone, Sandusky, Ohio. The trilobite is also reported from New York. The illustration is from Hall and Clarke (1888).

Odontochile phacoptyx (Hall and Clarke, 1888)
Type NYSM 4351

Only the pygidium is known from this Lower Devonian Oriskany trilobite. Delo (1940) also listed this trilobite from the lower Onondaga of Canada. The pygidium is triangular with a terminal spine. There are 13 axial rings and 16 pleural ribs. The surface is covered with spines or tubercles, which are more closely spaced on the pleura. There are tubercles of two sizes on the axial rings: the smaller are more abundant and the larger entirely obstruct the furrows.

Odontochile? sp. Plate 93

The specimen is from the Lower Devonian, upper Manlius Formation, Cobleskill, Schoharie County.

Phalangocephalus dentatus (Barrett, 1876) Figure 5.3D
Hypotypes NYSM 4310 to 4311

Type species. This species is characterized by the numerous triangular marginal spines. Pieces of this trilobite are so abundant in the Port Jervis Limestone southeast of Port Jervis in Orange County that the locality has been referred to as Trilobite Mountain. The cephalic border has 37 spines; the longest is medial and the shortest is on the flank of the genal spine.

Family Encrinuridae

Encrinurids are generally small pustulose trilobites. The glabella has near-parallel sides or expands forward. The thorax has 10 to 12 segments. The thoracopygidium is generally well tapered and may appear subtriangular. The small pygidium usually has very-well-defined pleural ribs, and the pygidial axis has significantly more lobes than there are pleural ribs.

Cybeloides ella (Raymond and Narraway)
Specimen MCZ 5093

The listed specimen is from the Middle Ordovician Black River Group at Buck's Quarry northeast of Poland, Herkimer County.

Cybeloides prima (Raymond, 1905) Plate 86B
Lectotype CM H1286 (Shaw 1968)

Cybeloides prima is a small encrinurid trilobite of the Middle Ordovician, lower and middle Chazy. It apparently is geographically restricted to the area of Valcour and the southern end of Valcour Island, Clinton County. The cephalon has long curved genal spines and a medial occipital spine. The pygidium is distinctive, as the anterior pleurae curve posteriorly and back toward the axis. Specimens described by Shaw are silicified remains. No articulated specimens are known.

Encrinurus cf. *E. raybesti* (Edgecombe and Chatterton, 1993) Plate 95
PRI 49628

Lower Silurian, Reynales Limestone, Niagara County. This limestone is much older than that containing *E. raybesti.* The poor state of preservation of the specimens available to Edgecombe and Chatterton made proper identification impossible. A specimen now at the PRI (PRI 49628), illustrated in Plate 95, suggests a new species may be involved (Greg Edgecombe 1999, private communication).

Encrinurus sp. Plate 96

The unidentified specimens in Plate 96 are from the Lockport Group of Wayne County.

Erratencrinurus vigilans (Hall, 1847) Plate 94
Lectotype AMNH 36070 (DeMott 1987), specimen NYSM 17008

Erratencrinurus vigilans is found in the lower half of the Middle Ordovician Trenton Group, particularly in Herkimer County. It is a small very pustulose trilobite with a highly tapered pygidium. *Erratencrinurus cybeleform* (Raymond 1921), listed by Fisher (1965) from the Middle Ordovician, lower Trenton Larrabee Limestone, and *E. trentonensis* (Walcott 1877), described with material from Wisconsin and Illinois, are junior synonyms of *E. vigilans.* There is unnumbered material in the USNM, all from the Middle Ordovician, lower Trenton, labeled *Encrinurus trentonensis.* Also the ROM has an entire specimen (ROM 18735) so labeled.

Physemataspis insularis (Shaw, 1968) Plate 86D
Holotype NYSM 12331

Physemataspis insularis is found in the Middle Ordovician, lower middle Chazy throughout the Champlain Valley. For location information, see Shaw (1968). It is a small trilobite described from silicified material with very good detail. The cranidium is very pustulose on the glabella and palpebral area, while the genal area is pitted. It has a spherical glabella and eyes raised on stalks well above the plane of the body.

Family Homalonotidae

The homalonotids as a family are known from the Lower Ordovician to the end of the Middle Devonian. All currently known New York species belong to the subfamily Homalonotinae that is represented from the Lower Silurian through the Middle Devonian. The homalonotins are characterized by a largely featureless cephalon and a rounded, subtriangular pygidium. There are facial sutures, but the cephalon on the New York homalonotids is always found intact, indicating that the free cheeks were fused and not lost during the molting process.

Brongniartella trentonensis (Simpson, 1890)
Specimen NYSM 17017

This is a Middle Ordovician Trenton age trilobite from Pennsylvania. Although Simpson reported it, Collie (1902) first described it. This trilobite is found locally abundant in the Salona Formation limestones at the Rte. 322 bypass at State College, Pennsylvania. The position of these limestones is possibly equivalent to the Kings Falls Formation of the Trenton Group in New York. The Pennsylvania limestones were deposited in deeper water than those of New York, indicating that the presence of this trilobite is facies controlled. There is no reason why *B. trentonensis* might not be an occasional migrant into the New York rocks.

Dipleura dekayi **(Green, 1832) Plates 97 and 98**
Hypotypes NYSM 4478 to 4487

Type species. *Dipleura dekayi* is most often found in Middle Devonian, Hamilton siltstone deposits. In some areas the cephala and pygidia are common. Whole, articulated specimens are far less common, but in certain central New York mudstones they are quarried successfully. This much-sought-after trilobite has been found in significant numbers from the famous "Cole Hill Road beds" in Madison County. The particular specimen shown in Plate 97 is from more western exposures of Kashong Shale and shows healed damage on the right side. There is no other Hamilton trilobite that can be confused or mistaken for *D. dekayi. Dipleura dekayi* can be found in specimens up to 15.2 cm (6 inches) or longer. In size, however, it cannot compete with its Lower Devonian cousin (i.e., *Homalonotus major).* The specimen illustrated in Plate 98 shows large, spar-filled thoracic perforations that extend through to the internal mold.

Homalonotus major **(Whitfield, 1885)**
Hypotype NYSM 4493

Two large, incomplete specimens of a homalonotid were discovered in the upper part of the Lower Devonian Oriskany Sandstone near Kingston, Dutchess County. The larger specimen is in the NYSM and the smaller was given to a local, Kingston museum. Whitfield, when describing the new species, calculated that the smaller of the two, if complete, would have measured 39.4 cm (15.5 inches) long by 14 cm (5.5 inches) wide. Hall and Clarke (1888) pictured this specimen as it might be complete, and added the additional comments that the smaller specimen was from the lower Oriskany and the larger, type specimen was from the upper. We are not aware of any additional specimens.

Trimerus delphinocephalus **(Green, 1832) Plate 99**
Hypotypes NYSM 4490 to 4492

Type species. *Trimerus delphinocephalus* is a well-known trilobite of the Lower Silurian Rochester Shale. Gillette (1947), in his faunal list, restricted it in New York to the upper Clinton Group, and it is found across the state in outcrops of this age. In the Rochester Shale it is mostly found in the upper, deeper-water facies (Tetreault, 1994). This large inflated trilobite has a triangular cephalon and small eyes. The axial lobe on the thorax is very wide and faint. The pygidium has a distinct axial lobe, triangular and tapering to a point. This trilobite is rarely found whole; however, disarticulated pieces are not uncommon. The specimen shown in Plate 99 is accompanied by *Dalmanites limulurus. Trimeras delphinocephalus* is also found in the Silurian beds in England, suggesting that protaspis was pelagic.

Trimerus vanuxemi **(Hall, 1859)**
Type AMNH 2614/1,2

This large homalonotid is from the Lower Devonian, lower Helderberg and the Oriskany in Orange County. Only pieces are known, but Hall and Clarke's (1888) reconstruction 27.9 cm is (11 inches) long. It is not at all clear what differentiates this trilobite from *Homalonotus major.*

Family Phacopidae

Phacopids are the best known of all New York trilobites because of the widespread availability of *Eldredgeops rana* in New York Middle Devonian rocks. The family occurs from the Lower Silurian through the Upper Devonian, but the Lower and Middle Devonian trilobites are the best known. The glabella is rounded and expands forward. Phacopids have schizochroal eyes. The genal angle is usually rounded, but short spines have been observed on some of the Lower Devonian species. The facial sutures are fused, and molted cephala are complete. Much of the diagnosis of phacopids relies on cephalic features such as glabellar furrows and **prosopon**. The glabellar furrow S1 can be complete across the glabella, forming a preoccipital ring; it can extend back to the occipital ring (SO), cutting off the first glabellar lobe (L1) into a node; or it can be a crease that does not cross the glabella or join the SO. The S2 and more anterior glabellar furrows may be faintly expressed or absent. The thoracic pleurae have rounded ends and the pygidium is semicircular. Maximova (1972) erected a new genus, *Paciphacops*, and two new subgenera, *Paciphacops* and *Viaphacops*, with two New York phacopids, *Phacops logani* and *Phacops pipa*, as the respective type species. We have elected to use *Paciphacops* and *Viaphacops* as generic labels rather than subgeneric ones. Struve (1990) differentiated *Phacops rana* from the type species *Phacops latifrons* and erected the new genus *Eldredgeops* for the "*Phacops*" species of eastern North America; consequently the Middle Devonian *Phacops*? listed here may be referred to *Eldredgeops* when this genus is better defined for North America. Struve (1990) also reassigned one of the *Viaphacops* species to *Burtonops*. Tables 5.5 and 5.6 list the features of the Lower and Middle Devonian phacopids in New York.

Burtonops cristatus **(Hall, 1861) Figure 5.4F**
Lectotype NYSM 13883/1 (Eldredge 1973), hypotype NYSM 4612

Type species. This phacopid is from the Lower Devonian Schoharie Formation. Exposures are found in Schoharie County. The cephalon has large genal spines and usually an occipital spine. The thorax has axial spines. There are 14 dorsoventral rows

Table 5.5. Features of the Lower Devonian phacopids from New York

	Paciphacops logani	*Paciphacops hudsoniscus*	*Paciphacops clarkei*	*Paciphacops* subspecies A	*Phacops? clarksoni*	*Burtonops cristatus*
Cephalon						
Glabellar shape in "standard orientation"	High (tumid)	High	Flattened	Flattened	Flattened	Relatively tumid
Posterior region of the glabella	Deeply incised and connected 1S glabellar furrows; forms conspicuous intercalating ring with distal nodes and a median node, which is recurved slightly anteriorly	Deeply incised and connected 1S glabellar furrows; forms conspicuous intercalating ring with distal nodes and a median node, which is recurved slightly anteriorly	Deeply incised and connected 1S glabellar furrows; forms conspicuous intercalating ring with distal nodes and a median node, which is recurved slightly anteriorly	Deeply incised and connected 1S glabellar furrows; forms conspicuous intercalating ring with distal nodes and a median node, which is recurved slightly anteriorly	Glabellar lobes 1L are present distally as nodes; the intercalating either is wholly absorbed into the composite glabellar lobe or is depressed, with the medial position of glabellar furrow 1S generally obsolescent	Intercalating ring depressed distally; medial lobes reflected anteriorly and incorporated into composite grabellar lobe; glabellar furrow 1S confluent with occipital furrow distally, then reflected anteromedially, becoming obsolescent medially
Glabellar ornamentation	Ground mass of small granules that cover the large tubercles and spaces between them	Ground mass of small granules that cover the large tubercles and spaces between them	Small tubercles and ground mass of granulation much finer	?	Large simple tubercles, generally of different size classes	Large simple tubercles, generally of different size classes
Dorsoventral files in the eye	17, occasionally 18	15 or 16	18	15	17	14
Ornamentation of the cephalic doublure	Granules covering the entire surface proximal to the furrow	Granules covering the entire surface proximal to the furrow				Retains granules anteromedially but develops connected granules posterolaterally
Genal spine	Present	Absent	Present	Absent	Large	Large
Occipital spine	Node	?	Absent	?	Absent	Present
Thorax						
Thoracic axial spines	Absent	?	Absent	?	?	Spines or nodes
Axiallateral lobes	Well developed	?	?	?	?	Faint
Stratigraphic unit	Kalkberg and New Scotland Formations	Kalkberg and New Scotland Formations	Glenerie Formation	Becraft Formation	Schoharie Formation	Schoharie Formation

Source: From Eldredge, 1973; Campbell, 1977; Delo, 1940.

in the eyes, with a total of 48 to 81 lenses. The illustrated lectotype is from Eldredge (1972).

Eldredgeops crassituberculatus **(Stumm, 1953) Plates 100, 110D, and 111**
Holotype UMMP 25537

This Middle Devonian phacopid is found in great numbers along with *E. milleri* (Plates 110F, 111) in the famous Silica Shale quarries near Silvania, Ohio. Specimens of this trilobite were identified from the Solsville Member (now recognized as the Pecksport Member) of the Hamilton Group in Madison County (Eldredge 1972). This trilobite has 18 dorsoventral files, with six, rarely seven, lenses per file. The interlensar sclera is well developed, and the lenses are generally flush with the scleral surface. Cephalic tubercles are large and often elongated. Plate 100 shows a specimen from Ohio.

Table 5.6. Features of the Middle Devonian phacopids of New York

	Viaphacops bombifrons	*Viaphacops canadensis*	*Phacops? iowensis*	*Eldredgeops rana*
Cephalon				
Glabellar shape in "standard orientation"	Relatively tumid	Flattened	Flattened	Flattened
Posterior region of the glabella	Intercalating ring depressed distally; medial lobes reflected anteriorly and incorporated into composite grabellar lobe; glabellar furrow 1S confluent with occipital furrow distally, then reflected anteromedially, becoming obsolescent medially	Intercalating ring depressed distally; medial lobes reflected anteriorly and incorporated into composite grabellar lobe; glabellar furrow 1S confluent with occipital furrow distally, then reflected anteriomedially, becoming obsolescent medially	Deeply incised and connected 1S glabellar furrows; forms conspicuous intercalating ring with distal nodes and a median node, which is recurved slightly anteriorly	1S deeply incised
Glabellar ornamentation	Large simple tubercles, generally of different size classes	?	Large simple tubercles, generally of different size classes	Rounded tubercles becoming transversly elongate at the anterior margin of the glabella
Dorsoventral files in the eye	14, rarely 13, 15, 16	13 or 14	13	18 in *E. r. crassituberculata* 15, rarely 16 or 17 in *E. r. norwoodensis* 17 in *E. r. rana*
Ornamentation of the cephalic doublure	Retains granules anteromedially but develops connected granules posterolaterally	?	Shows only terrace lines	
Genal spine	Short	Large	Absent	Absent
Occipital spine	Absent	Absent	Absent	Absent
Thorax				
Thoracic axial spines	Sometimes nodes	Absent	Absent	Absent
Axial-lateral lobes	Faint	?	Absent	Absent
Stratigraphic unit	Onondaga Formation	"Lower Onondaga Limestone," Port Colborne, Ontario	Hamilton Group	Hamilton Group

Source: From Eldredge, 1972, 1973; Delo, 1940.

Eldredgeops rana **(Green, 1832) Plates 101, 102, 103, 104, 110C, and 111**

Holotype NYSM 4645

Eldredgeops rana is the single most common trilobite found and collected in New York, being widely distributed in the Middle Devonian Hamilton Group shales and limestones and reported up into the Upper Devonian Chemung rocks. The eye has 17 dorsoventral files. The most productive localities are in the Grabau Trilobite Beds in the Wanakah Shale. These remarkable beds are well exposed in Erie, Genesee, and Livingston Counties. The exuviae, cephala, and pygidia are ubiquitous in the Hamilton Group, and entire specimens are common. Plate 101 shows the holotype (positive) and its mold (negative). Plates 102 and 103 are of trilobite clusters found in trilobite-rich beds, all in western New York. Clusters such as these have been described as molting or breeding assemblages. Plate 104 illustrates an *E. rana* with color centers that may reflect underlying muscle attachment areas.

Eldredgeops norwoodensis **(Stumm, 1953) Plates 110E and 111**

Holotype UMMP 25524

This is a trilobite of the late Middle Devonian. In New York it is only found in the Tully Formation, Tompkins County. The eye with usually has 15, but rarely 16 or 17, dorsoventral files. This species is distinguished from *E. rana* by the narrower, less

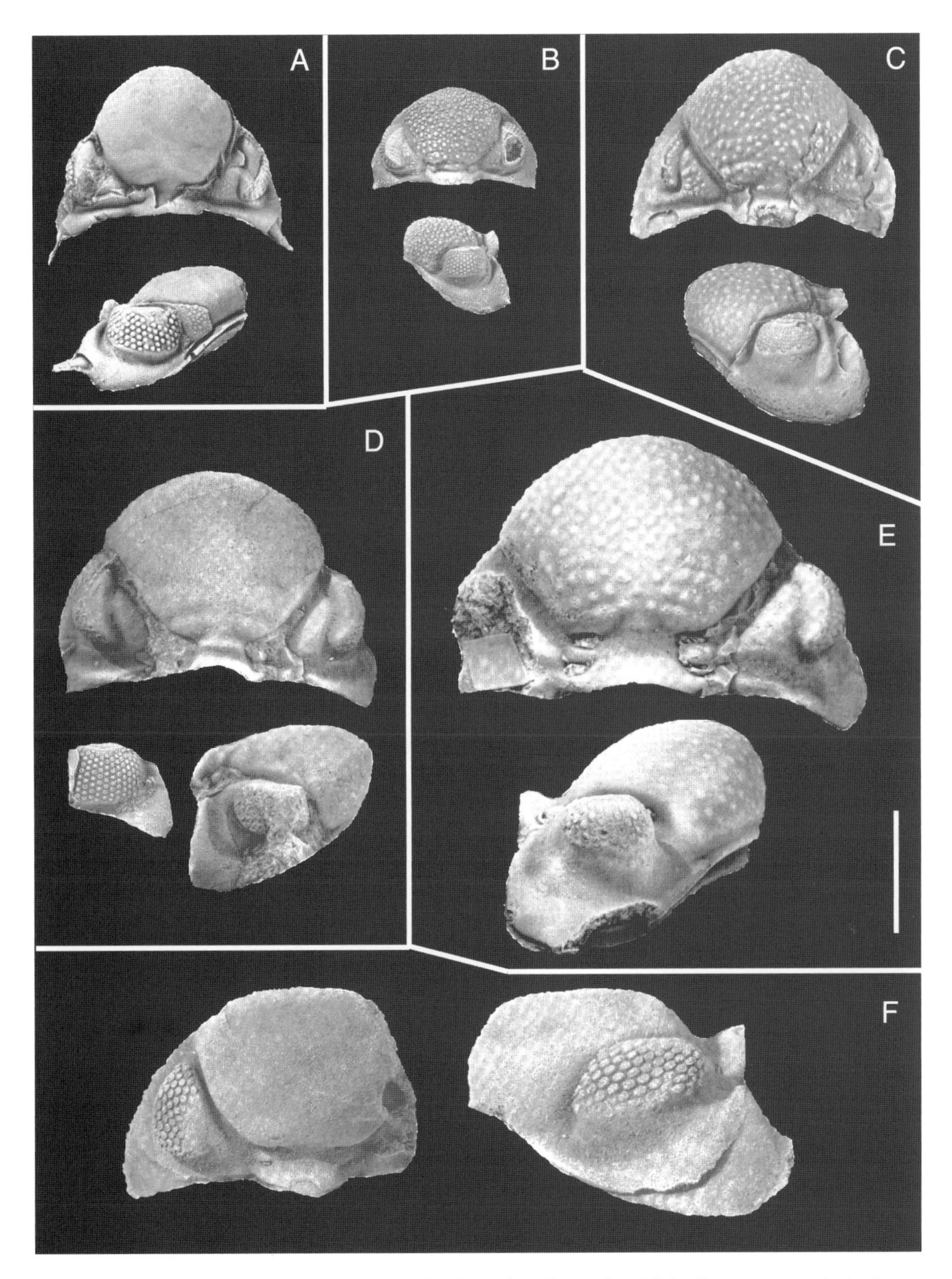

FIGURE 5.4. Lower Devonian and lower Middle Devonian phacopins. All the figures are cephala shown in dorsal and side views. The 1-cm scale bar is the same for all the images. A. *Phacops*? *clarksoni* from the Schoharie Formation, Schoharie County. The illustrated cephalon is AMNH 29244, the holotype. B. *Paciphacops hudsoniscus*, the small-eyed form from the Kalkberg–New Scotland Formations, Hudson, Columbia County. The illustration is AMNH 2613. C. *Paciphacops logani*, the large-eyed form from the New Scotland Formation near Schoharie, Schoharie County. D. *Paciphacops logani clarkei* from the Murailles Formation, Percé, Quebec. This subspecies is also found in the Glenerie Formation, Hudson, Columbia County. Illustrated is AMNH 29240, the holotype. E. *Viaphacops bombifrons* from the Onondaga Formation, western New York. The illustrated specimen is AMNH 4077/2A, the lectotype. F. *Burtonops cristatus* from the Schoharie Formation, New York. Illustrated is AMNH 13883/1, the lectotype. A–F are from Eldredge (1973), Reproduced with permission of the author.

expanding forward glabella and by the smooth pygidial pleura. The interpleural furrows on the pygidium are deeply incised. Although Eldredge (1972) did not comment on it, he illustrated what could be considered a large- and small-eyed form, with the New York specimen being small-eyed.

***Eophacops (Acernaspis?) trisulcatus* (Hall, 1843) Plates 105 and 106**
Holotype AMNH 1594

This trilobite is found in the Lower Silurian shales in Monroe County. Ludvigsen (1979b) figured a Silurian phacopid, referred to as an *Acernaspis* species, from the Cataract Group and the Thornloe Formation of Canada. The Canadian phacopid has a weakly inflated glabella with three pairs of thin, curving lateral furrows. The pygidium is small and lenticular, with few furrows. New York specimens are usually poorly preserved and are not inconsistent with this description. Further research is needed to properly classify the New York specimens. This phacopid is not uncommon in the Williamson or Sodus shales underlying the Irondequoit Limestone in Monroe County. Plates 105 and 106 are examples of this trilobite. Preservation of the exoskeleton is uniformly poor.

***Paciphacops hudsoniscus* (Hall, 1859) Figure 5.4B**
Type AMNH 2613

This trilobite is reported from the Lower Devonian Helderberg Group at "Becraft's Mountain" in Columbia County. Eldredge (1973) synonymized the species with *Paciphacops logani logani*, the small-eyed form. Ramsköld and Werdelin (1991) rejected this, and *P. hudsoniscus* is now considered a legitimate species.

***Paciphacops clarkei* (Eldredge, 1973) Figure 5.4D**
Holotype AMNH 29240

A phacopid of the Lower Devonian Glenerie Formation, Becraft Mountain, Columbia County. The glabella is somewhat flatter than in other *P. logani* subspecies. The genal spines are well developed. Glabellar furrow S1 generally is indistinct medially, and the intercalating ring tends to be incorporated medially into the glabellar lobe. The eyes have 17 or 18 dorsoventral files with eight or nine lenses per complete file. Ornamentation is fine tubercles over the entire test that are much smaller than those on the other subspecies.

***Paciphacops logani* (Hall, 1859) Figure 5.4C and Plates 107, 110A, and 111**
Lectotype NYSM 13885/2 (Hall and Clarke 1888; see Eldredge 1973)

Type species. This species is a widely distributed Lower Devonian, lower Helderberg trilobite. The occipital node is variably developed, and genal spines are usually present but may be absent, particularly in the small-eyed form. The nodes on the distal portions of the thoracic rings are large and distinct. Eldredge (1973) described a large-eyed and a small-eyed form. Further work has the large-eyed form, with 17 or 18 dorsoventral files and an average of 94 individual lens, described as *P. logani* (Plate 107), and the small-eyed form, with 15 or 16 dorsoventral files and with 46 to 48 lens, described as *P. hudsoniscus* (Ramsköld and Werdelin 1991). This species is reported from the Kalkberg and New Scotland Formations of Schoharie and Columbia Counties, and also in Quebec. Plate 107 shows a partially disarticulated specimen from the Helderberg Group. S1 extends across the glabella, leaving L1 as a ring on the posterior of the glabella just anterior to the occipital ring. The specimen also has acute genal angles. In addition, note the prominent circular lobes on the thorax where the axial and pleural lobes meet. Specimens illustrated by Hall (1861a,b) have genal angles from rounded to acute.

***Paciphacops logani* subsp. A (Eldredge, 1973)**
Specimen AMNH 29243

Lower Devonian Becraft Formation near Schoharie, Schoharie County. This species is based on a single cephalon with a distinctive eye. There are 15 dorsoventral files with a total of 73 lenses.

***Phacopina anceps* (Clarke, 1890)**
Holotype NYSM 4609

Middle Devonian Onondaga Limestone, Cayuga, Ontario. The holotype is a small, exfoliated cephalon. The presence in New York is suspected but not known.

***Phacopina? correlator* (Clarke, 1900)**
Holotype NYSM 13882/1

Only the holotype is known from the Glenerie Formation at Becraft Mountain. It is a mold of the left half of the cephalon.

***Phacops? clarksoni* (Eldredge, 1973) Figure 5.4A**
Holotype AMNH 29244

A species based on seven distinctive cephala from the Lower Devonian Schoharie Formation of the Schoharie Valley, Schoharie County. There are large genal spines and no occipital node. The eye has 17 dorsoventral files and more than 80 lenses. The lenses protrude farther beyond the interlensar sclera than is common for phacopids. The area under the visual surface is devoid of ornamentation. Eldredge believed that the specimens cannot be assigned to either *Paciphacops* or *Viaphacops*.

***Phacops? iowensis alpenensis* (Stumm, 1953) Plates 108, 110F, and 111**
Holotype UMMP 25516

Eldredge (1972) found specimens in the Middle Devonian Pecksport Member. Later Steve Pavilski found some in the Windom Member of the Hamilton Group in Madison County.

The S1 glabellar furrow is deeply incised. Moderately conical tubercles are all over the exoskeleton. The facial suture over the ocular platform is moderately deeply incised. This trilobite is relatively common in Michigan and very uncommon in New York. Plate 108 is from the upper Windom beds in Livingston County.

Viaphacops bombifrons **(Hall and Clarke, 1888) Figure 5.4E and Plates 109, 110B, and 111**
Lectotype AMNH 4071/2A (Eldredge 1973)

This species is widely distributed in the Middle Devonian Onondaga Limestones and their equivalent in New York, Ontario, and Ohio. The eye usually has 14 dorsoventral files, although 13 to 16 are known. Four or five lenses per row are also the norm. The area under the visual surface is coarsely tuberculate. There are short genal spines and sometimes an occipital node but no occipital spine. Normally there are no thoracic axial nodes or spines.

Family Pliomeridae

The only pliomerid from New York is *Pliomerops canadensis*. The cephalon is semicircular with a prominent glabella. The glabella expands forward uniformly. The S1 and S2 glabellar furrows are deep and extend about one-third of the way across the glabella. The thorax tapers to the rear with 18 segments. The pygidium is semicircular and curves sharply down all around. There are five pygidial axial rings and five pleurae separated by deep interpleural furrows.

Pliomerops canadensis **(Billings, 1859) Plates 112, 113, and 114**
Lectotype GSC 1101 (Whittington 1961)

Type species. *Pliomerops canadensis* is found throughout the Middle Ordovician Chazy in New York (for locations, see Shaw 1968) and in Chazy age rocks in Quebec, Tennessee, and Virginia. Articulated specimens are not rare in New York and may reflect this species's exceptionally robust exoskeleton, as this is generally not the case with other Chazy trilobites. In most of the specimens, the medial area is very rounded and the distal parts of the pleurae are perpendicular to the matrix. Plate 112 shows an articulated specimen. The very robust exoskeleton contributed to its preservation in a high-energy environment. Plate 113 is a side view of the same specimen, showing the vaulted pleural area and the cephalic structure, which is difficult to see from the top.

Family Pterygometopidae

The pterygometopids, in New York, are found in the Middle Ordovician. The glabella expands forward, and the eyes are schizochroal. The genal angle is often rounded to acute, but *Achatella* and *Eomonorachus* species have genal spines. The thoracic pleurae are rounded on the ends. The pygidium is somewhat triangular with 8 to 13 axial rings. *Calyptaulax* species at first glance appear to be phacopids, and the *Achatella* species resembles a dalmanitid. The family was reviewed by Delo (1940) and more recently by Ludvigsen and Chatterton (1982).

Achatella achates **(Billings, 1860) Plate 115**
Hypotype NYSM 4257

Type species. The holotype of this Middle Ordovician trilobite is from Ottawa, Canada. The general outline resembles dalmanitids. The cephalon is semicircular in outline with long genal spines. The glabella expands forward and the La lobe is pustulose. The eyes are schizochroal and elevated well above the plane of the cephalon. The body tapers toward the rear, ending with a pointed termination on the pygidium. Whole articulated specimens are found in the middle Trenton, Walcott-Rust Quarry, Herkimer County. It is also reported from Pennsylvania and Minnesota. Plate 115 shows a specimen from the Walcott-Rust Quarry. This trilobite is most often found as cephala on bedding planes where the high, schizochroal eyes are distinctive.

Calyptaulax annulata **(Raymond, 1905)**
Lectotype CM H-1293 (Shaw 1968); hypotypes NYSM 12368 to 12392

Calyptaulax annulata is found throughout the Middle Ordovician Chazy Group. In general shape and characteristics, it closely resembles *C. callicephalus* from the Trenton. Shaw illustrated specimens with and without short genal spines. The protaspids and meraspids are also known (Shaw 1968). It is the only pterygometopid known from the New York Chazy.

Calyptaulax callicephalus **(Hall, 1847) Plates 116 and 117**
Holotype AMNH 848

Calyptaulax callicephalus is found throughout the Middle Ordovician Trenton Group of New York and is also reported in Ontario, Manitoba, New Jersey, Kentucky, and Minnesota. It is especially common in the Kings Falls Limestone at Kings Falls on the Deer River, Lewis County. At first glance this trilobite appears to be a phacopid. The main differences are the cephalon, which in *C. callicephalus* is not rounded but rather comes to a blunt point in front, and the pygidium, which is subtriangular rather than rounded as in most phacopids. The pygidium could be mistaken for that of calymenids found in the same horizons, but once again the pygidium of *C. callicephalus* is more triangular and has many more axial rings and pleurae. A specimen in the MCZ is labeled with this name from the Black River Limestone Quarries near Poland, Herkimer County. Ruedemann (1926) reported a trilobite resembling *C. callicephalus* from the Upper Ordovician Indian Ladder Shales. Plates 116 and 117 illustrate the dorsal and ventral exoskeletal anatomy of this trilobite.

Calyptaulax eboraceous **(Clarke, 1894)**
Holotype NYSM 4767

A Middle Ordovician, lower Trenton (Glen Falls Limestone) trilobite from the quarries at Rawlins Mills, Saratoga County. It is also reported from the Trenton pebbles of the Rysedorph Conglomerate (Ruedemann 1901). It differs from *Eomonorachus intermedius* in that the S2 and S3 glabellar furrows do not isolate the L2 and L3 lobes. It has genal spines.

Chasmops? (or Sceptaspis) bebryx **(Billings, 1860) Plate 118**
Holotype? PRI 49631, hypotype GSC 13262, specimen USNM 26361

The trilobite is listed as Trenton age from Ottawa, Canada. Billings's illustration, however, was from a specimen owned by Colonel Jewett. This specimen is now at PRI and is designated as the holotype?. It is from Trenton age rocks at Jacksonburg, Montgomery County. The USNM specimen, labeled *Monorachus bebryx*, is from near Watertown, Lewis County, collected by the USGS in 1897. Delo (1940) illustrated the PRI specimen, by copying Billings, and designated it as the type of *Chasmops bebryx* but does not list it in the text of his paper. Plate 118 illustrates the holotype?. *Chasmops? bebryx* differs from *Eomonorachus convexus* (Plate 119) in the longer palpebral lobes, the much more triangular pygidium, and the much less strongly incised S1 on *C.? bebryx*. The specimen is very similar to the genus *Sceptaspis*, which is widely reported in the Midwest and Canada as *Sceptaspis lincolnensis*. If this is the same species, then the species name *bebryx* has priority.

Eomonorachus convexus **(Ulrich and Delo in Delo, 1940) Plate 119**
Holotype USNM 89987, specimen PRI 49632

The holotype is an internal mold labeled from "probably" the Middle Ordovician Trenton Falls, Herkimer County. There are two exfoliated cranidia on the small rock with the specimens. The left cheek and genal area are complete on one of them. It has short, thin genal spines. The eyes are missing. The matrix of the rock is a jumble of *Prasopora*. On the same rock are *Flexicalymene* and *Ceraurus* cranidia. The species was synonymized with *Eomonorachus intermedius* by Ludvigsen and Chatterton (1982) without comment. Delo (1940) stated that the species differs from *E. intermedius* in the rounded front margin, smaller second glabellar lobes, larger eyes, and coarser ornamentation. Different specimens are USNM 79015 from Rathbun Brook, Oneida County, and USNM 79014 from the lower Trenton quarries at Rawlins Mills, Saratoga County. A specimen with this name in the MCZ is labeled as from the basal Trenton near Poland, Herkimer County. Specimens collected from the Glens Falls Limestone near Amsterdam, Montgomery County, PRI 49632 (Plate 119), show the following characteristics: The cephalon is proparian and the free cheeks are missing. The glabella is pustulose. The palpebral lobe is crescent shaped, with a groove separating it from the fixed cheek. The occipital ring has a median node. The pygidium is subtriangular. The tapered axis extends nearly to the margin, with tapering more strongly for the most anterior five axial rings; there are at least 14 axial rings; the first 10 are strongly expressed. The axial rings have faint medial nodes. There are eight pygidial ribs on the pleura, with one or two thin raised areas near the medioposterior area. The first four most anterior ribs have weak intrapleural grooves. The interpleural grooves do not reach the margin.

Family Synphoriidae

The family Synphoriidae was erected from the subfamily Synphoriinae (Delo 1935, 1940) of the Dalmanitidae by Lespérance (1975) because of recognition of a number of new trilobites within the then subfamily. The family was earlier restricted to the Devonian and was almost exclusively North American. Campbell (1977) and Holloway (1981) substantially modified the concept of the Synphoriidae. Holloway extended the family into the Silurian. The family is characterized by the distance between the cephalic apodemes, S1 and S2, which is more than 1.5 times the distance between SO and S1 (Figure 5.2).

Anchiopella anchiops **(Green, 1832) Plate 120**
Holotype NYSM 4264

Type species. This trilobite from the Lower Devonian Schoharie Grit in Albany County is also referred to as *Anchiopsis*; however, the *Treatise* supports *Anchiopella*. The cephalon has slender genal spines, the glabella is sparsely covered with well-rounded tubercles of two sizes, and the occipital ring has a spine and large eyes. The pygidium is triangular with a sharp, upturned terminal spine. It has seven or eight pairs of rounded, unfurrowed pleural lobes and eight or nine axial rings.

Anchiopella anchiops sobrinus **(Hall and Clarke, 1888)**
Holotype NYSM 4243

This trilobite of the Lower Devonian Schoharie Grit, Albany County, differs from *A. anchiops* in having a more-rounded anterior margin and in lacking genal and occipital spines. Only the holotype cephalon is known, according to Delo (1940).

Coronura aspectans **(Conrad, 1841) Plates 121 and 122**
Type AMNH 4065/1, hypotype NYSM 4316

Type species. This species, from the Middle Devonian Onondaga Limestone, is characterized by furrowed ribs and regular, rather fine, ornamentation on the pygidium. A whole specimen, Plate 121, in NYSM is from Leroy, Genesee County. It is a latex pull from the external mold, as it retains better features than the counterpart.

Coronura helena **(Hall, 1861) Plate 127B**
Holotype AMNH 4250/1

This species from Middle Devonian Onondaga Limestone is characterized by weak to obsolete tuberculation, long slender lateral pygidial spines, and horizontally bifid terminal pygidial spines.

Coronura myrmecophorus (Green, 1835) **Figure 5.3F**
Hypotypes NYSM 4335, 4350

This trilobite of the Middle Devonian Onondaga Limestone, Genesee County, is characterized by its very large size, unfurrowed ribs, scattered tuberculose ornamentation, and erect terminal spines (Delo 1940).

Odontocephalus aegeria (Hall, 1861) **Plate 123**
Hypotype NYSM 4258

This species from the Middle Devonian Onondaga Limestone, Erie County, is characterized by 11 spatulate denticles on the anterior cephalic border and sharp, slender genal spines that extend back to the fourth thoracic segment. There are 10 pygidial pleurae.

Odontocephalus bifidus (Hall, 1861) **Plate 124**
Hypotype NYSM 4275

The type specimen is from quarries in the Onondaga Limestone, Genesee County. It is a pygidium characterized by a broad flat area posterior to the axial lobe and two stout, rapidly tapering, and close together, terminal axial spines. There are seven to eight pleurae. An entire specimen from the same quarries is illustrated in Plate 124. This is the only known whole specimen from New York. The cephalon has nine spatulate denticles and short genal spines.

Odontocephalus coronatus (Hall, 1861)
Holotype AMNH 4064, hypotype NYSM 4309

This trilobite is from the Middle Devonian Onondaga Limestone of central New York, Cayuga County. The cephalon is unknown. It is characterized by the pygidium, which has no border and eight pleurae. There are no terminal spines, but where they are on other odontocephalids, there are two rounded pygidial extensions.

Odontocephalus humboltensis (Sargent, 1953)
Specimen BMS E-16674-5

This Middle Devonian Onondaga species was discovered during road construction in Buffalo, Erie County. It is characterized by having 11 denticles on the anterior cephalic border, short genal spines, and a distinct border on the pygidium with no terminal pygidial spines.

Odontocephalus selenurus (Eaton, 1832) **Plate 125**
Hypotypes NYSM 4359 to 4368

Type species. This trilobite is reported from the Middle Devonian Onondaga limestones across the state. The species is in need of revision. It has been characterized as having nine spatulate denticles on the cephalon. Hall and Clarke (1988, Plate 12) illustrated specimens that may belong to two different species or morphotypes. One form illustrated in Plate 125 has a cephalon with ridges on the denticles and a slight forward protrusion of the anterior cephalic border. The pygidium is oval with short pygidial spines. The other form, illustrated in Plate 126, has a cephalon with flat denticles, a rounded anterior cephalic border, and very short genal spines. An associated pygidium is more triangular with relatively long pygidial spines. There are at least five species in this genus. The characteristics are listed in Table 5.7.

Odontocephalus sp. **Plate 126**
This specimen is from the Middle Devonian Moorehouse Member of the Onondaga Limestone in the LeRoy Quarry, Genesee County.

Schoharia emarginata (Hall, 1876)
Holotype NYSM 4318

Type species. Only the pygidium of this trilobite is definitely known from the Lower Devonian Schoharie Formation, Schoharie County. It is characterized by a posterior "notch," within which the most posterior pleural ribs terminate. There are

Table 5.7. Features of the genus *Odontocephalus* of New York

	O. aegeria	*O. bifidus*	*O. coronatus*	*O. huboltensus*	*O. selenurus*
Cephalon					
Length to width	1:2		Not known	1:2	1:2
Preglabellar dentations	11	9		11	9
Genal spines	Long	Short		Short	Short
Pygidium					
Border	None	None	None	Distinct	Distinct
Terminal spines	2, short, well separated	2, short, close together, parallel	None	None	2, short, well separated
Axial rings	9	8	7 (plus one indistinct)		8 or 9
Pleura	9 or 10	7	8	12	8 or 9

approximately 17 axial rings and 15 nonfurrowed pleural ribs. Ornamentation consists of scattered pustules.

Schoharia sp.
Specimen NYSM 4309

A single pygidium illustrated by Lespérance (1975) is from the Middle Devonian Onondaga Limestone, Schoharie County. It is characterized by a shallow posterior notch, a pseudo postaxial ridge, and low flattened ribs. Ornamentation consists of very few scattered pustules.

Synphoria? concinnus **(Hall, 1876)**
Types NYSM 4307 to 4308

This species of Lower Devonian, Schoharie Grit, Albany County, is known only from pygidia. The pygidium is subtriangular and very convex. The border is obscure laterally but widens posteriorly into a short, blunt, triangular terminus. The surface is smooth. There are seven or eight axial rings and seven or eight wide flat pleurae, the most anterior of which is indistinctly furrowed.

Synphoria sopita **(Lespérance, 1975)**
Paratype AMNH 29247

This species is from the Lower Devonian Oriskany/Glenerie Formation, Columbia County. The cephalon is triangular in outline with protruding genal angles. The anterior border has at least 15 crenulations, with perhaps up to 19 more. The pygidium has 14 or 15 axial rings and 9 or 10 pleural ribs. It is terminated with a blunt spine upturned at about 30 degrees.

Synphoria stemmata compacta **(Lespérance and Bourque, 1971)**
Holotype NYSM 4371

This trilobite, from the Lower Devonian Oriskany/Grenerie Formations of Becraft Mountain, Columbia County, resembles *S. stemmata stemmata* except the anterior cephalic border has 13 to 19 crenulations. The pygidium is nonspinose and well rounded, with 13 axial rings and nine pleural ribs.

Synphoria stemmata stemmata **(Clarke, 1900) Figure 5.3E**
Lectotype NYSM 4372 (Lespérance and Bourque, 1971)

Type species. This species is from the Lower Devonian Oriskany/Glenerie Formations, Columbia County. The cephalon is large, is slightly convex, has seven crenulations on the anterior border, and has no genal spines. The pygidium is nonspinose and bluntly rounded, with 13 axial rings and nine pleural ribs.

Synphoroides dolphi **(Clarke, 1893)**
Holotype NYSM 4317

Helderbergian at Port Jervis, Orange County.

Trypaulites calypso **(Hall and Clarke, 1888) Figure 5.3G**
Type AMNH 4249/1

Hall and Clarke (1888) listed this trilobite as Lower Devonian from Ohio and Schoharie, Schoharie County. None of the illustrations in Hall and Clarke's publication are specimens from New York, and all of the specimen listings from Lespérance (1975) are from Ohio, Kentucky, Michigan, and Illinois. One must question if this species is truly found in New York. The cephalon has short genal spines and large eyes. There are 11 thoracic segments. The pygidium is rounded with a well-defined border and has 11 axial rings with a rounded node and 10 or 11 broad, flat, furrowed pleurae.

Trypaulites erinus **(Hall, 1861) Plate 127A**
Holotype NYSM 4320

From the Middle Devonian Onondaga Group of western New York in Ontario, Genesee, and Erie Counties, this trilobite is known only from the pygidium. It has 12 (13?) slightly furrowed pleural ribs and 16 axial rings with a postaxial ridge. There is a border. The surface is covered with granules and the termination is slightly upturned. Pygidia of this species have been collected from the Bois Blanc near LeRoy, Genesee County, New York.

Trypaulites macrops **(Hall, 1861)**
Holotype NYSM 4327

This trilobite is known from a single incomplete, internal cephalic mold from the Onondaga Limestone of Schoharie, Schoharie County. The eye is much larger than that of *T. calypso*, with 37 dorsoventral files of 15 lenses each.

5.6 ORDER PROETIDA

This post-Cambrian order lasted until the final extinction of trilobites at the end of the Permian. The trilobites are generally oval to suboval in outline, with a medium-sized pygidium. Genal spines are commonly present.

Family Aulacopleuridae

Thomas and Owens (1978) combined two trilobite families, Aulacopleuridae and Otarionidae, into one family, Aulacopleuridae, with two subfamilies Aulacopleurinae and Scharyiinae. No members of the subfamily Scharyiinae are reported from New York. The aulacopleurids are small trilobites with a forward-tapering glabella. The L1 glabellar lobe is separated from the glabella by a groove. In many respects, except for size, they resemble proetids and are a member of the order Proetida. Thomas and Owens did not attempt to reclassify the large number of North American species represented by this family, so most of the names listed are traditional. Adrain and Chatterton (1995) revised the genera *Harpidella* and *Maurotarion*. Species referred to as *Otarion* in New York are in need of revision.

***Cyphaspis* sp.**
Specimen BMS E-11035

Stumm (1967, Plate II, no. 21) illustrated a trilobite identified as an *Otarion?* species. An articulated specimen collected in the Middle Devonian Deep Run Shale in Livingston County has been identified as a *Cyphaspis* species (Jon Adrain 1999, private communication).

***Harpidella craspedota* (Hall and Clarke, 1888) Plate 128**
Lectotype NYSM 4236

Harpidella craspedota is a small, relatively common trilobite from certain of layers the Centerfield Limestone in Middle Devonian Hamilton exposures in Ontario and Livingston Counties in western New York. Specimens of *Harpidella* that may belong to this species have been collected from the Mount Marion Formation and the Stafford Limestone. This trilobite is characterized by long genal spines and 12 thoracic segments. There is a short, axial spine on the fourth thoracic segment and a long one on the sixth. Plate 128 shows a small cluster of specimens.

***Harpidella spinafrons* (Williams in Cooper and Williams, 1935) Plates 129 and 130**
Holotype USNM 89751, specimen USNM 89980 (labeled *H. craspedota*)

This species is from the Middle Devonian Tully Formation, West Brook Member near Sherburne, Chenango County. Other specimens are listed as 4.8 km (3 miles) south of Sherburne. This trilobite has a double row of anterior spines along the cephalic border, a juvenile characteristic retained in the holaspid. Plate 129 shows an entire but damaged specimen. The base of the thoracic spine on the sixth segment is clearly shown. Plate 130 is of a well-preserved cephalon of the same species.

***Harpidella stephanophora* (Hall and Clarke, 1888)**
Hypotypes NYSM 4253 to 4256

This Middle Devonian trilobite is from "decomposed chert boulders in the neighborhood of Canandaigua, New York" (Hall and Clarke 1888). These boulders are Onondaga Limestone glacially transported into Ontario County. Adrain and Chatterton (1995) assigned *Cyphaspis stephanophora* to *Harpidella*.

***Harpidella* sp. Plate 131**

The trilobite in Plate 131 is an as yet, undescribed species from the Onondaga Limestone.

***Maurotarion* sp. Plate 132**

Specimens have been found in the Upper Silurian Lockport Group near Sodus, Wayne County. They cannot be identified with any known New York species.

***Otarion? (Maurotarion) coelebs* (Hall and Clarke, 1888)**
Plastotype NYSM 4233

Otarion? coelebs is a Lower Devonian trilobite from the Lower Helderberg of Albany and Schoharie Counties. This is likely a *Maurotarion* species (Adrain and Chatterton 1995).

***Otarion? diadema* (Hall and Clarke, 1888)**
Plastotype NYSM 4238

This Middle Devonian trilobite is from "decomposed chert boulders in the neighborhood of Canandaigua, New York" (Hall and Clarke 1888). These boulders, in Ontario County, are glacially transported Onondaga Limestone.

***Otarion? hudsonicum* (Ruedemann, 1901)**
Holotype NYSM 4236

This trilobite is from the Ordovician, upper Utica Shale of Green Island near Albany, Albany County. It was assigned to a new species because of the rarity of aulacopleurids in the Utica Shale.

***Otarion? hybrida* (Hall and Clarke, 1888)**
Holotype NYSM 4240

This Middle Devonian trilobite is from "decomposed chert boulders in the neighborhood of Canandaigua," Ontario County. These boulders are glacially transported Onondaga Limestone.

***Otarion? laevis* (Hall, 1876)**

Otarion? laevis is known from one cephalon from the Upper Devonian "Chemung Formation", Chemung County.

***Otarion? matutinum* (Ruedemann, 1901)**
Syntypes NYSM 4241 to 4242

This trilobite was identified from material in Middle Ordovician Trenton age pebbles within the Rysedorph Conglomerate, Rensselaer County. Shaw (1968) believed that *O. matutinum* and *O. hudsonicum* might not be aulacopleurids because of the non-separate nature of the basal lobes.

***Otarion? (Maurotarion) minuscula* (Hall, 1876)**
Hypotypes NYSM 4243 to 4249

Otarion? minuscula is a trilobite of the Schoharie Grit, Albany County, and Onondaga Limestone, Ontario, Genesee, and Erie Counties, which places it in the upper Lower Devonian or lower Middle Devonian. The specimen illustrated in the work by Hall and Clarke (1888) shows no evidence of the axial thoracic spine often found in this genus. This is likely a *Maurotarion* (Adrain and Chatterton 1995).

***Otarion? spinicaudatum* (Shaw, 1968)**
Holotype NYSM 12243

Otarion spinicaudatum is from the Middle Ordovician, lower Middle Chazy beds and was identified from silicified specimens.

It is considered to be one of the oldest members of this family. For locations, see Shaw (1968).

Family Bathyuridae

Bathyurid trilobites are found in the Lower and Middle Ordovician of New York—most, if not all, below the middle Trenton. They have a robust, nearly semicircular cephalon with stout genal spines. The glabella is prominent and expands forward or is parallel sided. Glabellar furrows are faint. The thorax has 9 or 10 segments and no conspicuous pleural spines. The pygidium is nearly semicircular, sometimes with an axial pygidal spine.

Acidiphorus whittingtoni **(Brett and Westrop, 1996)**
Holotype USNM

This bathyurid is from the Lower Ordovician Scotia Limestone Member of the Fort Cassin Formation, Washington County. One should refer to the original publication for the description.

Bathyurellus platypus **(Fortey, 1979)**
Holotype GSC 56847

This bathyurid is from the Lower Ordovician Scotia Limestone Member of the Fort Cassin Formation, Washington County. A cranidium and two pygidia were judged identical to material from Newfoundland (Brett and Westrop 1996).

Bathyurus **cf. *B. angelina*** **(Billings, 1859)**
Holotype GSC 1084c, specimen MCZ 3790

The holotype is from the Lower Ordovician of Quebec. The MCZ specimen was collected by C. D. Walcott in 1867 from a drift block near Trenton Falls in either Herkimer or Oneida County. This species was identified by P. Raymond in 1905.

Bathyurus extans **(Hall, 1847) Plate 133**
Lectotype NYSM 4139 (Whittington 1953)

Type species. *Bathyurus extans* is a characteristic trilobite from the Lowville Limestone Member of the Middle Ordovician Black River Group. It is found wherever these rocks are exposed in New York and is also reported from Minnesota. Plate 133 is a specimen from the Black River of Lewis County.

Bathyurus johnsoni **(Raymond, 1913)**
Specimen USNM

A trilobite with this name, but with no inventory number, is in the USNM and is listed from the Lowville Member of the Middle Ordovician Black River Group, Utica, Oneida County. There are no Black River rocks in the immediate Utica area, so this is probably from the Lowville Member at Newport, Herkimer County, in an area on West Canada Creek.

Bathyurus taurifrons **(Dwight, 1884)**
Syntypes NYSM 9634 to 9636

Bathyurus taurifrons is reported from the Lower Ordovician Roachdale Limestone in the Wappinger Valley in Dutchess County.

Benthamaspis striata **(Whitfield, 1897)**
Holotype AMNH 35823

This species is from the Lower Ordovician Fort Cassin Formation in Vermont. In all probability it is also found in New York (Brett and Westrop 1996).

Bolbocephalus seelyi **(Whitfield, 1889)**
Holotype AMNH 396

Type species. *Bolbocephalus seelyi* is a trilobite from the lower part of the Spellman Formation (Lower Ordovician, lower Beekmantown Group) at Beekmantown, Clinton County. The cephalon is convex, narrowing anterior to the occipital ring and then expanding forward. There are no glabellar furrows.

Grinnellaspis **cf. *G. marginiata*** **(Billings, 1865)**
Lectotype GSC 646 (Whittington 1953)

Brett and Westrop (1996) tentatively identified specimens from the Lower Ordovician Scotia Limestone Member of the Fort Cassin Formation, Washington County, as this species.

Raymondites ingalli **(Raymond, 1913)**
Holotype GSC 4328

Type species. *Raymondites ingalli* is a trilobite usually associated with the Middle Ordovician Kirkfield Formation of Ontario. Fisher (1962) listed it as characteristic of the lower Trenton Larrabee Limestone. DeMott (1987) stated that he can only place the holotype in this species with certainty. All other specimens he is aware of are *R. spiniger.*

Raymondites longispinus **(Walcott, 1877)**
Holotype MCZ 107237

The holotype of this Middle Ordovician, upper Black River/lower Trenton trilobite is from one of the quarries operated for building stone near Poland, Herkimer County, in the mid to late 1800s. Kay (1953) listed *R. longispinus* from the lower Trenton Rockland (Napanee) Member. The *Raymondites* genus differs from *Bathyurus* in that it has one pair of lateral glabellar furrows rather than two, and the surface of the cephalon is pustulose.

Raymondites spiniger **(Hall, 1847)**
Holotype AMNH 854

DeMott (1987) reported the holotype as GSC 4318, stating it is the specimen figured by Hall. Hall clearly stated that his figured specimen, and holotype, is from New York but did not give a specific location other than the Mohawk Valley. He did report another similar specimen belonging to a Mr. Logan from Montreal, which Logan obtained in Canada. (Logan is Sir William Logan, the first director of the Geological Survey of Canada.) This may be the GSC specimen. *Raymondites spiniger* is reported from the Middle Ordovician, upper Black River Chaumont Limestone and the lower Trenton Rockland (Napanee) Limestone. Geographically it is widely distributed into Ontario, Quebec, Kentucky, and Illinois.

Strigigenalis cassinensis **(Whittington, 1953)**
Holotype location unknown; specimens NYSM 15384 to 15390

Brett and Westrop (1996) described this species from the Lower Ordovician Scotia Limestone Member of the Fort Cassin Formation, Washington County.

Strigigenalis caudatus **(Billings, 1865)**
Holotype GSC 635

Raymond (1913) *Strigigenalis caudatus* reported from the Lower Ordovician Beekmantown at Ticonderoga, Essex County. The holotype is from Newfoundland.

Uromystrum brevispinum **(Raymond, 1905)**
Lectotype CM H-1287 (Shaw 1968)

Uromystrum brevispinum is a trilobite of the Middle Ordovician, upper Chazy Group known only from a poorly preserved cephalon, cranidia, and pygidia. For locations, see Shaw (1968). *Uromystrum brevispinum* has a long unfurrowed glabella.

Uromystrum minor **(Raymond, 1905)**
Holotype CM H-1284

Uromystrum minor is similar to *U. brevispinum* but is known only from cranidia and pygidia. It is from the Middle Ordovician, middle Chazy Group. The *U. minor* differs in that it has a low, indistinct glabella. For locations, see Shaw (1968).

Family Brachymetopidae

Brachymetopids are small trilobites in the Lower Silurian and Devonian in New York. The cephalon is nearly semicircular in outline, with genal spines with a distinct border. The glabella is short and tapers forward. The basal glabellar lobes (L1) stand alone, being cut off by glabellar furrow S1. The thorax has 10 segments. The pygidium is semicircular with numerous axial rings. The pygidiual margin may be smooth or spined. The entire dorsal surface of the trilobite is covered with tubercles.

Australosutura gemmaea **(Hall, 1876) Plate 134**
Lectotype NYSM 4217 (Stumm 1967)

This species is found in the Jaycox, Deep Run, Kashong, and Windom Shales. The illustrated specimen, in Plate 134 is from Livingston County. Specimens reported from the Onondaga are an undescribed species. The genus assignment is from an abstract (Scatterday 1986).

Cordania becraftensis **Clarke, 1900**
Lectotype NYSM 4210 (Whittington 1960)

Cordania becraftensis is from the Lower Devonian Glenerie Limestone, Becraft Mountain, Hudson, Columbia County.

Cordania cyclurus **(Hall and Clarke, 1888)**
Lectotype NYSM 4215 (Whittington 1960)

Cordania cyclurus is a trilobite of the Lower Devonian New Scotland Limestone in Albany County.

Mystrocephala arenicolus **(Hall and Clarke, 1888)**
Plastotype NYSM 4206

Mystrocephala arenicolus is found in the upper Lower Devonian Schoharie Grit, Schoharie County.

Mystrocephala ornata **(Hall, 1876)**
Hypotype NYSM 4250

This trilobite from the Middle Devonian Hamilton shales, in Ontario County, is probably different from *A. gemmaea.*

Mystrocephala ornata baccata **(Hall and Clarke, 1888)**
Syntypes NYSM 4251, 4252

This species is reported to be from the Centerfield Limestone in Ontario County.

Mystrocephala varicella **(Hall and Clarke, 1888)**
Lectotype NYSM 4223 (Whittington 1960)

Mystrocephala varicella is from the Middle Devonian Onondaga Limestone in Ontario County.

Radnoria **sp. Plate 135**

This trilobite was formerly referred to as *Proetus stokesii.* It is found in the Lower Silurian Rochester Shale. The specimen in Plate 135 is from Orleans County. This trilobite is being described (Jon Adrain 2000, private communication).

Family Dimeropygidae

Only one representative of this family is known in New York. These small trilobites, 1 cm or less in length, are mostly known from silicified remains etched from limestone blocks. The cephalon is convex and pustulose. The thorax has eight segments. The pygidium is convex with three to six segments. Whole, articulated specimens are very rare in eastern North America, probably because they inhabited shallower, high-energy environments where preservation is poor. For this same reason they, would be difficult to find unless silicified.

Dimeropyge clintonensis **(Shaw, 1968)**
Holotype NYSM 12357

Dimeropyge clintonensis is a small trilobite of the Middle Ordovician, middle Chazy and the only representative of the family in New York. For locations, see Shaw (1968). It is known from silicified maraspids and holaspids. The cephalon has long genal spines, and the transitory pygidium has a prominent axial spine from the second axial ring. This pygidial spine is apparently lost in the holaspis. *Dimeropyge clintonensis* closely resembles *D. virginiensis,* from Whittington and Evitt (1954). No articulated specimens of the New York species are known. An articulated *Dimeropyge* specimen is known from the Trenton of Canada (Ludvigsen 1979b).

Family Proetidae

Proetids existed over more geologic time than any other trilobite family. According to Fortey and Owens (1997) the family arose in the Early Ordovician and lasted until trilobites became extinct during the great Late Permian extinction event (about 250 million years ago). The order Proetida has members from the Late Cambrian. Proetids have a rounded, inflated glabella that tapers forward. They have a well-developed cephalic border and usually medium-length genal spines. The thoracic pleurae have rounded ends. The pygidium is semicircular and often has a border. The first New York species are found in the Middle Ordovician and are undoubtedly immigrants from European trilobite populations. These early representatives did not last past the Silurian but were replaced by another wave of proetids in the Early Devonian (Lieberman 1994). The proetids are nowhere as abundant as *Isotelus, Dalmanites,* and phacopids from their respective periods, but their presence is known from their exuviae. Whole, articulated specimens are uncommon, both because of low populations and because the exoskeleton was not very robust. In the Middle Devonian, however, there are locations in the Centerfield Limestone in Livingston County where groups of whole *Pseudodechenella rowi* have been found. In 1994 Lieberman revised the proetids of the Devonian of eastern North America including those in New York. The Silurian and Ordovician proetids need some attention. The features of the described proetids from New York are in Table 5.8.

Basidechenella? hesionea (Hall, 1861)
Holotype AMNH 2898

This Lower Devonian species from the Schoharie Grit, Schoharie County, is known from two pygidia. The pygidium is semicircular, with 13 axial rings and nine pleural segments. There are no tubercles on the axial rings.

Coniproetus angustifrons (Hall, 1861)
Lectotype AMNH 2900 (Stumm 1953b)

This species is from the Lower Devonian Schoharie Grit of Albany County. Apparently it is only positively known from cranidia. Pygidia assigned to this species by Hall and Clarke are not accepted by other authors. The species is distinguished by the larger than usual sagittal width of the cephalic border. There is no medial occipital node.

Coniproetus conradi (Hall, 1861)
Lectotype AMNH 39321 (Lieberman 1994), hypotype NYSM 4706

The lectotype is from the Lower Devonian Schoharie Grit in Albany County. The NYSM specimen is from Glenerie Formation, Becraft Mountain, Hudson, Columbia County. The cephalic features are very similar to those of *C. angustifrons.* The pygidium is rounded and 1.8 times wide as long. There are 10 axial rings. The pleural furrows are faint and become indistinct posteriorly. There is a border.

Coniproetus folliceps (Hall and Clarke, 1888) Plate 136
Lectotype NYSM 4722 (Lieberman 1994)

This species is from the Middle Devonian, upper Onondaga Limestone near LeRoy, Genesee County. The cephalon is semicircular in outline without genal spines. There is a narrow border. The surface is covered with low, coarse granules. The occipital ring has a faint axial node. The lateral occipital lobes are connected posteriorly to the occipital ring. There are no axial thoracic nodes. The pygidium has at least 10 axial rings and eitht or more pleurae. There is a narrow border. The surface is granulose. Plate 136 illustrates the lectotype.

Cornuproetus beecheri (Ruedemann, 1926)
Holotype YPM, hypotype YPM 27810

This trilobite is part of Beecher's collections from Beecher's Trilobite Bed in the Upper Ordovician Frankfort Shales north of Rome, Oneida County, the best specimens of which are in YPM. None of the New York Ordovician proetids have been redescribed. The assignment to *Cornuproetus* by Cisne (1973) must be considered tentative.

Crassiproetus brevispinus (Fagerstrom, 1961)
Specimens AMNH 39329, 39337, 44700, 44707 to 44713

From the Middle Devonian Onondaga Limestone and its equivalent in Ontario and western New York and Albany County, this species is known from cranidia and pygidia. It differs from *C. crassimarginatus* by having a narrow cephalic border and, a glabella that is wider and semicircular in the dorsal anterior view. There is a small median occipital node. The pygidium has 18 axial rings and 15 pleural segments.

Crassiproetus crassimarginatus (Hall, 1843)
Lectotype AMNH 39328 (Lieberman 1994)

The lectotype is from Williamsville, Erie County. This proetid is widely distributed from the Middle Devonian Onondaga Limestone and age-equivalent rocks of New York, Ontario, Michigan, Ohio, and Kentucky. Specimens have been reported from the Lower Devonian Schoharie Grit in Schoharie County and the Middle Devonian Onondaga Limestone in Genesee County. The cephalon is semicircular and has a wide border and no genal spines. The glabellar outline in dorsal anterior view is parabolic

Table 5.8. Features of the described proetids of New York

NAME	*AGE, STRATA*	*LATERAL OCCIPITAL LOBE*	*OCCIPITAL NODE*	*GENAL SPINES*	*THORACIC AXIAL NODES*	*PYGIDIAL AXIAL NODES*	*PYGIDIAL AXIAL RINGS*	*PYGIDIAL PLEURAL SEGMENTS*
Cornuproetus beecheri (Ruedemann, 1926)	U. Ord., Frankfort Shale	SEP	Yes	Yes	No	No	7	5
Proetus (Proetus) clelandi (Raymond, 1905)	M. Ord., Chazy?	No	Yes					
Proetus parviusculus (Hall, 1860)	Ord.							
Proetus spurlocki (Meek, 1872)	Ord.							
Proetus undulostriatulus (Hall, 1847)	M. Ord., Snake Hill Shale		No					
Hedstroemia pachydermata (Barrett, 1878)	Sil., Decker Ferry Limestone						15	
Proetus artiaxis (Howell and Sanford, 1947)	U. Sil., Oak Orchard Dolostone					No	3?	
Proetus tenuisulcatus (Howell and Sanford, 1947)	U. Sil., Oak Orchard Dolostone	SEP	No				7	Faint
Basidechenella hesionea (Hall, 1861)	L. Dev., Schoharie Grit					No	13	9
Coniproetus angustifrons (Hall, 1861)	L. Dev., Schoharie Grit	Faint	No	Yes	No		7 or 8	5 or 6
Coniproetus conradi (Hall, 1861)	L. Dev., Glenerie Formation	Faint	No	Yes	No		10	4 or 5
Crassiproetus schohariensis (Lieberman, 1994)	L. Dev., Schoharie Grit					No	14	12
Crassiproetus stummi (Lieberman, 1994)	L. Dev., Schoharie Grit					No	13	10
Gerastos protuberans (Hall, 1859)	L. Dev., New Scotland Limestone	No	No	No	No	No	8 or 9	4 Faint
Proetus hesione (Hall, 1861)	L. Dev., Schoharie Grit						10 or 11	8
Basidechenella canaliculata (Hall, 1861)	M. Dev., Onondaga Limestone	SEP	Faint			Yes	10	

Table 5.8. *Continued*

NAME	*AGE, STRATA*	*LATERAL OCCIPITAL LOBE*	*OCCIPITAL NODE*	*GENAL SPINES*	*THORACIC AXIAL NODES*	*PYGIDIAL AXIAL NODES*	*PYGIDIAL AXIAL RINGS*	*PYGIDIAL PLEURAL SEGMENTS*
Basidechenella clara (Hall, 1861)	M. Dev., Onondaga Limestone	SEP	No	Yes	No	No	11	4 or 5
Coniproetus folliceps (Hall and Clarke, 1888)	M. Dev., Onondaga Limestone	CON	Faint	No	No	No	10	8
Crassiproetus brevispinus (Fagerstrom, 1961)	M. Dev., Onondaga Limestone		Faint	Yes			18	15
Crassiproetus crassimarginatus (Hall, 1843)	M. Dev., Onondaga Limestone		No	No?	No	No	16	14
Crassiproetus neoturgitus (Lieberman, 1994)	M. Dev., Onondaga Limestone					No	13	10
Monodechenella halli (Stumm, 1953)	M. Dev., Onondaga Limestone					Yes	9	9
Proetus microgemma (Hall and Clarke, 1888)	M. Dev., Onondaga Limestone					Yes	11	7 or 8
Pseudodechenella arkonensis (Stumm, 1953)	M. Dev., Hamilton Group	SEP	Yes	Yes	Yes	Yes	13	9
Pseudodechenella rowi (Green, 1838)	M. Dev., Hamilton Group	SEP	Faint	Yes	No	No to faint	11	8
Dechenella haldemani (Hall, 1861)	M. Dev., Hamilton Group	SEP	No	No	No	No to faint	17	10
Monodechenella macrocephala (Hall, 1861)	M. Dev., Hamilton Group	SEP	No	Yes	Faint	Yes	14	12
Proetus jejunus (Hall and Clarke, 1888)	M. Dev., Hamilton Group					Yes	10	8
Proetus marginalis (Conrad, 1839)	M. Dev., Hamilton Group				No			

SEP, separated by a groove from the occipital ring; CON, connected to occipital ring.

or subsemicircular. There are about 16 pygidial axial rings and about 14 pleural segments. There is a pygidial border.

Crassiproetus neoturgitus **(Lieberman, 1994)**
Holotype AMNH 44720

This species was described using two exfoliated pygidia from the Middle Devonian Onondaga Limestone in Schoharie County. There are 13 axial rings and ten pleural segments. A pygidial border is present.

Crassiproetus schohariensis **(Lieberman, 1994)**
Holotype AMNH 44699

The holotype exfoliated pygidium is from the Lower Devonian Schoharie Grit of Orange County. Other pygidia assigned to this species are either from the Schoharie Grit or from the Middle Devonian Onondaga of Albany and Schoharie Counties. There are 14 axial rings and 12 pleural segments. No pygidial border is evident on the illustrations in Lieberman (1994).

Crassiproetus stummi **(Lieberman, 1994)**
Holotype UMMP 29519

This species, known only from pygidia, is found in the Onondaga Limestone and Schoharie Grit of Schoharie County. The holotype is from the Bois Blanc Formation of Michigan. There are 13 axial rings and ten pleural segments. There are no axial nodes, and a pygidial border is present.

Cyphoproetus cf. C. wilsonae **(Sinclair) Plate 137**
Specimen PRI 49633

This small proetid is from the Middle Ordovician, lower Trenton Group. The genus is characterized by the large L1 lobes isolated by the S1 furrow. A Canadian specimen figured by Ludvigsen (1979b, Figure 31C) differs slightly from the PRI trilobite. Another characteristic listed by Owens (1973, p. 27), lateral lobes on the occipital ring, is not shown on the Ludvigsen illustration but can just be made out on the left side of the PRI specimen in Plate 137. The Canadian specimen appears to have a narrow preglabellar field. This is not evident on the PRI trilobite. *Cyphoproetus* is a Silurian genus somewhat different from this species. These differences are under investigation (Jon Adrain 1999, private communication).

Dechenella haldemani **(Hall, 1861) Plate 138**
Lectotype AMNH 5504 (Lieberman 1994)

This trilobite is restricted to the Middle Devonian Marcellus Formation, lower Hamilton Group. Specimens have been found in Otsego and Onondaga Counties. The illustration from Hall and Clarke (1888) is of a specimen from Pennsylvania, which was clearly designated to be the holotype. The selection of a lectotype may be invalid. The cephalon is semicircular in outline with short genal spines. The cephalic border is broad, and there is a narrow preglabellar field. The glabella narrows anteriorly and the glabellar furrows are deeply incised. There is no median occipital node and the lateral occipital lobes are connected posteriorly to the occipital ring. There are no axial thoracic nodes. The pygidium has 17 axial rings and 12 pleural segments. There is a pygidial border, which widens posteriorly. Plate 138 is a whole specimen from the Union Springs Member of the Marcellus.

Decoroproetus corycoeus **(Conrad, 1842) Plates 139 and 140**
Lectotype AMNH 1829 (Holloway 1980), hypotype NYSM 4707

Found in the lower Rochester Shale and reported from the Lockport limestone, this trilobite is suboval and has long genal spines reaching the sixth thoracic segment. The eyes are large and reniform. Plate 140 shows the same specimen as Plate 139 but shows an associated ophiuroid (brittlestar *Furcaster echinatus*).

Gerastos protuberans **(Hall, 1859)**
Lectotype AMNH 35239 (Lieberman 1994)

The lectotype is a pygidium from the Lower Devonian New Scotland Limestone. A partial cephalothorax was also figured by Hall (1859b). The pygidium has eight axial rings and four or five pleural segments. Pygidia with this description are known from the Onondaga of Ontario County and the Bois Blanc Formation of Michigan.

Hedstroemia pachydermata **(Barrett, 1878)**

This genus is widespread in northern Laurentia, England, and Baltica. It was first reported in eastern North America as *H. pachydermata* in the Deckers Ferry Formation in Orange County and in New Jersey.

Mannopyge halli **(Stumm, 1953)**
Holotype AMNH 4074

Type species. The holotype of this trilobite is an incomplete thorax and complete pygidium from the Edgecliff Member of the Onondaga Limestone, Williamsville, Erie County. The pygidium is very distinctive. There are nine pygidial axial rings and nine pleural segments. A groove across the distal end of the pleural segments separates the distal parts into nodes. These nodes are rounded and appear as part of an unformed border. No other proetid from New York has these characteristics. This trilobite has also been reported from Ontario. The genus was erected by Ludvigsen (1987).

Monodechenella macrocephala **(Hall, 1861) Plate 141**
Lectotype AMNH 4734 (Lieberman 1994)

Type species. The lectotype is from Ontario County. A proetid of the Middle Devonian Hamilton Group and its equivalents, *M. macrocephala* is often found in the same horizons with *Pseudodechenella rowi*, although in significantly reduced numbers. However, in the Deep Run and Kashong Shales it is the most common proetid. The cephalon is semicircular in outline with short genal spines. The glabella is inflated and pustulose and in dorsal view extends anteriorly over the wide cephalic border. S1 is deeply incised and extends perpendicular toward the median line, with a sharp dog-leg bend posteromedially, nearly to SO. The occipital ring is wide, with lateral lobes separated by a shallow groove. There is no median occipital node. The pygidium is longer than wide and has 14 axial rings and 12 pleural segments. There is a pygidial border. The entire test is covered with granulations, coarsest on the glabella.

***Proetus artiaxis* (Howell and Sanford, 1947)**
Holotype USNM 488127

This trilobite is known from a single pygidium found in the Upper Silurian Eramosa Member of the Lockport Group. The pygidium has a narrow axial lobe, which is almost parallel sided. The axis is rather highly convex and has three well-defined segments in its anterior half.

***Proetus (Proetus) clelandi* (Raymond, 1905)**
Lectotype Cornell Museum 5678a (Shaw 1968)

This trilobite was part of the Jewett Collection purchased by Cornell University in 1868 (Shaw 1968). The description accompanying the only known specimen was "Chazy Limestone, Chazy Village, New York" in Clinton County. Inasmuch as both Chazy and Trenton Limestones are found in the area and since no other specimens have been found in the Chazy, the exact stratigraphic horizon is in doubt.

***Proetus jejunus* (Hall and Clarke, 1888)**

The species is known from a single exfoliated pygidium from the Middle Devonian "sandy shales" of the Hamilton Group, Albany County. The pygidium is wider than long and has ten axial rings and eight pleural segments. There are prominent axial nodes.

***Proetus marginalis* (Conrad, 1839)**
Hypotypes NYSM 4752, 4753

This proetid trilobite was found in a Devonian boulder near Ithaca, Tompkins County. "This has a much less prominent front than the rowi, a deeper groove between the eye and middle lobe, and the tubercle which nearly joins the lower angle of the eye is much smaller" (Conrad 1839). Hall and Clarke (1888) ascertained that the boulder referred to was from the Middle Devonian Tully Formation. A number of other proetids had been assigned to the species, and Hall synonymized them all with *Pseudodechenella rowi*. The location of the type is unknown. The hypotypes are from the Tully Formation near Ovid, Seneca County.

***Proetus microgemma* (Hall and Clarke, 1888)**
Syntype NYSM 4740

This species is known from two pygidia from the Middle Devonian Onondaga Limestone of Ontario County. It quite possibly is not a proetid.

***Proetus parviusculus* (Hall, 1860)**
Specimen AMNH 1071

Proetus parviusculus is an Upper Ordovician, Ohio proetid. The specimen mentioned is from Floyd, Oneida County, near Rome. Both Middle Ordovician Utica and Upper Ordovician Frankfort Shales are exposed in the area. A specimen in the USNM (USNM 34490) is a proetid from (probably Upper Ordovician) Utica Shale, Holland Patent, Oneida County, and is part of the Rust collection.

***Proetus spurlocki* (Meek, 1872)**
Specimen USNM 92544

Proetus spurlocki is an Ohio Upper Ordovician proetid. It is included here because of a small pygidium in the National Museum from Floyd, Oneida County, labeled with this name (Hurlburt Collection). Floyd is a town northeast of Rome, Oneida County. Both the Middle Ordovician Utica and Upper Ordovician Frankfort Shales are exposed in the area.

***Proetus tenuisulcatus* (Howell and Sanford, 1947)**
Holotype USNM 488134

This trilobite is from the Upper Silurian Eramosa Member of the Lockport Group in Monroe County. The glabella is subtriangular and reaches all the way to the brim in front. The glabellar furrows are very faint. The fixed cheeks are narrow, and the posterolateral portions are narrow and end in sharp points. The pygidium is moderately convex. The axis is narrow; it tapers only a little and rises well above the pleural lobes. There are seven well-defined axial segments. The pleural lobes are smooth, with only faint traces of ribs.

***Proetus undulostriatulus* (Hall, 1847)**
Holotype? AMNH 30101, plastoholotype NYSM 9855

Hall first described this proetid from the Middle Ordovician Snake Hill Beds, Saratoga County. Additional specimens from the same area have been described as *Cyphaspis*. The Snake Hill Beds are an eastern New York equivalent of the lower Utica Shale.

Proetid sp. Plate 142
Specimens USNM, MCZ 111714

Small unprepared, unnumbered, and undescribed Middle Ordovician proetids are in collections from the Walcott-Rust Quarry in the Rust Limestone Member of the Trenton Group in Herkimer County. Both the MCZ and the USNM have specimens. Plate 142 shows a prepared specimen from the Walcott-Rust Quarry now in the MCZ.

***Pseudodechenella arkonensis* (Stumm, 1953)**
Holotype UMMP 25541

This trilobite is described from the Middle Devonian Hamilton age rocks of southwestern Ontario. Lieberman (1994) reported this species from the Ludlowville Member of the Hamilton Group in central New York. Stumm described it as similar to *Pseudodechenella rowi* except that there is no trace of glabellar furrows, the genal spines reach to the sixth or seventh thoracic segment, and there is a distinct occipital node. There are axial nodes on the thorax, more strongly expressed on the more posterior segments. The pygidium has a prominent axial node on the most anterior axial ring and lacks tuberculation. In addition, the pygidium has 13 axial rings and nine pleural segments.

***Pseudodechenella canaliculata* (Hall, 1861)**
Holotype AMNH 4253, specimen YPM 33773

Lieberman (1994) assigned specimen YPM 33773 to this species. It is an exfoliated pygidium labeled as coming from Cherry Valley, Otsego County, and is tentatively assigned to the Onondaga. This is the only specimen from New York. The holotype is from the Jeffersonville Limestone at the Falls of the Ohio, Kentucky. The anterior cephalic border has a medial transverse groove that is raised on each side, giving the appearance of two transverse ridges on the border. One must question the identity of YPM 33773 until more information is available.

***Pseudodechenella clara* (Hall, 1861) Plate 143**
Lectotype AMNH 39326 (Stumm 1953)

This trilobite is described from the Middle Devonian Onondaga Limestone at Stafford in Genesee County. A good specimen is also in the USNM (USNM 25884). The species has a subconical, faintly constricted glabella with faint glabellar furrows. The cephalic border is wide, flat, or slightly concave and has a distinctive raised outer margin. The occipital ring has no medial node and has lateral lobes separated by a shallow furrow. The pygidium has 11 axial rings. The illustrated (Plate 143) specimen is from the Moorehouse Member of the Onondaga Limestone in the Honeoye Falls Quarry, Monroe County. This trilobite is widely distributed in the Onondaga of New York and age-equivalent rocks of Ontario, Michigan, Ohio, and Kentucky.

***Pseudodechenella rowi* (Green, 1838) Plates 144, 145, 146, and 147**
Plastotype NYSM 4750

Type species. This trilobite is found throughout the Middle Devonian Hamilton Group, with the exception of the deeper-water black shales of the group. It is the most commonly found proetid of New York, with numerous specimens found from Erie County to east of Livingston County. The cephalon is semicircular in outline with genal spines. The cephalic border has a narrow raised edge. There is no preglabellar field. The glabella is vaulted and tapers slightly forward. Glabellar tubercles are faint to absent. The occipital ring has a very slight to absent medial node. The lateral occipital lobes are separated by a groove. There are no axial nodes on the smooth thorax. The pygidium has 11 axial rings and eight pleural segments. Plate 144 shows a small group of individuals from the Windom Shale in Erie County and Plate 145 is of a *P. rowi* together with *Bellacartwrightia whiteleyi*. The species has also been reported in Michigan. Plate 147 shows a cluster of individuals from the Tully Formation, revealing some morphological differences from those lower in the Hamilton. Pillet (1972) erected the new genus *Pseudodechenella*, with *P. rowi* as the type species. Basse (1997) agreed with Pillet that the New World "*Basidechenella*" species are different from those from Europe.

5.7 ORDER ASAPHIDA

Family Asaphidae

Asaphids are the most numerous specimens represented in the New York Ordovician. Although some may be suspect by synonymy, this will not displace them from "the most numerous" title. We have found a total of 41 asaphids listed from the Lower, Middle, and Late Ordovician of New York. Asaphids have planktic protaspids, which makes for easy and widespread dispersal. The meraspids and early holaspids usually have long genal spines, which often carries through to the adult. When mature, the asaphids have a smooth, robust, unornamented body, which is well suited for plowing through and feeding in the surface mud and debris of the sea bottom. The holaspid has eight thoracic segments. The cephalic sutures are opisthoparian. The cephalon and pygidium are often similar in size and shape, an observation that led to the genus name *Isotelus*.

The oldest asaphids in New York are found in the Lower Ordovician Beekmantown Group, and then more or less are found continuously upward through the Chazy, Black River, Trenton, and Lorraine Groups. Asaphids geographically are very widely distributed in eastern North America; therefore, it is not likely that any straightforward evolutionary sequences will be found within the state. The family Asaphidae does not go beyond the Ordovician.

******Basilicus romingeri* (Walcott, 1877)**
Holotype MCZ, hypotype NYSM 12918

Basilicus romingeri specimens from the Middle Ordovician Black River Group, Herkimer County, are in the NYSM as NYSM 12918 (hypotype) and the USNM as USNM 61277. The genus *Basilicus* differs from *Basiliella* in that the pygidium is longer than wide. DeMott (1987) synonymized *Basilicus romingeri* and *B. wisconsensis* with *Basiliella barrandi*.

***Basilicus ulrichi* (Clarke, 1894)**
Specimen USNM 61278

The USNM specimen is from the Middle Ordovician, upper Black River Limestone, 1.6 km (1 mile) north of Poland, Herkimer County.

***Basilicus? vetustus* (Hall, 1847)**
Type AMNH 613

The type of this Middle Ordovician Black River trilobite is from "the compact Birdseye Limestone of the Mohawk Valley." In the USNM a similarly labeled pygidium (USNM 4779) has at least 14 axial rings.

Basilicus (Basiliella) whittingtoni **(Shaw, 1968)**
Holotype NYSM 12535

This species, from the Middle Ordovician Chazy Limestone, is only known from cranidia and pygidia but is unquestionably *Basilicus* and possibly *Basiliella*. Shaw equated this trilobite with *Asaphus marginalis* (abandon). The illustration by Raymond (1905) is a reconstruction of the former *A. marginatus*. The pygidium has paired, rounded projections. For location information, see Shaw (1968).

Basiliella barrandi **(Hall, 1851)**
Specimen MCZ 100948

Type species. Specimens of *B. barrandi* in the MCZ are from the Middle Ordovician Black River Limestones near Poland and Newport, Herkimer County. Members of the genera *Basiliella* and *Basilicus* are the only asaphids "with marginal position of the facial suture in front of glabella" (*Treatise*). Both have genal spines, a cephalic border with a convex rim, and a concave, well-defined border on the pygidium. *Basiliella barrandi* has a near semicircular pygidium. See also DeMott (1987).

Bellefontia gyracanthus **(Raymond, 1910)**
Lectotype? (Westrop, Knox, and Landing 1993)

Westrop et al. listed the family for genus *Bellefontia* from Lower Ordovician as uncertain. The *Treatise* placed it with the asaphids. There are specimens in the MCZ labeled *Asaphus gyracanthus* from Tribes Hill, Canajoharie, Montgomery County. They are MCZ 3478 and 3479.

? Ectenaspis homalonotoides **(Walcott, 1877) Plate 148**
Syntypes FMNH 12324a, 12341b

? Ectenaspis homalonotoides, originally described by Walcott based on specimens from Illinois, was said by Raymond and Narraway (1910) to be fairly common in the Middle Ordovician Black River Limestones at Pattersonville, Schenectady County. Later, Raymond (1925) withdrew this identification of the Black River specimens and equated the original with Ontario material from just above the base of the Trenton. This later publication also assigned it to his new genus *Ectenaspis*. *Ectenaspis homalonotoides* has a long, tapering projection to the front of the cephalon and a triangular pygidium. Plate 148 illustrates a specimen from Ontario identified as *Isoteloides homalonotoides*. See also DeMott (1987).

Homotelus stegops **(Green, 1832)**
Hypotypes NYSM 9757 to 9765

This trilobite is found in the Upper Ordovician Whetstone Gulf Shale of the Lorraine Group, Oneida County. The cephalon has no trace of a border, and the pygidium has a poorly defined one. The axis of the pygidium is not as prominent as the illustration from the *Treatise* suggests, but the posterior end of it is a well-defined rounded node. *Homotelus* is considered a questionable genus and may be assigned to *Isotelus*.

Hyboaspis depressa **(Raymond, 1925)**
Lectotype MCZ 101138 (Shaw 1968)

Only the pygidium of this Middle Ordovician Chazy species is known. However, Raymond (1925) considered these pygidia as fairly common. There is some question of whether it is an asaphid or an illaenid. See Shaw (1968) for location information.

Isoteloides angusticaudus **(Raymond, 1905)**
Holotype CM 1285

Shaw (1968) referred to this Middle Ordovician Chazy trilobite as "*Isotelus* sp." because of its close similarity to *Isotelus harrisi* and the fact that only the pygidium, which was the original figured specimen, can be found. An entire specimen later figured by Raymond (1910c) is lost. The *Isoteloides* species has well-defined, flattened borders on the cephalon and pygidium. See Shaw (1968) for location information.

Isoteloides canalis **(Whitfield, 1886) Plate 149**
Lectotype AMNH 35276 (Brett and Westrop 1996)

The species is from the Lower Ordovician Scotia Limestone Member of the Fort Cassin Formation in Washington County. The diagnosis from Brett and Westrop is "a species of *Isoteloides* with a relatively short axis that occupies up to about 80 percent of pygidial length and, consequently, relatively long border. Anterior branches of facial sutures are moderately divergent, so that maximum frontal area width is equal to, or slightly greater than glabellar length." The name is unfortunate. The first record of *Asaphus canalis* is in Hall's publication (1847). It is attributed to Conrad and refers to material from the Chazy group. Whitfield (1886b) used the name to refer to material from the Fort Cassin Formation. Later authors abandoned the name *Isotelus canalis* for the Chazy material because of poor definition of the type specimens. Thus, one might argue that *Isoteloides whitfieldi* is now the proper name.

Isoteloides peri **(Fortey, 1979)**
Holotype GSC 56799

This species is differentiated from *I. canalis* by a relatively longer pygidial axis, 85 to 90% of the pygidial length. The species is from the Lower Ordovician Scotia Limestone member of the Fort Cassin Formation, Washington County. The species is also reported from Newfoundland and Pennsylvania (Brett and Westrop 1996).

Isotelus

The genus *Isotelus* has not been properly revised, and species identification, particularly in older museum specimens, can be poor. Rudkin and Tripp (1985, 1987) of the ROM have done significant work with the Trenton and Upper Ordovician *Isotelus* species that occur in Ontario. Generally speaking, *I. platycephalus* is found in the Black River age rocks; *I. gigas* and *I. walcotii*, in the middle Trenton; *I. latus* and another isotelid informally known as "*I. mafritzae*" are found in the upper Trenton and lower Upper Ordovician; and *I. brachycephalus* and *I. maximus*, in the Upper Ordovician. Other unnamed species have been identified, but similar careful examination on the New York species has not been done. Mature specimens of *I. platycephalus*, *I. gigas*, and *I. latus* do not have genal spines. Genal spines are found on mature specimens of *I. walcotti*, "*I. mafritzae*," *I. brachycephalus*, and *I. maximus*. Although "*I. mafritzae*" is not a formally published name, specimens from the Lindsay Formation in southern Ontario are commercially available with this name (Rudkin and Tripp 1985, 1987, 1989; David Rudkin 1997, private communication).

Isotelus annectans **(Raymond, 1920)**
Holotype CM?

The holotype was originally described as *Ectenaspis homalnotoides* from New York but on further review was assigned to a new species. Raymond referred to this species as a link between *Ectenaspis* and *Isotelus*. It is described as fairly common in the Glens Falls Member of the Middle Ordovician Trenton Group at Pattersonville, Schenectady County, and Smiths Basin, Washington County. The borders on the cephalon and pygidium are poorly defined. Both the cephalic and pygidial axes are wide and poorly defined.

Isotelus giganteus **(Raymond, 1931)**
Holotype MCZ 104982

The holotype, and only specimen, is a huge hypostome from the upper Chazy Limestone at Tiger Point, Valcour Island, Clinton County. The hypostome measures 80 mm across the posterior points and about 100 mm at its widest. From the size of the hypostome and by knowledge of the relative size of the hypostoma of other species of the genus, Raymond believed this specimen was 600 to 675 mm (24 to 26 inches) long, making it a truly giant trilobite. It is not, however, likely to be a distinct species but a large specimen of the large asaphids already known in the Chazy.

Isotelus gigas **(DeKay, 1824) Plates 150, 151, 152, and 153**
Neotype MCZ 100938 (Rudkin and Tripp 1989)

Type species. *Isotelus gigas* is a very well-known New York trilobite because of the numbers of specimens, which were found in the Middle Ordovician, middle Trenton Limestones of the Trenton Falls area, Herkimer County. Immature holaspids have genal spines. These spines disappear as the trilobite grows, becoming shorter with each successive molt. The size at which the genal spines disappear is different for *I. gigas* from different localities. For example, *I. gigas* from the Lake Simcoe area of Ontario retains its spines longer (i.e., at larger sizes) than those of the Mohawk Valley. Plate 150 shows the neotype from the Walcott-Rust Quarry and is in the MCZ. Plate 151 is of a large, near-perfect specimen from the same location as the neotype. Plate 152 shows a cluster from Trenton Falls. Plate 153 is the ventral view of a specimen from Quebec.

Isotelus harrisi **(Raymond, 1905)**
Lectotype YPM 23297 (Shaw 1968)

Isotelus harrisi is a fairly common asaphid in some Middle Ordovician Chazy exposures. See Shaw (1968) for location information. It is a large trilobite but no articulated specimens are reported.

Isotelus jacobus **(Clarke, 1894)**
Holotype NYSM 4512 (lost)

The type is listed from Crown Point, Essex County. A large pygidium in the USNM, USNM 13642, is from this locality. It is also listed from Middleville, Herkimer County, a Middle Ordovician, lower Trenton area.

Isotelus latus **(Raymond, 1913)**
Holotype GSC; specimens NYSM 17009, 17010

The NYSM specimens are from Middle Ordovician, upper Black River or lower Trenton areas at Middleville and Newport, Herkimer County, respectively. One is a very large semicircular pygidium, 120 mm wide by 90 mm long. Given that *I. latus* is from higher up in the Trenton/Upper Ordovician, these specimens are most likely *I. platycephalus*.

Isotelus **cf. *I. maximus*** **(Locke, 1838) Plates 154, 155, 156, and 157**
Hypotypes NYSM 4313, 4315

Ruedemann (1901) found material in the Middle Ordovician Rysedorph Conglomerate, Rensselaer County, that he assigned to the Upper Ordovician trilobite, found in Ohio and Ontario. NYSM 4313 is a small, negative mold, which is not diagnostic. On the same matrix are two pygidia and a cranidium of a pterygometopid. He believed the asaphid resembled *I. maximus* from Ohio, which has a cephalon and pygidium with a very rounded outline. The genal spines on Ohio specimens are long and thin. They are kept through maturity. The New York specimens are a questionable assignment. Plate 154 illustrates a specimen of *I. maximus* from Ohio and Plates 155 and 156 are the ventral view of a pyritized specimen from Ohio, with the posterior appendages possibly pyritized. Plate 157 is of a molt with the cephalon overturned.

Isotelus platycephalus **(Stokes, 1824)**
Lectotype BM (Darby and Stumm 1965)

The lectotype is from the Black River Group of Ontario and is found in association with *Basiliella barrandei* and *Bumastoides milleri.* It differs from *I. gigas* in that both the cephalon and the pygidium are more rounded and less triangular in outline. Black River *Isotelus* species should be compared to *I. platycephalus* when one is considering their validity.

***Isotelus pulaskiensis* (Ulrich, 1926)**
Hypotype NYSM 9767

The type of this species is from Pennsylvania. The specimen in the USNM, USNM 23628, bearing this name is from the Pulaski drift (Upper Ordovician) in the Trenton Falls, Herkimer County, area and is from the Rust collection in a 1886 purchase. A small free cheek has long genal spines, longer than what is seen on immature *I. gigas* specimens.

***Isotelus simplex* (Raymond and Narraway, 1910)**
Holotype CM 1441

DeMott (1987) listed this Middle Ordovician Black River trilobite, originally described from Pennsylvania, as from Buck's Quarry near Poland, Herkimer County. It differs from *I. gigas* in that the shields are less triangular and lack concave borders. Some asaphids from the Black River Buck's Quarry near Poland, Herkimer County, bear the label *I. iowensis* (MCZ 426). These are probably specimens of *I. simplex.*

***Isotelus walcotti* Ulrich, in (Walcott, 1918) Plate 158**
Lectotype USNM 61261a (Rudkin and Tripp 1989)

Isotelus walcotti is from the Walcott-Rust Quarry, the source of all the specimens known to us. The quarry is located in the lower Rust Limestone of the Middle Ordovician Trenton Group, Herkimer County. The lectotype is figured in Plate 158. The trilobite bears a strong resemblance to *I. iowensis* from the Upper Ordovician of the north-central United States. The species differs from the *I. gigas* found in the same rocks in that it keeps its genal spines throughout its growth, and the anterior and posterior ends of the cephalon and pygidium, respectively, are rounded in *I. walcotti* while they have a more pointed appearance in *I. gigas.* Larger specimens of *I. walcotti* are rare. Only one approaching the size of a mature *I. gigas* is in the MCZ, MCZ 422, and it still has its genal spines.

***Nileoides perkinsi* (Raymond, 1910)**
Lectotype UVM 2-66 (Shaw 1968)

Nileoides perkinsi is from the Middle Ordovician, upper Chazy of Vermont. Raymond (1925, p. 97) stated that he found a specimen from the Chazy of Clinton County.

***Pseudogygites latimarginatus* (Hall, 1847) Plate 159**
Holotype AMNH 1070

Hall described this species from two isolated pygidia from an unspecified location near Watertown, Jefferson County. A similar species from Canada, found in abundance in the upper Middle Ordovician, was named *Pseudogygites canadensis.* Fieldwork established that *P. latimarginatus* can be found in the Hillier Limestone member of the topmost Middle Ordovician Trenton and in the base of the overlying shale. This work also established the fact that *P. latimarginatus* and *P. canadensis* are the same, and *P. canadensis* was abandoned as a species. Both the cephalon and pygidium have well-defined borders. There are short genal spines. The pygidium has numerous, prominent pleurae that almost reach the margin. Plate 159 is a NYSM specimen from Oneida County.

***Vogdesia bearsi* (Raymond, 1905)**
Neotype NYSM 12538 (Shaw 1968)

Type species. This trilobite is known from one area of the Middle Ordovician Chazy where it is considered abundant, Sloop Bay, Valcour Island, Clinton County. Strangely enough, similar age exposures a short distance away do not yield this trilobite. The eyes are on short stalks. The pygidium has a concave border, and the pleural field is flat. The thorax, doublure, and hypostome of *V. bearsi* are unknown.

***Vogdesia? obtusus* (Hall, 1847)**
Neotype NYSM 12524 (Shaw 1968)

This fairly common trilobite is found throughout areas of the Middle Ordovician Chazy. The fragmentary nature of most of the material prevents a more exact placement within the asaphids. Additional specimens are MCZ 3601 through 3609. Raymond (1925, p. 90) also reported this species from Vermont and Virginia. See Shaw (1968) for location information.

Family Idahoiidae

***Saratogia (Saratogia) calcifera* (Walcott, 1879) Plate 175B**
Lectotype USNM 58555 (Ludvigsen and Westrop 1983)

Upper Cambrian, Hoyt Limestone, Saratoga County.

Family Pterocephaliidae

***Cameraspis convexa* (Whitfield, 1878)**
Specimens NYSM 14159, 14160

Upper Cambrian, Galway Formation, Saratoga County. See Ludvigsen and Westrop (1983) for synonymies. This species is also found in Wisconsin, Minnesota, Pennsylvania, Oklahoma, Missouri, Montana, and Wyoming.

***Cameraspis* sp.**
Specimen MCZ 4810

Potsdam Sandstone, north edge of Battle Hill, near Fort Ann, Washington County.

Family Ptychaspididae

***Conaspis whitehallensis* (Walcott, 1912)**
Holotype USNM 58579

Upper Cambrian, Potsdam Sandstone, Whitehall, Washington County.

Idiomesus sp. **(Bird and Rasetti, 1968)**

Upper Cambrian trilobite from a limited exposure in the East Chatham quadrangle of Columbia County.

Keithiella depressa **(Rasetti, 1944)**

Holotype GSC 71161

Upper Cambrian, Hoyt Limestone, Saratoga County. See Ludvigsen and Westrop (1983) for synonymies. This species is also found in Quebec and Newfoundland.

Family Raphiophoridae

All of the raphiophorids in New York are found in the Middle Ordovician. All are assigned to the genus *Lonchodomas* and are characterized by a glabellar medioanterior spine projecting straightforward. There are also genal spines. The thorax has five segments. The pygidium is small and semicircular and has a narrow border.

Lonchodomas chaziensis **(Shaw, 1968)**

Holotype NYSM 12265

Lonchodomas chaziensis is a trinucleoid characterized by a long, four-sided glabellar spine projecting straight forward. It is found in the Lower and Middle Chazy. See Shaw (1968) for location information.

Lonchodomas halli **(Billings, 1865)**

Topotype NYSM 12279

In New York *L. halli* is found in the Upper Chazy. It is also found in Chazy age rocks in Virginia, Quebec, and Vermont. Shaw (1968) differentiated the two *Lonchodomas* species from the Chazy as follows: "*L. halli* differs from *L. chaziensis* by possession of the following characters: (1) glabellar profile more convex, with a more definite keel present. (2) Glabellar furrows are extremely faint to absent. (3) Glabella tapers more rapidly into glabellar spine, thus the glabella itself does not extend far beyond the fixed cheeks. (4) S-shaped genal spines. (5) Possession of a sight flat area on fixed cheek proximal to anterior portion of facial sutures." See Shaw (1968) for location information.

Lonchodomas hastatus **(Ruedemann, 1901)**

Syntypes NYSM 4128 to 4135

Lonchodomas hastatus was found in Trenton age pebbles in the Rysedorp Hill Conglomerate, Rensselaer County.

Family Remopleurididae

Remopleuridids are known from the Upper Cambrian to the Middle Ordovician in New York. They are not common anywhere. The family is characterized by long curved eye lobes. All species have genal spines and small pygidia.

Apatokephaloides sp. **(Bird and Rasetti, 1968)**

An Upper Cambrian species, from Columbia County.

Hypodicranotus striatulus **(Walcott, 1875) Plates 160, 161, and 162**

Lectotype MCZ 100986 (Raymond 1925)

Type species. *Hypodicranotus striatulus* is a very interesting trilobite that was first described from the Middle Ordovician, upper Trenton, Walcott-Rust Quarry, Herkimer County. In addition to the remopleurid cranidium, the species has long genal spines running parallel to spinelike extensions of the occipital ring, giving the appearance of a dual genal spine. The hypostome is like no other. It is shaped like a tuning fork and extends to the end of the thorax. The pygidium is very small. This species has been reported from New York, Ontario, Missouri, Quebec, and the District of Mackenzie, Canada. Plate 160 shows a specimen from the Walcott-Rust Quarry. Plate 161 is of another specimen from the Walcott-Rust Quarry that split longitudinally, showing the long hypostome. Plate 162 shows the small pygidium and the unusual, bifurcate lateral genal area. The genus has been found in Minnesota, Wisconsin, Oklahoma, Nevada, Scotland, the central Ural Mountains, and far northeastern Russia (see Ludvigsen and Chatterton 1991). Because of its shape it has been suggested the trilobite might be pelagic, and its wide distribution goes along with this.

Remopleurides canadensis **(Billings, 1865)**

Holotype GSC 1760 (Shaw 1968)

Remopleurides canadensis is found throughout the Middle Ordovician Chazy. Remopleurids have a distinctive cranidium, which can not easily be mistaken for any other genus. See Shaw (1968) for location information. This species is also reported from Ottawa and Virginia.

Remopleurides linguatus **(Ruedemann, 1901)**

Hypotypes NYSM 4770 to 4774

Middle Ordovician Trenton age pebbles in the Rysedorph Conglomerate.

Remopleurides tumidus **(Ruedemann, 1901)**

Holotype NYSM 4775

Middle Ordovician Trenton age pebbles in the Rysedorph Conglomerate, Rensselaer County.

Richardsonella sp. **(Bird and Rasetti, 1968)**

Upper Cambrian, Columbia County.

Robergiella cf. *R. brevilingua* **(Fortey, 1980)**

Brett and Westrop (1996) reported a cranidium and free cheek very similar to those of this species from the Lower Ordovician Scotia Limestone member of the Fort Cassin Formation, Washington County.

Family Saukiidae

***Hoytaspis speciosa* (Walcott, 1879)**

Lectotype USNM 58563 (Resser 1942a)

Type species. This species is known from the Upper Cambrian, Hoyt Limestone, Saratoga County, and also the Potsdam Sandstone drift near Trenton Falls, Herkimer County (MCZ 3858).

***Prosaukia briarcliffensis* (Lochman, 1946)**

Holotype YPM 17391

Upper Cambrian from rocks in Dutchess County referred to as "Hoyt Dolostone."

***Prosaukia hartti* (Walcott, 1879) Plate 175A**

Holotype USNM 58571

Upper Cambrian, Hoyt Limestone, Saratoga County. This species is also found in Wisconsin.

***Prosaukia tribulis* (Walcott, 1912)**

Holotype USNM 58578

Upper Cambrian, Hoyt Limestone, Saratoga County.

Family Symphysurinidae

This family, once considered a subfamily of Asaphidae, is represented by two species, and no general family description is given. The family is not listed in volume 1 of *Treatise* (revised).

***Symphysurina convexa* (Cleland, 1900)**

Holotype PRI 5072

This trilobite is from the Tribes Hill Formation of the Lower Ordovician in Montgomery County. The glabella is vaulted, longer than wide, and the palpebral lobes are about midpoint. The pygidium is semicircular to parabolic, rounded with a well-defined axis. Axial segments are defined but not prominent. *Symphysurina convexa* is characterized by a pitted prosopon or ornamentation on external surfaces. See Westrop, Knox, and Landing (1993).

***Symphysurina* cf. *S. woosteri* Ulrich, in (Walcott, 1924)**

Specimens NYSM 15209, 15210

Type species. This trilobite is from the Lower Ordovician Tribes Hill Formation, Montgomery County. It is based on two pygidia with stout medial spines. More material is necessary for positive identification. See Westrop, Knox, and Landing (1993).

Family Trinucleidae

Hughes, Ingham, and Addison (1975) reviewed and revised trinucleids. The trinucleids are represented in New York by five species or subspecies: *Tretaspis reticulata*, *T. diademata*, *Cryptolithus tessellatus*, *C. lorettensis*, and *C. bellulus*. The trinucleids are a family originating in the Southern Trilobite Province Zone of Whittington (1966) in the lower Arenig. They first appeared in North America during the latter part of the Middle Ordovician (early Caradoc), likely due to the closing of the Iapetus Ocean and associated changes in currents and distances that allowed the pelagic larvae to reach North America. Shaw and Lespérance (1994) argued that all the *Cryptolithus* species represented in New York are part of a long-lived interbreeding population and as such do not hold species rank but should be considered "morphs" of *Cryptolithus tessellatus*. Since these trilobites are clearly separated by time and stratigraphy in New York, we chose to continue to list them as separate species. The mature *Cryptolithus* trilobites are eyeless, while the tretaspids have lateral eye tubercles. Both are considered to be benthic, living and feeding by plowing through the soft bottom and ingesting the microfauna or organic detritus. All have a characteristic, radially pitted, cephalic brim or fringe. The pits are considered to have a sensory function, but they also strengthen the fringe significantly over a smooth surface. The brim is divided horizontally into an upper part of the dorsal cephalic exoskeleton and a lower portion, which comes free, probably during ecdesys. Because of this laminar nature, the upper brim and lower brim are called *lamellae*. The lower lamella has the genal spines.

***Cryptolithus bellulus* (Ulrich, 1878) Figure 5.5C, D and Plate 163**

Holotype USNM 41876

Cryptolithus bellulus geographically is a widely distributed Late Ordovician trilobite found in both limestones and shales. In New York State it is found in the Lorraine Group shales in Oneida County. *Cryptolithus bellulus* with preserved appendages was found along with the more famous appendaged *Triarthrus eatoni* in the Frankfort Shales of Beecher's Trilobite Bed. All we know of the soft tissue anatomy of the cryptolithids is from the 13 pyritized, partial specimens known from this site. *Cryptolithus bellulus* is very similar to *C. lorettensis* with the exception that the pits in E_1 and I_1 are radially aligned in *C. lorettensis* and alternating in *C. bellulus*. There are also statistical differences in the pit counts of the concentric rows, but this is not diagnostic with small sample sizes. Because of the observed co-occurrence of *C. bellulus* and *C. lorettensis* in Canada, it is questionable that these are distinct species. In New York, however, there is no intermix. One must go outside the state to observe *Cryptolithus* species through the time sequence between the middle Trenton and the Lorraine. Plate 163 shows a specimen from the Upper Ordovician shales in Oneida County.

***Cryptolithus lorettensis* (Foerste, 1924) Figure 5.5A, E and Plates 164 and 165**

Holotype GSC 10790

This trilobite is found in the upper Sugar River Limestone, Rathbun Member, or its equivalent limestones in New York. It is found near the top of the member where it replaces *C. tessellatus*. In some areas in Oneida County there is a narrow horizon where both may be found, but this is not common. In each of the USNM and NYSM (E-920) collections there is a box of mixed *C. tessellatus* and *C. lorettensis* specimens labeled from Trenton Falls, Herkimer County. In the Mohawk Valley, *C. tessellatus* is found in the middle Sugar River Limestone, but in the Rathbun or upper member of the Sugar River, *C. lorettensis* is the only *Cryptolithus* species (Vere 1972). The primary distinguishing feature of *C. lorettensis* versus *C. tessellatus* is that *C. lorettensis* has an additional row of pits, I_3, anterior to the glabella and genal area. This row of pits is not always complete in front of the glabella but always extends farther toward the median line than with *C. tessellatus*. Additionally, on large specimens of *C. lorettensis*, the exterior, median genal area is reticulated, and this area is smooth on large specimens of *C. tessellatus*.

Cryptolithus tessellatus **(Green, 1832) Figure 5.5B and Plates 166 and 167**
Holotype lost, neotype MCZ 106492 (Shaw and Lespérance 1994)

Type species. The Middle Ordovician *C. tessellatus* in New York is considered an index fossil of the Sugar River Limestone of the Trenton Group because it apparently first appears in New York in this limestone and is rarely, if ever, found in the overlying limestones. It is common in some horizons and locations in the Sugar River in Oneida County but is not uniformly distributed anywhere. Although small, this trilobite has a distinctive and robust cephalon. The glabella is vaulted and smooth, with only two shallow pits anterior to the occipital ring for glabellar furrows. The cephalon is rounded in outline and about twice as long as wide, including the long genal spines. Cephala are found with and without genal spines. Those without genal spines have lost the lower lamella. The occipital ring has a central spine extending backward over the first few thoracic segments. The most characteristic feature of the trilobite is the anterior brim or fringe, which has uniform rows of pits following the curvature of the anterior edge. These pits, their number, and arrangement are major factors in distinguishing the various species. Pit rows E_1 and I_1 have a thickened area called a *girder*, seen only on the ventral surface, that separates them. *Cryptolithus tessellatus* differs from the other New York *Cryptolithus* species in that there are only three rows of pits (E_1, I_1, I_2) immediately anterior to the genal area. The thorax has six thoracic segments, and the pygidium is rounded, being much broader than it is long. Whole specimens from New York are rare. Plate 166 shows an uncommon, whole specimen from the Sugar River Formation of Herkimer County. Plate 167 is included because it is listed from Trenton Falls, an area not usually included as exposing the Sugar River rocks.

Tretaspis diademata **(Ruedemann, 1901)**
Holotype NYSM 4819

Tretaspis diademata is known from one specimen in the Middle Ordovician part of the Rysedorph Conglomerate, Rensselaer County. It differs from *T. reticulata* in that the rows of pits are arranged radially (aligned) and not just concentric as they are in *T. reticulata*.

Tretaspis reticulata **(Ruedemann, 1901)**
Hypotypes NYSM 4820 to 4822

Tretaspis reticulata was described from remains in Middle Ordovician Trenton age pebbles in the Rysedorph Conglomerate, Rensselaer County. The genus *Tretaspis* differs from *Cryptolithus* primarily in the glabella. Both have a bulbous, forward-expanding glabella, but *Tretaspis* species have three pairs of distinct glabellar furrows, which are absent in *Cryptolithus*. There are three rows of pits anterior to the glabella. The surface of the glabella is pitted and reticulated. There is no occipital spine. The pygidium is small and widely triangular. The number of thoracic segments is unknown for *T. reticulata*, but six may be expected by analogy. For a more complete discussion of the trinucleids of New York, particularly *Cryptolithus*, see Whittington (1941, 1968) and Vere (1972).

5.8 ORDER PTYCHOPARIIDA

This order includes trilobites primarily of the Cambrian and Ordovician. In New York these trilobites are only represented in these periods. This diverse order does not have any easily recognized defining features. Within the individual Ordovician families, however, there are distinctive features.

Family Conocoryphidae

Atops trilineatus **(Emmons, 1846)**
Plastotype NYSM 4204

Type species. Lower Cambrian.

Conocoryphe verrucosa **(Whitfield, 1884)**
Syntypes AMNH 000280, 035978-035983

Upper Cambrian, Potsdam Sandstone, Ausable Chasm, Keesville, Essex County.

Family Dokimocephalidae

Sulcocephalus saratogensis **(Resser, 1942)**
Holotype NYSM 10511

Lower Cambrian, Theresa Formation.

Family Elviniidae

Calocephalites **cf.** ***C. minimus*** **(Kurtz, 1975)**

Upper Cambrian, Galway Formation, Saratoga County. See Ludvigsen and Westrop (1983).

Dellea? landingi **(Ludvigsen and Westrop, 1983)**
Holotype NYSM 14138

Upper Cambrian, Hoyt Limestone, Saratoga County. See Ludvigsen and Westrop (1983).

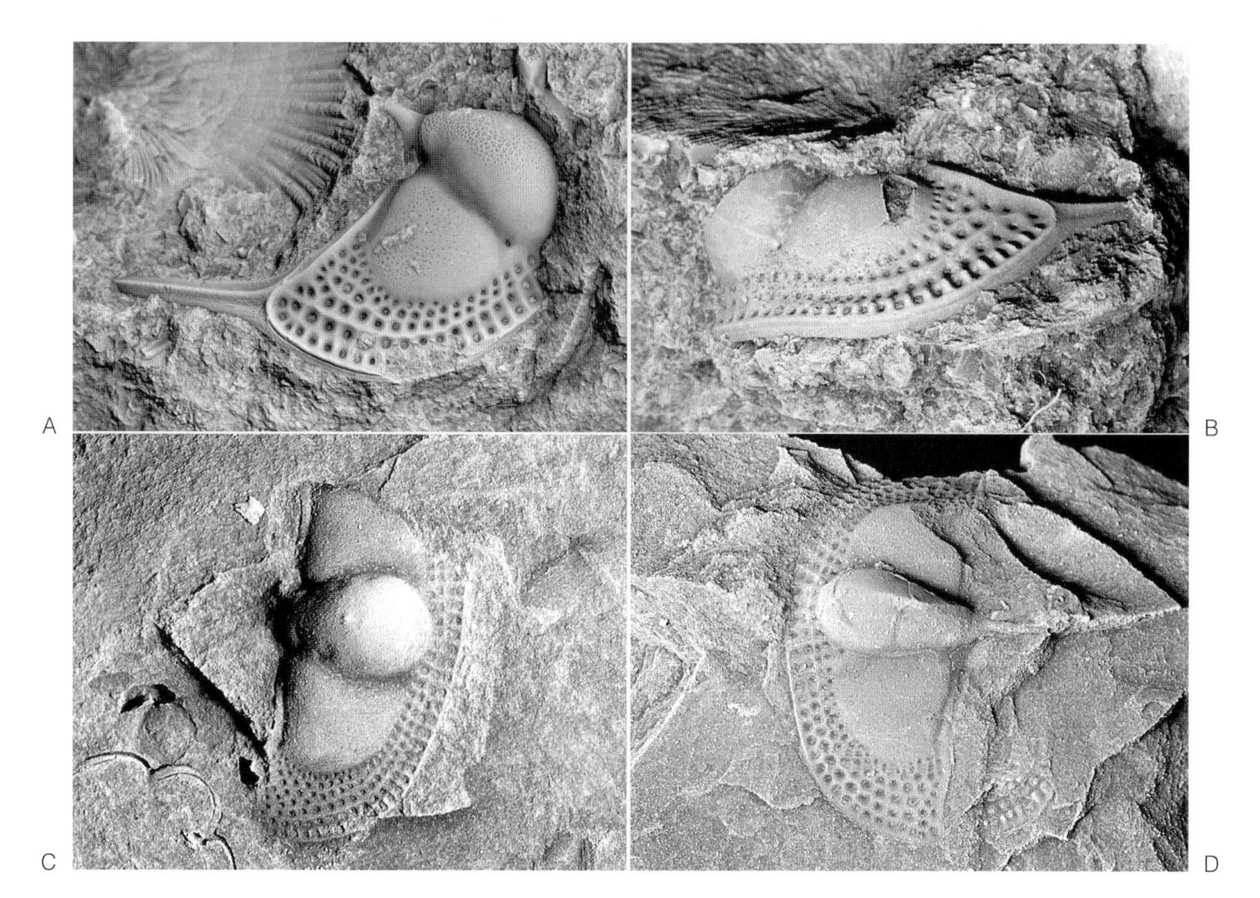

FIGURE 5.5. Features of New York *Cryptolithus*. The specimens are all about the same size. For scale, see the plates. A. *Cryptolithus lorettensis* from the upper Sugar River Formation of the Trenton Group, Jefferson County. Note the four rows of pits anterior to the cheek area (NYSM). B. *Cryptolithus tessellatus* from the lower Sugar River Formation of the Trenton Group, Herkimer County. Note that there are three rows of pits anterior to the cheek area (TEW collection). C. *Cryptolithus bellulus* from the Pulaski Member of the Lorraine Group, Oneida County. There are four rows of pits anterior to the cheek area in this specimen. D. *Cryptolithus bellulus* from the Pulaski Member of the Lorraine Group, Lewis County. There are four rows of pits anterior to the cheek area. The radial alignment of the pits in rows E_1 and I_1 are not quite the same as in C. The specimens in C and D are internal molds and not as clearly defined as those in A and B. The numbering system for the pit rows in *Cryptolithus* are as follows: R_1, R_2, R_3, . . . are designations for the radially aligned rows starting with the medial row; E_1 is the first circumferal row from the front edge of the cephalon; I_1, I_2, I_3, . . . designate the circumferal rows after E_1.

Dellea saratogensis **(Resser, 1942)**
Holotype NYSM 10511

Upper Cambrian, Galway Formation, Saratoga County. See Ludvigsen and Westrop (1983).

Drabia **cf.** ***D. curtoccipita*** **(Wilson, 1951)**

Upper Cambrian, Galway Formation, Saratoga County. See Ludvigsen and Westrop (1983).

Drabia cf. D. menusa **(Wilson, 1951)**

Upper Cambrian, Galway Formation, Saratoga County. See Ludvigsen and Westrop (1983).

Elvinia granulata **(Resser, 1942)**
Holotype USNM 108815

Upper Cambrian, Galway Formation. Saratoga County. See Ludvigsen and Westrop (1983). The holotype is from Nevada. The species is also found in Utah and Indiana.

Family Glaphuridae

Glaphurids are best known from the two genera in eastern North America. Long considered trilobites of the Middle Ordovician Chazy Limestones, they are currently known to have a longer time range (Shaw 1968). Glaphurids are smallish trilobites found in reef facies. Their exoskeleton is subelliptical, tapering from the large cephalon to the small semicircular pygidium. The surface is either very pustulose or spiny. The glabella is suboval and prominent.

Glaphurina lamottensis (Ulrich, 1930) Plate 168B
Holotype USNM 80553

Type species. Shaw (1968) considered *G. lamottensis* from the upper part of the Middle Ordovician, in the lower and the lower part of the middle Chazy, to be a considerably different trilobite from *G. pustulosus*. The exoskeletal prosopon is widely spaced tubercles, compared to the spines on *G. pustulosus*, and there are 12 thoracic segments instead of the 10 in *G. pustulosus*.

Glaphurus pustulosus (Walcott, 1877) Plate 168A
Lectotype MCZ 7566B (Shaw 1968)

Glaphurus pustulosus is a trilobite of upper middle and basal upper Chazy Group limestones. Disarticulated parts can be very abundant locally, but whole articulated specimens are rare. The one articulated and two partially articulated specimens reported by Shaw exhibit spines radiating from the front of the cephalon, the genal angle, the axis of the thoracic segments, and the ends of the pleura. The single articulated specimen is 3.5 mm long and could be a late meraspid. There are ten thoracic segments on this specimen.

Family Harpidae

Harpids are nearly impossible to mistake for any other family. The cephalon is semicircular in the anterior portion, with a very wide fringe. The lateral fringe sweeps back posteriorly to the rear of the cephalon, forming a wide platform, too wide to call a genal spine. This genal platform extends close to the posterior end of the trilobite. The fringe has a laminar structure that is separated into upper and lower lamellae by a suture around the entire lateral edge. Separation is parallel to the plane of the fringe surface. During ecdysis the lamellae separate along this suture, and the trilobite crawls forward, leaving the molted exoskeleton behind. The eyes are small tubercles with two lenses each. The thorax has 12 to 29 segments. The pygidium is small with few segments. Three of the six known harpids in New York are from the Middle Ordovician Chazy limestones.

Dolichoharpes sp. (Shaw, 1968)
Specimens YPM 23294, NYSM 12294 to 12300

This trilobite is from the Middle Ordovician, middle and upper Chazy. Dolichoharpids have a narrower fringe that has three raised, serrated rows along the lateral edge of the upper fringe. For location information, see Shaw (1968).

Hibbertia ottawaensis (Billings, 1865) Plate 169C
Holotype GSC 329, hypotype NYSM 9736, specimen MCZ 12856

The holotype is from the Middle Ordovician of Ontario. The NYSM specimen is from the Snake Hill beds, Snake Hill, Saratoga County. The MCZ material is two specimens from the Middle Ordovician Trenton Limestone of Port Jackson, Clinton County. The specimen in Plate 169C is from a negative of a metal replica of the holotype.

Eoharpes pustulosus (Hall, 1847)
Holotype AMNH 841

Eoharpes pustulosus is a trilobite of the Middle Ordovician, upper Black River Group in the Watertown area, Jefferson County.

Hibbertia valcourensis (Shaw, 1968) Plate 169A, B
Holotype NYSM 12293

Hibbertia valcourensis from the Middle Ordovician, lower Chazy was described by Shaw from material previously described from the Chazy by Raymond (1905b, 1910b) as *Harpina antiquatus* and *Eoharpes ottawaensis* (both described by Billings (1859, 1865) from Canadian specimens). The genus *Hibbertia* differs from *Dolichoharpes* in the width of the fringe area and the ornamentation. For location information, see Shaw (1968).

Hibbertia sp. (Shaw, 1968)
Specimen NYSM 12301

This trilobite is from the Middle Ordovician, lower Chazy Group. For location information, see Shaw (1968).

Scotoharpes cassinensis (Whitfield, 1897) Plate 169D
Holotype USNM 35817

Scotoharpes cassinensis is a Lower Ordovician trilobite first reported from the Beekmantown Group of Vermont. Fisher (1962) reported it also from New York.

Family Hystricuridae

Hystricurids are only found in the Lower Ordovician in New York. They are small pustulose trilobites with opisthoparian cephalic sutures. All have genal spines. Articulated specimens are unknown from New York.

Hystricurus conicus (Billings, 1859)
Holotype GSC 516

Type species. The holotype is from the Lower Ordovician Beekmantown of Quebec. It is also from the Lower Ordovician Spellman Formation.

Hystricurus crotalifrons (Dwight, 1884)
Syntype NYSM 9766

From the Lower Ordovician Roachdale Limestone, Dutchess County.

Hystricurus ellipticus (Cleland, 1900)
Holotype PRI 5073

This species is known from numerous cranidia and pygidia from the Lower Ordovician Tribes Hill Formation, Montgomery County. See Westrop, Knox, and Landing (1993).

Hystricurus cf. *H. oculilunatus* (Ross, 1951)
Specimens NYSM 15230, 15231

Tribes Hill Formation, Montgomery County. See Westrop, Knox, and Landing (1993).

Family Kingstoniidae

Kingstonia seelyi (Walcott, 1912)
Cotypes USNM 58582 to 58584

Upper Cambrian, Potsdam Sandstone. See Resser (1942).

Family Komaspididae

Komaspidids are small opisthoparian trilobites with forward-narrowing glabella. The eyes are characteristic as they are quite long, extending more than half the length of the glabella. There are genal spines. The pygidium is semicircular and roughly the same size as the cranidium. There are only two or three axial rings and pygidial pleurae. Interpleural furrows do not reach the margin. The family is normally found from the Upper Cambrian to the Lower Ordovician (*Treatise*). There is some question whether the below-mentioned trilobite is in the correct family. "*Carrickia* is probably a proetacean; it is definitely not a komaspidid" (R. Ludvigsen 1997, private communication).

"Carrickia" setoni (Shaw, 1986)
Holotype NYSM 12246

"*Carrickia*" *setoni* was described from silicified material from the Middle Ordovician, lower Middle Chazy. Additional material was found in the upper lower Chazy as well. Most of the material is very small silicified specimens. The glabella tapers forward and from the dorsal view is parabolic to semicircular. There are no glabellar furrows. The eyes are set wide apart and extend from about one-third back on the cranidium to nearly the posterior cephalic border. The thorax is unknown. The pygidium is semicircular with a well-defined axis with two axial rings. There are two pleurae. The pleural furrows do not quite reach the margin. This trilobite is the only representative of this normally upper Middle Cambrian to Lower Ordovician family. For location information, see Shaw (1968).

Family Lonchocephalidae

Lonchocephalus (Bradley, 1860)
Plesiotypes USNM 58567 to 58570

Upper Cambrian, Potsdam Sandstone, Keesville, Essex County.

Family Marjumiidae

Modocia punctata (Rasetti, 1967)
Holotype USNM 156687

Middle Cambrian, Stockport Station, Columbia County. See Bird and Rasetti (1968, p. 26).

Family Menomoniidae

Bolaspidella fisheri (Rasetti, 1967)
Holotype USNM 156682

Middle Cambrian, Stockport Station, Columbia County. See Bird and Rasetti (1968, p. 26).

Family Olenidae

Olenids are known in the Ordovician of New York from at least three, possibly five, species of *Triarthrus*. These are small trilobites with a semicircular cephalon. The glabella is rectangular, curving gently in front. The S1 and S2 glabellar furrows are distinct and curve toward the rear. S3 is usually faint. The free cheeks are "yoked," meaning they are attached together anteriorly to the front center of the glabella and are lost as a single unit during the molting process. The fixed cheeks flare outward from the rear end of the palpebral lobes, crossing the genal angle. There is an occipital node. The thorax has 12 to 15 segments. *Triarthrus beckii* and the early *T. eatoni* have axial nodes on all the thoracic segments. *Triarthrus eatoni* from the (upper) Whetstone Gulf Shale does not have axial nodes on the most anterior segments. The pygidium is very small and can be distinguished from the thorax by the absence of intrapleural furrows. See Ludvigsen and Tuffnell (1983, 1994) for a complete review.

Triarthrus beckii (Green, 1832) Plate 170
Hypotype NYSM 9892

Type species. *Triarthrus beckii* is widely distributed in Middle Ordovician Trenton equivalent shales such as the Flat Creek, lower Utica, and Dolgeville Shales in Montgomery County. It has also been reported in some lower Trenton Limestones. *Triarthrus beckii* is primarily distinguished from *T. eatoni* by the position of the palpebral lobe. On *T. beckii* the lobe is short and the posterior end is opposite the S2 glabellar furrow. On *T. eatoni* the palpebral lobe is significantly longer and extends to the S1 glabellar furrow. This distinction is usually clear. In addition, the glabella length and width are nearly equal with *T. beckii*, while the glabella of *T. eatoni* is longer than wide. On some specimens, however, such as those from the Holland Patent site (see *T. spinosus*), both characteristics of *T. beckii* and *T. eatoni* can be found. Ludvigsen and Tuffnell (1994) indicated that the ranges of *T. beckii* and *T. eatoni* significantly overlap. Plate 170 enables one to see the short palpebral length compared to *T. eatoni* (Plates 171 and 172).

Triarthrus eatoni (Hall, 1838) Plates 171 and 172
Hypotypes NYSM 9893, 9894

Triarthrus eatoni is a common trilobite of the Upper Ordovician, upper Utica Shales and overlying Lorraine Group shales. Many, if not most, of the specimens labeled as *T. eatoni* from the Middle Ordovician shales are *T. beckii*. A complete meraspid growth series of this trilobite is known from collections made by Walcott and Rust in the Utica Shale overlying the Steuben Limestone near Holland Patent, Oneida County. Because of its presence in Beecher's Trilobite Bed with superbly pyritized appendages, a great deal has been written concerning the structure of this trilobite (see Whittington and Almond (1987) and the references therein). The protaspid of this trilobite reported from these beds, however, may be that of a proetid (G. Edgecombe 1990, private communication). *Triarthrus eatoni* lived in deep, poorly oxygenated waters. Plate 172 shows a Beecher specimen with the pyritized appendages. The specimen in Plate 171 has all the characteristics of *T. eatoni* except it lacks the axial nodes on the eight most-anterior thoracic segments.

***Triarthrus glaber* (Billings, 1859)**
Syntypes GSC 1936e, h; specimen USNM

The syntypes are from the Upper Ordovician of Quebec. Unnumbered specimens in the USNM from Lorraine, Jefferson County, bear this name.

***Triarthrus spinosus* (Billings, 1857) Plates 173 and 174**
Holotype USNM 96235, hypotypes NYSM 9897 to 9900

Ruedemann (1926) found this normally Upper Ordovician Canadian trilobite in collections made by Walcott and Rust. It was found in the Utica Shales overlying the Stueben Limestone near Holland Patent, Oneida County. This olenid is distinguished by having genal spines, an occipital spine, and thoracic spines from the eighth, ninth, and tenth thoracic segments. There are no axial nodes on the thoracic segments. The specimen in plate 173 is the holotype. Plate 174 shows a specimen from Ottawa, Ontario.

Family Plethopeltidae

***Plethometopus knopfi* (Lochman, 1946)**
Holotype YPM 17394

Upper Cambrian, from rocks in Dutchess County referred to as "Hoyt Dolostone."

***Plethopeltis granulosa* (Resser, 1942)**
Holotype USNM 58561

Upper Cambrian, Hoyt Limestone, Saratoga County. See Ludvigsen and Westrop (1983).

***Plethopeltis saratogensis* (Walcott, 1890) Plate 175C**
Holotype USNM 58558

Type species. This species is from the Upper Cambrian, Hoyt Limestone, Lester State Park, Saratoga County. See Ludvigsen and Westrop (1983).

Family Ptychopariidae

***Ptychoparia minuta* (Bradley, 1860)**
Specimen MCZ 1045

Upper Cambrian, Potsdam Sandstone, High Bridge, Keesville, Essex County.

***Ptychoparia matheri* (Walcott, 1912)**
Types USNM 58585 to 58587

Upper Cambrian, Potsdam Sandstone, Whitehall, Washington County.

Family Shumardiidae

This is a family of very small, blind trilobites without facial sutures and oval in outline. The cephalon is semicircular with deep lateral glabellar furrows. The occipital ring is prominent. The pygidial shape is variable, as are the number of axial rings (four to seven). Their family relationship is in doubt.

***Shumardia pusilla* (Sars, 1835)**
Hypotype NYSM 9857

Lower Ordovician Deepkill Shales.

Family Solenopleuridae

***Rimouskia typica* (Resser, 1938)**
Plesiotype USNM 156694

Type species. This species is from the Lower Cambrian, Ashley Hill and Judson Point, Columbia County. See Bird and Rasetti (1968, p. 28). Rasetti (1967) classified *Rimouskia* as "family undetermined." The *Treatise* placed it within this family.

Family Uncertain

***Clelandia parabola* (Cleland, 1900)**
Holotype PRI 5070

From the Lower Ordovician Tribes Hill Formation. Westrop, Knox, and Landing (1993) reviewed the trilobite. The species is also found in Idaho and Oklahoma.

Appendix A

Trilobites and Their Environments

Environment	***Ordovician Chazy***	***Ordovician Black River***	***Ordovician Trenton***	***Ordovician Lorraine***
Lagoonal		***Lowville*** *Bathyurus* *Bumastoides*		
Reef/shoals	**Reef limestone** *Thaleops* *Glaphurus* *Glaphurina* *Paraceraurus* *Niloides* *Uromystrum* *Plattillaenus* *Hyboaspis* *Isotelus* *Bumastus*	***Watertown*** *Raymondites* *Bathyurus* *Bumastoides* *Illaenus* *Isotelus* *Eoharpes*	**Steuben/*Kings Falls*** *Calyptaulax* *Primaspis* *Hemiarges* *Flexicalymene*	
Shoal margins	**Calcarenite** *Plattillaenus* *Glaphurina* *Bumastoides* *Cybeloides* *Lonchodomas* *Calyptaulax*		**Sugar River/*Denley*** *Isotelus* *Primaspis* *Flexicalymene* *Ceraurus* *Cryptolithus*	
Proximal ramp	**Silty limestone** *Calyptaulax* *Cybeloides* *Lonchodomas* *Ceratocephala*		***Napanee*** *Isotelus* *Flexicalymene*	***Pulaski*** *Flexicalymene* *Cryptolithus*
Distal ramp			***Canajoharie*** *Gravicalymene* *Isotelus* *Triarthrus*	**Whetstone/ *Frankfort*** *Homotelus* *Triarthrus* *Cryptolithus*
Upper slope			***Dolgeville*** *Triarthrus*	
Lower slope			***Utica*** *Triarthrus*	
Basin			***Utica*** *Triarthrus*	

Information on trilobites and their environment from the work of Carlton Brett, Gerald Kloc, and Thomas Whiteley.

Silurian Clinton	*Silurian Lockport*	*Devonian Helderberg*	*Devonian Onondaga*	*Devonian Hamilton*
Neahga/*Sodus* *Eophacops* Calymenids		**Chrysler/*Thacher*** Eurypterids		***Tully (lower)*** *Scuttelum* *Pseudodechenella* *Harpidella*
U. Irondequoit *Dalmanites* *Scutellum* *Spathocalymene*	***Gasport*** Calymenids	**Coeymans/*Becraft*** *Odontochile* *Scutellum* *Proetus*	**Edgecliff *(western)*** *Maurotarion* *Mystrocephala* *"Calymene"* *Synphoria* *Trypaulites* *Terataspis*	***Tichenor*** *Otarion*
Lowest Rochester ***L. Irondequoit*** *Deiphon* *Bumastus* *Dicranopeltis* *Trochurus* *Liocalymene* *Encrinurus* *Eophacops*		**Kalkberg/*Becraft*** *Paciphacops* *Coniproetus* *Scutellum* *Synphoroides*	**Moorehouse/** ***Clarence*** *Harpidella* *Coronura* *Odontocephalus* *Viaphacops*	**Centerfield/*Kashong*** *Basidechenella* *Monodechenella* *Eldredgeops* *Harpidella* *Bellacartwrightia*
Lower Rochester Shales *Trimerus* *Calymene* *Dalmanites* *Radnoria* *Arctinurus* *Decoroproetus* *Bumastus*	**Eramosa *(basal)*** *Encrinurus* *Dalmanites* Calymenids	***New Scotland*** *Paciphacops* *Kettneraspis* *Dicranurus* *Oinchoe* *Acanthopyge* *Neoprobilium* *Dalmanites* *Cordania*	***Seneca*** *Viaphacops*	**Wanakah/*Windom*** *Eldredgeops* *Pseudodechenella* *Bellacartwrightia* *Greenops* *Dipleura* *Monodechenella*
Upper Rochester Shales ***Willowvale*** *Dalmanites* *Trimerus* *Liocalymene*		***Esopus***	***Nedrow*** *Viaphacops* *Odontocephalus* *Kettneraspis*	***Ledyard*** *Eldredgeops* *Greenops*
				Levanna *Eldredgeops* *Greenops*
Williamson *Liocalymene*				***Marcellus*** *Eldredgeops?*

Appendix B

The Photography

All of the photographs for this book were taken with a fairly standard, single-lens reflex, 35-mm camera equipped with a macro focusing lens. With few exceptions, the camera was placed on a copy stand, and electronic flash was always used for exposure. The specimens, preferably, were set on a "lab-jack" support so they could be raised or lowered for focusing independent of the camera once the proper composition was reached. The flash was as close to on-axis as the setup permitted. The exposure was determined through the lens by the electronics of the camera. An aperture of f11 was used for very flat specimens and f22, for three-dimensional specimens. The preferred films were color negative films selected for the highest sharpness available. Processing and printing were performed by commercially available sources.

Dark specimens usually benefited from whitening with ammonium chloride. This procedure was not available for specimens in museums.

With a Nikon LS-1000 35-mm film scanner, the resulting color negatives were scanned directly into a computer running Adobe Photoshop software. All of the scanning was at a resolution of 2700 pixels/inch (106 pixels/mm). The scanned images were adjusted with the autolevels command in Photoshop; treated with unsharp masking. Scratches on the negative were retouched in Photoshop; this was the only retouching done to the image. (There were rare exceptions to this "no retouching" statement. When glue bubbles in a repaired specimen were thought to be distracting, they were retouched away using the "clone" tool—the same tool used to eliminate scratches). The digital images were then stored as TIFF files.

The images in the book were adjusted for composition in Photoshop and the final images were stored as TIFF files at a resolution of 300 pixels/inch (11.8 pixels/mm).

Some Specifics

- Cameras. The two cameras used exclusively were Olympus OM-2N and OM-4T.
- Lenses. The macro lenses used were the Olympus 50-mm f3.5 and the Vivitar 90-mm f2.5. (The Vivitar was preferred because the longer focal length allowed better composition.)
- Flash. The flash used was an Olympus T-32, which couples with the exposure system of the camera to give proper exposure regardless of the aperture.
- Film. The preferred film was Kodak 25-speed color negative film because of its very high resolution. When it became no longer available, Kodak Gold 100 was used, as it is the next sharpest available color negative film. Color negative film was preferred over black and white film (which can be sharper) and slide film (which has poor exposure latitude and is not as sharp). The color negative films are fine enough in grain so graininess is not a problem and the sharpness of the photographic image is well above that of the image when digitized (i.e., it is not a limiting factor in image quality). High-quality color negative processing is available anywhere in the country (world?), and the color print is very useful for cataloging.
- Exposure. Electronic flash was preferred because (1) it stops all motion and (2) most color films are balanced for daylight exposure. There is a measurable, and theoretically known, falloff in sharpness as a camera lens is stopped down beyond about f11. Apertures between f11 and f22 were chosen to give the necessary depth of field. Aperture 22 was only used when the specimen had a lot of three-dimensional character.
- Scanning. The photographic images were scanned into the computer, as digital images are much easier to manipulate than are optical images in the traditional darkroom. Once one gets through the learning curve with digital images, the time difference between using the computer and the traditional darkroom is orders of magnitude. Scanning was done at 2700 pixels/inch in order to have the maximum amount of information to work with in the computer. Images can always be sized down with little loss in quality but sizing up always introduces potential problems. Occasionally the prints, if good enough, were scanned into the computer with a flat bed scanner. One should use the maximum optical resolution of the scanner.
- Software. Adobe Photoshop is the most-used image-processing program available. A limited edition that is now on the market for significantly less money has all the capability most needed for this type of work. The levels command evaluates the image and adjusts it to occupy the gray-scale space that Photoshop can handle. This is important. The autolevels function

generally does an excellent job, but occasionally one will wish to additionally adjust the image for lightness or darkness. To do this, use the levels command and move the center slider to get the preferred image. Unsharp masking enhances edges and is also important. Anytime an image goes through a lens, it loses sharpness. This happens both in the camera and in the scanners. Unsharp masking will regain the edge sharpness desired in the image.

- Stored images. Most publishers want images that have not been compressed. All compression algorithms throw some information away as redundant. For a onetime compression and decompression and printing, it is unlikely that anyone will notice changes in the image. Compressing images that previously have been compressed and then modified can result in compression artifacts that reduce image quality. It is inescapable. By far the safest mode is to keep images in an uncompressed mode (TIFF) and only compress them when no further work is going to be done on them. Since a 35-mm TIFF color image can be about 25 megabytes in size, one should consider storing images on a CD-ROM, as CD "burners" are readily available and writable CDs are inexpensive.
- Printing. The ink-jet printers on the market can give near-photographic-quality prints in either color or black and white. Various consumer groups and computer magazines rate the printers. Their advice is usually sound. The price range is large and is driven more by printing speed than by final print quality. In other words, within one manufacturer's printers, there will be little if any difference in print quality between those selling for $150 and those selling for $300 but the printing speed may be significantly different. If you do not mind waiting 5 to 10 minutes for an 8 × 10 color print, buy the less expensive model. No specific recommendations can be given because the technology is moving too fast and any recommendations will be obsolete within a year.

Glossary

For the geological terms we were assisted by the use of Keary (1996) and for the trilobite terms, by Whittington and Kelly (1997).

abathochroal. Having schizochroal eyes with less lenses and lacking stellar projections.
abaxial. Away or farther from the axial line.
accretionary wedge. A complex of sediments with multiple thrust faults resulting from the scrape-off of sediment, onto the overridding plate, from the oceanside plate during subduction.
acritarchs. A group of microfossils consisting of hollow, organic-walled, unicellular vesicles, mostly 20 to 150 μm in size. Known from the Precambrian to the recent, they are used in stratigraphic correlation.
adaxial. Toward or closer to the axial line.
aerobic. Able to live only where free oxygen is available.
allochthonous. A term describing rock units that have been transported from the site of their original deposition.
amalgamated. Closely joined, inseparable rock units.
anaerobic. Able to live or can only live where free oxygen is absent.
anoxic. A term describing an environment where the concentration of oxygen is too low to support oxygen-breathing life.
antennae (singular, antenna). Modified appendages that are jointed and extend from the front, ventral part of the cephalon.
anterior. Toward the front or in front of.
anterior cephalic border. The border on the forward part of the cephalon generally bounded by the anterior cephalic sutures.
apodemes. Paired projections on the inner surface of the exoskeleton along the axial groove. There is one pair for each thoracic segment and a pair for each cephalic and pygidial appendage. Apodemes are the attachment points for the appendages to the exoskeleton.
apomorphy. A derived characteristic.
appendifer. See *apodeme.*
arenites. Sandstones with grain sizes from 0.1 to 2 mm.
arthropod. An animal with an exoskeleton and jointed appendages.
autochthonous. A term describing rock units in place where they were deposited (i.e., not transported).
axis. The longitudinal center line of an organism.
ball-and-pillow-type structures. Rounded sandstone structures resulting from the soft sediment settling of sandstone into underlying mud.
basis (plural, bases). The most proximal part of the trilobite, appendage structure. The basis attaches to the trilobite, and the other two parts of the appendage branch off from it.
benthic (benthonic). Living on the seafloor.
bentonites. Clay layers consisting of altered volcanic ash useful in tracing stratigraphic horizons.
biofacies. Rocks with distinctive and characteristic fossils.
bioherms. An organic mound or reef.
biomeres. A term used to describe periods of time in the Cambrian when there was an extinction event with nearshore trilobites and the area was repopulated from deeper-water species that evolved to occupy the now-vacant ecological niches.
biramous. Having appendages that branch into two branches with differing functions.
border. For the trilobite, a distinctive area around the margin of the cephalon or pygidium.
boundstones. Limestones in which the grains were bound together by organisms.
breccias. Rock units with included angular pieces, or clasts, of other broken-up units.
brecciated. Containing breccia or distinctive inclusions from other units.
cephalic border. The border around the margin of the cephalon.
cephalon (plural, cephala). The head of the trilobite.
chamositic. Containing the iron mineral chamosite.
chelicerate. A subphylum of arthropods without antennae and with the first pair of appendages modified as chelicerae (pinchers).
chert. Microcrystalline quartz found within limestones.
cladistics. A taxonomic methodology studying relationships using shared characteristics.
class. In biology, a grouping just after the phylum of plants or animals having a similar basic structure. Trilobites are in the phylum Arthropoda and class Trilobita.

clasts. Pieces of broken-up rock units.
cleavage. Preferred direction of splitting of rocks.
comminuted. Broken down into fine particles.
conodonts. Phosphatic teeth of an extinct marine animal with chordate characteristics; widely used in stratigraphic correlation.
conterminant. A condition where the hypostome is under the anterior glabella and attached to the anterior doublure by a suture.
coordinated stasis. Long-term evolutionary stability for an entire fossil community.
coquina. Limestones with a high density of accumulated skeletal material.
cotype. When two or more specimens are used to define a species (term is no longer used).
coxa. The part of an arthropod appendage attached to the body.
cranidium (plural, cranidia). The central portion of the trilobite cephalon, bounded by the facial sutures; basically the fixed cheeks and the glabella.
cross-stratification. Inclined bedding structure resulting from water- or wind-deposited grains.
Cruziana. A trace fossil attributed to trilobite tracks or trails.
cuticle. The exoskeleton of the trilobite.
degree. A term used to describe the number of thoracic segments in trilobite meraspid.
dendroid. A highly branched inclusion often seen on bedding planes.
desiccation cracks. Cracks in fine-grained sediments resulting from aerial drying out; usually seen as polygon-shaped structures.
diachronous. A stratigraphic term describing a unit that varies in age over its geographic range.
diagenesis. Process by which physical, biological, and chemical changes alter a sediment or fossil from the time of deposition to the present.
Diplichnites. A fossil trace believed to be tracks from the appendages of arthropods.
distal. Away from a reference point.
dorsal. Toward the upper surface of an animal while in its life position.
doublure. A continuation, by folding under, of the dorsal exoskeleton on the ventral side.
down warp. The side of a fault displaced downward.
dysaerobic. Able to live in low-oxygen conditions.
dysoxic. A term used to describe a low-oxygen environment that poorly supports oxygen-breathing organisms.
ecdysis. The process of shedding the exoskeleton in order to grow larger.
encrinal. Consisting primarily of crinoid material.
endites. Inwardly directed projections on a segment, podomere, of a trilobite's walking leg.
endopod. The lower or walking branch of a trilobite's appendage.
epifaunal. A term describing organisms that live on the sea bottom, either moving freely or attached.
epipelagic. Living at or very close to the surface of the sea.
etymology. The origin and development of a word or name.
exfoliated. Loss of the actual fossil, leaving an internal mold.
exite. The outer or brachial branch of a trilobite's appendage.
exopod. The upper or brachial, breathing, branch of a trilobite's appendage.
exoskeleton. The external support structure for arthropods.
facial suture. The line for opening or separation of the free cheek from the cranidium.
families. Grouping of organisms with very similar characteristics. In taxonomy, family is the grouping after order and before genus.
firmgrounds. Sea bottoms that have consolidated enough to support some forms of direct attachments.
fixigenae (singular, fixigena). The fixed cheeks; portions of the cheek area not lost by the opening of the facial suture.
flooding surfaces. Sediment-starved areas at the interface of two very different rock units due to a rapid rise in sea level or a transgression event.
fluidize. To convert a particulate substrate with stable, semisolid characteristics into a fluid that can flow.
flute. A groovelike physical marking on the base of a rock unit resulting from erosion in the underlying unit.
foreland basin. A basin or deepening of a sea between a stable craton and mountain building due to plate tectonics.
genae (singular, gena). The cheeks of a trilobite; the areas lateral from the central area of the glabella.
genal angle. The angle formed where the anterior margin of the cephalon turns toward the axis.
genal spine. A spine directed posteriorly from the genal angle.
genus (plural, genera). The taxonomic grouping after family that denotes an increasingly similar evolutionary relationship. It is the grouping just before the actual species.
glabella. The central, axial, part of a trilobite's cephalon.
glabellar furrow. A furrow along the sides of the glabella paralleling the axis of the trilobite.
gnathobases. The most proximal projections on the podomeres of a trilobite's appendages.
gonatoparian. The condition where the posterior end of the facial suture emerges on the genal angle.
gradient currents. Underwater currents flowing downslope, often carrying sediment into deeper water.
grainstones. Granular limestones with no included muds.
graptolites. Extinct animals with a characteristic, often sawtooth, shape, used in stratigraphic correlation.
graywacke. A sandstone with greater than 15% clay minerals.
groove-casts. Grooves on an underlying layer caused by the movement of a stone along the bottom. These grooves are expressed as casts on the underside of the overlying rock.

hemipelagic. A term describing a deposit of fine-grained sediment that has flowed downslope within a gradient current and come to rest, settling out.

holaspid. The final growth phase of a trilobite; in most cases, signifies that no more thoracic segments are added during molting.

holochroal. Having compound eyes with a smooth external surface.

holotype. The single specimen used to define a species.

hypersalinity. Salt content above what one measures in the ocean.

hypostome (plural, hypostoma). A mineralized tergite on the ventral cephalon under the glabella.

hypotype. A specimen described in the literature.

Iapetus. An ocean lying between proto North America and the African craton. The closing of this ocean resulted in major mountain building, the Caladonian Orogeny.

impendent. Having a condition where the anterior edge of the hypostome is posterior to the anterior edge of the glabella.

infaunal. A term describing organisms that live just under the surface of the sea bottom.

instar. Recognizable phase during the protaspid phase of trilobite growth.

Isopach maps. Maps showing lines of equal thickness of rock units.

isostatic. A term describing buoyancy forces.

karstic. Having surface characteristics due to subsurface solution of rocks. Major cave-forming areas often have karstic surfaces.

Konservat-Lagerstätten. Fossil deposits with very unusually good preservation of the organisms.

lag. Deposit in which the finer particles have been winnowed or washed away.

lamella (pleural, lamellae). The horizontally separable part of the cephalon, of trinucleids and harpids. A suture runs around the outer edge of the cephalon, separating it into an upper and lower lamella.

lappets. Short, rounded projections from the ends of the pleura.

lateral glabellar furrows. Symmetrical furrows in the glabella that are roughly perpendicular to the axis and end or start, from the glabellar furrow.

lateral glabellar lobes. Bilateral lobes on the glabella separated by the lateral glabellar furrows.

Laurentia. A continental land mass composed of the Canadian shield, Greenland, parts of western Europe, and other parts of what becomes North America.

lectotype. The single specimen selected to define the species when, in the original definition, a group of specimens, syntypes or cotypes, were used.

librigenae (singular, librigena). Free cheeks.

load casts. Depressions in an underlying sediment from overloading by the overlying layers; expressed as rounded protrusions on the base of the overlying layer.

LO, L1, . . . Sequential designation for the lateral glabellar furrows, starting with the occipital furrow, LO.

lowstand. The lowest sea level following a regression.

lutites. Very-fine-grained limestones.

MaBP. Million of years before present.

magmas. Molten rocks forced to the surface from underground chambers.

marginal lappets. Lappets radially projecting from the pygidial margin.

marginal spines. Spines radially projecting from the pygidial margin.

meraspid. A growth phase of trilobites from when the cephalon and the protopygidium are separated to when no further thoracic segments are added during molting.

meraspis. An individual specimen in the meraspid phase.

metamorphic. A term to describe rocks altered by heat and pressure.

micritic. Containing very-fine-grained limestones that form the matrix to fossil deposits; see also *lutites.*

mineralized. When inorganic materials become part of a basically organic structure; used here to indicate the mineral structure of a trilobite's exoskeleton.

morphological. Having to do with physical shape or morphology.

mudstones. Rocks composed of consolidated muds with few larger particles.

natent. A condition where the hypostome is below the anterior glabella but unattached to the doublure.

neotype. The single specimen used to define a known species when the original type is lost or poorly defined.

nodes. Small circular or oval raised areas on the exoskeleton.

obducted. A term used to describe rocks that have been forced over and above when two plates collide.

occipital ring. Raised area at the posterior of the glabella running transverse across the axis and bounded by the glabellar furrows.

ontogenetic. Distintive phases of an animal's growth.

ontogeny. The processes of the growth of an organism.

ooids. Small particles composed of concentric layers.

oolitic. Containing oolites.

opisthoparian. The condition where the posterior end of the facial suture emerges on the posterior cephalic margin (i.e., toward the axis from the genal angle).

order. The taxonomic grouping after class.

orogenesis. The building of mountains.

orogenic. Resulting from mountain building.

ossicles. The name given to individual echinoderm plates.

packstones. Granular limestones with mud between the granules.

palpebral area. Portion of the fixed cheek that is immediately adaxial to the eye.

paralectotypes. When a lectotype is chosen from a type group, syntypes or cotypes, the remaining specimens are paralectotypes.

paratypes. Specimens from the same locations and horizon as the holotype that have been used in the original species definition.

parsimonious. Used in cladistics to describe the closest relationship or fit in a grouping.

pelagic. Living in the open ocean or deep sea.

phaselus. A minute oval structure believed to be a pre-protaspid phase in a trilobite's life cycle.

phylum (plural, phyla). The first taxonomic division of plants or animals; indicates a basic shared characteristic.

planktic. Floating free in the surface of the sea.

planktonic. See *planktic.*

plastotype. A cast or mold of the original type specimen.

plesiomorphic. Having shared primitive characteristics, meaning they are primitive to the group (e.g., biramous appendages are plesiomorphic as they represent a common ancestor to arthropods not exclusive to trilobites).

plesiotype. A specimen very similar to the type (term is not used today).

pleura (plural, pleural). Lateral portion of each thoracic segments, or the raised area separated by a groove on the pleural area of the pygidium.

pleural spines. Spines projecting from the thoracic pleura.

podomeres. The sections of the trilobite walking leg.

posterior. Toward the rear or rearward.

preglabellar field. An area immediately in front of the glabella separated from the cephalic margin by a groove.

progradation. The buildup and extension of deltaic sediments within a basin as the result of sediment washed from the land. The weight of the sediments can cause the basin axis to move away from the sediment source.

proparian. The condition where the posterior end of the facial suture emerges in front of the genal angle.

prosopon. Exoskeletal ornamentation.

protaspid. First unequivocal stage of a trilobite's life cycle.

protaspis. An individual trilobite at the protaspid stage.

Proterozoic. The late Precambrian.

protocephalon. The anterior part of the meraspid that represents the future cephalon.

Protopangea. Precambrian supercontinent (Rodinia).

protopygidium. The posterior-most part of the trilobite during the meraspid stage.

proximal. Near, close to.

punctuated equilibria. Periods of long-term evolutionary stability followed by abrupt evolutionary changes.

pustule. Small rounded, raised area on the exoskeleton. When pustules are numerous, the specimen is often referred to as *pustulose.*

pygidium (plural, pygidia). The posterior or tail of a trilobite.

pyritized. Replaced by pyrite in diagenesis.

radiolarian. Marine, single-cell animal with an ornate silica shell.

regression. The lowering of sea level.

ridges. Elongated raised areas or sometimes elongated, abrupt raised areas as in terrace ridges.

rifting. The separation or pulling apart of continental plates.

ripple marks. Marks on a bedding plane due to preserved evidence of water rippling.

Rodinia. A proposed Precambrian supercontinent.

rostral plate. A part of the ventral, anterior border of a trilobite bounded by sutures; not present on all trilobites.

Rusophycus. A trace fossil attributed to trilobites; considered a trilobite resting pit; often seen on the base of sandstones as a wedge-shaped raised area.

schizochroal. Having eyes where the lenses are distinctly separate.

sclera. The area between the lenses of the schizochroal eye.

sclerites. Individual skeletal parts of complex fossils.

scoured. A sea bottom marked by current-moved particles; expressed as grooves or channels.

setae (singular, seta). Small hairlike projections from various parts of a trilobite.

silicified. A term describing exoskeletal material or appendages that has been replaced with silica.

soft sediment deformation. Uneven, wavy surfaces at rock-layer interfaces due to the settling of the upper layer into the still soft layer beneath it.

SO, S1, . . . Designations used to identify the lateral grooves or sulcus in the trilobite glabella.

speciation. The process of evolution of one species into another.

species. The smallest taxonomic division of life.

spines. Elongated, often pointed, exoskeletal projections. Spines can come off radially to parts of the exoskeleton or may project dorsally.

storm wave base. Maximum depth of the alteration of the sea bottom by storm waves.

stromatolites. Mound-shaped, laminar structures formed by algae and cyanobacteria.

stromatoporoids. Extinct reef-forming sponges with layered calcite skeletons.

subaerial. Exposed to the air.

subduction zone. The area during plate collision where an oceanic plate is forced under the continental plate.

subsiding. The deepening of a basin bottom due to tectonics or the weight of the overlying sediments.

sulcus. A furrow or groove.

supratidal. A term describing an area at the water's edge that is above normal high tides but submerged during exceptional tides.

suture. A line in the cephalon where the parts separate during the molting process.

synapomorphies. Shared derived characteristics.
synjacent. A term used to describe overlying and underlying rock units.
syntypes. Organisms used to define the species (a taxonomic term, no longer applied).
systematics. The orderly classification of organisms.
taphonomy. The study of processes involved with an organism from its death to its discovery.
tectophases. Phases during mountain formation and erosion from plate tectonics.
telepodite. The inner or walking branch of a trilobite's appendage.
tempestites. Sedimentary formations resulting from deposition of sediments transported as the result of storms.
tergite. A part of the dorsal exoskeleton that separates along sutures or articulations.
terminal axial spine. A spine on the pygidium that projects posteriorly from the axial region.
terrace lines. Elongated structures on the exoskeleton that are formed by microsteps.
terranes. Geographic areas inserted within larger areas during plate movement and consolidation.
terrigenous. Containing sediments derived from the erosion and wash-off from land.
thoracic segments. The articulated parts of a trilobite's thorax.
thoracopygidium. Trilobite fossil recovered without the cephalon, probably as a result of molting.
thorax. The central body of the trilobite between the cephalon and the pygidium.
topotype. A specimen from the type location.
trace fossils. Markings or structures in sedimentary layers believed to be due to the activity of living creatures.
transgression. Sea level rising.
transgressive lag. Concentrated deposit formed during times of low sediment deposition and erosion during transgression.
transitory pygidium. The most posterior unit of the trilobite meraspid phase.
Treatise. *Treatise on Invertebrate Paleontology: Part O, Anthropoda 1* (first edition) by Moore (1959).
trough and hummocky cross-stratification. Cross-stratification where the bounding surfaces of the stratification are both spoon shaped (lower) and raised in the center (upper).
trough cross-bedding. Cross-stratification where the lower surface of the stratification is curved upward on the ends.
tsunamis. Tidal waves.
tubercles. Small projections on the exoskeleton; often used to imply a structure larger than a pustule yet not a spine.
turbidites. Sediment layers from water-born sediments flowing downslope and coming to rest.
turbidity currents. Turbulent currents of sediment-laded water flowing downslope.
unconformity. A time gap in the stratigraphic column during which no sediment was deposited or other rocks were eroded.
ventral. The lower or under surface.
vugs. Hollows within rocks.
winnowed. A term describing deposits of coarse particulate material from which the fine sediments have been removed.
winnowing. The process of currents removing fine particles from around coarser ones.
XYZ. A series of rocks exhibiting low-to-high-to-low-energy deposition.

References

Adrain, J. M., and B. D. E. Chatterton. 1995. The otarionine trilobites *Harpidella* and *Maurotarion*, with species from northwestern Canada, the United States and Australia. Journal of Paleontology 69: 307–326.

Allison, P. A. 1986. soft-bodied fossils: the role of decay in fragmentation during transport. Geology 14: 979–981.

——. 1988a. Konservat-Lagerstätten: cause and classification. Paleobiology 14: 331–344.

——. 1988b. Phosphatized soft-bodied squids from the Jurassic Oxford Clay. Lethaia 21: 403–410.

——. 1990. Decay processes. In D. E. G. Briggs and P. R. Crowther (eds.), Paleobiology: a synthesis, pp. 213–216. Oxford: Blackwell Scientific.

Allison, P. A., and C. E. Brett. 1995. In situ bentho and paleo-oxygenation in the Middle Cambrian Burgess Shale, British Columbia, Canada. Geology 23: 1079–1082.

Allison, P. A., and D. E. G. Briggs. 1991a. The taphonomy of soft-bodied animals. In S. K. Donovan, The Processes of fossilization, pp. 120–140. London: Belhaven Press.

——. 1991b. Taphonomy: Releasing the Data Locked in the fossil record. New York: Plenum.

——. 1993. Exceptional fossil record: distribution of soft-tissue preservation through the Phanerozoic. Geology 21: 527–530.

Angelin, N. P. 1851. Palaeontologia Scandinavica: Pars I, Crustacea Formationis Transitionis. Holmiae, 24 p.

Anthony, J. G. 1838. *Crotocephala ceralepta*. American Journal of Science 34: 379.

Babcock, L. E. 1982. Original and diagenetic color patterns in two phacopid trilobites from the Devonian of New York. In B. Mamet and M. J. Copeland (eds.), Third North American Paleontological Convention, Proceedings, vol. 1, pp. 17–22. Toronto: Business and Economic Services.

——. 1993. Trilobite malformations and the fossil record of behavioral asymmetry. Journal of Paleontology 67: 217–229.

Babcock, L. E., and S. E. Speyer. 1987. Enrolled trilobites from the Alden pyrite bed, Ledyard Shale (Middle Devonian) of New York. Journal of Paleontology 61: 539–548.

Barrande, J. 1852. System Silurien du centre de la Boheme, 1: 226–230, 629.

Barrett, S. T. 1876. Description of a new trilobite, *Dalmanites dentata*. American Journal of Arts and Science, 3rd ser. 11: 200.

——. 1878. The coralline, or Niagara Limestone of the Appalachian system as represented at Nearpass's Cliff, Montaque, New Jersey. American Journal of Arts and Science, 3rd ser. 15: 370–372.

Barton, D. C. 1913. A new genus of the Cheiruridae, with descriptions of new species. Bulletin of the Museum of Comparative Zoology (Harvard) 54: 547–556.

Basse, M. 1997. Trilobiten aus Mittlerem Devon des Rhenohercynikums: II. Proetida (2), Ptychopariida, Phacopida (1). Palaeontographica A 246: 53–142.

Beecher, C. E. 1893a. Larval forms of trilobites from the Lower Helderberg Group. American Journal of Science, 3rd ser. 46: 142–147.

——. 1893b. A larval form of *Triarthrus*. American Journal of Science, 3rd ser. 46: 361–362.

——. 1893c. On the thoracic legs of *Triarthrus*. American Journal of Science, 3rd ser. 46: 467–470.

——. 1894a. The appendages of the pygidium of *Triarthrus*. American Journal of Science, 3rd ser. 47: 298–300.

——. 1894b. On the mode of occurrence, and the structure and development of *Triarthrus beckii*. American Geologist 13: 38–43.

——. 1895a. Structure and appendages of *Trinucleus*. American Journal of Science, 3rd ser. 49: 307–311.

——. 1895b. The larval stages of trilobites. American Geologist 16: 166–197.

——. 1896. The morphology of *Triarthrus*. American Journal of Science, 4th ser. 1: 251–256.

Bergström, J. 1969. Remarks on the appendages of trilobites. Lethaia 2: 395–414.

——. 1972. Appendage morphology of the trilobite *Cryptolithus* and its implications. Lethaia 5: 85–94.

——. 1973. Organization, life, and sytematics of trilobites. Fossils and Strata 2: 69 p.

——. 1990. Hunsrück slate. In D. E. G. Briggs and P. R. Crowther (eds.), Paleobiology: a synthesis, pp. 277–279. Oxford: Blackwell Scientific.

Bergström, J., and G. Brassel. 1984. Legs in the trilobite *Rhenops* from the Lower Devonian Hunsruck Slate. Lethaia 17: 67–72.

Berner, R. A. 1981a. Authigenic mineral formation resulting from organic matter decomposition in modern sediments. Fortschritte der Minerologie 59: 117–135.

——. 1981b. A new geochemical classification of sedimentary environments. Journal of Sedimentation Petrology 51: 359–365.

Bigsby. 1825. Description of a new species of trilobite. Journal of the Academy of Natural Science, Philadelphia 4: 365–368.

Billings, E. 1857. New species of fossils from the Silurian rocks of Canada. Geological Survey of Canada, report for 1856.

——. 1859. Descriptions of some new species of trilobites from the Lower and Middle Silurian rocks of Canada. Canadian Naturalist and Geologist 4: 367–383.

——. 1860. Description of some new species of fossils from the Lower and Middle Silurian rocks of Canada. Canadian Naturalist 5: 49–69.

——. 1863. Geology of Canada. Geological Survey of Canada (Montreal), 983 p.

——. 1865. Paleozoic fossils: vol. 1. Containing descriptions and figures of new or little known species of organic remains from the Silurian rocks. Geological Survey of Canada (Montreal), pp. 395–426.

Bird, J. M., and F. Rasetti. 1968. Lower, Middle and Upper Cambrian faunas in the Taconic Sequence of eastern New York: stratigraphic and biostratigraphic significance. Geological Society of America, special paper, 113. Boulder.

Bolton, T. E. 1966. Catalogue of type invertebrate fossils of the Geological Survey of Canada. vol. 3: 203 p. Ottawa: Geological Survey of Canada.

Bradley, F. H. 1860. Description of a new trilobite from the Potsdam Sandstone. American Journal of Science and Arts, 2nd ser. 30: 241–243.

Brandt, D. S. 1985. Ichnologic, taphonomic, and sedimentologic clues to the deposition of Cincinnatian shales (Upper Ordovician) Ohio, USA. In H. A. Curran (ed.), Biogenic Structures: There Use in Interpreting Depositional Environment. Society of Economic Paleontologists and Mineralogists special publication 35: 299–307.

——. 1996. Epizoans on *Flexicalymene* (Trilobita) and implications for trilobite paleoecology. Journal of Paleontology 70: 442–449.

Brett, C. E. 1978. Host-specific pit-forming epizoans on Silurian crinoids. Lethaia 11: 217–232.

——. 1985. *Tremichnus*: a new ichnogenus of circular-parabolic pits in fossil echiuoderms. Journal of Paleontology 59: 625–635.

Brett, C. E., and G. C. Baird. 1986. Comparative taphonomy: a key to paleoenvironmental interpretation based on the fossil preservation. Palaios 1: 207–227.

——. 1992. Coordinated stasis and evolutionary ecology of Silurian-Devonian marine biotas in the Appalachian Basin. Geological Society of America, Abstracts with Programs, 24: 139.

——. 1993. Taphonomic approaches to temporal resolution in stratigraphy: examples from Paleozoic marine mudrocks. In S. M. Kidwell and A. K. Behrensmeyer (eds.), Taphonomic approaches to time resolution in fossil assemblages, pp. 250–274. Short courses in paleontology 6, Knoxville, Tenn.: Paleontological Society.

——. 1995. Coordinated stasis and evolutionary ecology of Silurian to Middle Devonian faunas in the Appalachian basin. In D. H. Erwin and R. L. Anstey (eds.), New approaches to speciation in the fossil record, pp. 285–315. New York: Columbia University Press.

——. 2002. Revised stratigraphy and facies relationships of the upper part of the type area, central New York State: sedimentology and tectonics of a Middle Ordovician (Caradocian) shelf-to-basin succession. Physics and Chemistry of the Earth, in press.

Brett, C. E., L. C. Ivany, and K. M. Schopf. 1996. Coordinated stasis: an overview. Paleogeography, Paleoclimatology, Paleoecology 127: 1–20.

Brett, C. E., and A. Seilacher. 1991. Fossil Lagerstätten: a taphoanomic consequence of event sedimentation. In G. Einsele, W. Ricken, A. and Seilacher (eds.), Cycles and events in stratigraphy, pp. 283–297. Berlin: Springer-Verlag.

Brett, C. E., D. H. Tepper, W. M. Goodman, S. T. LoDuca, and B. H. Eckert. 1995. Revised Stratigraphy and Correlations of the Niagaran Provincial Series (Medince, Clinton, and Lockport Groups) in the Type Area of Western New York. U.S. Geological Survey Bulletin 2086, 66 p.

Brett, C. E., T. E. Whiteley, P. A. Allison, and E. Yochelson. 1997. The Walcott-Rust Quarry: a Middle Ordovician Konservat-Lagerstätte. 2nd International Trilobite Conference, St. Catherines, Ontario, Abstracts, p. 11.

——. 1999. The Walcott-Rust Quarry: a Middle Ordovician Konservat-Lagerstätten. Journal of Paleontology 73: 288–305.

Brett, K. D., and S. R. Westrop. 1996. Trilobites of the Lower Ordovician (Ibexian) Fort Cassin Formation, Champlain Valley region, New York State and Vermont. Journal of Paleontology 70: 408–427.

Briggs, D. E. G., S. H. Bottrell, and R. Raiswell. 1991. Pyritization of soft-bodied fossils: Beecher's Trilobite Bed, Upper Ordovician, New York State. Geology 19: 1221–1224.

Briggs, D. E. G., and P. R. Crowther. 1990. Paleobiology: a synthesis. Oxford: Blackwell Scientific. 583 p.

Briggs, D. E. G., and G. D. Edgecombe. 1993. Beecher's Trilobite Bed. Geology Today (May–June): 97–102.

Briggs, D. E. G., and A. J. Kear. 1993. Fossilization of soft tissue in the laboratory. Science 259: 1439–1442.

Bromley, R. G. 1990. Trace fossils biology and taphonomy. London: Unwin Hyman.

Burmeister, H. 1843. Die Organisation der trilobiten. Berlin: Georg Reimer, 147 p.

——. 1846. The organization of the trilobites. London: Ray Society, [An English translation of Burmeister (1843).]

Butterfield, N. J. 1990. Organic preservation of non-mineralizing organisms and the taphonomy of the Burgess Shale. Paleobiology 16: 272–286.

Campbell, K. S. W. 1975. The functional morphology of *Cryptolithus*. Fossils and Strata 4: 65–86.

——. 1977. Trilobites of the Haragan, Bois d'Arc and Frisco Formations (early Devonian) Arbuckle Mountains region, Oklahoma. Oklahoma Geological Survey Bulletin 123. Norman: Oklahoma Geological Survey, 227 p.

Canfield, D. E., and R. Raiswell. 1991a. Pyrite formation and fossil preservation. In P. A. Allison and D. E. G. Briggs (eds.), Taphonomy: releasing the data locked in the fossil record. pp. 337–388, New York: Plenum.

——. 1991b. Carbonate precipitation and dissolution: its relevance to fossil preservation. In P. A. Allison and D. E. G. Briggs (eds.), Taphonomy: Releasing the Data Locked in the Fossil Record, pp. 411–453. New York: Plenum.

Cassa, M. R., and D. L. Kissling. 1982. Carbonate facies of the Onondaga and Bois Blanc Formations, Niagara Peninsula, Ontario. In E. J. Buehler and P. E. Calkin (eds.), Geology of the northern Appalachian Basin, western New York, pp. 65–98. Field trips guidebook for New York State Geological Association 54th annual meeting.

Chatterton, B. D. E., G. D. Edgecombe, S. E. Speyer, A. S. Hunt, and R. A. Fortey. 1994. Ontogeny and relationships of Trinucleoidea (Trilobita). Journal of Paleontology 68: 523–540.

Chatterton, B. D. E., G. D. Edgecombe, N. E. Vaccari, and B. G. Waisfeld. 1997. Ontogeny and relationships of the Ordovician odontopleurid trilobite *Ceratocara*, with new species from Argentina and New York. Journal of Paleontology 71: 108–125.

Chatterton, B. D. E., D. J. Siveter, G. D. Edgecombe, and A. S. Hunt. 1990. Larvae and relationships of the Calymenina (Trilobita). Journal of Paleontology 64: 255–277.

Chatterton, B. D. E., and S. E. Speyer. 1990. Applications of the study of trilobite ontogeny. In D. G. Mikulic (ed.), Arthropod paleobiology, pp. 113–136. Short courses in paleontology 3. Knoxville, Tenn.: Paleontological Society.

——. 1997. Ontogeny. In R. L. Kaesler (ed.), Treatise on invertebrate paleontology: part O, Arthropoda 1, Trilobita, revised, pp. 173–247. Boulder, Col.: Geological Society of America.

Cisne, J. L. 1973. Beecher's Trilobite Bed revisited: ecology of an Ordovician deep water fauna. New Haven: Postilla, Peabody Museum, Yale University. 160 p.

——. 1975. Anatomy of *Triarthrus* and the relationships of the Trilobita. Fossils and Strata 4: 45–63.

——. 1981. *Triarthrus eatoni* (Trilobita): anatomy of its exoskeletal, skeletomuscular, and digestive systems. Palaeontographica Americana 9: 99–140.

Clarke, J. M. 1889. Eighth annual report of the New York State geologist, p. 57.

——. 1890. Archivos do Museu Nacional do Rio de Janeiro 9: 16.

——. 1891. 10th report of the state geologist of New York for 1890, 1891, p. 69, 71.

——. 1893. Report of the assistant paleontologist. Annual Report to the Regents, New York State Museum 46: 191–195.

——. 1894 The Lower Silurian trilobites of Minnesota. Geology of Minnesota, Paleontology 3(2): 695–759.

——. 1900. Oriskany fauna of Becraft Mountain. New York State Museum Memoir 3. 128 p.

——. 1908. Early Devonic History of New York and Eastern North America. New York State Museum Memoir 9: Plate 5, no. 1. Albany.

Clarkson, E. N. K. 1975. The evolution of the eye in trilobites. Fossils and Strata 4: 7–32.

——. 1979. Invertebrate paleontology and evolution. London: George Allen and Unwin.

Cleland, H. F. 1900. The calciferous of the Mohawk Valley. Bulletin of American Paleontology 3. 127 p.

Collie, G. L., 1902. Ordovician section near Bellefonte. Geological Society of America Bulletin 14: 418.

Conrad, T. A. 1839. Second annual report of T. A. Conrad on the Paleontological Department of the Survey. Geological report of New York, 1839–40, p. 56–66.

——. 1840. Annual report of the New York State Geological Survey.

——. 1841. Description of new genera and species of organic remains, Crustacea. New York Geological Survey, 5th report of the state paleontologist, p. 25–57.

——. 1842. Journal of the Academy of Natural Sciences, Philadelphia 8: 277.

——. 1843. Proceedings of the Academy of Natural Sciences, Philadelphia 1: 332.

Cooper, G. A., and J. S. Williams. 1935. Tully Formation of New York. Geological Society of America Bulletin 46: 781–868.

Dale, N. C. 1953. Geology and mineral resources of the Oriskany Quadrangle (Rome Quadrangle). New York State Museum Bulletin 345. 197 p. Albany.

Dalziel, I. W. D. 1997. Neoproterozoic-Paleozoic geography and tectonics: Review, hypothesis, environmental speculation. Geological Society of America Bulletin 109: 16–42.

Darby, D. G., and E. C. Stumm. 1965. A revision of the Ordovician trilobite *Asaphus platycephalus* Stokes. Contributions from the Museum of Paleontology, University of Michigan 20: 63–73.

Davidek, K., E. Landing, S. A. Bowring, S. R. Westrop, A. W. A. Rushton, R. A. Fortey, and J. M. Adrain. 2000. New uppermost

Cambrian U-Pb date from Avalonian Wales and age of the Cambrian-Ordovician boundary. Geological Magazine 137: 303–309.

DeKay, J. E. 1824. Observations on the structure of trilobites, and description of an apparently new genus. Annales of the Lyceum of Natural History, New York 1. p. 174.

Delo, D. M. 1934. The Fauna of the Rust Quarry, Trenton Falls, New York. Journal of Paleontology 8: 247–249.

——. 1940. Phacopid trilobites of North America. Geological Society of America, special publication 29.

——. 1935. A revision of the phacopid trilobites. Journal of Paleontology 9: 402–420.

DeMott, L. L. 1987. Platteville and Decorah trilobites from Illinois and Wisconsin. In R. E. Sloan (ed.), Report of investigations 35, Minnesota Geological Survey, pp. 63–98. St. Paul.

Dick, V. B., and C. E. Brett. 1986. Petrology, taphonomy and sedimentary environments of pyritic fossil beds from the Hamilton Group (Middle Devonian) of western New York. pp. 102–128. In C. E. Brett (ed.), Dynamic stratigraphy and depositional environments of the Hamilton Group (Middle Devonian) in New York State, Part 1. New York State Museum Bulletin 457. Albany: New York State Museum.

Donovan, S. K. (ed.). 1991. The process of fossilization. New York: Columbia University Press. 303 p.

Dwight, W. P. 1884. Recent explorations in the Wappinger Valley limestones of Duchess County, New York. American Journal of Science, 3rd ser. 27: 249–259.

Eaton. 1832. Geological textbook.

Edgecombe, G. D. 1992. Trilobite phylogeny and the Cambrian-Ordovician "event": cladistic reappraisal. In M. J. Novacek and Q. D. Wheeler (eds.), Extinction and phylogeny, pp. 144–177. New York: Columbia University Press.

Edgecombe, G. D. 1993. Silurian acastacean trilobites of the Americas. Journal of Paleontology 67: 535–548.

Edgecombe, G. D., and J. M. Adrain. 1995. Silurian calymenid trilobites from the United States. Palaeontographica (A) 235: 1–19.

Edgecombe, G. D., and B. D. E. Chatterton. 1993. Silurian (Wenlock-Ludlow) encrinurine trilobites from the Mackenzie Mountains, Canada and related species. Palaeontographica (A) 229: 75–112.

Einsele, G., W. Ricken, and A. Seilacher (eds.). 1991. Cycles and events in stratigraphy. Berlin: Springer-Verlag. 955 p.

Eldredge, N. 1972. Systematics and evolution of *Phacops rana* (Green, 1832) and *Phacops iowensis* Delo, 1935 (Trilobita) from the Middle Devonian of North America. American Museum of Natural History Bulletin 147: 45–114.

——. 1973. Systematics of Lower and Lower Middle Devonian species of the trilobite *Phacops* Emmrich in North America. American Museum of Natural History Bulletin 151: 286–335.

Eldredge, N., and S. J. Gould. 1972. Punctuated equlibria: an alternative to phyletic gradualism. In T. M. Schopf (ed.), Models in paleobiology, pp. 82–115. San Francisco: Freeman, Cooper.

Emmons, E. 1842. Natural History of New York. Geology of New York part 2. Comprising the Survey of the Second Geological District. Albany.

——. 1846. Natural history of New York, part 5, agriculture of New York, 1: 371 p.

——. 1855. The Taconic System. American Geologist 1: 236.

Ettensohn, F. R. 1987. Rates of relative plate motion during the Acadian Orogeny based on the spatial distribution of black shales. Journal of Geology 95: 630–639.

Ettonsohn, F. R., and C. E. Brett. 1995. Tectonic components in Silurian cyclicity: examples from the Appalachian Basin and global implications. In E. Landing and M. E. Johnson (eds.), Silurian Cycles: Linkages of Dynamic Stratigraphy with Atmospheric, Oceanic, and Tectonic Changes. New York State Museum Bulletin 491: 145–162. Albany.

Evitt, W. R. 1953. Observations on the Trilobite *Ceraurus.* Journal of Paleontology 27: 33.

Fagerstrom, J. A. 1961. The fauna of the Middle Devonian Formosa Reef Limestone of southwestern Ontario. Journal of Paleontology 35: 1–48.

Fisher, D. C. 1975. Swimming and burrowing in *Limulus* and *Mesolimulus.* Fossils and Strata 4: 281–290.

Fisher, D. W. 1953. Additions to the stratigraphy and paleontology of the lower Clinton of western New York. Buffalo Society of Natural Science Bulletin 21: 26 p.

——. 1957. Mohawkian (Middle Ordovician biostratigraphy of the Wells Outlier Hamilton County) New York. New York State Museum and Science Service Bulletin 359. Albany: University of the State of New York.

——. 1962. Correlation of the Ordovician Rocks in New York State. New York State Museum and Science Service Geological Survey, Map and chart series: No. 3. Albany.

Foerste, A. F. 1887. The Clinton Group of Ohio. part II. Bulletin of the Science Laboratories, Denison University 2: 89–110.

——. 1893. Fossils of the Clinton Group in Ohio and Indiana. Report of the Geological Survey of Ohio 7: 516–597.

——. 1909. Preliminary notes on Cincinnatian and Lexington fossils. Bulletin of the Science Laboratories, Denison University 14: 294.

——. 1910. Preliminary notes on Cincinnatian and Lexington fossils of Ohio, Indiana, Kentucky and Tennessee. Bulletin of the Science Laboratories, Denison University 16: 17–100.

——. 1917. Notes on Silurian fossils from Ohio and other central states. Ohio Journal of Science 17: 187–264.

——. 1920. The Kimmswick and Plattin Limestones of northeastern Missouri. Denison University Bulletin, Journal of the

Scientific Laboratories 19: 175–224.
——. 1924. Upper Ordovician faunas of Ontario and Quebec. Geological Survey of Canada, Memoir 138. Ottowa.
Ford, S. W. 1873. Remarks on the distribution of fossils in the Lower Potsdam rocks at Troy, New York, with descriptions of a few new species. American Journal of Science 106: 134–140.
——. 1876. On additional species of fossils from the primordial of Troy and Lansingburg, Rensselaer County, New York. American Journal of Science, 3rd ser. 11: 369–371.
——. 1877. On some embryonic forms of trilobites from the primordial rocks at Troy, New York. American Journal of Science, 3rd ser. 13: 265–273.
——. 1878. Note on the development of *Olenellus asaphoides*. American Journal of Science, 3rd ser. 15: 129–130.
——. 1878. Descriptions of two new species of primordial fossils. American Journal of Science, 3rd ser. 15: 124–127.
Fortey, R. A. 1975. Early Ordovician trilobite communities. Fossils and Strata 4: 331–352.
——. 1979. Early Ordovician trilobites from the Cartoche Formation (St. Georges Group), western Newfoundland. Geological Survey of Canada, Bulletin 321: 61–114.
——. 1980. The Ordovician trilobites of Spitsbergen. III. Remaining trilobites of the Valhallfonna formation. Norsk Polarinstitutt Skrifter 171: 1–163.
——. 1985. Pelagic trilobites as an example of deducing the life habits of extinct arthropods. Transactions of the Royal Society, Edinburgh 76: 219–230.
——. 1990a. Ontogeny, hypostome attachment and trilobite classification. Palaeontology 33: 529–576.
——. 1990b. Cladistics. In D. E. G. Briggs and P. R. Crowther (eds.), Paleobiology: a synthesis, pp. 430–434. Oxford: Blackwell Scientific.
——. 2001. Trilobite systematics: the last 75 years. Journal of Paleontology 75: 1141–1151.
Fortey, R. A., and B. D. E. Chatterton. 1988. Classification of the trilobite suborder Asaphina. Paleontology 31: 165–222.
Fortey, R. A., and S. F. Morris. 1978. Discovery of nauplius-like trilobite larvae. Palaeontology 21: 823–833.
Fortey, R. A., and R. M. Owens. 1997. Evolutionary history. In R. E. Kaesler (ed.), Treatise on invertebrate paleontology: part O, Arthropoda 1, Trilobita, revised, pp. 249–287. Boulder: Geological Society of America.
——. 1999. Feeding habits in trilobites. Palaeontology 42: 429–465.
Fortey, R. A., and J. N. Theron. 1995. A new Ordovician arthropod *Soomaspis*, and the agnostid problem. Palaeontology 37: 841–861.
Fortey, R. A., and H. B. Whittington. 1989. The Trilobita as a natural group. Historical Biology 2: 125–138.
Fortey, R. A., and N. V. Wilmot. 1991. Trilobite cuticle thickness in relation to paleoenvironment. Paleontologische Zeitschrift 65: 141–151.
Gillette, T. 1947. The Clinton of western and central New York. New York State Museum Bulletin 341. 191 p. Albany: New York State Museum.
Goldring, W. 1923. Devonian crinoids of New York. New York State Museum Memoir 16. 670 p. Albany: New York State Museum.
——. 1943. Geology of the Coxsackie Quadrangle, New York. New York State Museum Bulletin 332.
Golubic, S., R. D. Perkins, and R. J. Lucas. 1975. Boring microorganisms and microbarings in carbonate substrates. In R. W. Frey (ed.), The Study of Trace Fossils, pp. 229–259. New York: Springer-Verlag.
Goodwin, P. W., and E. J. Anderson. 1985. Punctuated aggradational cycles: a general hypothesis of episodic stratigraphic accumulation. Journal of Geology 93: 515–533.
Green, J. 1832. A monograph of the trilobites of North America, 93 p. Philadelphia: Joseph Brano.
——. 1835. A supplement to the monograph of the trilobites of North America, 30 p. Philadelphia: Joseph Brano.
——. 1837a. Descriptions of several new trilobites. American Journal of Science 32: 343–349.
——. 1837b. Description of a new trilobite. American Journal of Science and Arts 32: 167–169.
——. 1838. Description of a new trilobite. American Journal of Science 33: 406–407.
Griffing, D. H., and C. A. Ver Straeten. 1991. Stratigraphy and depositional environments of the lower part of the Marcellus Formation (Middle Devonian) in eastern New York State. In J. R. Ebert (ed.), New York State Geological Association Field Trip Guidebook 63, pp. 205–249. New York State Geological Association.
Hall, J. 1838. Descriptions of two species of trilobites, belonging to the genus *Paradoxides*. American Journal of Science 33: 139–142.
——. 1843. Natural history of New York. Part IV, Geology of New York, comprising the survey of the Fourth Geological District, 525 p. Albany.
——. 1847. Containing descriptions of the organic remains of the lower division of the New York System. Natural History of New York: Palaeontology, vol. I. 383 p.
——. 1851. Description of new and rare species of fossils from the Palaeozoic series. In Foster and Whitney, Report on the Lake Superior land district: part 2, the iron region together with the general geology: U. S. Congress, 32nd, Special Session, Senate Executive Document 4, pp. 203–231.
——. 1852. Containing descriptions of the organic remains of the lower middle divisions of the New York System. Natural History of New York: Palaeontology, vol. II. 362 p.
——. 1859a. New York State Cabinet of Natural History, Annual report no. 12, 72 p.
——. 1859b. Containing descriptions and figures of the organic remains of the Lower Helderberg Group and the Oriskany

Sandstone. Natural History of New York: Palaeontology, vol. III. 532 p. Albany.

——. 1860. Contributions to paleontology 1858 and 1859. 13th annual report to the regent of the University of the State of New York and the conditions of the State Cabinet of Natural History.

——. 1861a. Descriptions of new species of fossils from the upper Helderberg, Hamilton and Chemung Groups. New York State Cabinet of Natural History, annual report, 14.

——. 1861b. Preliminary notice of the trilobites and other crustaceans of the upper Helderberg, Hamilton and Chemung Groups. Report of the New York State Cabinet of Natural History, 15, pp. 27–113.

——. 1863. Notes and corrections. 16th annual report of New York State Cabinet of Natural History, pp. 223–226.

——. 1867. Account of some new or little known species of fossils from the rocks of the age of the Niagara Group. 20th annual report of New York State Cabinet of Natural History, pp. 305–401.

——. 1876. Illustration of Devonian fossils. Gastropoda, Pteropoda, Cephalopoda, Crustacea, and corals of the upper Helderberg, Hamilton and Chemung Groups, Albany, New York.

Hall, J., and J. M. Clarke. 1888. Containing descriptions of the trilobites and other crustacea of the Oriskany, upper Helderberg, Hamilton, Portage, Chemung and Catskill. Natural History of New York: Palaeontology, vol. VII. 236 p. Albany.

Hall, J., and R. P. Whitfield. 1875. Descriptions of invertebrate fossils, mainly from the Silurian System. Report Geology Survey Ohio, 2, part 2, pp. 65–161.

Henningsmoen, G. 1975. Moulting in trilobites. Fossils and Strata 4: 179–200.

Hesselbro, S. P. 1987. The biostratinomy of *Dikelocephalus* sclerites: implications for the use of trilobite attitude data. Palaios 2: 605–608.

Holloway, D. J. 1980. Middle Silurian trilobites from Arkansas and Oklahoma, U. S. A. Paleontographica Abt. A 170: 1–85.

——. 1981. Silurian dalmanitacean trilobites from North America and the origins of Dalmanitidae and Synphoriinae. Palaeontology 24: 695–731.

Holloway, D. J., and P. D. Lane. 1998. Effaced styginid trilobites from the Silurian of New South Wales. Palaeontology 41: 853–896.

Howell, B. F., and J. T. Sanford. 1946. Trilobites from the Silurian Irondequoit Formation of New York. Bulletin Wagner Free Institute of Science 21: 5–10.

——. 1947. Trilobites from the Silurian Oak Orchard Member of the Lockport Formation of New York. Bulletin Wagner Free Institute of Science 22: 33–42.

Hudson, J. A. 1982. Pyrite in ammonoid-bearing shales from the Jurassic of England and Germany. Sedimentology 27: 639–667.

Hughes, C. P., J. K. Ingham, and R. Addison. 1975. The morphology, classification and evolution of the Trinucleidae (Trilobita). Philosophical Transactions of the Royal Society 272: 537–604.

Hughes, N. C., and R. E. Chapman. 1995. Growth and variation in the Silurian proetide trilobite *Aulacopleura koninki* and its implications for trilobite paleobiology. Lethaia 28: 333–353.

Hughes, N. C., and D. L. Cooper. 1999. Paleobiologic and taphonomic aspects of the "*Granulosa*" trilobite cluster, Kope Formation (Upper Ordovician, Cincinnati region). Journal of Paleontology 73: 306–319.

ICZN. 1985. International code of zoological nomenclature, 3rd edition. Berkeley, Cal.: International Trust for Zoological Nomenclature, University of California Press. 338 p.

Irwin, M. L. 1965. General theory of epeiric clear water sedimentation. American Association of Petroleum Geologists Bulletin 49: 445–459.

Isachsen, Y. W., E. Landing, J. M. Lauber, L. V. Rickard, and W. B. Rogers (eds.). 1991. Geology of New York, a simplified account. New York State Museum/Geological Survey Educational Leaflet, 28. 284 p. Albany.

Jell, P. A. 1975. The abathochroal eye of *Pagetia*, a new type of trilobite eye. Fossils and Strata 4: 33–44.

Johnson, T. T. 1985. Trilobites of the Thomas T. Johnson collection: how to find, prepare and photograph trilobites. Dayton, Ohio: Litho-print. 178 p.

Kaesler, R. L. (ed.). 1997. Treatise on invertebrate paleontology: part O, Arthropoda 1, Trilobita, revised. Boulder, Col.: Geological Society of America. 530 p.

Kammer, T. W., C. E. Brett, D. R. Boardman II, and R. H. Mapes. 1986. Ecologic stability of the dysaerobic biofacies during the late Paleozoic. Lethaia 19: 109–121.

Kay, M. 1937. Stratigraphy of the Trenton Group. Geological Society of America Bulletin 48: 233–302.

——. 1943. Mohawkian series on West Canada Creek, New York. American Journal of Science 241: 597–606.

——. 1953. Geology of the Utica Quadrangle, New York. New York Museum Bulletin 347.

——. 1968. Ordovician formations in northwestern New York. Naturaliste Canadien 95: 1373–1378.

Keary, P. 1996. The new Penguin dictionary of geology. London: Penguin Books. 366 p.

Kidwell, S. M., and D. Bosence. 1991. Taphonomy and time-averaging of marine shelly faunas. In P. A. Allison and D. E. G. Briggs (eds.), Taphonomy: releasing the data locked in the fossil record, pp. 116–209. New York: Plenum.

Klapper, G. 1981. Review of New York Devonian conodont biostratigraphy. In W. A. Oliver and G. Klapper (eds.), Devonian biostratigraphy of New York, part 1, pp. 57–66. Washington, D.C.: International Union of Geological Sciences, Subcommittee on Devonian Stratigraphy.

Kloc, G. 1992. Spine function in the odontopleurid trilobites

Leonaspis and *Dicranurus* from the Devonian of Oklahoma. North American Paleontological Convention Abstracts and Program, Paleontological Society Special Publication 6: 167.

——. 1993. Epibionts on Selenopeltinae (Odontopleurida) trilobites. Geological Society of America, 1993 annual meeting, Abstracts with programs 25: 103.

——. 1997. Epibionts on *Dicranurus* and some related genera. In Second International Trilobite Conference, St. Catharines, Ontario, Abstracts with Program, p. 28.

Kobayashi, T. 1939. Journal of the Faculty of Science, Imperial University, Tokyo, sec. 2 5: 112.

Kurtz, V. E. 1975. Franconian (Upper Cambrian) trilobite faunas from the Elvins Group of southeast Missouri. Journal of Paleontology 49: 1009–1043.

Lambert, A. E. 1904. Description of *Dalmanites lunatus*. Bulletin of the Geological Society of America 15: 480–482.

Landing, E. 1988. Depositional tectonics and biostratigraphy of the western portion of the Taconic allochthon, eastern New York State. In E. Landing (ed.), The Canadian paleontology and biostratigraphy seminar, pp. 96–110. New York State Museum Bulletin 462: 96–110. Albany: New York State Museum.

Landing, E., S. A. Bowring, K. L. Davidek, A. W. A. Rushton, R. A. Fortey, and W. A. P. Wimbledon. 1998a. Cambrian-Ordovician boundary age and duration of the lowest Ordovician Tremadoc Series based on U-Pb zircon dates from Avalonian Wales. Geological Magazine 137: 485–494.

Landing, E., S. A. Bowring, K. L. Davidek, S. R. Westrop, G. Geyer, and W. Heldmaier. 1998b. Duration of the Early Cambrian: U-Pb ages of volcanic ashes from Avalon and Gonwanda. Canadian Journal of Earth Sciences 35: 329–338.

Lane, P. D., and A. T. Thomas. 1983. A review of the trilobite suborder Scutelluina. Palaeontological Association Special Papers in Palaeontology 30: 141–160.

Lask, P. R. B. 1993. The hydrodynamic behavior of sclerites from the trilobite *Flexicalymene meeki*. Palaios 2: 219–215.

Lespérance, P. J. 1975. Stratigraphy and paleontology of the Synphoriidae (Lower and Middle Devonian Dalmanitacean trilobites). Journal of Paleontology 49: 91–137.

Lespérance, P. J., and P.-A. Bourque. 1971. The Synphoriinae: an evolutionary pattern of Lower and Middle Devonian trilobites. Journal of Paleontology 45: 182–208.

Leutze, W. P. 1961. Arthropods from the Syracuse Formation, Silurian of New York. Journal of Paleontology 35: 49–64.

Levi-Setti, R. 1975. Trilobites: a photographic atlas. Chicago: University of Chicago Press. 213 p.

——. 1993. Trilobites, 2nd edition. Chicago: University of Chicago Press. 342 p.

Lieberman, B. E. 1994. Evolution of the trilobite subfamily Proetinae Salter, 1864, and the origin, diversification, evolutionary affinity, and extinction of the Middle Devonian proetid fauna of eastern North America. Bulletin American Museum Natural History 223. 176 p. New York: American Museum of Natural History.

Lieberman, B. E., and G. Kloc. 1997. Evolutionary and biogeographic patterns in the Asteropyginae (Trilobita, Devonian). Bulletin American Museum Natural History 232. 127 p. New York: New York Museum of Natural History.

Linnarsson, J. G. O. 1869. Om Vestergotlands Cambriskaoch Siluriska Aflagringar. K. Svenska Vetenskapakadamiens Handligar 8: 87.

Linsley, D. M. 1994. Devonian paleontology of New York. Paleontological Research Institution special publication 21. 472 p. Ithaca: Paleontological Research Institution.

Lochman, C. 1946. Two Upper Cambrian (Trempealeau) trilobites from Dutchess County, New York. American Journal of Science 244: 547–553.

Locke. 1838. 2nd annual report geology survey of Ohio. Columbia.

LoDuca, S. T. 1995. Thallophytic-algae-dominated biotas from the Silurian Lockport Group of New York and Ontario. Northeastern Geology and Environmental Sciences 17: 371–382.

Ludvigsen, R. 1977. The Ordovician trilobite *Ceraurinus* Barton in North America. Journal of Paleontology 51: 959–972.

——. 1979a. The Ordovician trilobite *Pseudogygites* Kobayashi in eastern and arctic North America. Life sciences contributions 120. Toronto: Royal Ontario Museum.

——. 1979b. Fossils of Ontario, part 1: the trilobites. Life sciences miscellaneous publications. Toronto: Royal Ontario Museum.

——. 1987. Reef trilobites from the Formosa Limestone (Lower Devonian) of southern Ontario. Canadian Journal of Earth Science 24: 676–688.

Ludvigsen, R., and B. D. E. Chatterton. 1982. Ordovician Pterygometopidae (Trilobita) of North America. Canadian Journal of Earth Science 19: 2179–2209.

——. 1991. The peculiar Ordovician trilobite *Hypodicranotus* from the Whittaker Formation, District of Mackenzie. Canadian Journal of Earth Science 28: 616–622.

Ludvigsen, R., and P. A. Tuffnell. 1983. A revision of the Ordovician olenid trilobite *Triarthrus* Green. Geological Magazine 120: 567–577.

——. 1994. The last olenacean trilobite: *Triarthrus* in the Witby Formation (Upper Ordovician) of southern Ontario. New York State Museum Bulletin 481: 183–212.

Ludvigsen, R., and S. R. Westrop. 1983. Franconian trilobites of New York State. New York State Museum Memoir 23: 83 p. Albany: New York State Museum.

——. 1986. Classification of the Late Cambrian trilobite *Idiomesus* Raymond. Canadian Journal of Earth Science 23: 300–307.

Martin, R. E. 1999. Taphonomy: a process approach. Cambridge: Cambridge University Press.

Matthew, G. F. 1896. Faunas of the *Paradoxides* beds of eastern North America. New York Academy of Science Transactions 15: 192–247.

Maximova, Z. A. 1972. New Devonian trilobites of the Phacopoidea. Paleontological Journal 1: 78–83.

McNamara, K. J., and D. Rudkin. 1984. Techniques of trilobite exuviation. Lethaia 17: 153–173.

Meek, F. B. 1872. Descriptions of new species of fossils from the Cincinnati Group of Ohio. American Journal of Science vol. 3, 3rd ser: 426.

Middleton, G. V., M. Rutka, and C. J. Salas. 1987. Geologic setting of the northern Appalachian Basin during the Early Silurian. In W. L. Duke (ed.), Sedimentology, stratigraphy, and ichnology of the Lower Silurian Medina Formation in New York and Ontario, Society of Economic Paleontologists and Mineralogists, Eastern Section. 1987 Annual Field Trip Guidebook. pp. 1–15.

Mikulic, D. G., and J. Kluessendorf. 2001. Moulting behavior and disarticulation patterns of Silurian calymenids: implications for interpreting trilobite ecology and taphonomy, In Abstracts of papers for the Third International Conference on Trilobites and their Relatives, pp. 22–23. New York: University of Oxford.

Miller, J. 1975. Structure and function of trilobite terrace lines. Fossils and Strata 4: 155–178.

Miller, M. F., and J. Rehmer. 1979. Trace fossils in interpretation of sharp lithologic boundaries; and example from the Devonian of New York. Geological Society of America, Abstracts with Program 11: 45.

Moore. R. C. (ed.). 1959. Treatise on invertebrate paleontology: part O, Arthropoda 1. Boulder, Col.: Geological Society of America and University of Kansas Press.

Morris, P., L. C. Ivany, K. M. Schopf, and C. E. Brett. 1995. The challenge of paleoecological stasis: reassessing sources of evolutionary stability. Proceedings of the National Academy of Sciences 92: 11269–11273.

Novacek, M. L., and Q. D. Wheeler (eds.). 1992. Extinction and phylogeny. New York: Columbia University Press. 253 p.

Osgood, R. G., Jr. 1970. Trace fossils of the Cincinnati area. Paleontographica America (v. 6) 41: 281–439.

Osgood, R. G., and W. I. Drennen. 1975. Trilobite trace fossils from the Clinton Group (Silurian) of east-central New York State. Bulletins of American Paleontology 67: 299–349.

Owen, A. W. 1985. Trilobite abnormalities. Transactions of the Royal Society, Edinburgh, Earth Science 76: 255–272.

Owens, R. M. 1973. British Ordovician and Silurian proetidae (Trilobita). Palaeontographical Society Monograph vol. 127, no. 535: 1–98.

Palmer, A. R. 1965. Biomere–a new kind of biostratigraphic unit. Journal of Paleontology 39: 149–153.

——. 1984. The biomere problem: evolution of an idea. Journal of Paleontology 58: 599–611.

Parsons, K. M., and C. E. Brett. 1991. Taphonomic processes and biases in modern marine environments: and actualistic perspective on fossil assemblage preservation. In S. K. Donovan (ed.). The processes of fossilization, pp. 22–65. London: Belhaven Press.

Phleger, F. B. 1937. New Lichidacea in the collection of the Museum of Comparative Zoology. Bulletin of the Museum of Comparative Zoology (Harvard) 80: 415–423.

Pillet, J. 1972. Les trilobites du Dévonian inférieur et du Dévonian moyen du sud-est du Massif armoricain. Société D'Études Scientifique de l'Anjou, Mémoire 1: 307 p., Angers France.

Plotnick, R. E. 1986. Taphonomy of a modern shrimp: implications for the arthropod fossil record. Palaios 1: 286–293.

Pratt, B. R. 1998. Possible predation on Upper Cambrian trilobites and its relevance for the extinction of soft-bodied Burgess Shale-type animals. Lethaia 31: 73–88.

Pribyl, A., J. Vanek, and I. Pek. 1985. Phylogeny and taxonomy of family Cheiruridae (Trilobita). Acta Universitatis Palackianae Olomucenis Facultas Rerum Naturalium Geographica-Geologica XXIV 83: 107–191.

Prouty, F. 1923. Systematic paleontology. Maryland Geological Survey: Silurian. Baltimore.

Ramsköld, L. 1991. Pattern and process in the evolution of the Odontopleuridae (Trilobita). The Selenopeltinae and Ceratcephalinae. Transactions of the Royal Society of Edinburgh: Earth Science 82: 143–181.

Ramsköld, L., and B. D. F. Chatterton. 1991. Revision and subdivision of the polyphyletic "*Leonaspis*" (Trilobita). Transactions of the Royal Society, Edinburgh: Earth Science 82: 333–371.

Ramsköld, L., and G. D. Edgecombe. 1991. Trilobite monophyly revisited. Historical Biology 4: 267–283.

——. 1996. Trilobite appendage structure—*Eoredlichi*a reconsidered. Alcheringa 20: 269–276.

Ramsköld, L., and L. Werdelin. 1991. The phylogeny and evolution of some phacopid trilobites. Cladistics 7: 29–74.

Rasetti, F. 1944. Upper Cambrian trilobites from the Lévis Conglomerate. Journal of Paleontology 18: 229–258.

——. 1945. New Upper Cambrian trilobites from the Lévis Conglomerate. Journal of Paleontology 19: 462–478.

——. 1946. Revision of some late Upper Cambrian trilobites from New York, Vermont and Quebec. America Journal of Science 244: 537–546.

——. 1948. Lower Cambrian trilobites from the conglomerates of Quebec (exclusive of the Ptychopariidae). Journal of Paleontology 22: 1–24.

——. 1952. Revision of the North American trilobites of the family Eodiscidae. Journal of Paleontology 26: 434–451.

——. 1966a. Revision of the North American species of the Cambrian trilobite genus *Pagetia*. Journal of Paleontology 39: 502–511.

——. 1966b. New Lower Cambrian trilobite faunule from the Taconic Sequence of New York. Smithsonian Miscella-

neous Collections 148. 52 p. Washington D.C.: Smithsonian Institution.
——. 1967. Lower and Middle Cambrian trilobite faunas from the Taconic Sequence of New York. Smithsonian Miscellaneous Collections 152. no. 4: 135 p. Washington D.C.: Smithsonian Institution.
Rasetti, F., and G. Theokritoff. 1967. Lower Cambrian agnostid trilobites of North America. Journal of Paleontology 41: 189–196.
Raymond, P. E. 1905a. The fauna of the Chazy Limestone. American Journal of Science, 4th ser. 20: 353–382.
——. 1905b. The trilobites of the Chazy Limestone. Annals of the Carnegie Museum 3: 328–386.
——. 1910a. Trilobites of the Chazy Formation in Vermont. 7th Report of the State Geologist, Vermont, pp. 213–248. Burlington.
——. 1910b. Notes on Ordovician trilobites. IV. New and old species of trilobites from the Chazy. Annals of the Carnegie Museum 7: 60–80.
——. 1910c. Notes on Ordovician trilobites. I. Asaphidae from the Beekmantown. Annals of the Carnegie Museum 7: 35–45.
——. 1913. Notes on some new and old trilobites in the Victoria Memorial Museum. Bulletin of the Victoria Memorial Museum, Geology Survey of Canada 1: 45. Ottawa.
——. 1914. Notes on the ontogeny of *Isotelus gigas* Dekay. Bulletin of the Museum of Comparative Zoology, Harvard 63: 247–269.
——. 1916. A new *Ceraurus* from the Chazy. New York State Museum Bulletin 189: 1–41. Albany.
——. 1920a. The appendages, anatomy and relationships of trilobites. Memoirs of the Connecticut Academy of Arts and Science 7: 169 p. New Haven, Connecticut.
——. 1920b. Some new Ordovician trilobites. Bulletin of the Museum of Comparative Zoology, Harvard 64: 273–296.
——. 1921. A contribution to the description of the fauna of the Trenton Group. Canadian Geological Survey Museum Bulletin 31. Ottawa.
——. 1925. Some trilobites of the Lower Middle Ordovician of eastern North America. Bulletin of the Museum of Comparative Zoology, Harvard 67 (1): 1–180.
——. 1931. An unusually large isoteloid hypostoma. Bulletin of the Museum of Comparative Zoology, Harvard 55: 205–209.
Raymond, P. E., and D. C. Barton. 1913. A revision of the American species of *Ceraurus*. Bulletin of the Museum of Comparative Zoology, Harvard 54 (20): 523–543.
Raymond, P. E., and J. E. Narraway. 1908. Notes on Ordovician trilobites: Illaenidae from the Black River Limestone near Ottawa, Canada. Annals of the Carnegie Museum 4: 242–255.
——. 1910. Notes on Ordovician trilobites. III. Asaphidae from the Lowville and Black River. Annals of the Carnegie Museum 7: 46–59.
Resser, C. E. 1938. Cambrian system (restricted) of the southern Appalachians. Geological Society of America Special Paper 15: 1–140. Boulder.
——. 1942a. New Upper Cambrian trilobites. Smithsonian Miscellaneous Collections 101: 1–58.
——. 1942b. New Upper Cambrian trilobites. Smithsonian Miscellaneous Collections 103: 1–136.
Rickard, L. V. 1962. Late Cayugan (Upper Silurian) and Helderbergian (Lower Devonian) stratigraphy in New York. New York State Museum Bulletin 386. Albany.
——. 1975. Correlation of the Silurian and Devonian rocks in New York State. New York State Museum, Map and Chart Ser. 24. Albany.
——. 1981. The Devonian System of New York State. In W. A. Oliver and G. Klapper (eds.), Devonian biostratigraphy of New York, part 1, pp. 5–21. Washington, D.C.: International Union of Geological Sciences, Subcommittee on Devonian Stratigraphy.
Ross, R. J. 1951. Stratigraphy of the Garden City Formation in northeastern Utah, and its trilobite fauna. Peabody Museum of Natural History Bulletin 6.
——. 1967. Calymenid and other Ordovician trilobites from Kentucky and Ohio. Geological Survey Professional Paper 583-B: B1–B19. Washington D.C.: United States Geological Survey.
——. 1975. Early Paleozoic trilobites, sedimentary facies, lithospheric plates, and ocean currents. Fossils and Strata 4: 331–352.
——. 1979. Additional trilobites from the Ordovician of Kentucky. Geological Survey Professional Paper 1066-D: D1–D13. Washington, D.C.: United States Geological Survey.
Rudkin, D. M., and R. P. Tripp. 1985. A reassessment of the Ordovician trilobite *Isotelus*, part I: North American species. Canadian Paleontology and Biostratigraphy Seminar, Quebec City, September 1985.
——. 1987. A reassessment of the Ordovician trilobite *Isotelus*, part II: Ontario species. Canadian Paleontology and Biostratigraphy Seminar, London, Ontario, September 1987.
——. 1989. The type species of the Ordovician trilobite genus *Isotelus*: *I. gigas* DeKay, 1824. Life Sciences Contribution 152. 18 p. Toronto: Royal Ontario Museum.
Rudkin, D. M., R. P. Tripp, and R. Ludvigsen. 1994. The Ordovician trilobite genus *Hemiarges* (Lichidae: Trochurinae) from North America and Greenland. New York State Museum Bulletin 481: 289–306.
Ruedemann, R. 1901. Trenton conglomerate of Rysedorph Hill Rensselaer Co., New York and its fauna. New York State Museum Bulletin 49: 3–114.
——. 1916a. Account of some little-known species of fossils, mostly from the Paleozoic rocks of New York. New York State Museum Bulletin 189: 1–112.
——. 1916b. The presence of a median eye in trilobites. New York State Museum Bulletin 189: 127–143.

——. 1926. The Utica and Lorraine Formations of New York. Part 2 systematic paleontology, no. 2 Mollusks, Crustaceans and Eurypterids. New York State Museum Bulletin 272: 1–227.

Sargent, J. D. 1953. A new trilobite from the Onondaga Limestone of New York. Buffalo Society Natural Science Bulletin 21, no. 2: 91–93.

Sars. 1835. Okens Isis, p. 333.

Scatterday, J. W. 1986. The Middle Devonian brachymetopid trilobite *Australosutura gemmaea* (Hall and Clarke, 1888), close relatives, descendants, and ancestors. Canadian Paleontology and Biostratigraphy Seminar, Abstracts, pp. 29–30.

Seilacher, A., W. E. Reif, and F. Westfal. 1985. Sedimentological, ecological, and temporal patterns of fossil Lagerstätten. Philosophical Transactions of the Royal Society of London B311: 5–23.

Sevon, W. D., and D. L. Woodrow. 1985. Middle and Upper Devonian stratigraphy within the Appalachian Basin. In D. L. Woodrow and W. D. Sevon (eds.), The Catskill Delta, Geological Society of America Special Paper 201. 246 p.

Shaw, F. C. 1968. Early Middle Ordovician Chazy trilobites of New York. New York State Museum and Science Service Memoir 17. 163 p. Albany.

Shaw, F. C., and P. J. Lespérance. 1994. North America biogeography and taxonomy of *Cryptolithus* (Trilobita, Ordovician). Journal of Paleontology 68: 808–823.

Shu, D.-G., G. Geyer, L. Chen, and X.-L. Zhang. 1995. Redlichiacean trilobites with preserved soft-parts from the Lower Cambrian Chengjiang fauna (South China). In E. Landing and G. Geyer (eds.), Morocco '95: The Lower-Middle Cambrian standard of western Gonwanda, pp. 203–242. Beringeria Special Issue 2. Würzburg, Germany: Institüt für Paläontologie der Universität Würzburg.

Signor, P. W., and C. E. Brett. 1984. The mid-Paleozoic precursor to the Mesozoic marine revolution. Paleobiology 10: 229–245.

Simpson. 1890. Transactions of the American Philosophical Society, n.s., 16: 460.

Sloss, S. L. 1963. Sequences in the cratonic interior of North America. Geological Society of America Bulletin 74: 93–113.

Smith, A. B. 1994. Systematics and the fossil record: documenting evolutionary patterns. Blackwell Scientific. 223 p.

Speyer, S. E. 1985. Moulting in phacopid trilobites. Transactions Royal Society, Edinburgh 76: 221–223.

——. 1987. Comparative taphonomy and paleoecology of trilobite Lagerstätten. Alcheringa 11: 205–232.

——. 1990a. Gregarious behavior and reproduction in trilobites. In A. J. Boucot (ed.), Evolutionary paleobiology and coevolution, pp. 405–409. Amsterdam: Elsevier.

——. 1990b. Trilobite molt patterns. In A. J. Boucot (ed.), Evolutionary paleobiology and coevolution, pp. 491–499. Amsterdam: Elsevier.

——. 1990c. Enrollment in trilobites. In A. J. Boucot (ed.), Evolutionary paleobiology and coevolution, pp. 450–456. Amsterdam: Elsevier.

Speyer, S. E., and C. E. Brett. 1985. Clustered trilobite assemblages in the Middle Devonian Hamilton Group. Lethaia 18: 85–103.

——. 1986. Trilobite taphonomy and Middle Devonian taphofacies. Palaios 1: 312–327.

——. 1988. Taphofacies models for epeiric sea environments: Middle Paleozoic examples. Palaeogeography, Palaeoclimatology, Palaeoecology 63: 225–262.

——. 1991. Taphofacies controls—background and episodic processes in fossil assemblage preservation. In P. A. Allison and D. E. G. Briggs (eds.), Taphonomy: releasing the data locked in the fossil record, pp. 501–545. New York: Plenum.

Stewart, G. A. 1927. Fauna of the Silica Shale of Lucas County, Ohio. Ohio Geology Survey Bulletin, 4th. ser. 32: 1–76.

Stokes, C. 1824. On a trilobite from Lake Huron. In J. J. Bigsby, Notes on the geography and geology of Lake Huron. Transactions Geological Society, ser. 2 1: 175–209.

Størmer, L. 1939. Studies on trilobite morphology. Part I. The thoracic appendages and their phylogenetic significance. Norsk Geology tidsskrift 19: 143–273.

——. 1951. Studies on trilobite morphology. Part III. The ventral cephalic structures with remarks on the zoological position of the trilobites. Norsk Geology tidsskrift 29: 108–158.

Struve, W. 1990. Palaozoologie III (1986–1990). Cour. Forsch.-Inst. Senckenberg 127: 251–279.

——. 1992. Neues zur stratigraphie und fauna des rhenotypen Mittel-Devouian. Sencken bergiana Lethaea. 71: 503–624.

Stumm, E. C. 1953a. Trilobites of the Devonian Traverse Group of Michigan. Contributions of the Museum of Paleontology, University of Michigan 10: 101–157.

——. 1953b. Lower Middle Devonian proetid trilobites from Michigan, southwestern Ontario, and northern Ohio. Contributions of the Museum of Paleontology, University of Michigan 11: 11–31.

——. 1967. Devonian trilobites from northwestern Ohio, northern Michigan and western New York. Contributions of the Museum of Paleontology, University of Michigan 21: 109–122.

Stürmer, W. 1970. Soft parts of cephalopods and trilobites: some surprising results of X-ray examination of Devonian slates. Science 170: 1300–1302.

Stürmer, W., and J. Bergström. 1973. New discoveries on trilobites by X-rays. Palaeontologische Zeitschrift 47: 104–141.

Swirydczuk, K., B. H. Wilkinson, and G. R. Smith. 1981. Synsedimentary lacustrine phosphorites from the Pliocene Glenns Ferry Formation of southwestern Idaho. Journal of Sedimentation Petrology 51: 1205–1214.

Taylor, W. L., and C. E. Brett. 1996. Taphonomy and paleoecology of Echinoderm Lagerstätten from the Silurian (Wenlockian) Rochester Shale. Palaios 11: 118–140.

Tetreault, D. K. 1992. Paleoecologic implications of epibionts on the Silurian trilobite *Arctinurus*. North American Paleontology Convention, abstracts and program, Paleontological Society special publication 6: 289.

——. 1994. Brachiopod and trilobite biofacies of the Rochester Shale (Silurian, Wenlockian Series) in western New York. New York State Museum Bulletin 481: 347–361.

Thomas, A. T., and D. S. Holloway. 1988. Classification and phylogeny of the trilobite order Lichida. Philosophical Transactions of the Royal Society, London B321: 179–262.

Thomas, A. T., and R. M. Owens. 1978. A review of the family Aulacopleuridae. Palaeontology 21: 65–81.

Tillman, C. G. 1960. Spathacalymene, an unusual new Silurian trilobite genus. Journal of Palaeontology 34: 891–895.

Titus, R. 1982. Fossil communities of the Middle Trenton Group (Ordovician) of New York State. Journal of Paleontology 56: 477–485.

——. 1986. Fossil communities of the Upper Trenton Group (Ordovician) of New York State. Journal of Paleontology 60: 805–824.

Towe, K. M. 1973. Trilobite eyes: calcified lens in vivo. Science 179: 1007–1009.

Treatise (revised). See Kaesler (1997).

Ulrich, E. O. 1878. Descriptions of some new species of fossils from the Cincinnati Group. Journal of the Cincinnati Society of Natural History 1: 99.

——. 1879. Description of a trilobite from the Niagara Group of Indiana. Journal of the Cincinnati Society of Natural History 2: 131–134.

——. 1926. In R. Ruedemann, The Utica and Lorraine Formations of New York: Part 2 systematic paleontology: no. 2 Mollusks, Crustaceans and Eurypterids. New York State Museum Bulletin 272: 128.

——. 1930. Ordovician trilobites of the family Telephidae and concerned stratigraphic correlations. United States National Museum Proceedings 76: 1–101.

Ulrich, E. O., and D. M. Delo. 1940. In D. M. Delo, Phacopid trilobites of North America. Geological Society of America special paper 29: 108. Boulder: Geological Society of America.

Vanuxem, L. 1842. Natural history of New York geology of New York. Part III. Comprising the survey of the Third Geological District, 306 p. Albany.

Vere, V. K. 1972. The biostratigraphy, evolution and paleoautecology of *Cryptolithus* (Trilobita) in New York State and western Vermont. Ph.D. dissertation, Syracuse University. 848 p.

Ver Straeten, C. A., and C. E. Brett. 2000. Bulge migration and pinnacle reef development, Devonian Appalachian foreland basin. Journal of Geology 108: 339–352.

Vogdes, A. W. 1879. A short notice upon the geology of Catoosa County, Georgia. American Journal of Science, 3rd ser. 18: 475–477.

Vogel, K., S. Gobulic, and C. E. Brett. 1987. Endolith associations and their relation to facies distribution in the Middle Devonian of New York State. Lethaia 20: 263–290.

Walcott, C. D. 1875a. Description of a new species of trilobite. Cincinnati Quarterly Journal of Science 2: 273.

——. 1875b. New species of trilobite from the Trenton limestone at Trenton Falls, New York. Cincinnati Quarterly Journal of Science 2: 347–349.

——. 1875c. Notes on *Ceraurus pleurexanthemus* Green. Annals of the Lyceum of Natural History, New York 11: 155–159.

——. 1875d. Description of the interior surface of the dorsal shell of *Ceraurus pleurexanthemus* Green. Annals of the Lyceum Natural History, New York 11: 159–162.

——. 1876. Preliminary notice of the discovery of the remains of the natatory and brachial appendages of trilobites. Advanced print, Dec. 1876. 28th Annual Report of the New York State Museum of Natural History 1879: 89–92.

——. 1877a. Description of new species of fossils from the Trenton Limestone. Advanced print, Jan. 1877. 28th Annual Report New York State Museum Natural History 1879: 93–97.

——. 1877b. Notes on some sections of trilobites from the Trenton limestone. Advanced print, Sept. 20, 1877. 31st Annual Report New York State Museum Natural History 1879: 61–63.

——. 1877c. Note on the eggs of the trilobite. Advanced print, Sept. 20, 1877. 31st Annual Report New York State Museum of Natural History 1879: 66–67.

——. 1877d. Descriptions of new species of fossils from the Chazy and Trenton limestones. Advanced print, Sept. 20, 1877. 31st Annual Report New York State Museum of Natural History 1879: 68–71.

——. 1879a. Description of new species of fossils from the Calciferous Formation. Advanced print Jan. 3, 1879. 32nd Annual Report New York State Museum Natural History 1879: 129–131.

——. 1879b. Fossils of the Utica Slate and the metamorphoses of *Triarthrus becki*. Advanced print, June 1879. Transactions of the Albany Institute 1883 10: 18–38.

——. 1881. The trilobite: new and old evidence relating to its organization. Bulletin of the Museum of Comparative Zoology, Harvard 8: 191–224.

——. 1890. Descriptions of new forms of Upper Cambrian fossils. United States National Museum Proceedings 13: 267–279.

——. 1891. The fauna of the Lower Cambrian or *Olenellus* zone. U.S. Geological Survey Tenth Annual Report, pp. 515–629.

——. 1912. Cambrian geology and paleontology II, New York Potsdam-Hoyt fauna. Smithsonian Miscellaneous Collections 57: 269.

——. 1918. Cambrian geology and paleontology, IV, appendages of trilobites. Smithsonian Miscellaneous Collections 67: 115–216.

——. 1921. Cambrian geology and paleontology, IV, notes on the structure of *Neolenus*, supplementary notes. Smithsonian Miscellaneous Collections 67: 377–456.

——. 1924. Cambrian and Lower Ozarkian trilobites. Smithsonian Miscellaneous Collections 75: 53–60.

Weller, S. 1903. New Jersey Geological Survey: Paleontology 3: 252.

——. 1907. The paleontology of the Niagaran Limestone in the Chicago area: the Trilobita. Chicago Academy of Sciences Bulletin 4, part II of the Natural History Survey, 281 p.

Westrop, S. R. 1983. The life habits of the Ordovician illaenine trilobite *Bumastoides*. Lethaia 16: 14–24.

Westrop, S. R., L. A. Knox, and E. Landing, 1993. Lower Ordovician (Ibexian) trilobites from the Tribes Hill Formation, central Mohawk Valley, New York State. Canadian Journal of Earth Sciences 30: 1618–1633.

Whiteley, T. E., C. E. Brett, and D. Lehmann. 1993. The Walcott-Rust Quarry; a unique Ordovician trilobite Konservat-Lagerstätte. Geological Society of North America, Northeastern Section, 28th Annual Meeting, Abstracts with programs, 25: 89.

Whiteley, T. E., and K. Smith. 2001. Fossil Lagerstätten of New York State 3: Caleb's Quarry: Part I, the trilobites. American Paleontologist 9, No. 1: 3–5.

Whitfield, R. P. 1884a. Notice of some new species of primordial fossils in the collections of the museum and correction of previously described species. American Museum of Natural History Memoir 1: 146.

——. 1884b. Notice of some new species of primordial fossils in the collections of the museum, and corrections of previously described species. American Museum of Natural History Bulletin 5: 149–153.

——. 1885. Notice of a very large species of *Homalonotus* from the Oriskany sandstone formation. American Museum of Natural History Bulletin 6: 193–195.

——. 1886a. Notice of geological investigations along the eastern share of Lake Champlain, conducted by Prof. H. Seely and Pres. Ezra Brainerd of Middlebury College, with descriptions of the new fossils discovered. American Museum of Natural History Bulletin 7: 336.

——. 1886b. Notice of geological investigations along the eastern shore of Lake Champlain conducted by Prof. H. M. Seely and Pres. Ezra Brainard of Middlebury College, with descriptions of the new fossils discovered. American Museum of Natural History Bulletin 8: 293–345.

——. 1889a. Observations on some imperfectly known fossils from the calciferous sandrock of Lake Champlain, and descriptions of several new forms. American Museum of Natural History Bulletin 11: 41–63.

——. 1889b. Additional notes on *Asaphus canalis*, Conrad. American Museum of Natural History Bulletin 11: 63–64.

——. 1897. Descriptions of new species of Silurian fossils from near Fort Cassin and elsewhere on Lake Champlain. American Museum of Natural History Bulletin 9: 177–184.

Whittington, H. B. 1941. The Trinucleidae—with special reference to North American genera and species. Journal of Paleontology 15: 21–41.

——. 1952. A unique remopleuridid trilobite. Breviora, Harvard 4: 1–9.

——. 1953. North American Bathyuridae and Leiostegiidae (Trilobita). Journal of Paleontology 27: 647–678.

——. 1957. Ontogeny of *Ellipsocephala, Paradoxides, Sao, Blainia*, and *Triarthrus* (Trilobita). Journal of Paleontology 31: 934–946.

——. 1959. Silicified Middle Ordovician trilobites: Remopleuridae, Trinucleidae, Raphiophoridae, Endymioniidae. Bulletin of the Museum of Comparative Zoology, Harvard 121: 371–496.

——. 1960. *Cordania* and other trilobites from the Lower and Middle Devonian. Journal of Paleontology 34: 405–420.

——. 1961a. Middle Ordovician Pliomeridae (Trilobita) from Nevada, New York, Quebec, Newfoundland. Journal of Paleontology 35: 911–922.

——. 1961b. A natural history of trilobites. Natural History 70 (August): 8–17.

——. 1963. Middle Ordovician trilobites from Lower Head, western Newfoundland. Museum of Comparative Zoology (Harvard) Bulletin 129: 1–118.

——. 1966. Phylogeny and distribution of Ordovician trilobites. Journal of Paleontology 40: 696–737.

——. 1968. *Cryptolithus* (Trilobita): specific characters and occurrence in Ordovician of eastern North America. Journal of Paleontology 42: 702–714.

——. 1971a. Silurian calymenid trilobites from the United States, Norway, and Sweden. Palaeontology 14: 455–477.

——. 1971b. The Burgess Shale: history of research and preservation of fossils. Proceedings of the First North American Paleontological Convention 1: 1176–1201.

——. 1980. Exoskeleton, moult stage, appendage with habits of the Middle Cambrian trilobite *Olenoides serratus*. Palaeontology 23: 171–204.

——. 1985. The Burgess Shale. New Haven: Yale University Press. 151 p.

——. 1992. Trilobites. Rochester, N.Y.: Boydell Press. 145 p., 120 plates.

——. 1993. Anatomy of the Ordovician trilobite *Placoparia*. Philosophical Transactions of the Royal Society, London 339: 109–118.

——. 1997a. Life mode, habits and occurrence. In R. L. Kaesler (ed.), Treatise on invertebrate paleontology: part O, Arthropoda 1, Trilobita, revised, pp. 152–158. Boulder: Geological Society of America.

——. 1997b. Illaenidae (Trilobita): morphology of thorax, classification, and mode of life. Journal of Paleontology 71: 878–896.

Whittington, H. B., and J. E. Almond. 1987. Appendages and habits of the Upper Ordovician trilobite *Triarthrus eatoni*. Philosophical Transactions of the Royal Society, London B317: 1–46.

Whittington, H. B., and W. R. Evitt. 1954. Silicified Middle Ordovician trilobites. Geological Society of America Memoir 59: 137.

Whittington, H. B., and C. P. Hughes. 1972. Ordovician geography and faunal provinces deduced from trilobite distribution. Philosophical Transactions of the Royal Society, London 263: 235–278.

Whittington, H. B., and S. R. A. Kelly. 1997. Morphological terms applied to Trilobita. In R. L. Kaesler (ed.), Treatise on invertebrate paleontology: part O, Arthropoda 1, Trilobita, revised, pp. 313–329. Boulder, Col.: Geological Society of America and the University of Kansas.

Wilson, A. E. 1947. Trilobita of the Ottawa Formation of the Ottawa-St. Lawrence lowland. Geological Survey Bulletin 9. Canadian Department of Mines and Resources. 86 p. Ottawa.

Wilson, J. L. 1951. Franconian trilobites of the central Appalachians. Journal of Paleontology 25: 617–654.

Witzke, B. J. 1990. Palaeoclimate constraints for Palaeozoic palaeolatitudes of Laurentia and Euroamerica. In W. S. McKerrow and C. R. Scotese (eds.), Palaeozoic palaeogeography and biogeography, pp. 57–73. Memoir 12. London: Geographical Society.

Yochelson, E. L. 1987. Walcott in Albany, New York: James Hall's "special assistant." Earth Sciences History 6: 86–94.

Yochelson, E. L., and M. A. Fedonkin. 1993. Paleobiology of *Climactichnites*, an enigmatic late Cambrian fossil. Smithsonian Contributions to Paleobiology 74. pp. 1–74.

Young, F. P. 1943a. Black River stratigraphy and faunas, part I. American Journal of Science 241: 141–166.

——. 1943b. Black River stratigraphy and faunas, part II. American Journal of Science 241: 209–240.

Trilobite Index: Order-Family-Genus-Species

The arrangement of the orders in this index reflects that in the trilobite *Treatise* (revised) (Kaesler 1997). Families, genera, and species are alphabetical.

ORDER	FAMILY	GENUS	SPECIES	PAGE	PLATE
Agnostida	Diplagnostidae	*Baltagnostus*	*angustilobus*	118	
		Baltagnostus	*stockportensis*	56, 118	
	Peronopsidae	*Peronopsis*	*evansi*	56, 118	
		Peronopsis	*primigenea*	56, 118	
	Ptychagnostidae	*Ptychagnostus*	*gibbus*	56, 118	
		Ptychagnostus	*punctuosus*	56, 118	
		Ptychagnostus	*elegans*	56, 118	
	Spinagnostidae	*Eoagnostus*	*acrorachis*	56, 118	
		Eoagnostus	*primigeneus*	118	
		Hypagnostus	*parvifrons*	56, 118	
	Calodiscidae	*Calodiscus*	*agnostoides*	55, 119	
		Calodiscus	*fissifrons*	55, 119	
		Calodiscus	*lobatus*	55, 119	
		Calodiscus	*meeki*	55, 119	
		Calodiscus	*occipitalus*	55, 119	
		Calodiscus	*reticulatus*	56, 119	1E
		Calodiscus	*schucherti*	56, 119	
		Calodiscus	*theokritoffi*	56, 119	
		Calodiscus	*walcotti*	56, 119	
		Chelediscus	*chathamsis*	56, 119	
	Eodiscidae	*Pagetia*	*bigranulosa*	56, 119	
		Pagetia	*connexa*	56, 119	
		Pagetia	*clytioides*	56, 119	
		Pagetia	*erratic*	56, 119	
		Pagetia	*laevis*	119	
		Pagetides	*amplifrons*	56, 119	
		Pagetides	*elegans*	56, 119	
		Pagetides	*leiopygus*	56, 119	
		Pagetides	*minutus*	56, 119	
		Pagetides	*rupestris*	56, 119	
	Hebediscidae	*Hebediscus*	*marginatus*	119	
		Neopagetina	*taconica*	56, 119	
	Weymouthiidae	*Acidiscus*	*birdi*	55, 119	
		Acidiscus	*hexacanthus*	55, 119	1H
		Acimetopus	*bilobatus*	55, 119	1A
		Analox	*bipunctata*	119	
		Analox	*obtusa*	119	
		Bathydiscus	*dolichometopus*	55, 119	
		Bolboparia	*elongata*	55, 119	1F

ORDER	FAMILY	GENUS	SPECIES	PAGE	PLATE
		Bolboparia	*superba*	55, 119	1E
		Leptochilodiscus	*punctulatus*	56, 120	
		Litometopus	*punctulatus*	120	1G
		Mallagnostus	*desideratus*	120	
		Microdiscus	*connexus*	120	
		Oodiscus	*binodosus*	56, 120	
		Oodiscus	*longifrons*	120	
		Oodiscus	*subgranulatus*	56, 120	
		Serrodiscus	*griswoldi*	56, 120	
		Serrodiscus	*latus*	120	
		Serrodiscus	*speciosus*	56, 120	1B, C, D, K
		Serrodiscus	*spinulosus*	56, 120	
		Serrodiscus	*subclovatus*	56, 120	
		Stigmadiscus	*gibbosus*	56, 120	
		Stigmadiscus	*stenometopus*	56, 120	
		Weymouthia	*nobilis*	56, 120	
Redlichiida	Holmiidae	*Elliptocephala*	*asaphoides*	56, 118	1K, 2
Corynexochida	Dolichometopidae	*Athabaskiella*	*proba*	118	
		Bathyuriscidella	*socialis*	118	
		Bathyuriscus	*eboracensis*	56, 118	
		Bathyuriscus	*fibriatus*	118	
		Corynexochides	*expansus*	56, 118	
	Dorypygidae	*Fordaspis*	*nana*	55, 56, 118	
		Kootenia	*fordi*	55, 56, 120	
		Olenoides	*stissingensis*	120	
		Olenoides	*stockportensis*	56	
	Illaenidae	*Bumastoides*	*aplatus*	120	
		Bumastoides	*bellevillensis*	121	
		Bumastoides	*billingsi*	66, 121	
		Bumastoides	*comes*	62	
		Bumastoides	*gardenensis*	62, 121	
		Bumastoides	*globosus*	62, 121	3
		Bumastoides	*holei*	75, 121	4, 5
		Bumastoides	*milleri*	66, 121	6
		Bumastoides	*porrectus*	71, 75, 121	7
		Bumastoides	*trentonensis*	71, 121	
		Illaenus	*arcturus*	121	
		Illaenus	*consimilis*	122	
		Illaenus	*crassicauda*	62	
		Illaenus	*latidorsata*	71	
		Nanillaenus	*americanus*	75, 122	8
		Nanillaenus	*conradi*	71, 122	
		Nanillaenus	*latiaxiatus*	66, 122	9
		Nanillaenus	*punctatus*	62, 122	
		Nanillaenus	*raymondi*	62, 122	10A
		Thaleops	*longispina*	62, 122	10B
		Thaleops	*ovata*	62, 66, 122	
	Styginidae	*Bumastus*	*ioxus*	89, 123	11
		Eobronteus	*lunatus*	78, 123	12
		Eobronteus	sp.	62	

ORDER	FAMILY	GENUS	SPECIES	PAGE	PLATE
		Failleana	*indeterminata*	66, 123	13
		Illaenoides	cf. *trilobita*	89, 123	14, 15
		Platillaenus	*erastusi*	62, 123	
		Platillaenus	*limbatus*	123	
		Scutellum	*barrandi*	123	
		Scutellum	*niagarensis*	85, 123	16
		Scutellum	*pompilius*	98, 123	
		Scutellum	*rochesterense*	88, 89, 123	26B
		Scutellum	*senescens*	116, 123	17
		Scutellum	*tullius*	112, 123	18
		Scutellum	*wardi*	93, 124	
	Zacanthoididae	*Prozacanthoides*	*eatoni*	56, 124	
Lichida	Lichidae	*Acanthopyge*	*consanquinea*	98, 124, 125	
		Acanthopyge	*contusa*	107, 124, 125	
		Amphilichas	*conifrons*	75, 124	
		Amphilichas	*cornutus*	75, 124	19
		Amphilichas	*minganensis*	62, 124	
		Amphilichas	*trentonensis*	71, 124	
		Arctinurus	*boltoni*	1, 18, 89, 126	20, 21
		Autoloxolichas	*inconsuetus*	71, 126	
		Ceratolichas	*dracon*	107, 125, 126	
		Ceratolichas	*gryps*	107, 125, 126	
		Dicranopeltis	*fragosa*	126	
		Dicranopeltis	*nereus*	27, 89, 126	22, 23
		Echinolichas	*eriopsis*	107, 126	24
		Echinolichas	*hispidus*	105, 107, 125, 126	
		Hemiarges	*paulianus*	69, 71, 126	25
		lichid	sp.	126	26A
		Oinochoe	*bigsbyi*	98, 125, 126	
		Oinochoe	*pustulosus*	98, 127	27
		Richterarges	*ptyonurus*	127	28
		Terataspis	*grandis*	17, 105, 107, 127	29
		Trochurus	*bulbosa*	127	
		Trochurus	*halli*	89, 127	30, 31
		Trochurus	*phlyctainoides*	127	
	Odontopleuridae	*Apianurus*	*narrawayi*	62, 127	
		Ceratocara	*shawi*	127	
		Ceratocephala	*triacantheis*	62, 128	
		Diacanthaspis	*parvula*	75, 128	32
		Dicranurus	*elegantus*		33
		Dicranurus	*hamatus*	128	
		Kettneraspis	*callicera*	105, 107, 128	34
		Kettneraspis	sp.	128	36
		Kettneraspis	*tuberculata*	6, 98, 128	35
		Meadowtownella	*trentonensis*	75, 128	37, 38
		Odontopleura	*cerelepta*	80, 128	
		Odontopleura	sp.	128	
		Primaspis?	*crosota*	128	39

ORDER	FAMILY	GENUS	SPECIES	PAGE	PLATE
Phacopida	Acastidae	asteropygin		129	40
		Astropyginae aff. *Greenops*			49
		Bellacartwrightia	*calderonae*	112, 129	41
		Bellacartwrightia	*calliteles*	129	
		Bellacartwrightia	*jennae*	112, 129, 130	42, 43
		Bellacartwrightia	*phyllocaudata*	112, 129, 130	44
		Bellacartwrightia	*pleione*	107, 130, 131	
		Bellacartwrightia	*whiteleyi*	112, 131	45
		Bellacartwrightia	sp.	112, 131	46, 47, 48
		Greenops	*barberi*	112, 130, 131	50
		Greenops	*boothi*	130, 131	51
		Greenops	*grabaui*	11, 112, 131	52, 53, 54
		Greenops	sp.	11, 131	55, 56
		Kennacryphaeus	*harrisae*	112, 131	
	Calymenidae	*Calymene*	*camerata*	132	57
		Calymene	*conradi*	80, 132	
		Calymene	*niagarensis*	9, 89, 132	58, 59
		Calymene	*platys*	105, 107, 132	60
		Calymene	*singularis*	93, 132	61
		calymenid	sp.	8, 27, 132	62, 63
		Diacalymene	*rostrata*	85, 132	
		Diacalymene	sp.	132	64, 65
		Diacalymene	*vogdesi*	132	
		Flexicalymene	*granulosa*	80, 133	
		Flexicalymene	*meeki*	25, 80, 133	66
		Flexicalymene	*senaria*	14, 69, 71, 75, 123	67, 68, 69, 70
		Gravicalymene	*magnotuberculata*	71, 133	71, 72
		Liocalymene	*clintoni*	89, 133	73
		Liocalymene	*cresapensis*	133	
	Cheiruridae	*Acanthoparypha*	sp.	62, 134	
		Acanthoparypha	*trentonensis*	134	
		Ceraurinella	*latipyge*	62, 134	86C
		Ceraurinella	*scofieldi*	66, 134	74
		Ceraurinus	*marginatus*	80, 134	75
		Ceraurinus	sp.	134	
		Cerauropeltis	*ruedemanni*	62, 134	84C to E
		Ceraurus	*montyensis*	134	
		Ceraurus	*pleurexanthemus*	9, 12, 19, 20, 22, 71, 134	76, 77, 78
		Ceraurus	sp.	125	79, 80
		Cheirurus	sp.	135	81
		Deiphon	*pisum*	89, 135	84B
		Forteyops	*approximus*	135	
		Gabriceraurus	*dentatus*	71, 75, 135	82, 83
		Gabriceraurus	*hudsoni*	62, 135	
		Heliomera	*akocephala*	62, 135	
		Kawina	*chazyensis*	62, 135	
		Kawina	*vulcanus*	62, 135	
		Nieszkowskia	*satyrus*	62, 135	
		Sphaerexochus	*parvus*	62, 135	
		Sphaerocoryphe	*goodnovi*	62, 135	86B

ORDER	FAMILY	GENUS	SPECIES	PAGE	PLATE
		Sphaerocoryphe	*major*	78, 135	
		Sphaerocoryphe	*robusta*	75, 135	85
		Staurocephalus	sp.	136	91B
	Dalmanitidae	*Corycephalus*	*pygmaeus*	107, 136	
		Corycephalus	*regalis*	105, 137, 138	
		Dalmanites	*aspinosus*	137	87
		Dalmanites	*bisigmatus*	101, 137	
		Dalmanites	*limulurus limulurus*	18, 27, 89, 137	88, 89, 99
		Dalmanites	*limulurus lunatus*	137	
		Dalmanites	*pleuroptyx*	98, 137	90
		Dalmanites	sp.	137	91A
		Forillonaria	*russelli*	137	
		Neoprobilium	*nasutus*	98, 137, 138	
		Neoprobilium	*tridens*	98, 137, 138	
		Odontochile	*litchfieldensis*	137	
		Odontochile	*micrurus*	98, 137	92
		Odontochile	*phacoptyx*	101, 139	
		odontochilid	sp.	139	93
		Phalangocephalus	*dentatus*	138, 139	
	Encrinuridae	*Cybeloides*	*ella*	139	
		Cybeloides	*prima*	139	86B
		Encrinurus	cf. *raybesti*	85, 93, 139	95
		Encrinurus	sp.	139	96
		Erratencrinurus	*vigilans*	69, 71, 139	94
		Physemataspis	*insularis*	139	86D
	Homalonotidae	*Brogniartella*	*trentonensis*	139	
		Dipleura	*dekayi*	11, 140	97, 98
		Homalonotus	*major*	17, 101, 140	
		Trimerus	*delphinocephalus*	89, 93, 140	99
		Trimerus	*vanuxemi*	101, 140	
	Phacopidae	*Burtonops*	*cristatus*	102, 105, 140, 141, 143	
		Eldredgeops	*crassituberculatus*	112, 141	100
		Eldredgeops	*rana*	2, 5, 34, 112, 142	101, 102, 103, 104
		Eldredgeops	*rana norwoodensis*	112, 142	
		Eophacops	*trisulcatus*	85, 144	105, 106
		Paciphacops	*hudsoniscus*	141, 143, 144	
		Paciphacops	*clarkei*	101, 141, 143, 144	
		Paciphacops	*logani*	98, 107, 141, 143, 144	107
		Paciphacops	*logani* subsp. A	141, 144	
		Phacopina	*anceps*	144	
		Phacops?	*clarksoni*	105, 141, 143, 144	
		Phacops?	*iowensis*	112, 142, 144	108
		Viaphacops	*bombifrons*	9, 107, 142, 143, 145	109
	Pliomeridae	*Pliomerops*	*canadensis*	62, 145	112, 113, 114
	Pterygometopidae	*Achatella*	*achates*	23, 75, 78, 145	115

ORDER	FAMILY	GENUS	SPECIES	PAGE	PLATE
		Calyptaulax	*annulata*	62, 145	
		Calyptaulax	*callicephalus*	9, 71, 75, 78, 145	116, 117
		Calyptaulax	*eboraceous*	71, 75, 78, 145	
		Calyptaulax	sp.	71, 146	
		Chasmops?	*bebryx*	71, 146	118
		Eomonorachus	*convexus*	71, 146	119
	Synphoriidae	*Anchiopella*	*anchiops*	102, 105, 136, 146	120
		Anchiopella	*sobrinus*	105, 146	
		Coronura	*aspectans*	107, 146	121, 122
		Coronura	*helena*	107, 146	127B
		Coronura	*myrmecophorus*	17, 107, 138, 147	
		Odontocephalus	*aegeria*	107, 147	123
		Odontocephalus	*bifidus*	107, 147	124
		Odontocephalus	*coronatus*	107, 147	
		Odontocephalus	*humboltensis*	107, 147	
		Odontocephalus	*selenurus*	107, 147	125
		Odontocephalus	sp.	147	126
		Schoharia	*emarginata*	105, 147	
		Schoharia	sp.	148	
		Synphoria	*concinnus*	105, 107, 148	
		Synphoria	*sopita*	101, 148	
		Synphoria	*stemmata compacta*	101, 148	
		Synphoria	*stemmata stemmata*	101, 138, 148	
		Synphoroides	*dolphi*	148	
		Trypaulites	*calypso*	138, 148	
		Trypaulites	*erinus*	105, 107, 148	127A
		Trypaulites	*macrops*	107, 148	
Proetida	Aulacopleuridae	*Cyphaspis*	sp.	112, 149	
		Harpidella	*craspedota*	149	128
		Harpidella	sp.	149	131
		Harpidella	*spinafrons*	112, 149	129, 130
		Harpidella	*stephanophora*	107, 149	
		Maurotarion	*minuscula*	102, 105, 107, 149	
		Maurotarion	sp.		132
		Otarion?	*coelebs*	149	
		Otarion?	*diadema*	107, 149	
		Otarion?	*hudsonicum*	80, 149	
		Otarion?	*hybrida*	107, 149	
		Otarion?	*laevis*	116, 149	
		Otarion?	*matutinum*	78, 149	
		Otarion?	*spinicaudatum*	62, 149	
	Bathyuridae	*Acidiphorus*	*whittingtoni*	61, 150	
		Bathyurellus	*platypus*	61, 150	
		Bathyuropsis	*schucherti*	71	
		Bathyurus	cf. *B. angelina*	150	
		Bathyurus	*extans*	150	133
		Bathyurus	*johnsoni*	66, 150	

ORDER	FAMILY	GENUS	SPECIES	PAGE	PLATE
		Bathyurus	*perkinsi*	61	
		Bathyurus	*taurifrons*	150	
		Benthamaspis	*striata*	61, 150	
		Bolbocephalus	*seelyi*	61, 150	
		Grinnellaspis	cf. *G. marginiata*	61, 150	
		Raymondites	*ingalli*	71, 150	
		Raymondites	*longispinus*	66, 150	
		Raymondites	*spiniger*	66, 150	
		Strigigenalis	*cassinensis*	61, 151	
		Strigigenalis	*caudatus*	61, 151	
		Uromystrum	*brevispinum*	62, 151	
		Uromystrum	*minor*	62, 151	
	Brachymetopidae	*Australosutera*	*gemmaea*	107, 112, 151	134
		Cordania	*becraftensis*	101, 151	
		Cordania	*cyclurus*	98, 151	
		Mystrocephala	*arenicolus*	105, 151	
		Mystrocephala	*ornata*	151	
		Mystrocephala	*baccata*	112, 151	
		Mystrocephala	*varicella*	107, 151	
		Radnoria	sp.	151	135
	Dimeropygidae	*Dimeropyge*	*clintonensis*	62, 151	
	Proetidae	*Basidechenella*	*hesionea*	105, 152, 153	
		Coniproetus	*angustifrons*	105, 152, 153	
		Coniproetus	*conradi*	105, 152, 153	
		Coniproetus	*folliceps*	107, 152, 154	136
		Cornuproetus	*beecheri*	78, 80, 152, 153	
		Crassiproetus	*brevispinus*	107, 152, 154	
		Crassiproetus	*crassimarginatus*	107, 152, 154	
		Crassiproetus	*neoturgitus*	107, 154, 155	
		Crassiproetus	*schohariensis*	101, 153, 155	
		Crassiproetus	*stummi*	107, 153, 155	
		Cyphoproetus	*wilsonae*	71, 155	137
		Dechenella	*haldemani*	112, 154, 155	138
		Decoroproetus	*corycoeus*	89, 155	139, 140
		Gerastos	*protuberans*	153, 155	
		Hedstroemia	*pachydermata*	153, 155	
		Mannopyge	*halli*	107, 154, 155	
		Monodechenella	*macrocephala*	9, 112, 154, 155	141
		Proetus	*artiaxis*	93, 153, 156	
		Proetus	*clelandi*	62, 153, 156	
		Proetus	*hesione*	153	
		Proetus	*jejunus*	112, 154, 156	
		Proetus	*marginalis*	154, 156	
		Proetus	*microgemma*	154, 156	
		Proetus	*parviusculus*	153, 156	
		Proetus	*spurlocki*	80, 153, 156	
		Proetus	*tenuisulcatus*	93, 153, 156	
		Proetus	*undulostriatulus*	153, 156	
		proetid	sp.	156	142
		Pseudodechenella	*arkonensis*	112, 154, 156	
		Pseudodechenella	*canaliculata*	107, 153, 157	

ORDER	FAMILY	GENUS	SPECIES	PAGE	PLATE
		Pseudodechenella	*clara*	107, 154, 157	143
		Pseudodechenella	*rowi*	112, 154, 157	144, 145 146, 147
Asaphida	Asaphidae	*Basilicus*	*romingeri*	157	
		Basilicus	*ulrichi*	66, 157	
		Basilicus	*vetustus*	66, 157	
		Basilicus	*whittingtoni*	17, 62, 158	
		Basiliella	*barrandi*	66, 158	
		Bellefontia	*gyracanthus*	61, 158	
		Ectenaspis?	*homalonotoides*	66, 158	148
		Homotelus	*stegops*	80, 158	
		Hyboaspis	*depressa*	62, 158	
		Isoteloides	*angusticaudus*	62, 158	
		Isoteloides	*canalis*	61, 62, 158	149
		Isoteloides	*peri*	61, 158	
		Isoteloides	*whitfieldi*	61	
		Isotelus	*annectans*	159	
		Isotelus	*beta*	62	
		Isotelus	*giganteus*	17, 62, 159	
		Isotelus	*gigas*	1, 12, 14, 17, 23, 71, 75, 159	150, 151, 152, 153
		Isotelus	*harrisi*	62, 159	
		Isotelus	*jacobus*	71, 159	
		Isotelus	*latus*	71, 159	
		Isotelus	*maximus*	9, 17, 78, 159	154, 155, 156, 157
		Isotelus	*platycephalus*	1, 159	
		Isotelus	*pulaskiensis*	80, 160	
		Isotelus	*simplex*	66, 160	
		Isotelus	*walcotti*	3, 75, 160	158
		Niloides	*perkinsi*	62, 160	
		Pseudogygites	*latimarginatus*	80, 160	159
		Vogdesia	*bearsi*	62, 160	
		Vogdesia	*obtusus*	62, 160	
	Idahoiidae	*Saratogia*	*calcifera*	51, 160	175B
	Pterocephaliidae	*Cameraspis*	*convexa*	51, 160	
		Cameraspis	sp.	160	
	Ptychaspididae	*Conaspis*	*whitehallensis*	160	
		Idiomesus	sp.	161	
		Kiethiella	*depressa*	51, 161	
	Raphiophoridae	*Lonchodomas*	*chaziensis*	62, 161	
		Lonchodomas	*halli*	62, 161	
		Lonchodomas	*hastatus*	78, 161	
	Remopleurididae	*Apatokephaloides*	sp.	161	
		Hypodicranotus	*striatulus*	23, 75, 161	160, 161, 162
		Remopleurides	*canadensis*	62, 161	
		Remopleurides	*linguatus*	78, 161	
		Remopleurides	*tumidus*	78, 161	
		Richardsonella	sp.	161	
		Robergiella	*brevilingua*	61, 161	

ORDER	FAMILY	GENUS	SPECIES	PAGE	PLATE
	Saukiidae	*Hoytaspis*	*speciosa*	51, 162	
		Prosaukia	*briarcliffensis*	162	
		Prosaukia	*hartti*	51, 162	175A
		Prosaukia	*tribulis*	51, 162	
	Symphysurinidae	*Symphysurina*	*convexa*	61, 162	
		Symphysurina	*woosteri*	61, 162	
	Trinucleidae	*Cryptolithus*	*bellulus*	20, 23, 28, 78, 80, 162	163
		Cryptolithus	*lorettensis*	11, 28, 69, 71, 162	164, 165
		Cryptolithus	*tessellatus*	28, 69, 71, 163	166, 167
		Tretaspis	*diademata*	78, 163	
		Tretaspis	*reticulata*	78, 163	
Ptychopariida	Conocoryphidae	*Atops*	*trilineatus*	55, 163	
		Conocoryphe	*verrucosa*	163	
	Dokimocephalidae	*Sulcocephalus*	*saratogensis*	163	
	Elviniidae	*Calocephalites*	*minimus*	51, 163	
		Dellea	*landingi*	51, 163	
		Dellea	*saratogensis*	51, 164	
		Drabia	*curtoccipita*	51, 164	
		Drabia	*menusa*	51, 164	
		Elvinia	*granulata*	51, 164	
	Glaphuridae	*Glaphurina*	*lamottensis*	62, 165	168B
		Glaphurus	*pustulosus*	62, 165	168A
	Harpidae	*Dolichoharpes*	sp.	62, 165	
		Eoharpes	*pustulosus*	66, 165	
		Hibbertia	*ottawaensis*	165	167C
		Hibbertia	sp.	62, 165	
		Hibbertia	*valcourensis*	62, 165	169A, B
		Scotoharpes	*cassinensis*	61, 165	169D
	Hystricuridae	*Hystricurus*	*conicus*	61, 165	
		Hystricurus	*crotalifrons*	61, 165	
		Hystricurus	*ellipticus*	61, 165	
		Hystricurus	*oculilunatus*	61, 166	
	Kingstoniidae	*Kingstonia*	*seelyi*	166	
	Komaspididae	*Carrickia*	*setoni*	62, 166	
	Lonchocephalidae	*Lonchocephalus*	sp.	166	
	Marjumiidae	*Modocia*	*punctata*	166	
	Menomoniidae	*Bolaspidella*	*fisheri*	56, 166	
	Olenidae	*Triarthrus*	*beckii*	71, 76, 166	170
		Triarthrus	*eatoni*	14, 19, 20, 22, 23, 78, 80, 166	171, 172
		Triarthrus	*glaber*	80, 167	
		Triarthrus	*spinosus*	80, 167	173, 174
	Plethopeltidae	*Plethometopus*	*knopfi*	167	
		Plethopeltis	*granulosa*	51, 167	
		Plethopeltis	*saratogensis*	51, 167	175C
	Ptychopariidae	*Ptychoparia*	*minuta*	167	
		Ptychoparia	*matheri*	167	
	Shumardiidae	*Shumardia*	*pusilla*	61, 167	
	Solenopleuridae	*Rimouskia*	*typica*	56, 167	
Uncertain	Uncertain	*Clelandia*	*parabola*	167	

Index

Abrasion, corrosion, and encrustation, 37
Acadian Orogeny, 49, 97, 99, 100, 105, 107, 110, 112, 115, 116
Agnostida, 117
Agnostina, 117
Alden pyrite beds, 111
Algonquin Arch, 85, 86
Alsen Formation, 99, 100, 101
Ammonoosuc island arc, 61, 62
Amsterdam Formation, 68, 70
apodemes, 12
appendages, 17, 19, 20
Arthrophycus, 85
articulated remains, 33–35
Asaphida, 157
Ashokan Formation, 110
Austin Glen Formation, 55, 64, 66, 81
Avalonia, 49, 100, 116
axial lobe, 5

Bald Hill bentonite, 100
Baltica, 49, 83
basis, 19, 20
Beaver Dam Shale, 102
Becraft Limestone, 99, 100, 101
Beecher's Trilobite Bed, 17, 28, 40, 78
Beekmantown Group, 57, 61
Bertie Group, 83, 95
bioherm, 62, 88, 90, 91, 92, 105
biostratinomy, 32
Black River Group, 59, 64, 67
Bloomsburg Formation, 94
Bois Blanc Formation, 102
brachial appendage, 19, 20
Bridgewater Member, 109
Bridport Member, 61
Brown's Creek bed, 41, 111
Browns Pond formation, 53, 55
Burgess Shale, 17
Burleigh Hill Member, 89
Butternut Member, 110

Cabot Head Shales, 83
Caladonian Orogeny, 83, 100
Cambrian allochthon, 51–56
Cambrian autochthon, 49–51
Cambrian Period, 46–56
Cambrian stratigraphic charts, 52, 55
Camillus Formation, 109
Canajoharie Arch, 69
Canajoharie Shale, 76
Cardiff Formation, 109
Carlisle Center Sandstone, 102
Catskill facies, 115
Cedar Valley Limestone, 112
cephalic border, 8
cephalic spine, 6
cephalic sutures, 9
cephalon, 5, 8
Centerfield Member, 110
Chazy Group, 59, 62, 63
Chazy stratigraphic chart, 63
Chemung facies, 115
Chenango Member, 110
Cherokee Unconformity, 80, 81, 83, 84
Cherry Valley Limestone, 109
Chestnut Street bed, 107
Chittenango Shale Member, 60
City Brook bed, 71
cladistics, 30–31
Clinton Group, 83, 84, 85
Clinton iron ores, 85
Cobleskill Limestone, 95
Coeymans Member, 96, 98, 101
Cole Hill tongue, 110
Columbus Limestone, 105
Corynexochida, 118
cranidium, 7
Creek phase (Tippecanoe Supersequence), 62, 64, 68
Crown Point Formation, 62
Cruziana, 102

Daedalus, 85
Dawes Sandstone, 86
Day Point Formation, 62
death, 32
Death, decay and disarticulation, 32, 35
DeCew Formation, 90
Decker Formation, 95
Deep Kill Formation, 55, 61
Deep Run Member, 111
Deer River Shale, 76, 78
Delphi Station Member, 110
Denley Formation, 70, 71, 72, 73, 74
Denmark Formation, 71
Devonian Period, 95–116
Devonian stratigraphic charts, 96, 97, 103, 108, 115
disarticulation, 34
Dolgeville Formation, 76, 77, 78
doublure, 12

Early Devonian, 96
Early Silurian, 83–85
Eastern Trenton equivalents and Utica Shale, 76–78
Edgecliff Member, 105
eggs, 13
endites, 19, 21
Eodiscina, 118
Eramosa Formation, 93
Esopus Formation, 101, 102
evolution, 30
exites, 19
exoskeleton, 4–6
exoskeletons with attached fauna, 18
eyes, 7

Findley Arch, 85
fixed cheek, 8
Flat Creek Formation, 76
Fonda Member, 60
Fort Cassin Formation, 60, 61
fossil diagenesis, 32, 37–40

Fragmentation, and biased preservation, 36–37
Frankfort Formation, 78, 80, 81
free cheek, 8
Frontenac Arch, 57

Galway Formation, 51
Gasport Formation, 90, 92
genal angle, 10
genal spine, 6
Genesee Formation, 115
Geneseo Shale, 112, 114, 115
glabella, 7
glabellar furrows, 8
glabellar lobes, 8
Glens Falls Limestone, 69
Glenerie Limestone, 101
Glenmark Shale, 90
gnathobases, 19
Goat Island Formation, 90
gonatoparion cephalic suture, 9
Gondwana, 96, 100
Granville Shale, 53
"Grabau Trilobite Beds", 111
Grenville Orogeny, 43
Grenville rocks, 45, 46, 47, 48
Grimsby Formation, 83
Gull River Formation, 64
gut, 22

Halihan Hill beds, 109
Hamilton Group, 107, 108
Hannacroix Member, 98
Hasenclever Member, 78
Hatch Hill Formation, 55, 56, 61
heart, 22
Helderberg Group, 96
hepatopancreatic organ, 22
Herkimer Formation, 87, 89
High Falls Ash, 73, 74
Hillier Limestone, 75, 79
holaspid, 13, 14
holochroal eye, 7, 9
holotype, 3
House Creek Formation, 64
Hoyt Formation, 51, 54
hypostome, 12
hypotype, 3

Iapetus Ocean (Protoatlantic), 46, 49, 61
Ilion Formation, 90, 93
Indian Castle Formation, 76, 77
Indian River Formation, 55, 61, 63, 78
internal anatomy, 21, 22
Irondequoit Limestone, 87, 88, 91
Ivy Point Member, 111

Jamesville Member, 98
Jaycox Member, 111
Jeffersonville Limestone, 105
Joshua Coral bed, 111

Kalkberg Formation, 96, 98, 101
Kashong Member, 111
Kaskaskia Supersequence, 100
Keefer Formation, 89
Keyser Formation, 95
Kings Falls Limestone, 69, 70
Kirkland Formation, 89
Knox Unconformity, 57, 60, 61, 62, 63, 64
Kodak Sandstone, 84, 85

Lagerstätten, 17, 40–42
lappets, 10
Late Devonian strata, 115–116
Late Early Devonian, 96–100
Late Ordovician, 78–81
Late Silurian, 93–95
Laurentia, 45, 49, 56, 61, 100, 116
Ledyard Member, 110, 111
Leicester Pyrite, 112
LeRoy bed, 109
Levanna Shale, 110
Lewiston Member, 89
Lichida, 124
life-mode, 21–30
Little Falls Formation, 50, 51
Lockport Group, 83, 90–93
Lorraine Group, 78
Lowville Formation, 64, 65
Ludlowville Formation, 108, 110, 113

Mackenzie Formation, 93
Manitoulin Dolostone, 83
Manlius Formation, 96, 101
Manorkill Formation, 110
Maplewood Shale, 84, 85
Marcellus Formation, 105, 109
Medina Group (Sandstone), 82, 83, 84, 85
Menteth Limestone, 111, 113
meraspid, 13, 14
Middle and Upper Trenton Group, 71–76
Middle Granville Shale, 55
Middle Devonian, 105–114
Middle Ordovician, 62–78
Middle Ordovician allochthonous rocks, 62–64
Middle Silurian, 85–93
molting, 14, 15
Moorehouse Member, 105
Moscow Formation, 111
Mottville Member, 110
Mount Marion Formation, 106, 109
Mount Merino Formation, 55, 63
Moyer Creek Member, 78
Murder Creek Beds, 42
muscles, 22

Napanee Formation, 67, 68, 70
Nassau Formation, 53
Neahga Shale, 85
Nedrow Member, 104, 105
neotype, 3
New Scotland Limestone, 98, 101
Niagara Gorge, 88
nodes, 6
Normanskill Group, 63

Oak Orchard Member, 93
Oatka Creek Shale, 109
obrution, 34
Ogdensburg Formation, 60
Olney Member, 98
Onondaga Group, 102, 103, 105, 106
Onondaga Indian Nations Ash, 105
Ontogeny, 13–17
opisthoparion cephalic suture, 19
Ordovician allochthonous rocks, 61
Ordovician Period, 56–81
Ordovician stratigraphic chart, 57
Oriskany Sandstone, 101
Oswego Formation, 78, 79, 80
Otisco Shale, 111
Otsquago Formation, 86

Palatine Bridge Member, 60
Paleozoic geology, 43–116
Paleozoic stratigraphic charts, 44, 45
palpebral lobe, 5, 7
Pamelia Formation, 64
paratype, 3
Pecksport Shale, 109
Penfield Formation, 90
Penn Dixie Quarry, 42
perforations, 11
Phacopida, 128
phaselus, 13
Phytopsis, 64, 65, 69
pits, 11
plastotype, 3
Plattekill Formation, 110
pleural lobes, 5

podomeres, 19, 21
Poland Member, 71, 72, 73
Pompey Member, 110
Pools Brook Member, 98
Port Ewen Formation, 99, 100, 101
Port Jervis Formation, 100
Portage facies, 115
Potsdam Sandstone, 50, 51, 53
Poulteney Formation, 61
Power Glen Shale, 83, 84
Prasopora, 71, 76
preglabellar field, 10
Prelude to the Paleozoic, 43–46
Proetida, 148
proparion cephalic suture, 9
protaspid, 13, 14
Ptychopariida, 163
Pulaski Member, 76, 79, 80
pustules, 6
pygidial spine, 6
pygidium, 5, 7, 10
pyritization, 9, 21, 37, 38, 39

Queenston Formation, 78, 80, 84

Rafinesquina, 75
Rathbun Member, 69, 71
Ravenna Member, 98
Redlichiida, 118
Retsof beds, 111
Richmond Group, 80
Rickard Hill Member, 102
Rochdale Formation, 61
Rochester Shale, 41, 87, 88, 89, 91
Rockway Formation, 86, 91
Rodinia, 43, 48
Rondout Group, 81, 83, 95
Rose Hill Shale, 86
Rusophycus, 83, 85, 89
Russia Member, 71, 72, 74
Rust Formation, 72, 73, 74, 78
Rysedorph Hill conglomerate, 76

Salina Group, 83, 93, 94
Salinic Orogeny, 83, 94
Sauk Supersequence, 49, 60, 62
Sauk Unconformity, 61
Sauquoit Shale, 81, 86
Schenectady Formation, 78, 79, 80
schizochroal eye, 7, 9
Schodack Limestone, 55
Schoharie Formation, 102
Schoharie Grit, 102
Selinsgrove Formation, 107
setae, 19
Shadow Lake Formation, 64
Shawangunk, 85
Silurian Period, 81–95
Silurian stratigraphic chart, 82
Skunnemunk Outlier, 100, 101
Sloss Supersequences, 62
Smoke Creek beds, 42
Snake Hill Formation, 70, 76
Sodus Shale Formation, 85
soft body parts, 17
soft tissue decay, 33
Solsville Sandstone, 109
Sonyea Group, 116
Stafford Member, 110
Staghorn Point, 111
Steuben Formation, 75, 78
Stockbridge Member, 60
stomach, 22
Stony Hollow Member, 109
stromatolites, 51, 54, 94, 112
Sugar River Formation, 69
syntype, 3
Syracuse Formation, 95

Taconic allocthon, 61, 62, 68
Taconic Orogeny, 53, 56, 61, 76, 82
taphofacies, 40
terminal spine, 10
Tetradium, 64, 69
Thatcher Member, 96
Theresa Dolostone, 51, 60
thoracic segments, 10
thoracic spines, 6
thorax, 5
Thorold Sandstone, 85
Tichenor Member, 111, 113
Tioga bentonite, 105
Tippecanoe Supersequence, 62, 64, 68, 83, 96
Tonoloway Formation, 95
trace fossils, 24, 25
Transport and reorientation, 35–36
Trenton Group, 68, 70
Tribes Hill Formation, 57, 60
trilobite injury, 18, 26, 27
trilobite Lagerstätten in New York, 17, 40
Trilobite Mountain, 100
trilobite names, 2
trilobite preservation, 34
trilobite shape, 23
trilobite size, 17
Tristates Group, 96, 101–106
Trocholites, 71
Truthville Shale, 53, 55
Tully Limestone, 112, 114
Tuscarora Formation, 80, 83, 85
Tutelo Phase (Tippecanoe Supersequence), 83, 96

Union Springs Formation, 106, 107, 109
Upper Clinton Group, 86–90
Upper Ordovician Oswego and Queenston formations, 80
Utica Shale Group, 76, 78

Valcour Formation, 62
Valley Brook Shale, 76
ventral anatomy, 12, 28
Vernon Formation, 94
volcanic island arc, 56

Walcott-Rust Quarry beds, 17, 41, 74, 75
Wallbridge Unconformity, 101, 102, 104
Wanakah Member, 111
Ward Siltstone Member, 61
Watertown Formation, 64, 66, 67, 68
West Falls Group, 115
Westmoreland, 86
Whetstone Gulf Member, 78
Whirlpool Sandstone, 83, 84
Whitehall Formation, 54, 60
Williamson Formation, 90
Williamson-Willowvale shales, 86
Windom Member, 112, 113, 114
Winchell Creek Member, 60
Wintergreen Flat beds, 76
Wolcott Formation, 86
Wolf Hollow Member, 60

Zoophycos, 102

Plates

The specimens on the following pages were photographed and reproduced with the permission of their owners.
The scale bars are 1 cm except where noted.

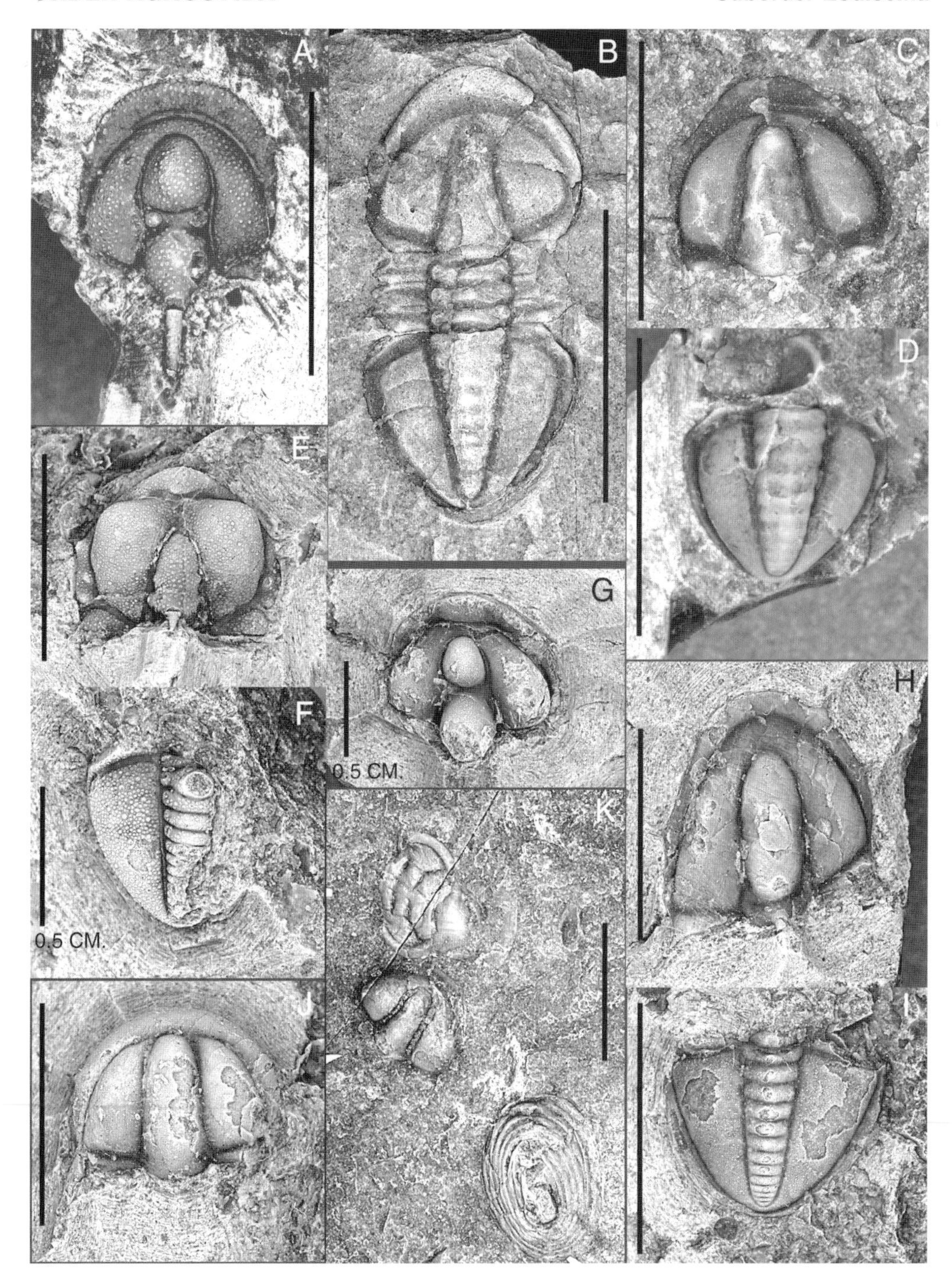

PLATE 1. Eodiscid trilobites from the Early Cambrian allochthonous rocks of New York. A. *Acimetopus bilobatus* cephalon from Columbia County. NYSM 17020. B. *Serrodiscus speciosus* from Columbia County. USNM 156592. Articulated specimens are rare in these rocks. C. *Serrodiscus speciosus* cephalon from Rensselaer County. NYSM 17018. D. *Serrodiscus speciosus* pygidium from Rensselaer County. NYSM 17019. E. *Bolboparia superba* cephalon from Columbia County. USNM 145998 (holotype). F. *Bolboparia elongata* pygidium from Columbia County. USNM 146002b (paratype). G. *Calodiscus reticulatus* cephalon from Columbia County. USNM 146006 (holotype). H. *Acidiscus hexacanthus* cephalon from Columbia County. USNM 145989 (holotype). I. *Acidiscus hexacanthus* pygidium from Columbia County. USNM 156574. J. *Litometopus longispinus* cephalon from Columbia County. USNM 146012. K. Cephala of *Serrodiscus speciosus* and *Elliptocephala asaphoides* from Rensselaer County. NYSM 17021.

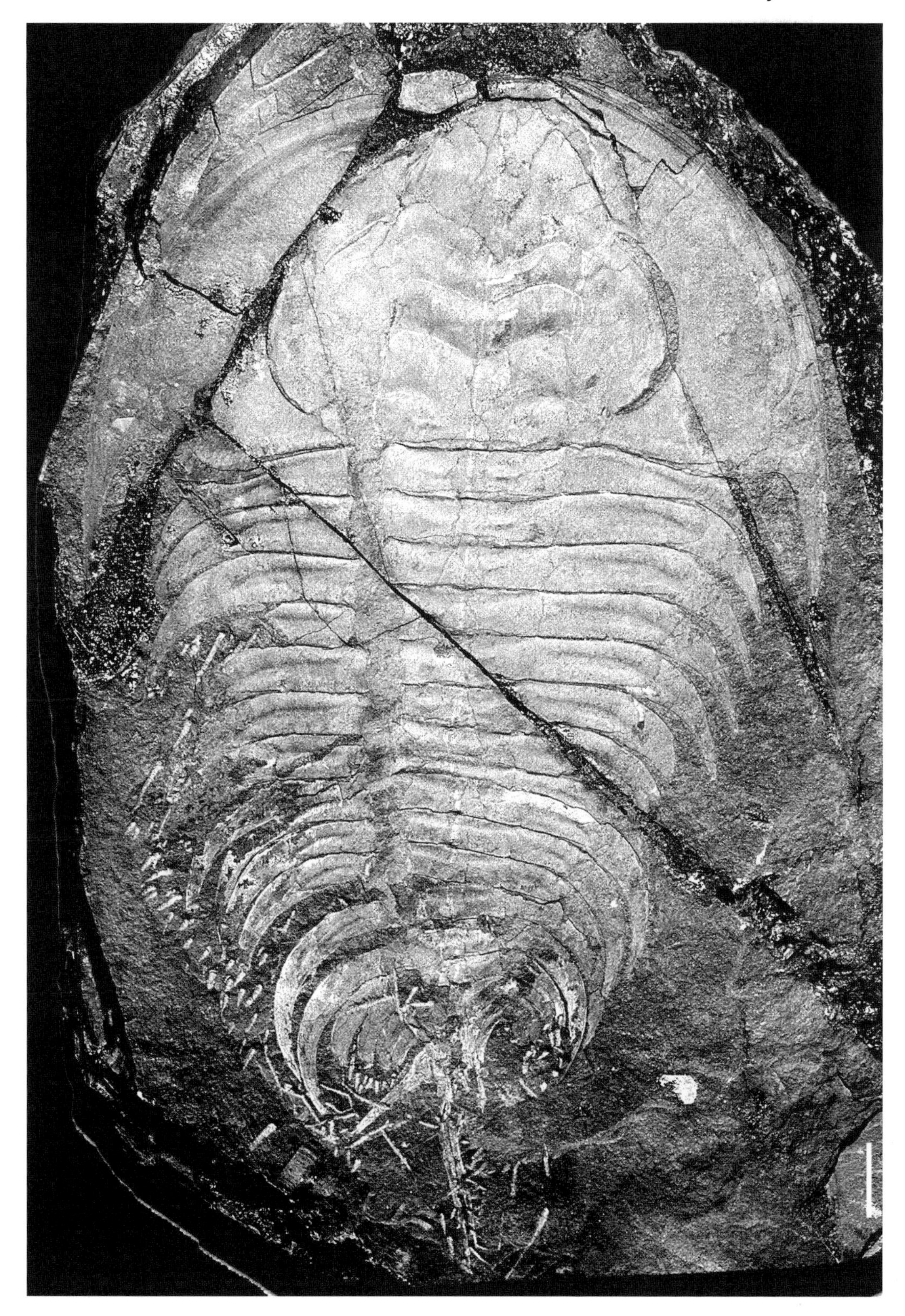

PLATE 2. *Elliptocephala asaphoides* from the Lower Cambrian Schodack Formation in Washington County. The cephalon is 100 mm wide. Articulated specimens of this trilobite are very rare. USNM 18350.

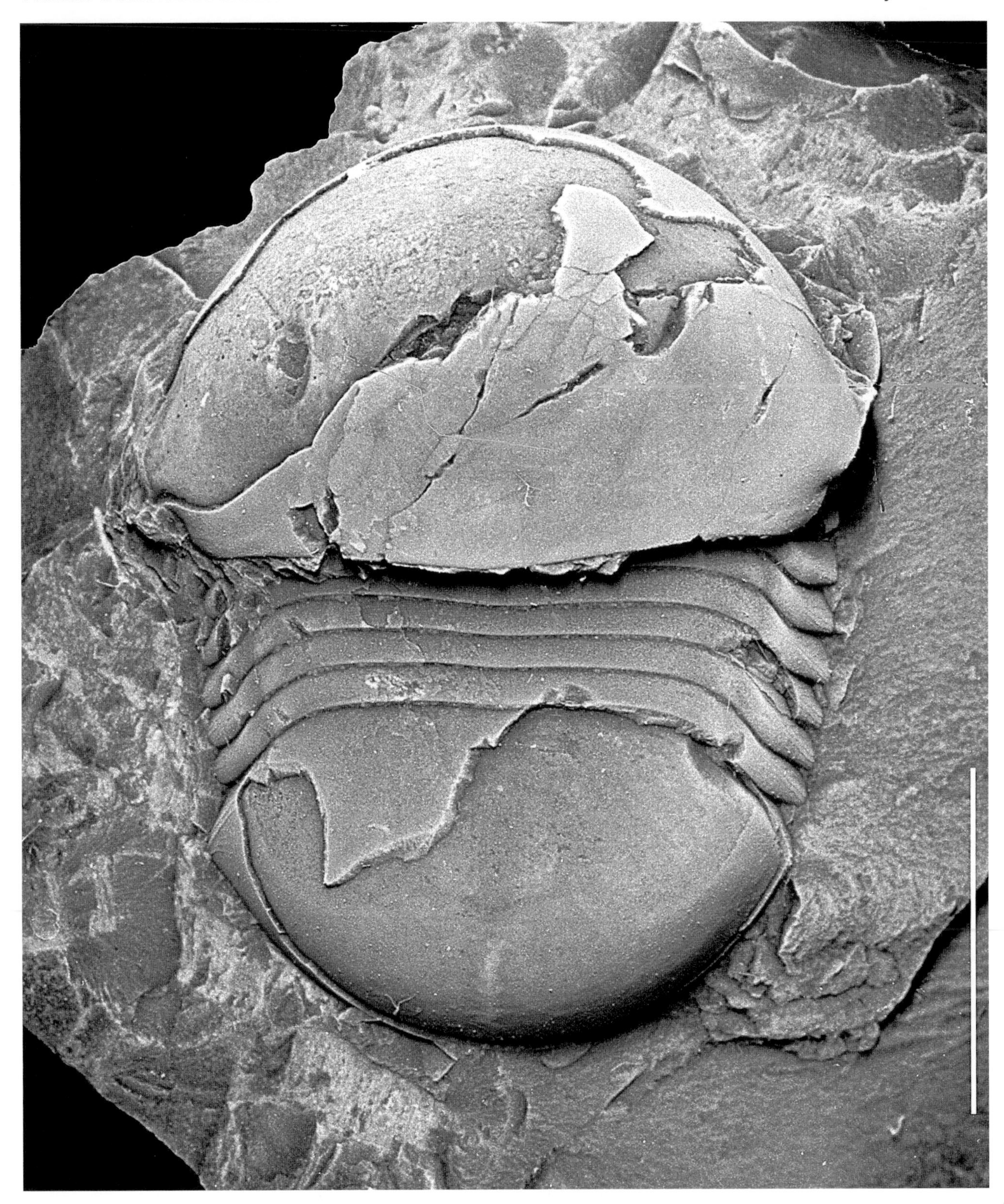

PLATE 3. *Bumastoides globosus* from the Middle Ordovician Chazy Group, Valcour Formation, Valcour Island, Clinton County. The trilobite is 21 mm long. NYSM 12481.

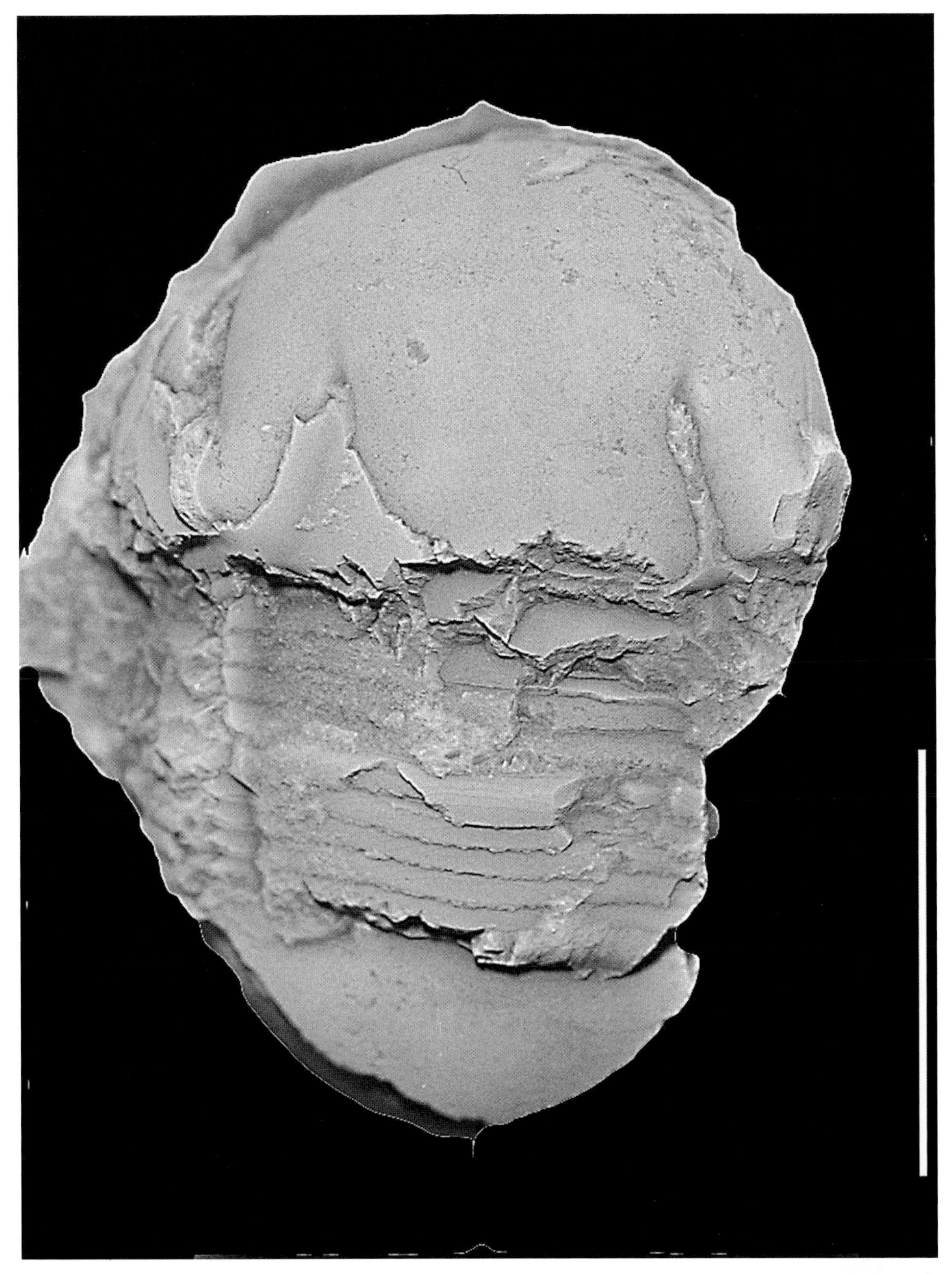

PLATE 4. *Bumastoides holei* from the Middle Ordovician Rust Formation, Trenton Group, Walcott-Rust Quarry, Herkimer County. The trilobite is 26 mm long. MCZ 3756.

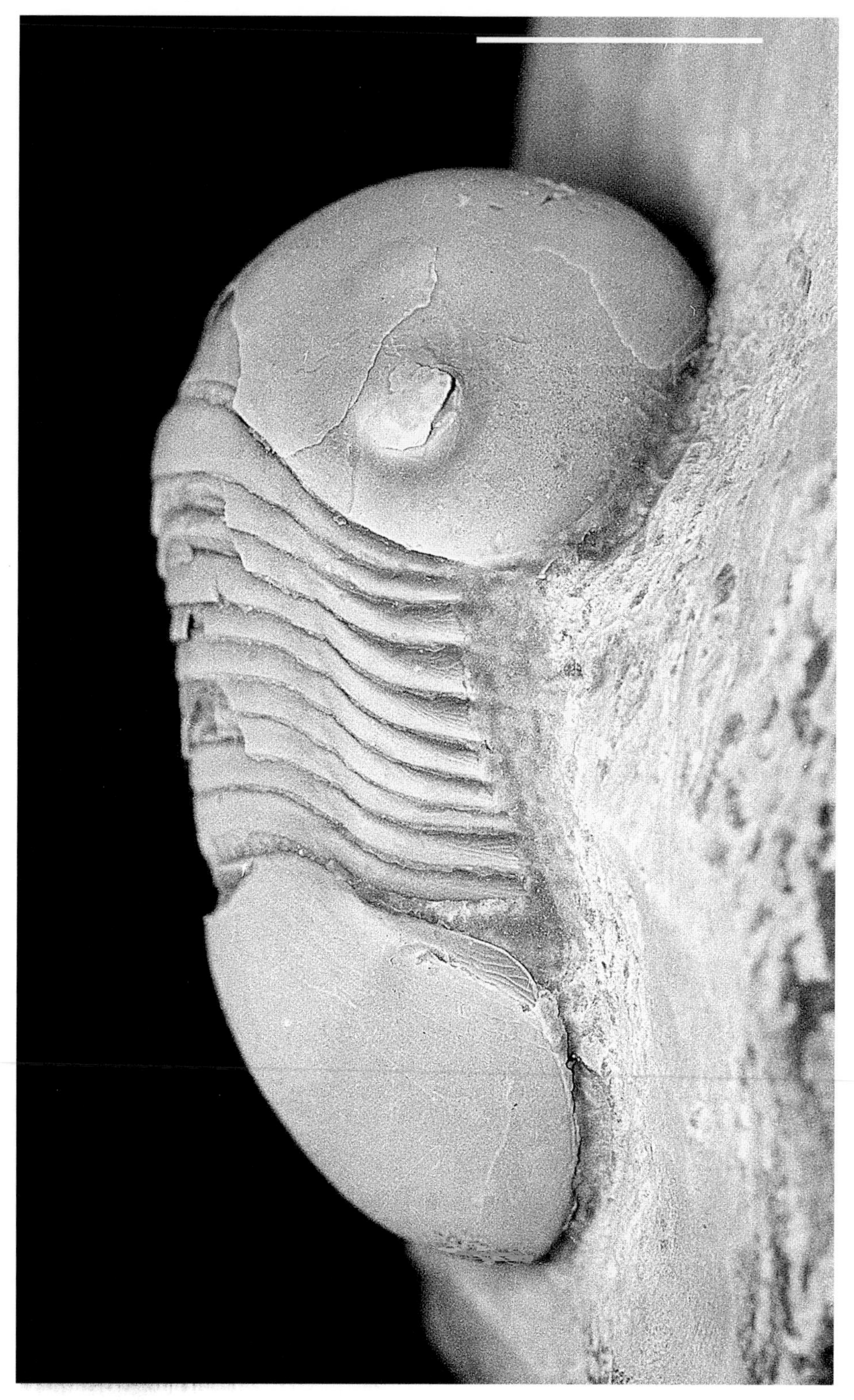

PLATE 5. *Bumastoides holei* from the Middle Ordovician Rust Formation, Trenton Group, Walcott-Rust Quarry, Herkimer County. The trilobite is 37 mm long. MCZ 115909.

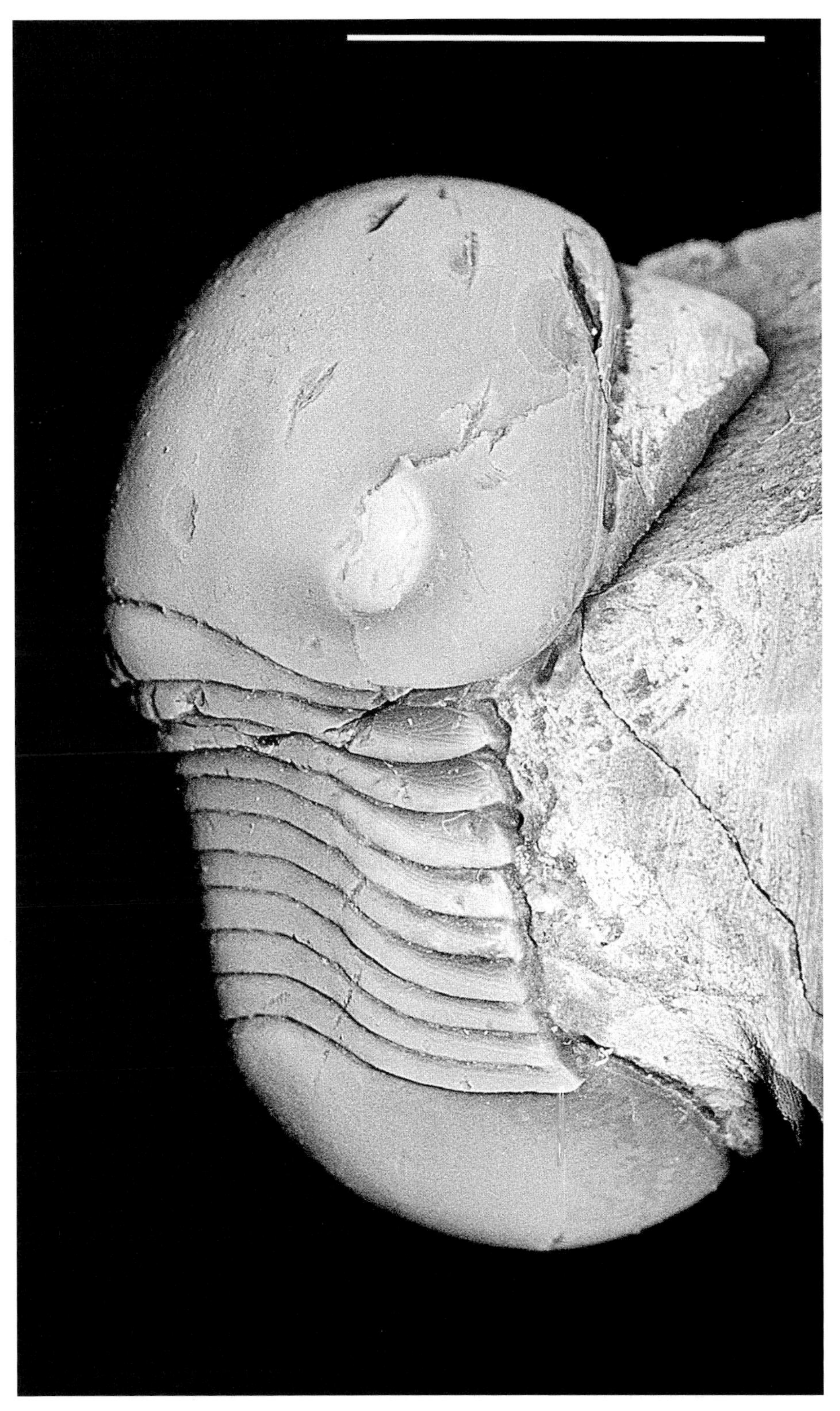

PLATE 6. *Bumastoides milleri* from the Middle Ordovician Trenton Group in Watertown, Jefferson County. The specimen is 25 mm long. PRI 49621.

PLATE 7. *Bumastoides porrectus* from the Middle Ordovician Rust Formation of the Trenton Group, Walcott-Rust Quarry, Herkimer County. The trilobite is 23 mm long. MCZ 101147.

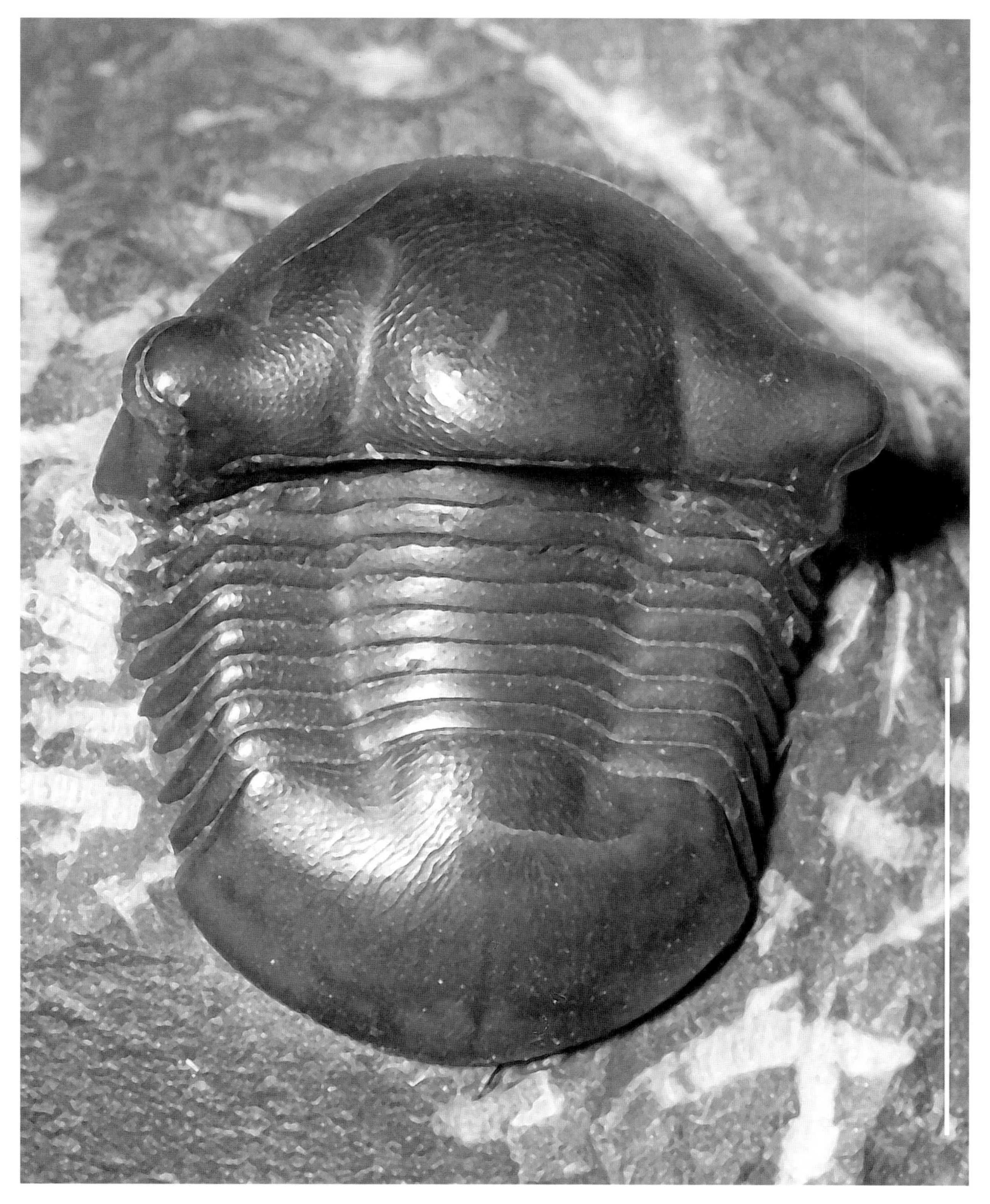

PLATE 8. *Nanillaenus americanus* from the Middle Ordovician Rust Formation of the Trenton Group at the Walcott-Rust Quarry, Herkimer County. The trilobite is 20 mm long. MCZ 707.

PLATE 9. *Nanillaenus latiaxiatus* from the Middle Ordovician in the Bowmanville Quarry, Ontario. This trilobite is reported in New York. The trilobite is 36 mm long. In the collection of William Pinch.

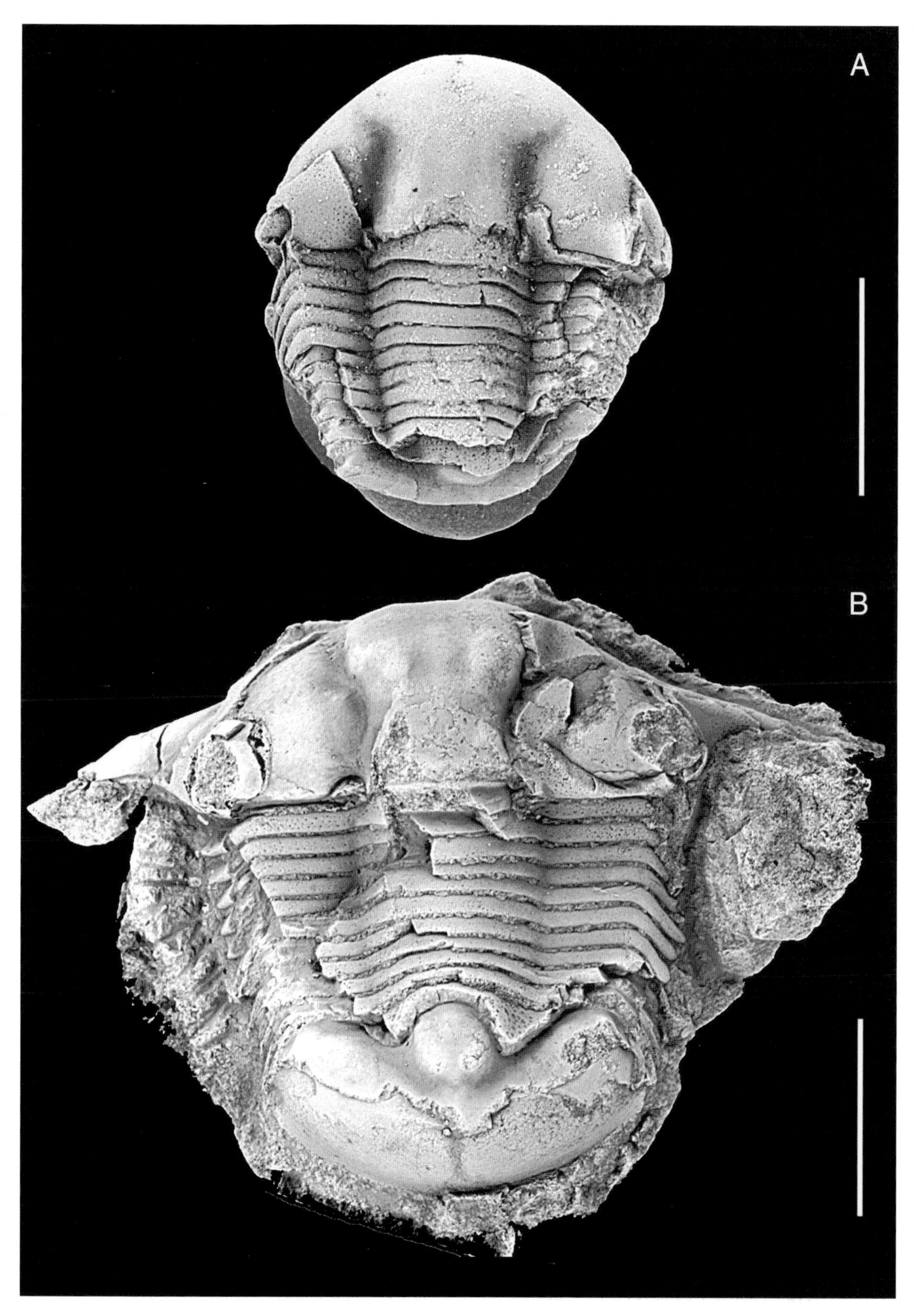

PLATE 10. Illaenids from the Middle Ordovician Chazy Group. A. *Nanillaenus ? raymondi* from the Day Point Formation, Valcour, Clinton County. NYSM 12491 (holotype). B. *Thaleops longispina* from the Crown Point Formation, Essex County. NYSM 12922 (holotype).

PLATE 11. *Bumastus ioxus* from the Lower Silurian Rochester Shale in a commercial quarry near Middleport, Orleans County. The trilobite is 91 mm long. In the collection of Kent Smith.

PLATE 12. *Eobronteus lunatus* from the Middle Ordovician Snake Hill Shale, Snake Hill, Saratoga, Saratoga County. The trilobite is 45 mm long. NYSM 17001.

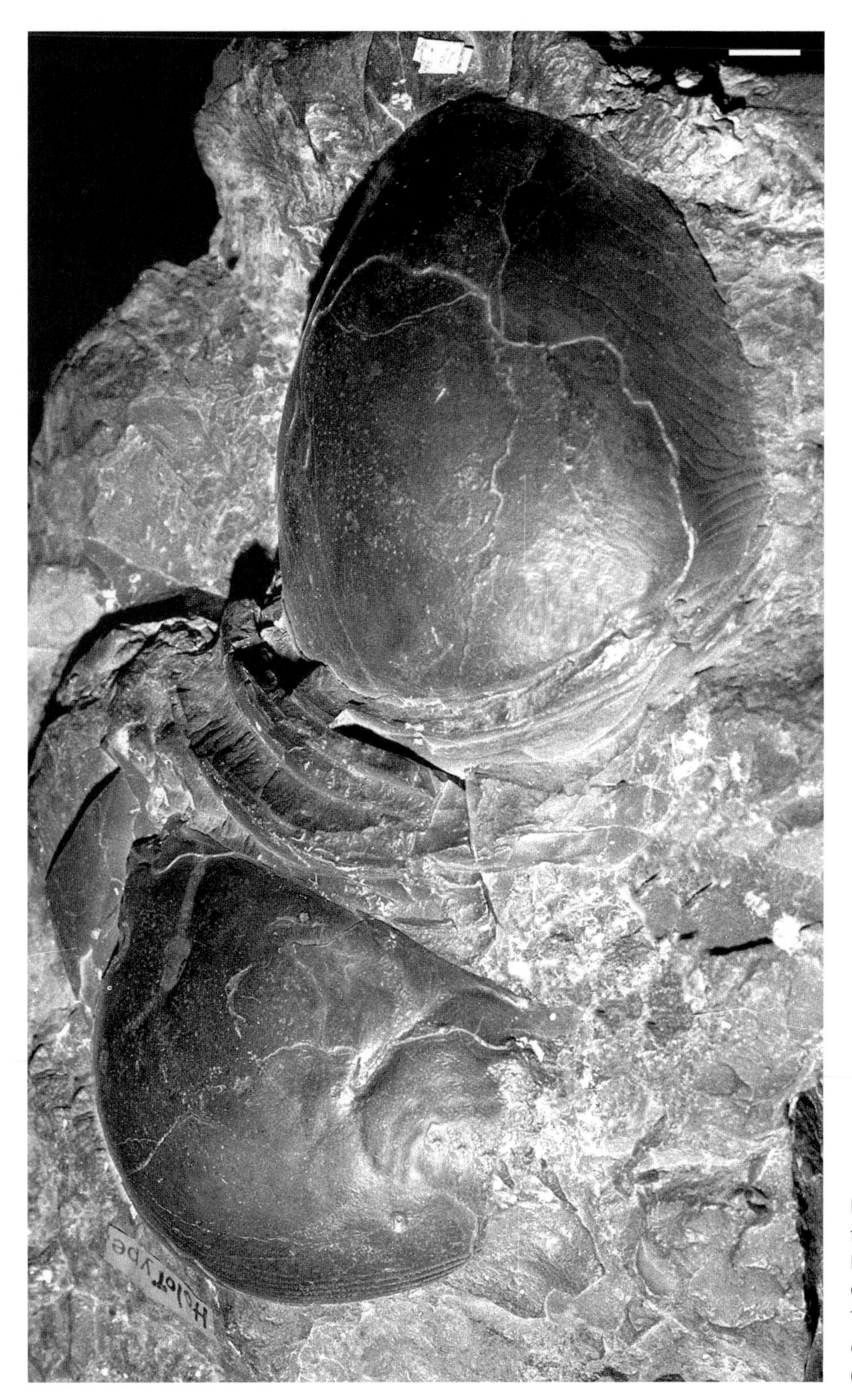

PLATE 13. *Failleana indeterminata* from the Middle Ordovician Black River Group limestones at Buck's Quarry, Poland, Herkimer County. The trilobite is somewhat disarticulated. MCZ 104928 (holotype).

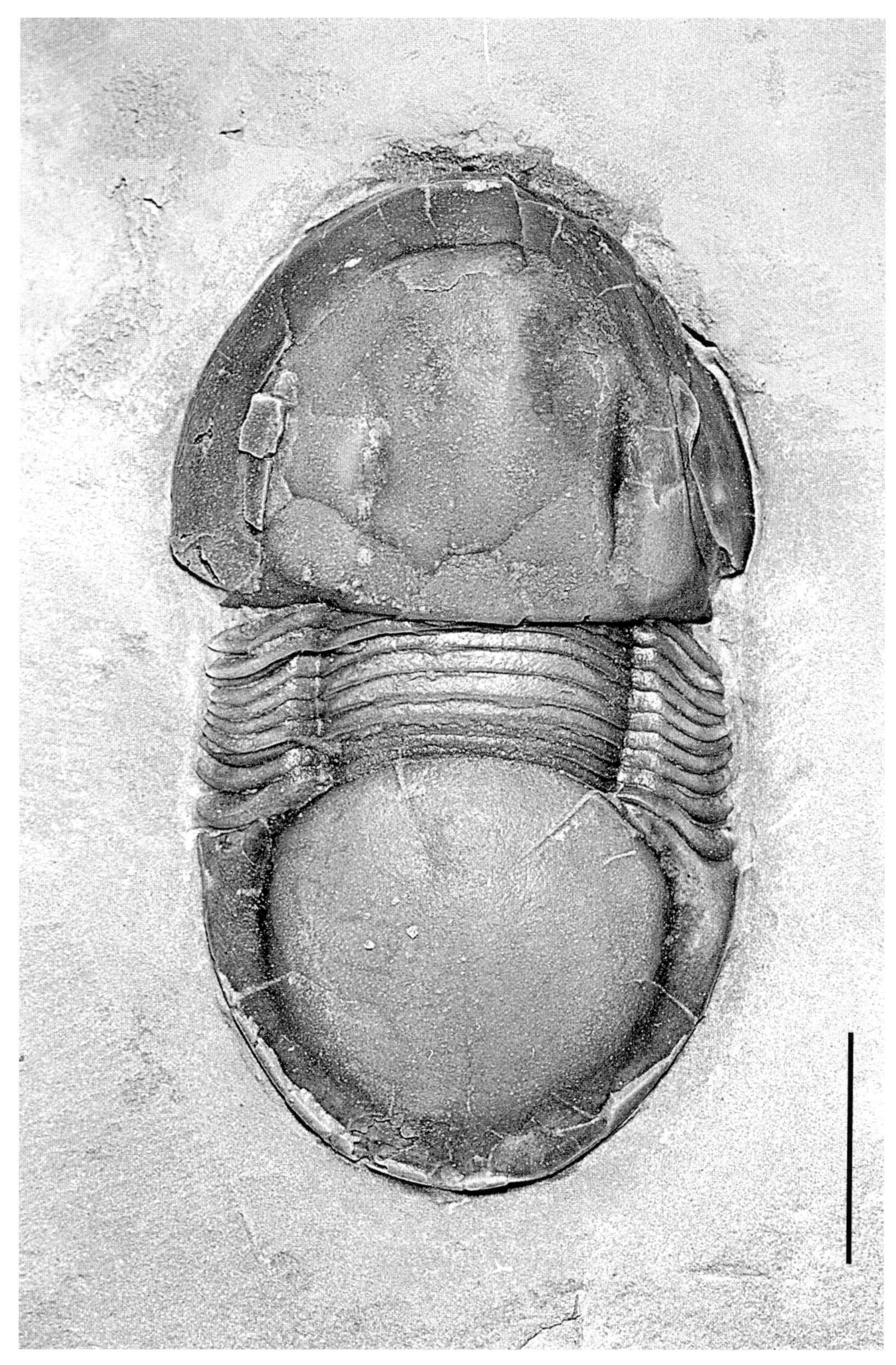

PLATE 14. *Illaenoides* cf. *I. trilobita* from the Lower Silurian Rochester Shale in a commercial quarry near Middleport, Orleans County. The trilobite is 45 mm long. In the collection of Kent Smith.

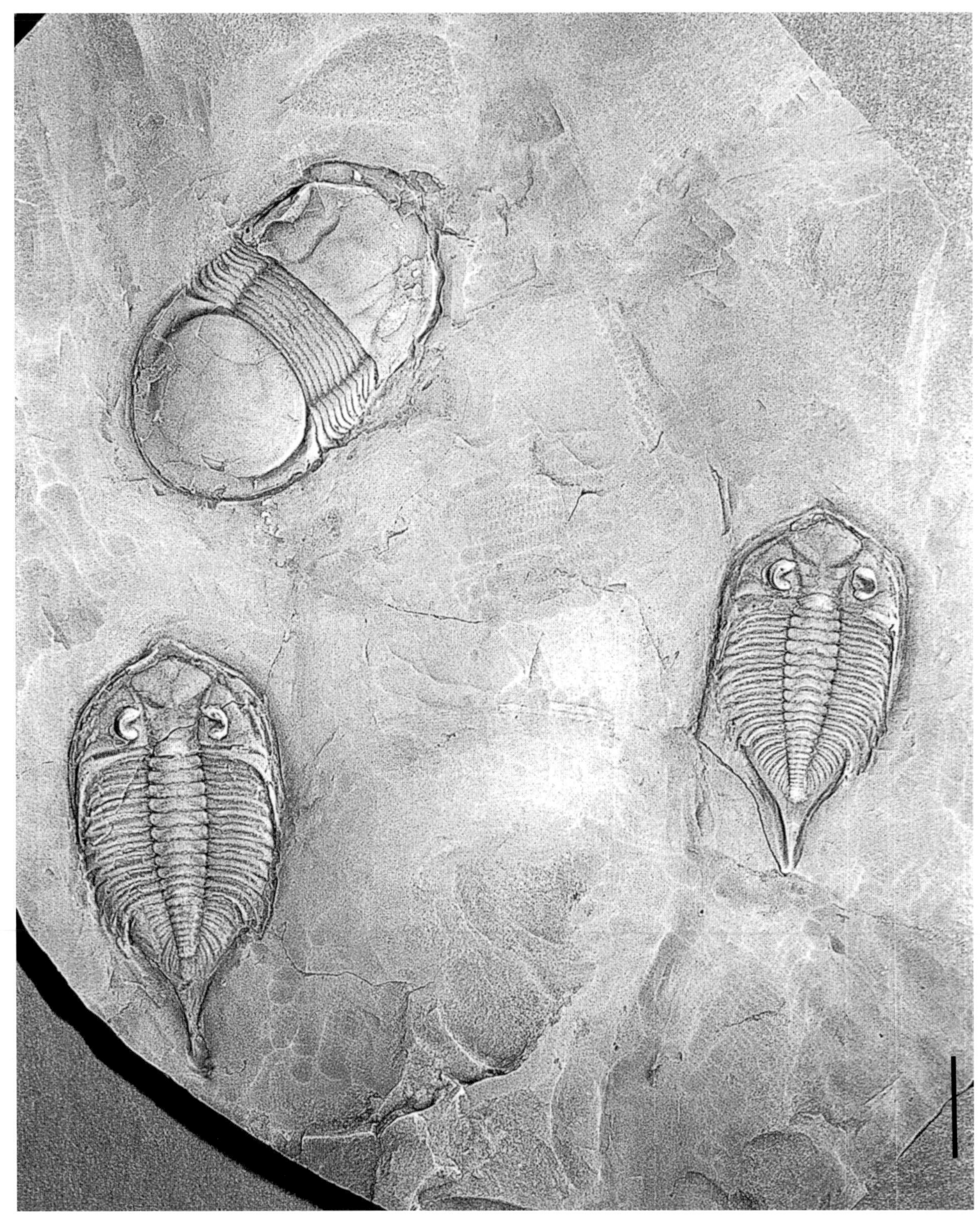

PLATE 15. *Illaenoides* cf. *I. trilobita* and *Dalmanites limulurus* from the Lower Silurian Rochester Shale in a commercial quarry near Middleport, Orleans County. The *Illaenoides* species is 51 mm long. In the collection of Kent Smith.

PLATE 16. *Scutellum niagarensis* from drift block of the Lower Silurian Reynales Limestone in Clarendon, Orleans County. The cephalon is 17 mm long and the pygidium is 22 mm long. NYSM 17023 (cephalon) and NYSM 17024 (pygidium).

PLATE 17. *Scutellum senescens* from the Upper Devonian Chemung beds near Avoca, Chemung County. Trilobites are rare in the Upper Devonian in New York. The specimen is about 63 mm long. NYSM 4152 (hypotype).

PLATE 18. *Scutellum tullius* from the Middle Devonian Tully Limestone at Borodino, Cayuga County. The illustrated cephala in A are 12 and 8 mm long. PRI 49622. B. Associated pygidium of the same species. C. Reconstruction of the trilobite by Professor Wells, formerly at Cornell University.

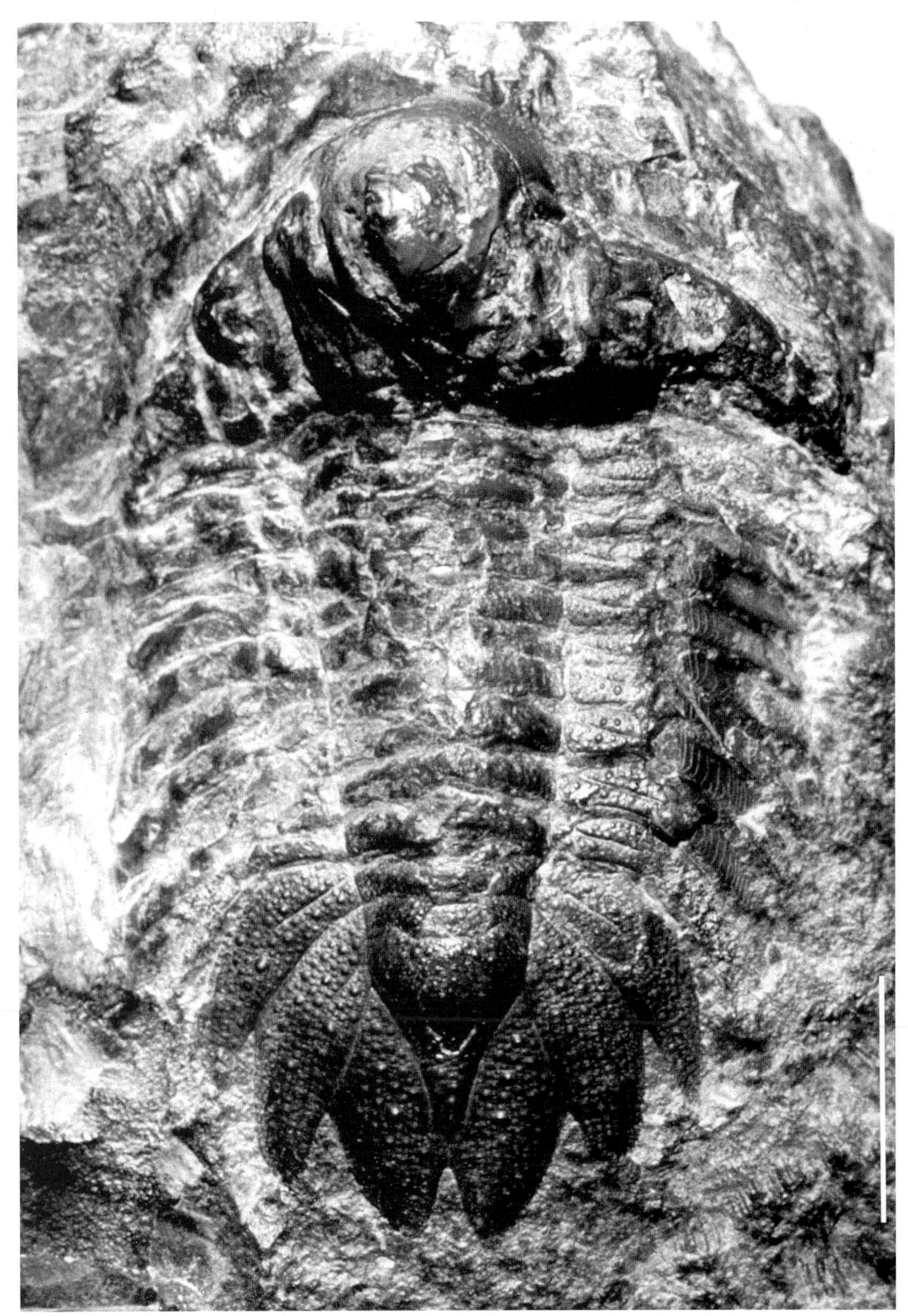

PLATE 19. *Amphilichas cornutus* from the Middle Ordovician Rust Formation, Trenton Group, Walcott-Rust Quarry, Herkimer County. The trilobite was damaged probably during attempts to prepare it. The specimen is 36 mm long. NYSM 4533 (holotype).

PLATE 20. *Arctinurus boltoni* from the Lower Silurian Rochester Shale at Middleport, Niagara County. This specimen shows attached brachiopods. The trilobite is 150 mm long. USNM 499453.

PLATE 21. *Arctinurus boltoni* from the Lower Silurian Rochester Shale in a commercial quarry near Middleport in Orleans County. This specimen shows healed damage, perhaps from a predator. The trilobite is 127 mm long. PRI 42095.

PLATE 22. *Dicranopeltis nereus* from the Lower Silurian Rochester Shale in a commercial quarry near Middleport in Orleans County. This trilobite is uncommon and good specimens are rare. The trilobite is 70 mm long. In the Kent Smith collection.

PLATE 23. *Dicranopeltis nereus* from the Lower Silurian Rochester Shale in a commercial quarry near Middleport in Orleans County. This specimen shows considerable damage, perhaps by a predator. The trilobite is 59 mm long. PRI 49623.

PLATE 24. *Echinolichas eriopsis* from the Lower Devonian of eastern New York. NYSM 4537 (type).

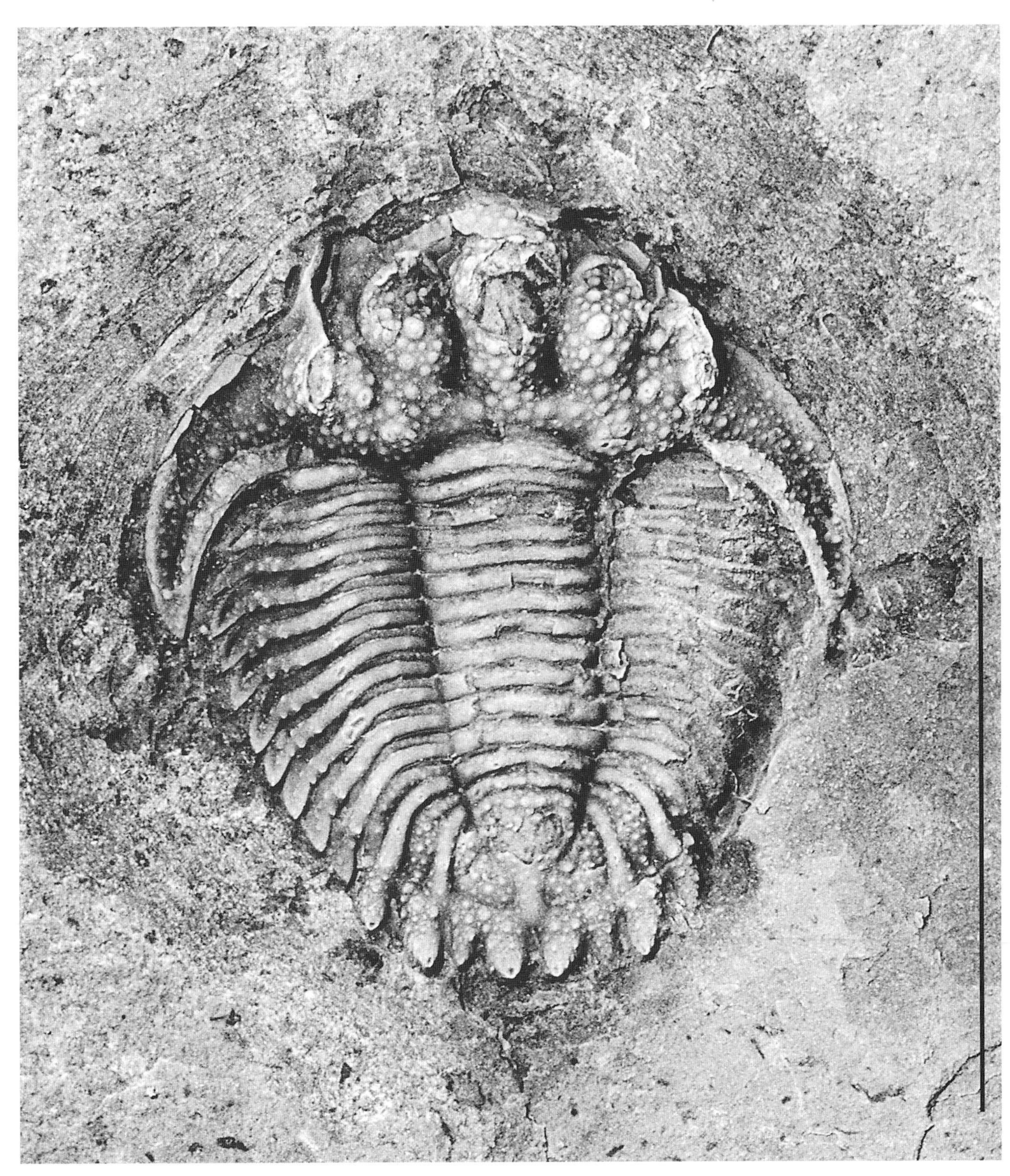

PLATE 25. *Hemiarges paulianus* from the Middle Ordovician Bobcaygeon Formation at the Carden Quarry, Lake Simcoe area, Ontario. This small lichid is fairly common in the Kings Falls Limestone of the lower Trenton Group. The trilobite is 14 mm long. PRI 49637.

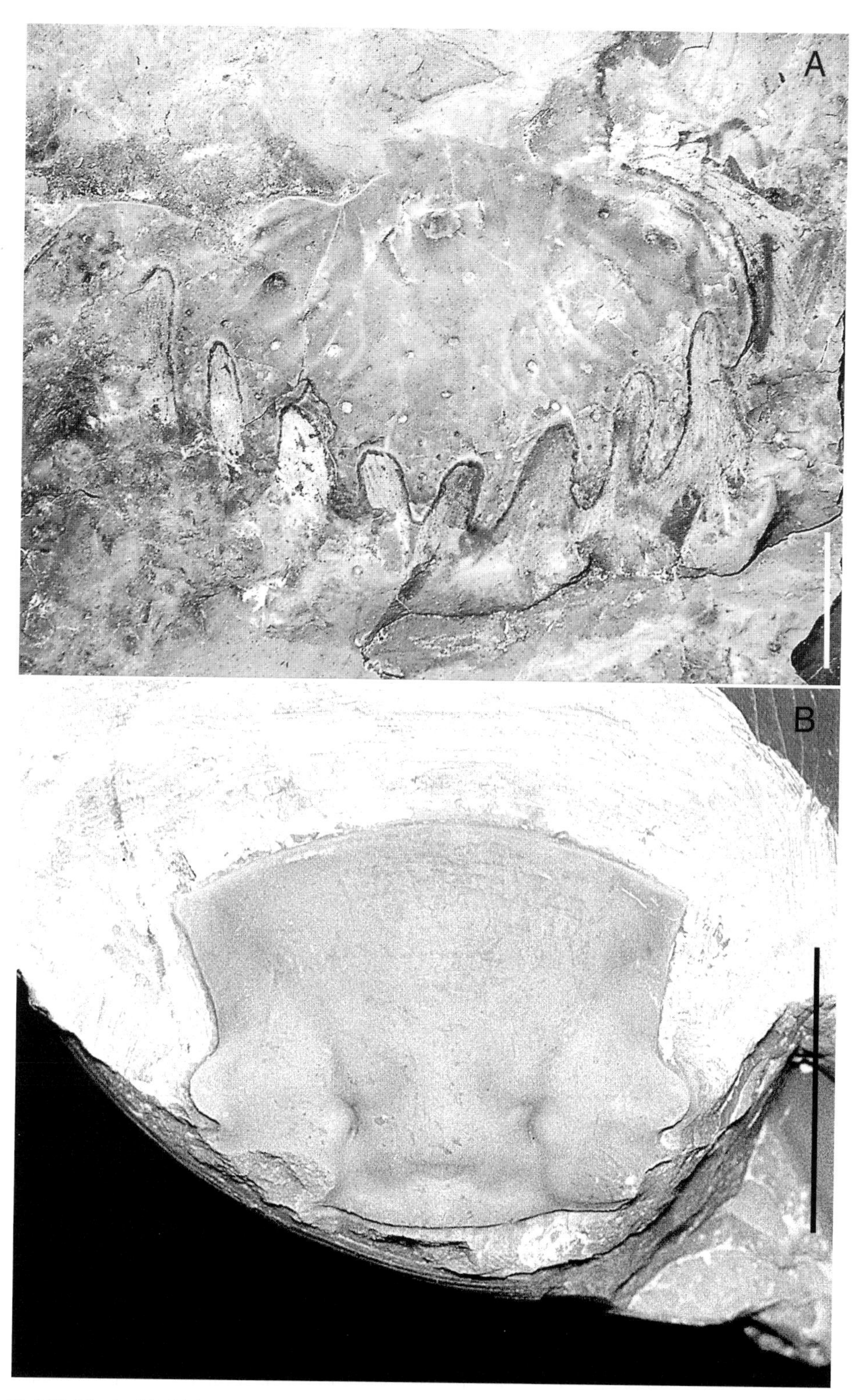

PLATE 26. A. A lichid species from the Middle Devonian Onondaga Limestone, Cheektawaga Quarry, Erie County. The pygidium is 30 mm long. In the collection of Lee Tutt. B. *Scutellum rochesterense* (family Styginidae) from the Lower Silurian Irondequoit Limestone; exact locality is unknown. This cranidium is associated with closely packed pygidia and cranidia of *Bumastus ioxus*, a situation characteristic of bioherms in the Irondequoit. NYSM 17016.

PLATE 27. Cranidium of *Oinochoe pustulosus* from the Lower Devonian New Scotland Limestone in eastern New York. NYSM 17002.

PLATE 28. *Richterarges ptyonurus* from the Upper Silurian Cobleskill Limestone in Schoharie County. The figured cranidium is 7 mm long and the pygidium is 10 mm wide. NYSM 4555.

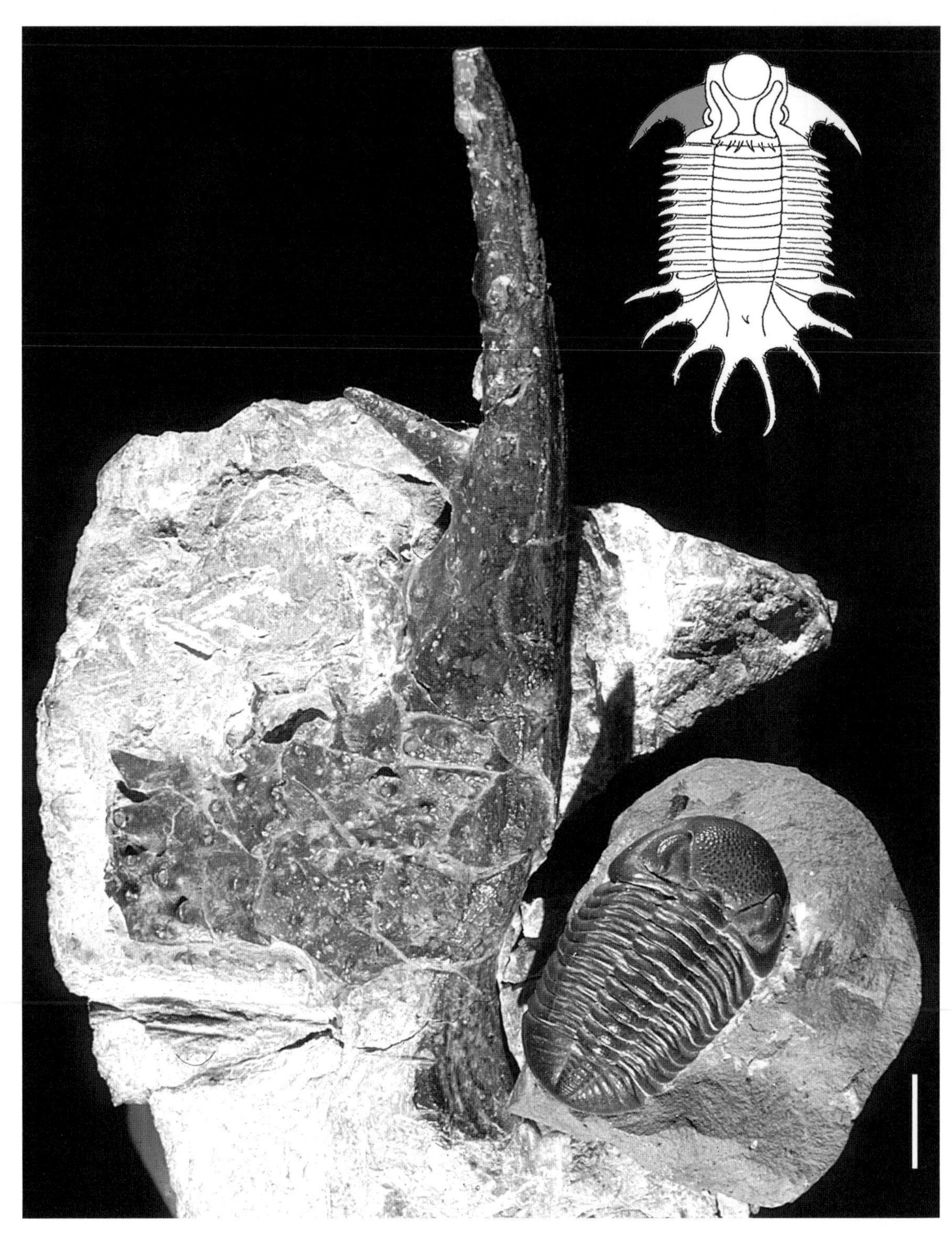

PLATE 29. *Terataspis grandis* from the Middle Devonian Williamsville Quarry, Erie County, in the Onondaga Limestone. This very large genal spine came from a trilobite about 580 mm (23 inches) long. PRI 49638. The reconstruction drawing in the upper corner shows where this genal spine was on the trilobite and gives some idea of the scale of the complete specimen. The specimen of *Eldredgeops milleri* from the Middle Devonian of Arkona, Ontario, Canada, is also included for scale. The *Eldredgeops* is 42 mm long. PRI 49639.

PLATE 30. *Trochurus halli* from the Lower Silurian Rochester Shale at Brockport, Monroe County. This specimen, which is missing the cranidium, is a molt 30 mm long. In the collection of Paul Krohn.

PLATE 31. *Trochurus halli* from the Lower Silurian Rochester Shale in an unnamed creek in the Sodus area, Wayne County. This cranidium gives some idea of what the trilobite from Plate 30 must have looked like whole. The cranidium is 28 mm wide. In the collection of Gerald Kloc.

PLATE 32. *Diacanthaspis parvula* from the Middle Ordovician Rust Formation, Trenton Group, Walcott-Rust Quarry, Herkimer County. This is a rare trilobite in New York, perhaps because of its small size. The specimen is 8 mm long. MCZ 4372.

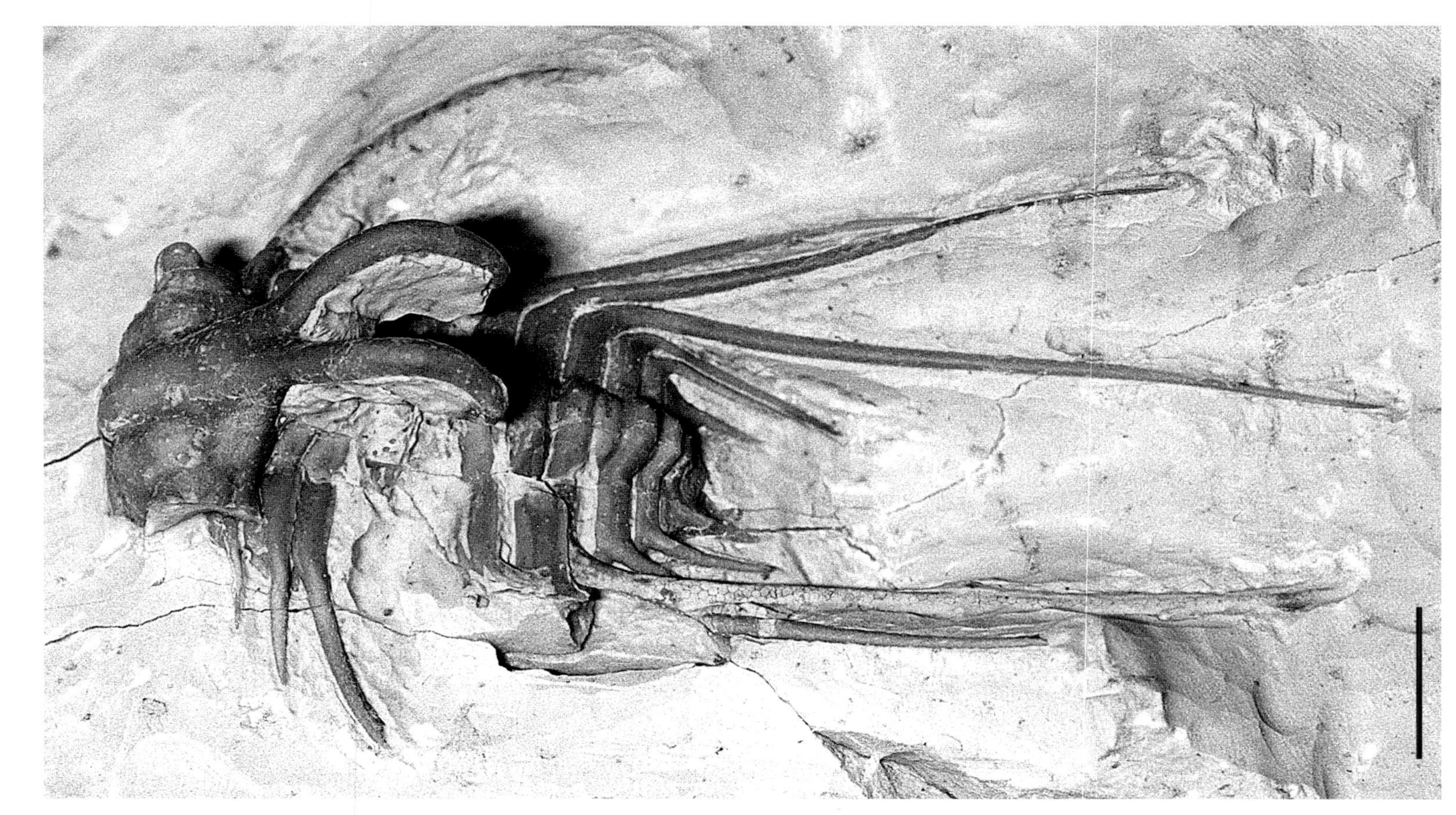

PLATE 33. *Dicranurus elegantus* from the Lower Devonian Haragan Formation in Oklahoma. This specimen is illustrated to give an idea of how the *Dicranurus hamatus* from the Lower Devonian of New York looked like as an articulated trilobite. The illustrated specimen is 38 mm long. In the collection of Gerald Kloc.

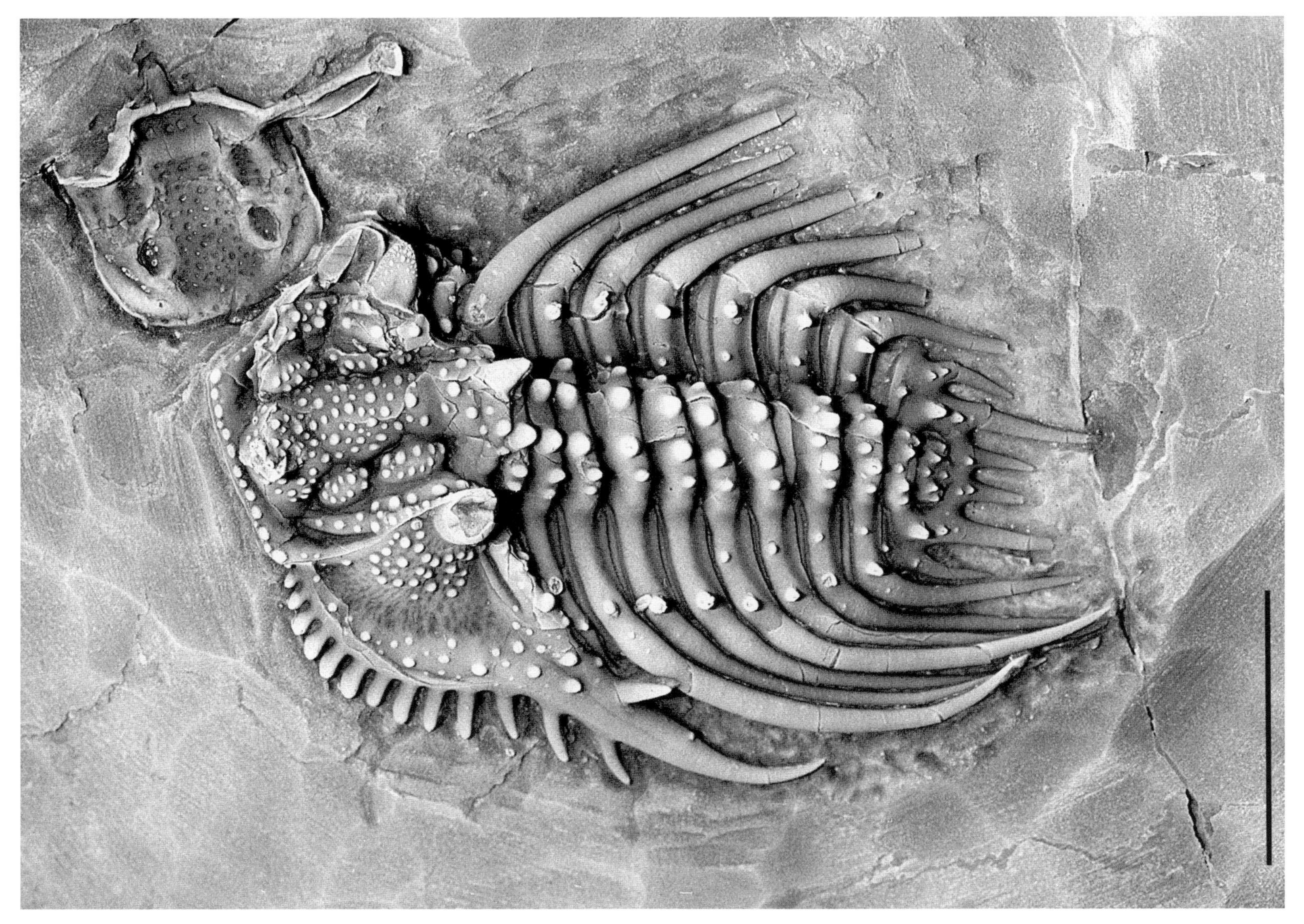

PLATE 34. *Kettneraspis callicera* from the Middle Devonian Onondaga Limestone at the Benchmark Quarry, Oak Corners, Ontario County. The specimen is 26 mm wide. In the collection of Gerald Kloc.

PLATE 35. *Kettneraspis tuberculata* from the Lower Devonian New Scotland Limestone of the Helderberg Group. The specimen is 28 mm long and is found in the Blue Circle Quarry, Ravenna, Albany County. In the collection of Gerald Kloc.

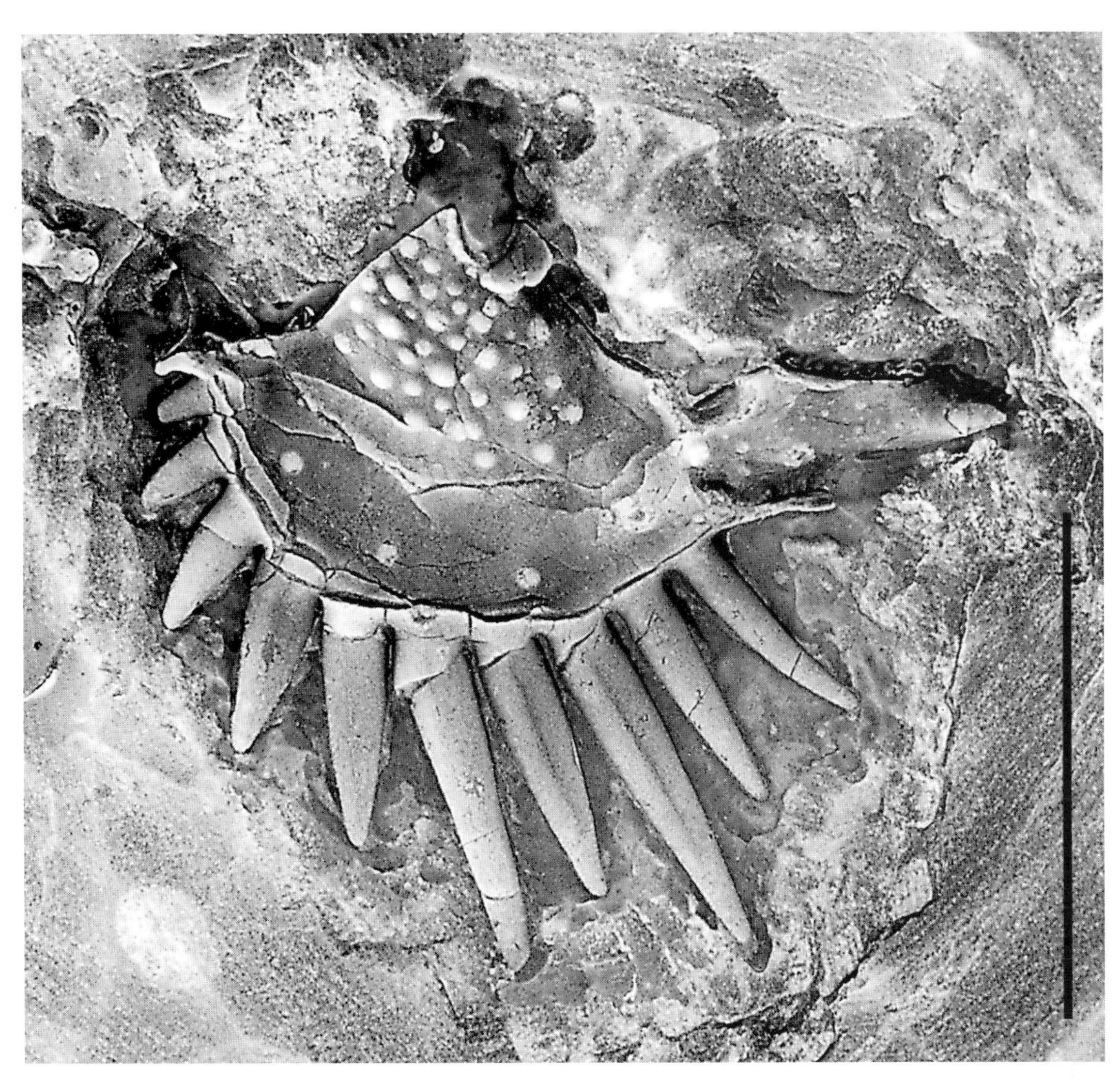

PLATE 36. *Kettneraspis?* species, with a free cheek obviously quite different from that in the odontopleurid in Plate 35. This species is also found in the Lower Devonian New Scotland Limestone, Helderberg Group, at the Blue Circle Quarry, Ravenna, Albany County. The specimen is 18 mm across at the widest point. In the collection of Gerald Kloc.

PLATE 37. *Meadowtownella trentonensis* from the Middle Ordovician Rust Formation, Trenton Group, Walcott-Rust Quarry, Herkimer County. This is a fairly common trilobite in the quarry but is not often reported from other Trenton rocks, possibly owing to its small size. The specimen is 12 mm long and shows the ventral exoskeletal anatomy. MCZ 111717.

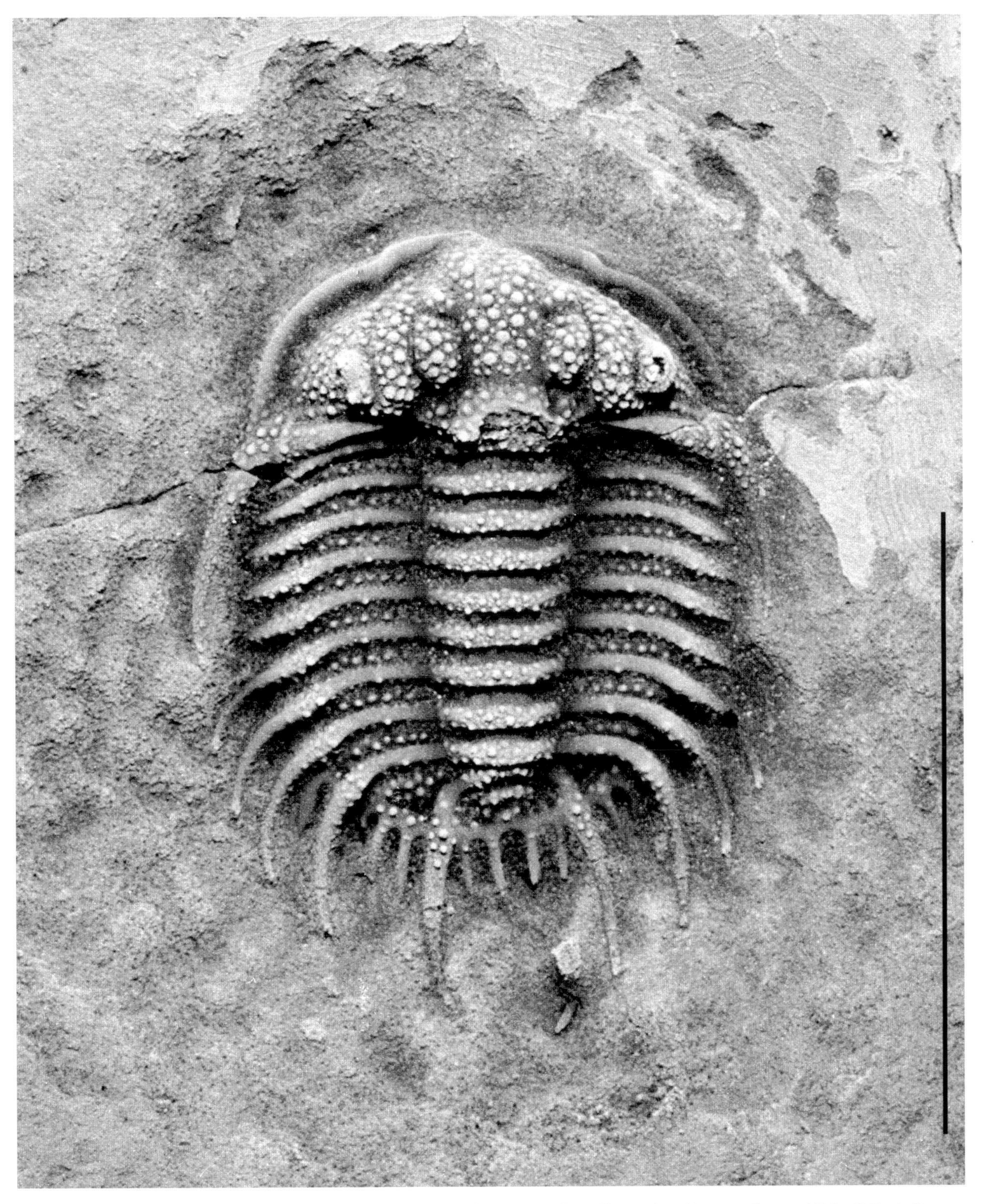

PLATE 38. *Meadowtownella trentonensis* from the Middle Ordovician Rust Formation, Trenton Group, Walcott-Rust Quarry, Herkimer County. A dorsal view of the same trilobite in Plate 37. MCZ.

PLATE 39. *Miraspis crosota* from the Upper Ordovician Frankfort Member of the Lorraine Group at Rome, Oneida County. The glabella of this partial specimen is 1 mm long. USNM 23600.

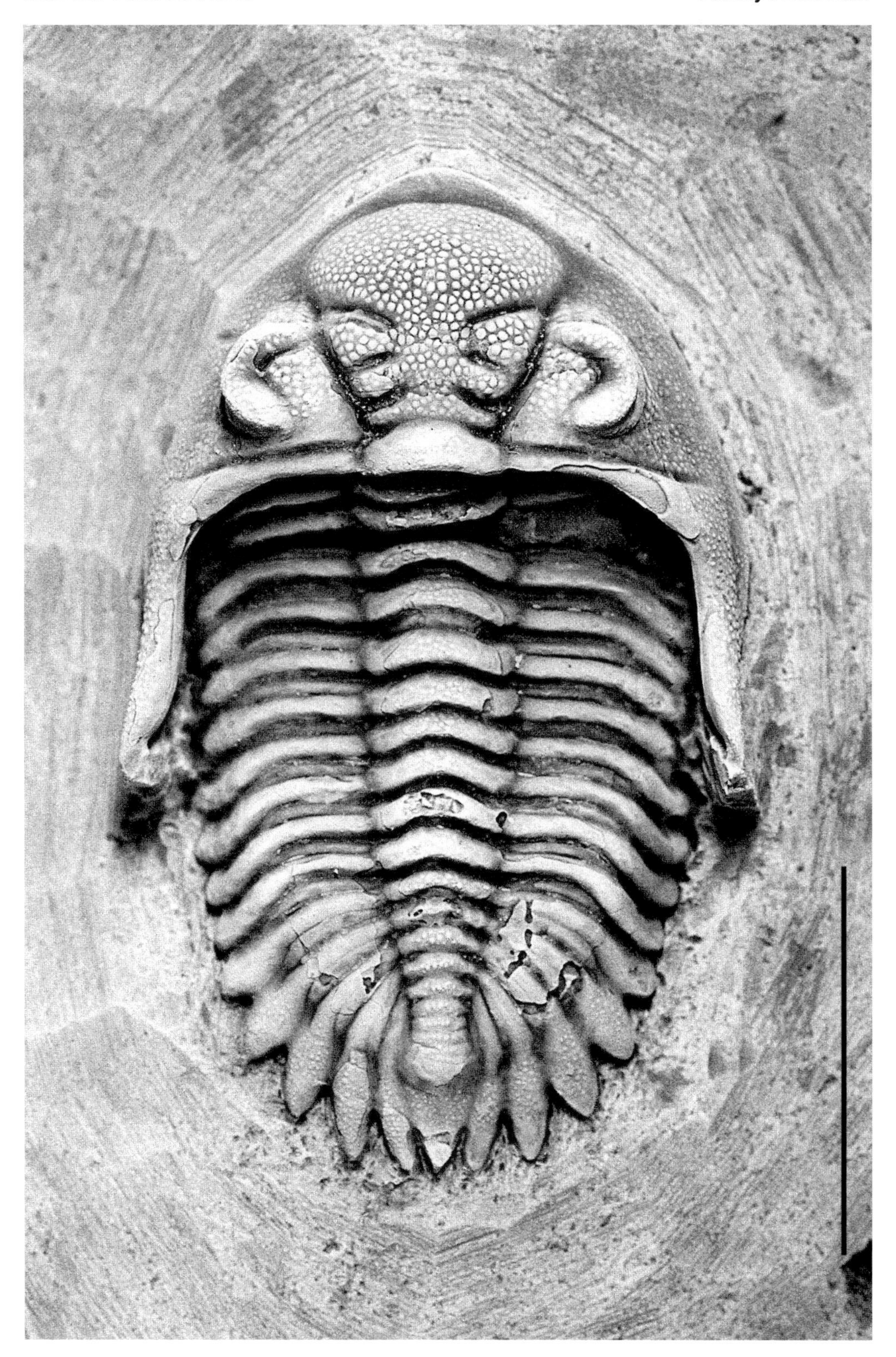

PLATE 40. An unidentified asteropygin from the upper Middle Devonian, Tully Limestone in Groves Creek off Cayuga Lake, Seneca County. The trilobite is 26 mm long. In the collection of Gerald Kloc.

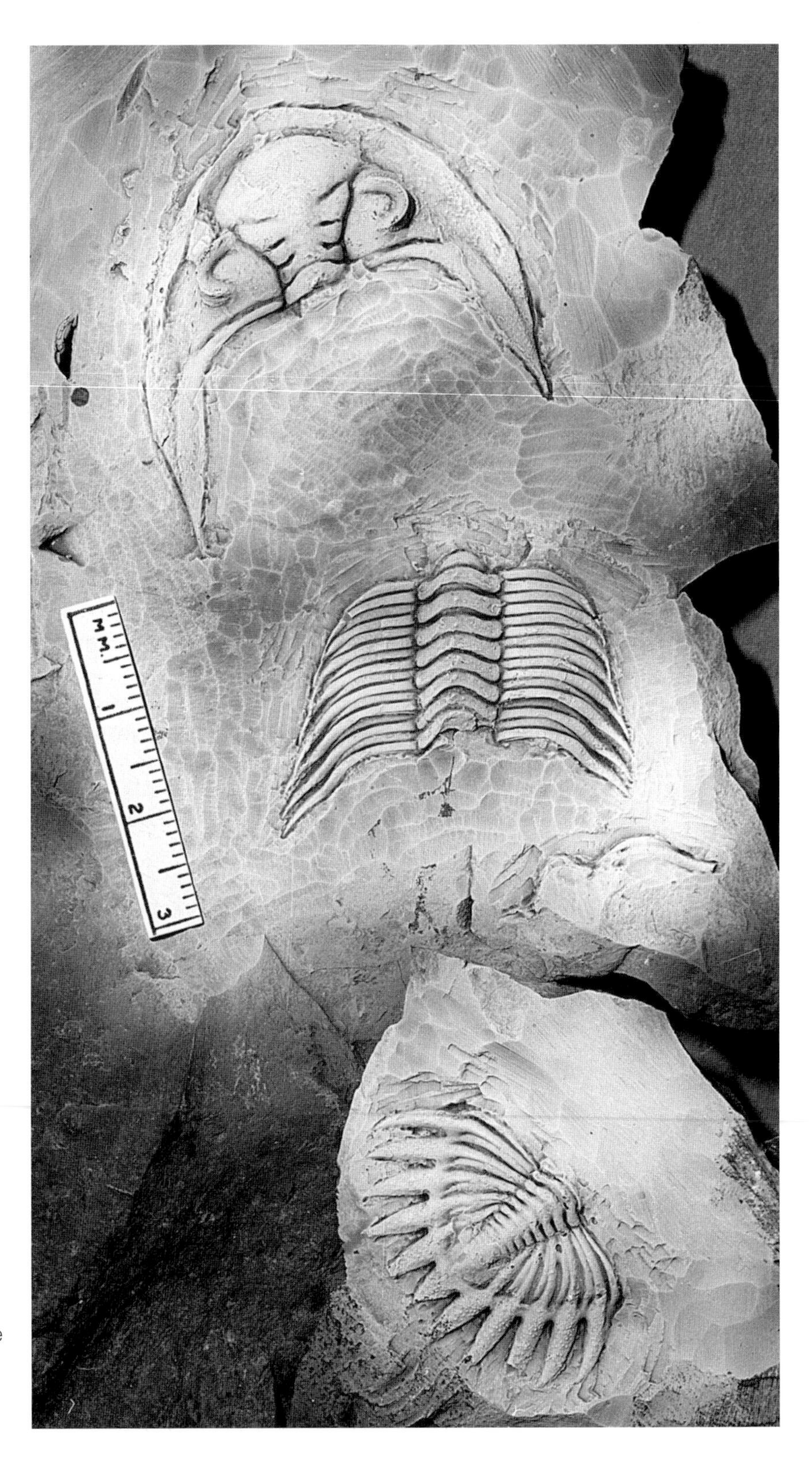

PLATE 41. Disarticulated specimen of *Bellacartwrightia calderonae* from the Middle Devonian Windom Shale, Hamilton Group, Kashong Glen, Ontario County. AMNH 45273 (holotype).

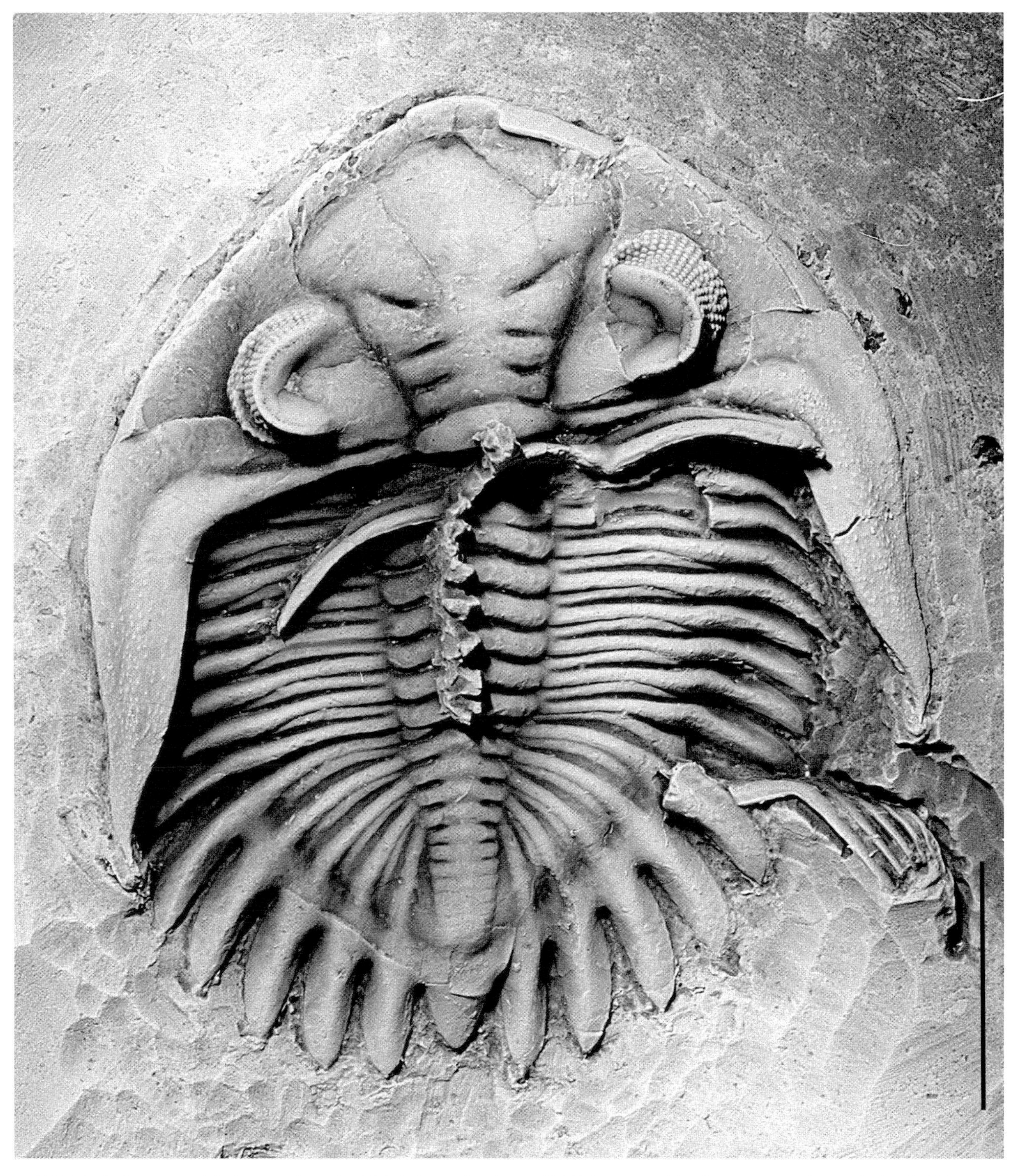

PLATE 42. *Bellacartwrightia jennyae* from the Middle Devonian Centerfield Limestone, Hamilton Group, at Brown's Creek, Livingston County. The specimen is 38 mm long. AMNH 45310 (paratype).

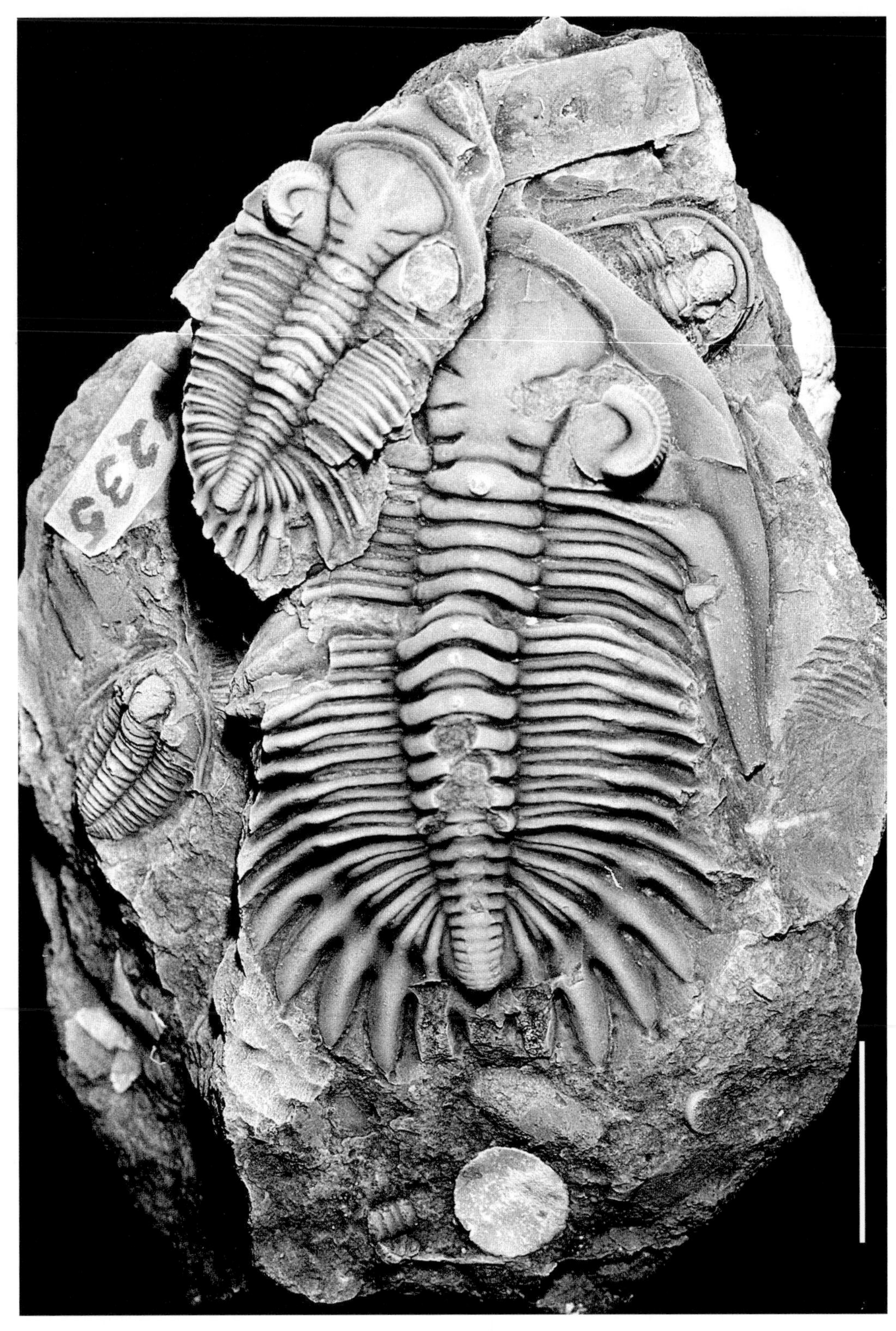

PLATE 43. *Bellacartwrightia jennyae* and *Harpidella craspedota* from the Middle Devonian Centerfield Limestone, Hamilton Group, at Centerfield, Ontario County. The largest *Bellacartwrightia* is 40 mm long. NYSM 4235.

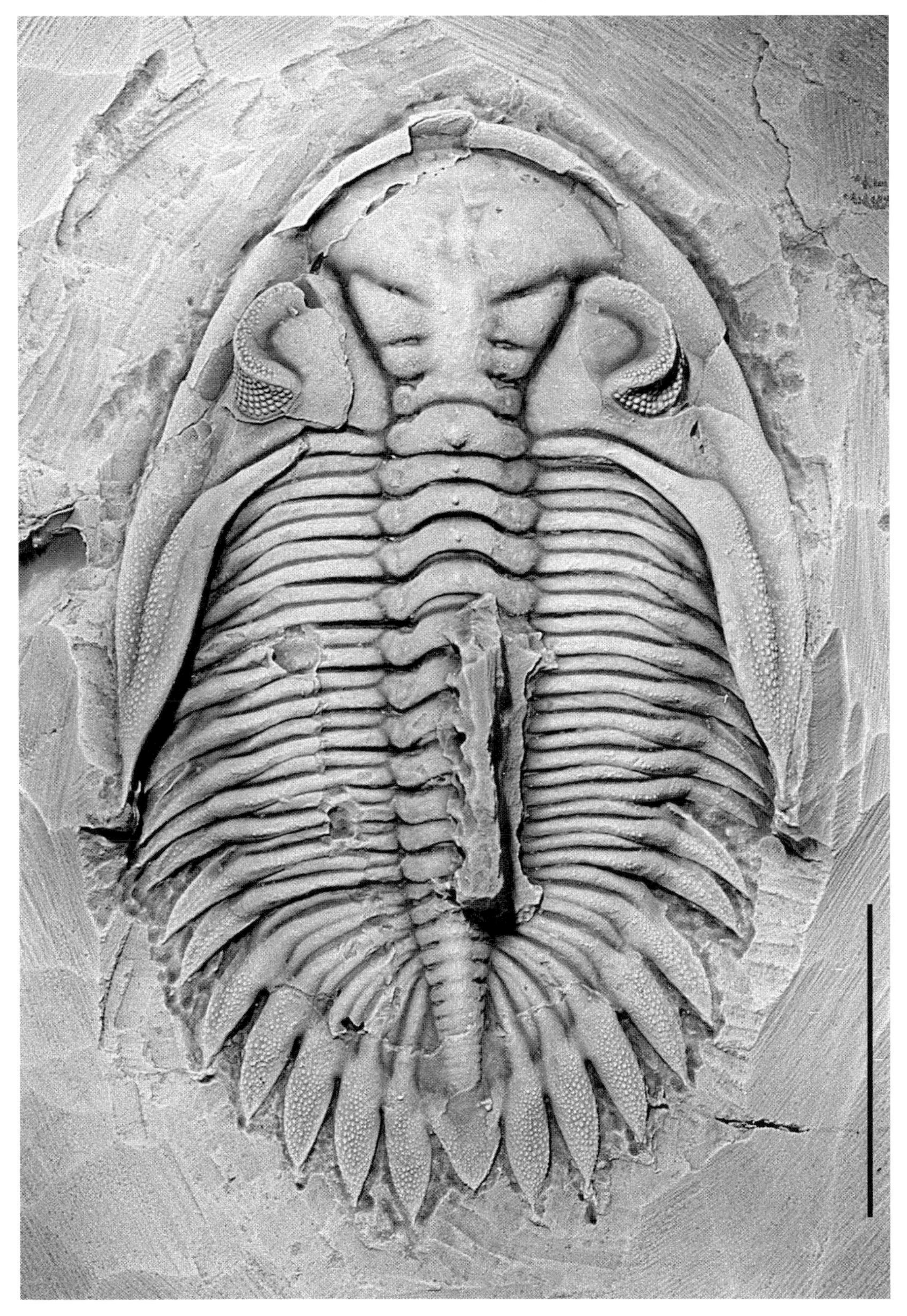

PLATE 44. *Bellacartwrightia phyllocaudata* from the Middle Devonian Deep Run Member, Hamilton Group, near Geneseo, Livingston County. The specimen is 34 mm long. AMNH 45230 (holotype).

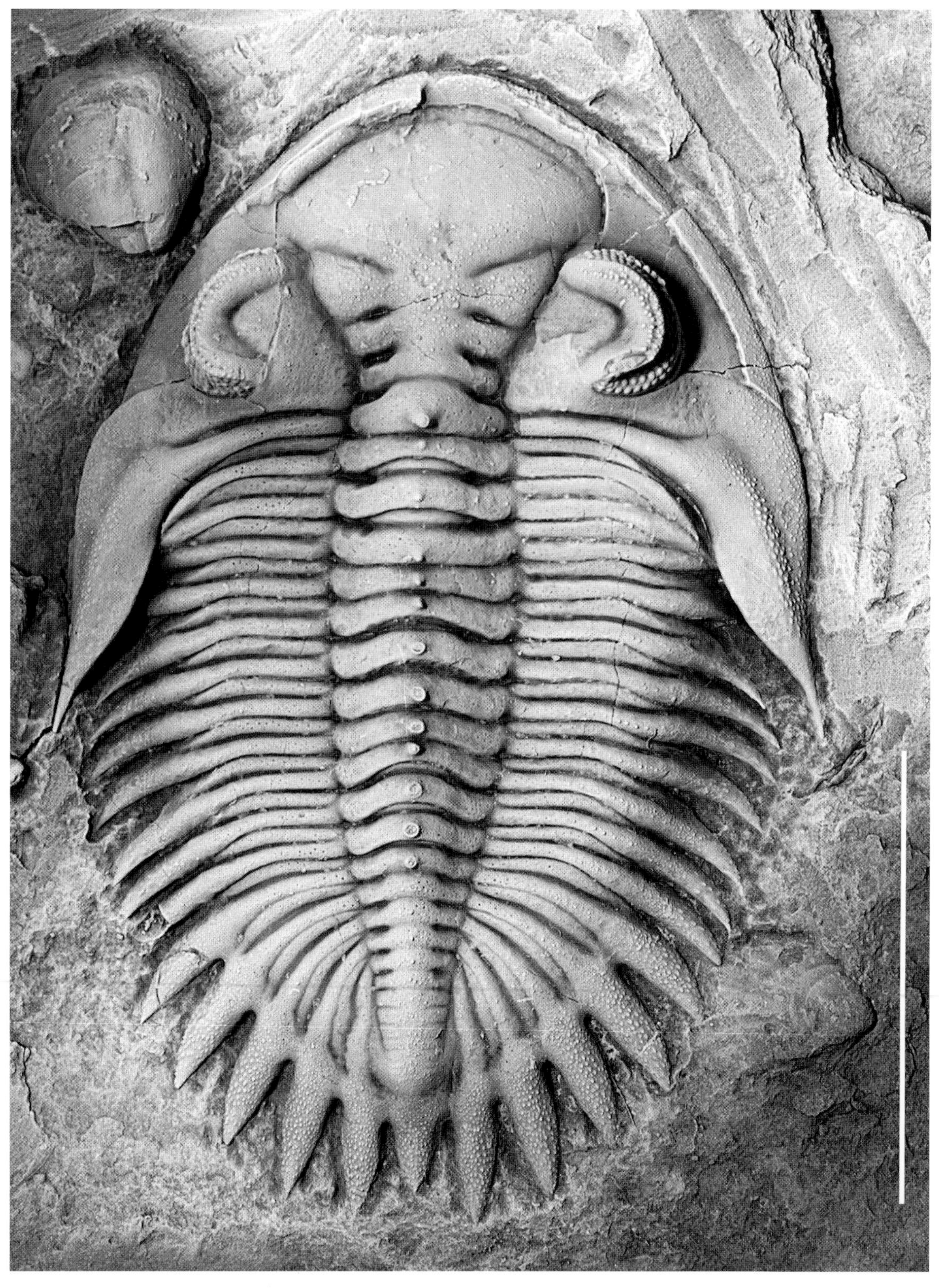

PLATE 45. *Bellacartwrightia whiteleyi* from the Middle Devonian Wanakah Member, Hamilton Group, Darien, Genesee County. The trilobite is 25 mm long. AMNH 45314 (paratype).

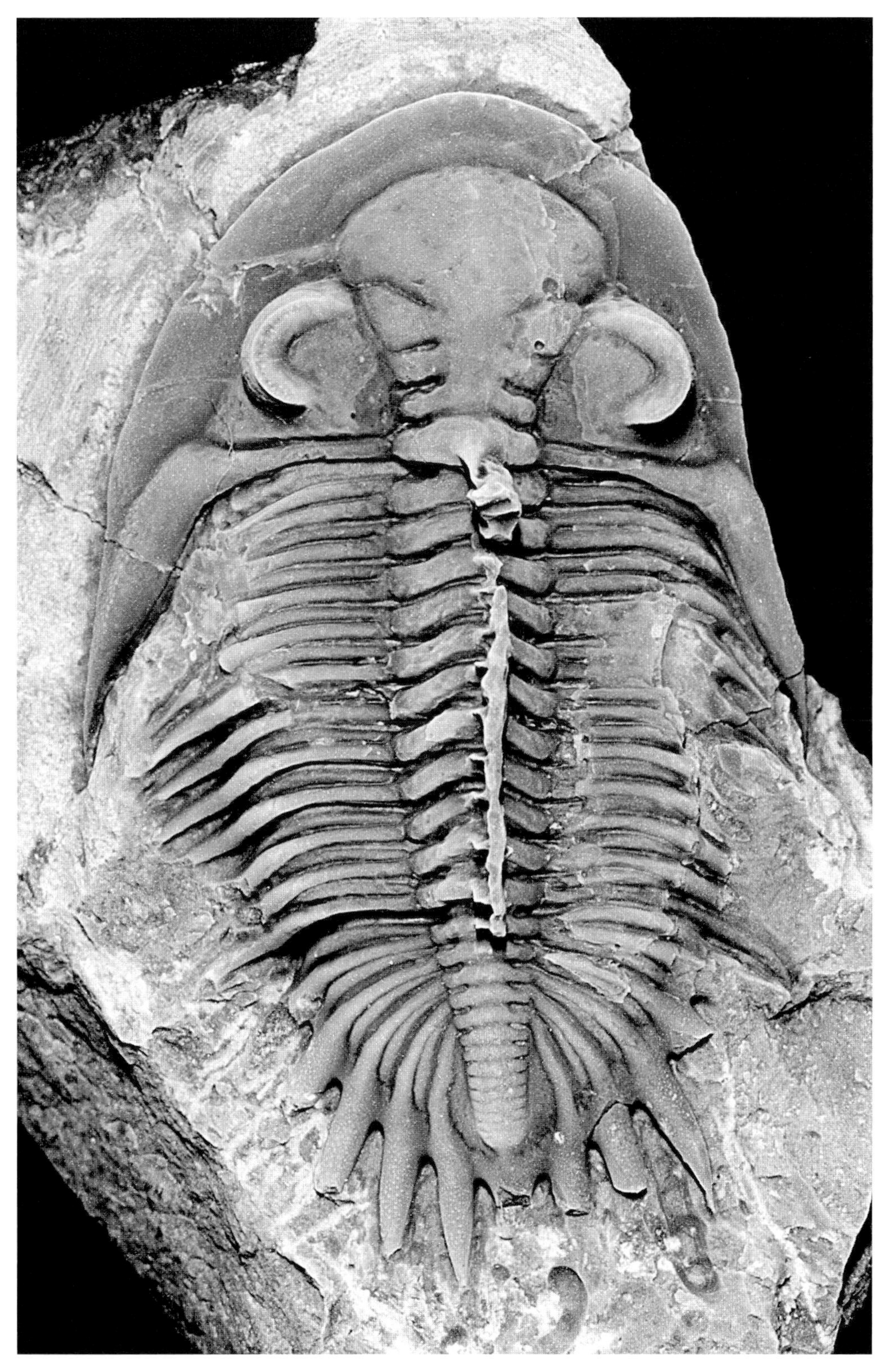

PLATE 46. *Bellacartwrightia* species from the upper Middle Devonian, Tully Limestone in Madison County. USNM 89959.

PLATE 47. *Bellacartwrightia* species from the Middle Devonian Windom Shale, Hamilton Group, on the Geneseo College campus, Geneseo, Livingston County. The ventral exoskeletal features of the *Bellacartwrightia* are shown. The specimen is 41 mm long. In the collection of James Scatterday.

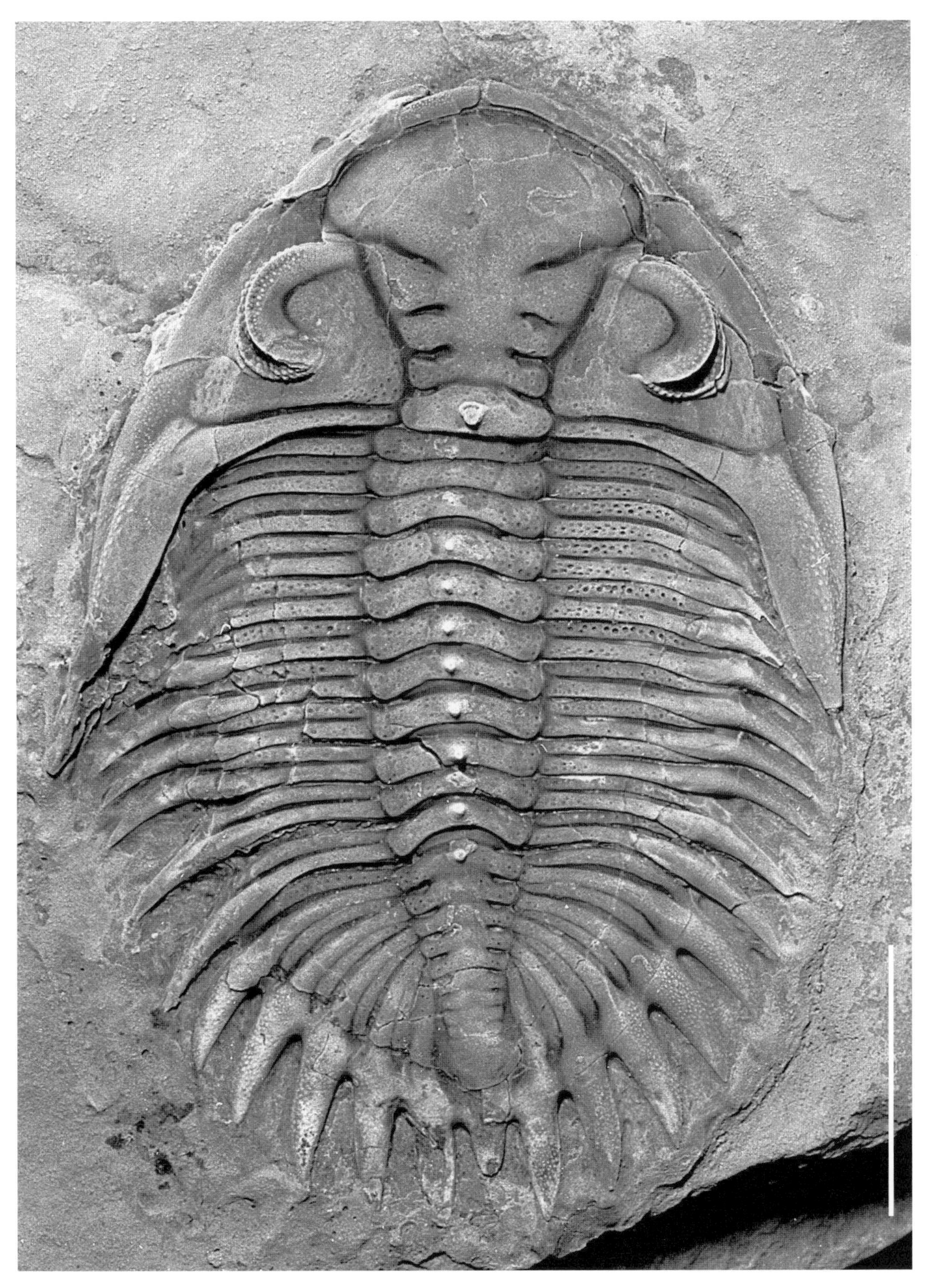

PLATE 48. *Bellacartwrightia* species from the Middle Devonian Windom Shale, Hamilton Group, at the Penn-Dixie Quarry, Hamburg, Erie County. The specimen is 40 mm long. In the collection of Gregory Jennings.

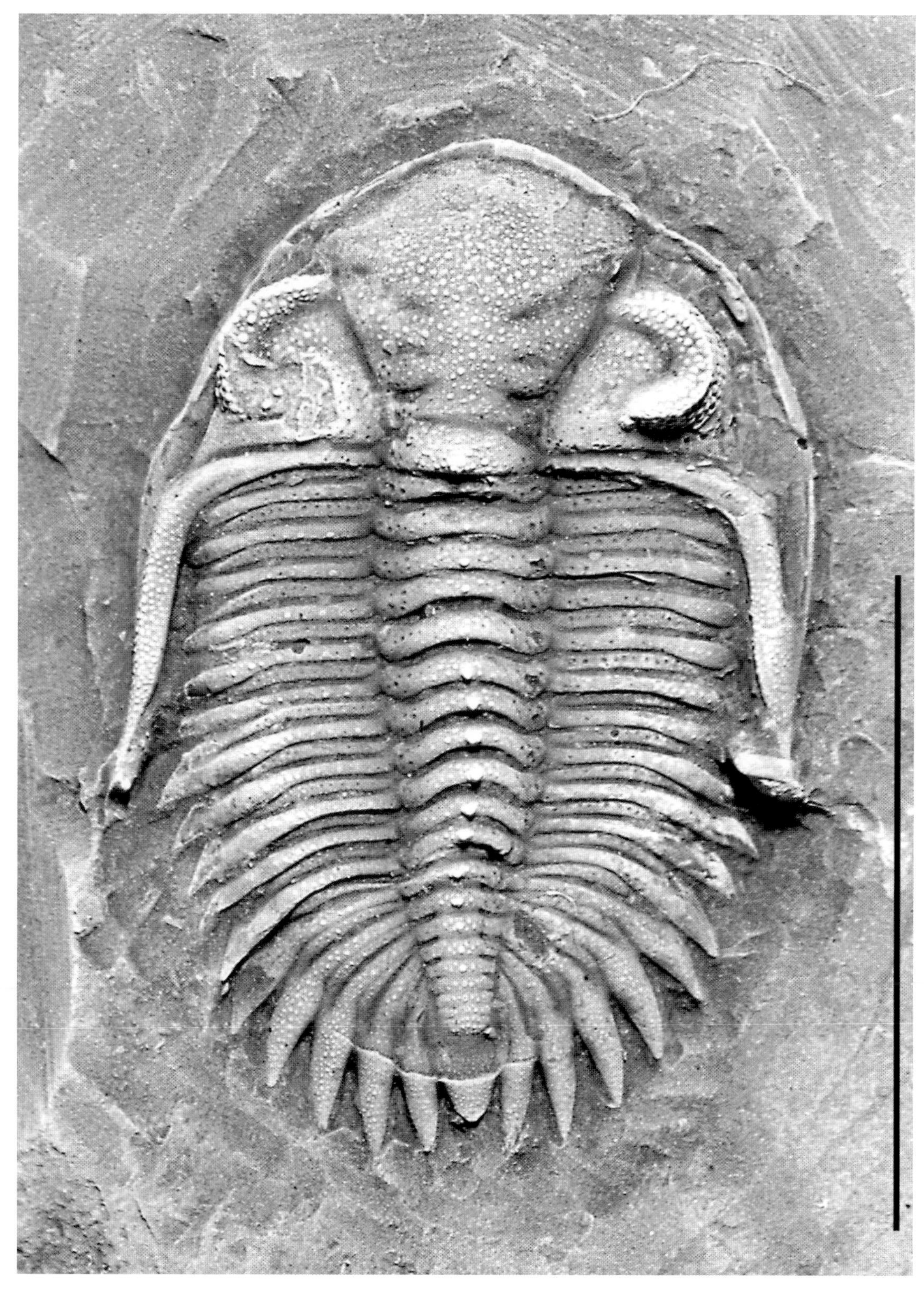

PLATE 49. Astropyginae aff. *Greenops* from the Middle Devonian Pompey Member, Hamilton Group, at Rockefeller Road, Cayuga County. The specimen is 15 mm long. In the collection of Gerald Kloc.

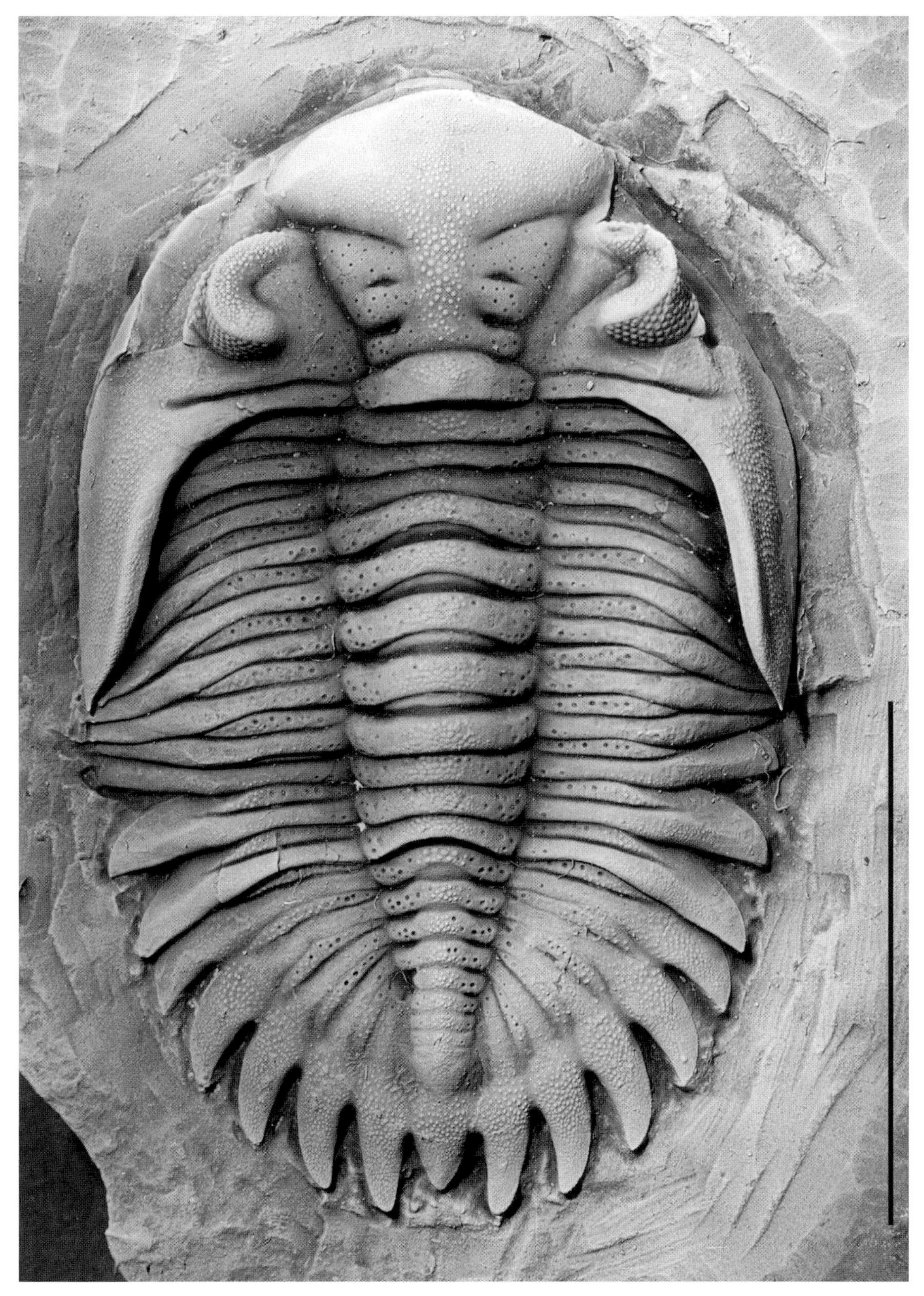

PLATE 50. *Greenops barberi* from the Middle Devonian Windom Member, Hamilton Group, Buffalo Creek, Erie County. The specimen is 21 mm long. AMNH 45277 (paratype).

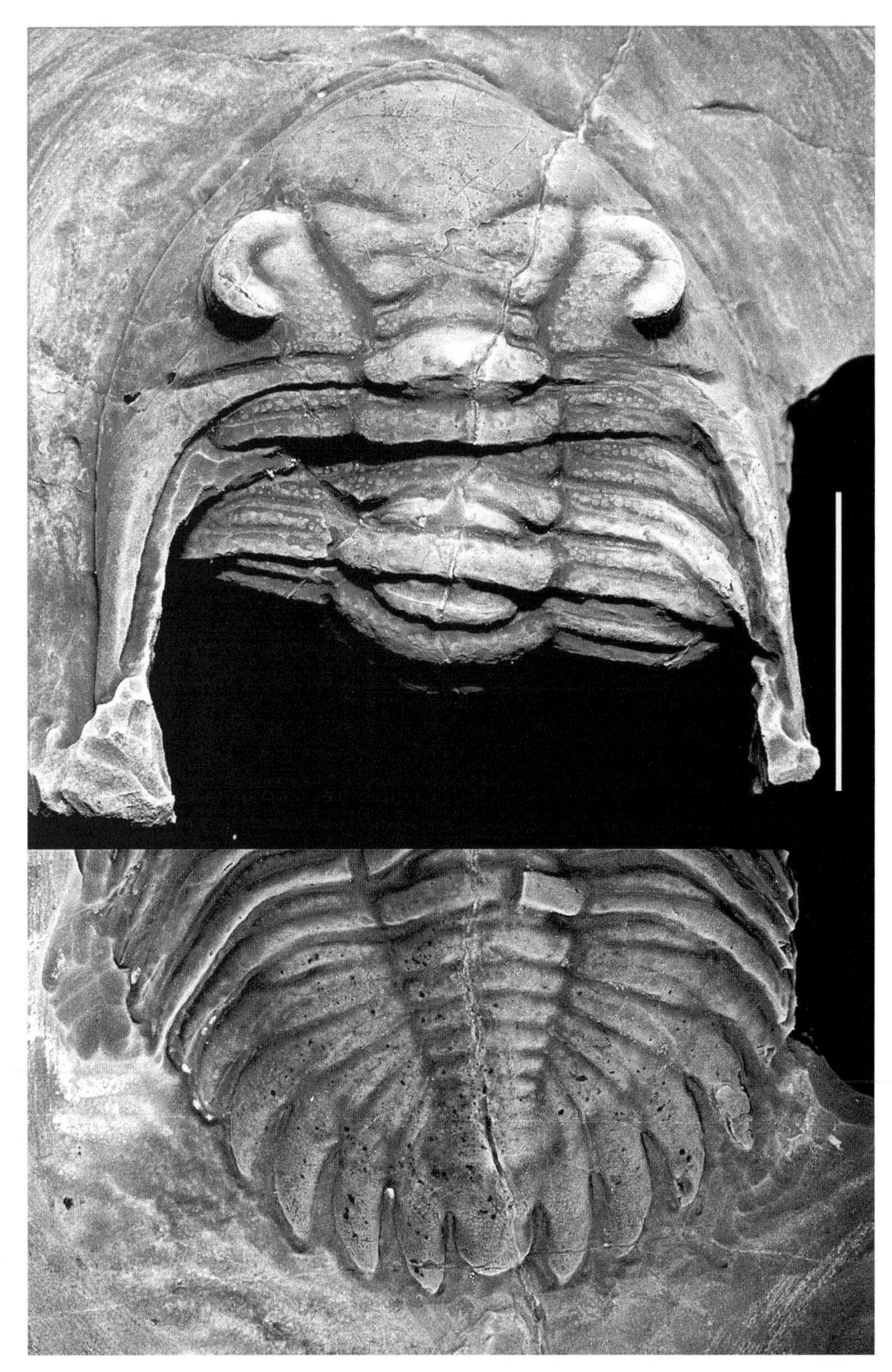

PLATE 51. *Greenops boothi* from the Middle Devonian Frame Member, Mahantango Formation, Huntington, Pennsylvania. This is the area where the original Green specimen came from and is the type area for the genus. The specimen is coiled. The cephalon is 24 mm wide. YPM 35807 (neotype).

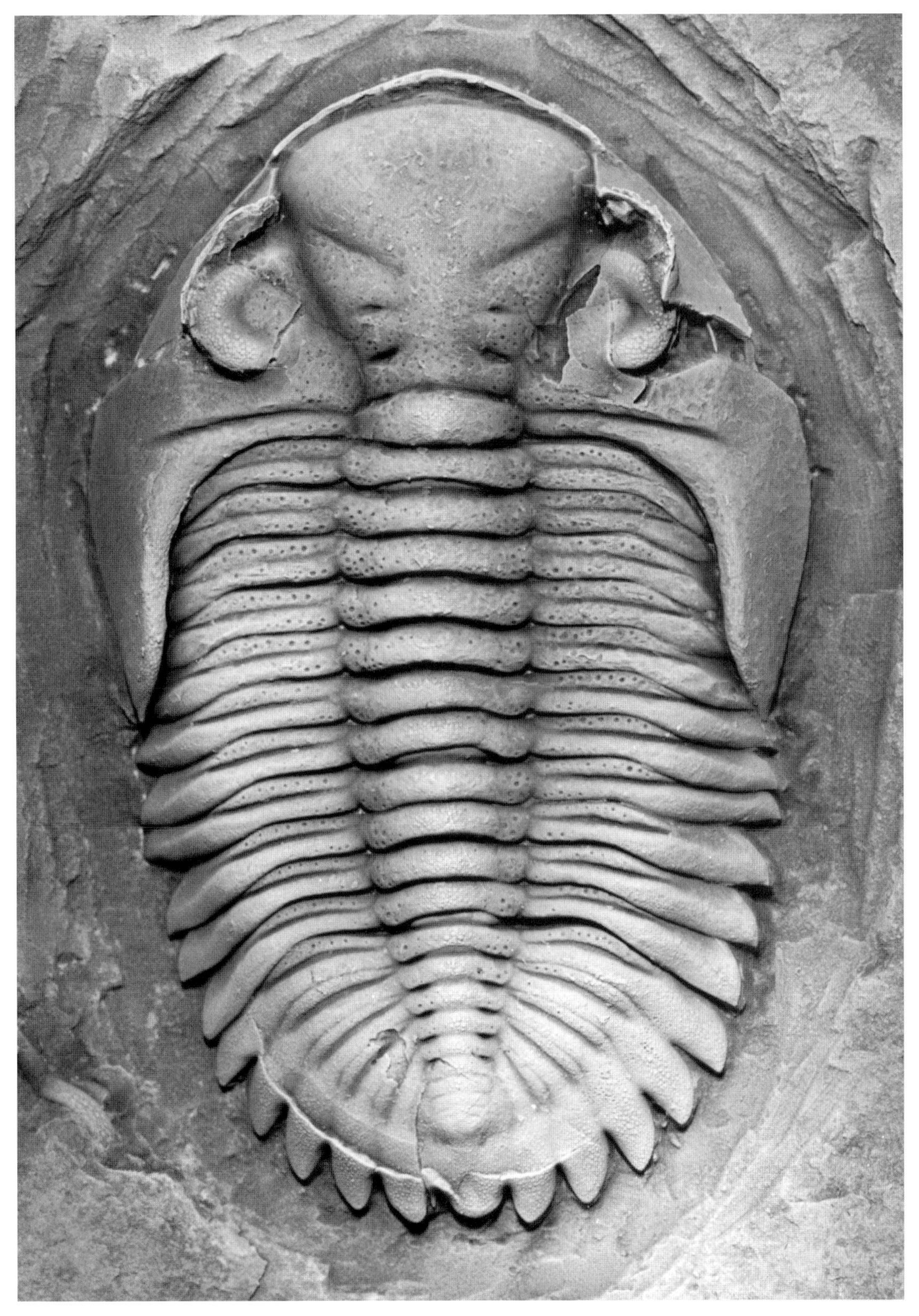

PLATE 52. *Greenops grabaui* from the Middle Devonian Wanakah Member, Hamilton Group, Erie County. In the collection of Fred Barber.

PLATE 53. *Greenops grabaui* from the Middle Devonian Wanakah Member, Hamilton Group, Lake Erie shore, Erie County. This meraspid is 10 mm long. In the collection of Kym Pocius.

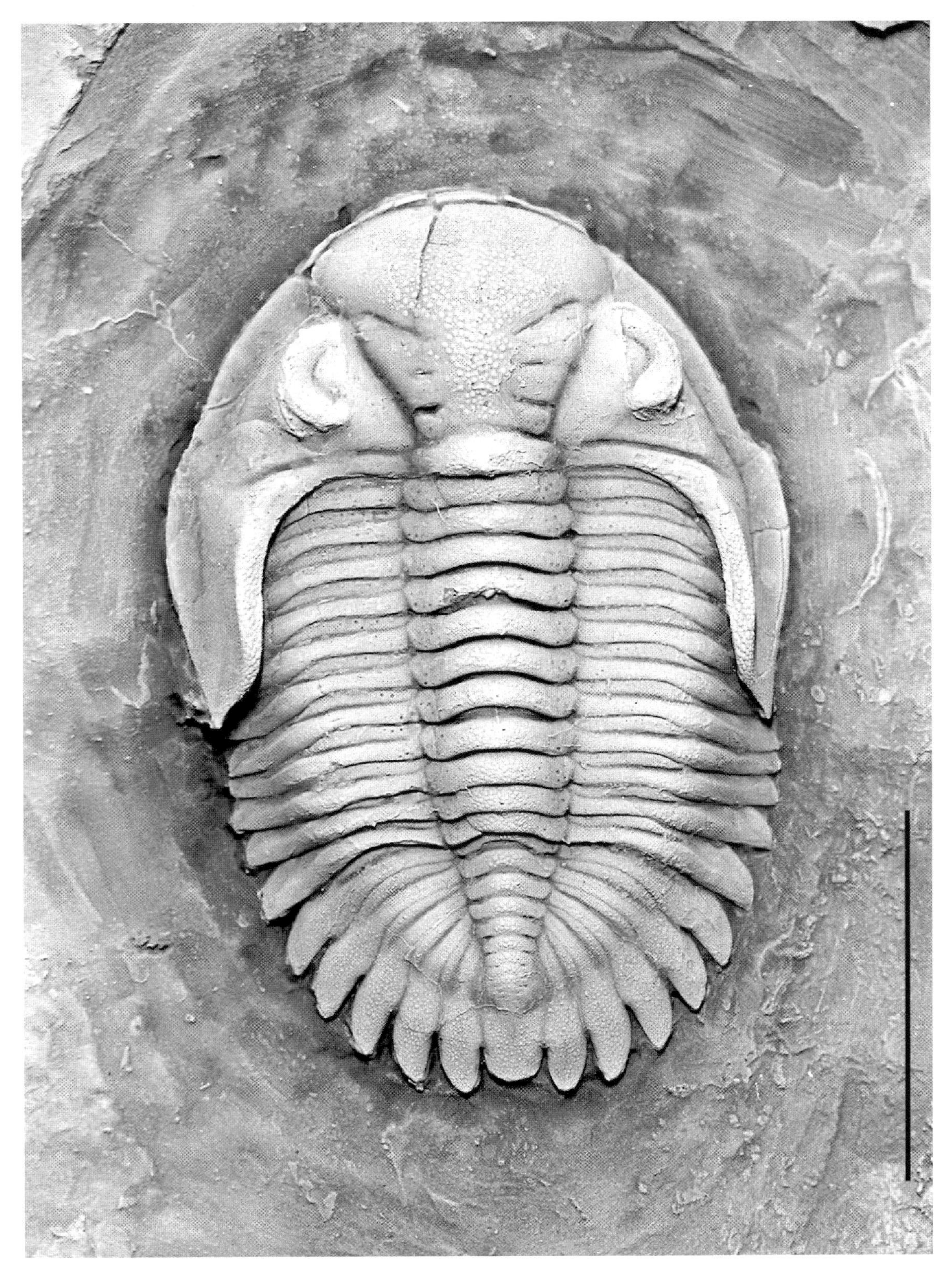

PLATE 54. *Greenops grabaui* from the Middle Devonian Wanakah Member, Hamilton Group, Darien, Genesee County. The specimen is 24 mm long. PRI 49640.

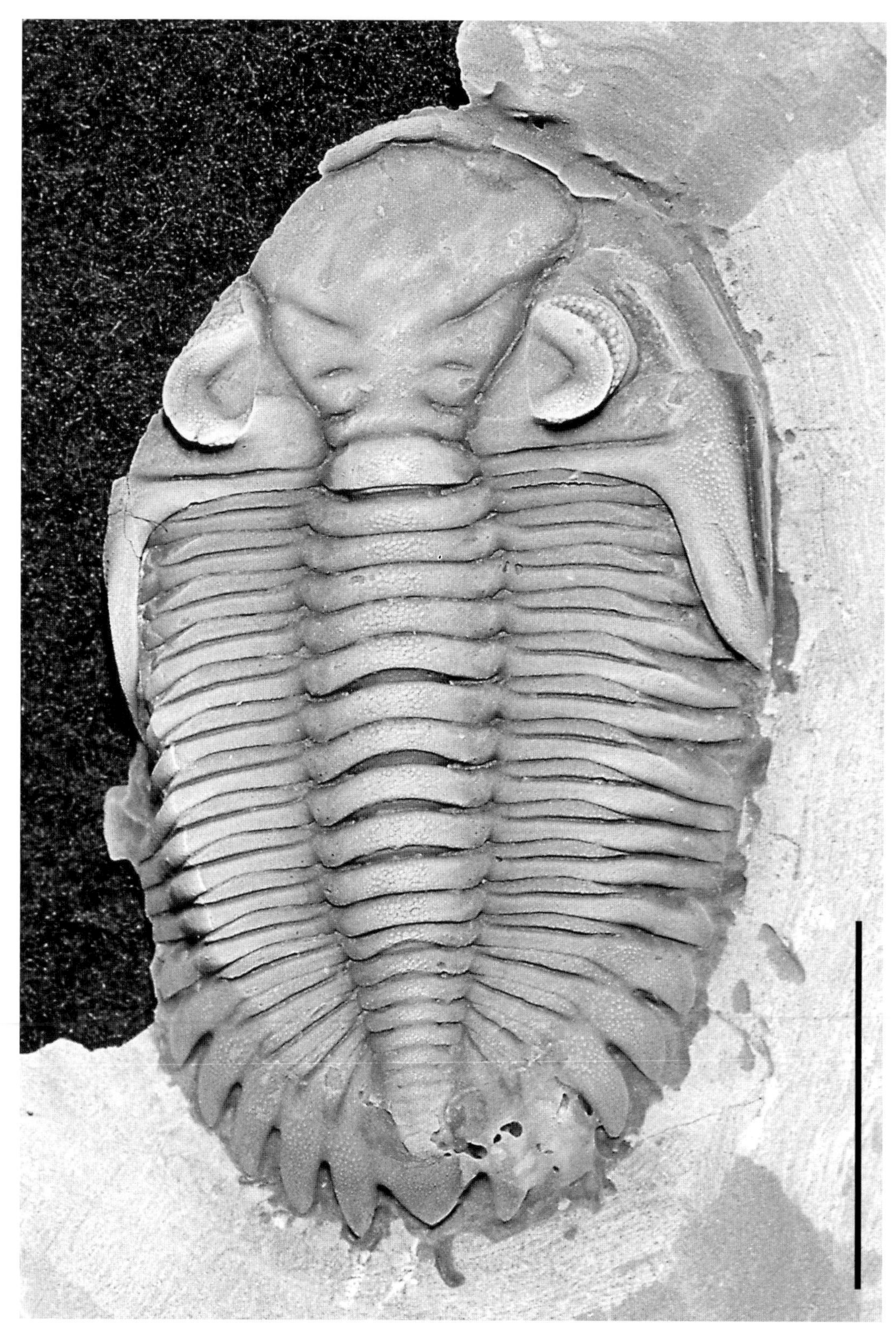

PLATE 55. *Greenops* species from the Middle Devonian, upper Windom Member, Hamilton Group, Groves Creek, Seneca County. The specimen is 30 mm long. In the collection of Gerald Kloc.

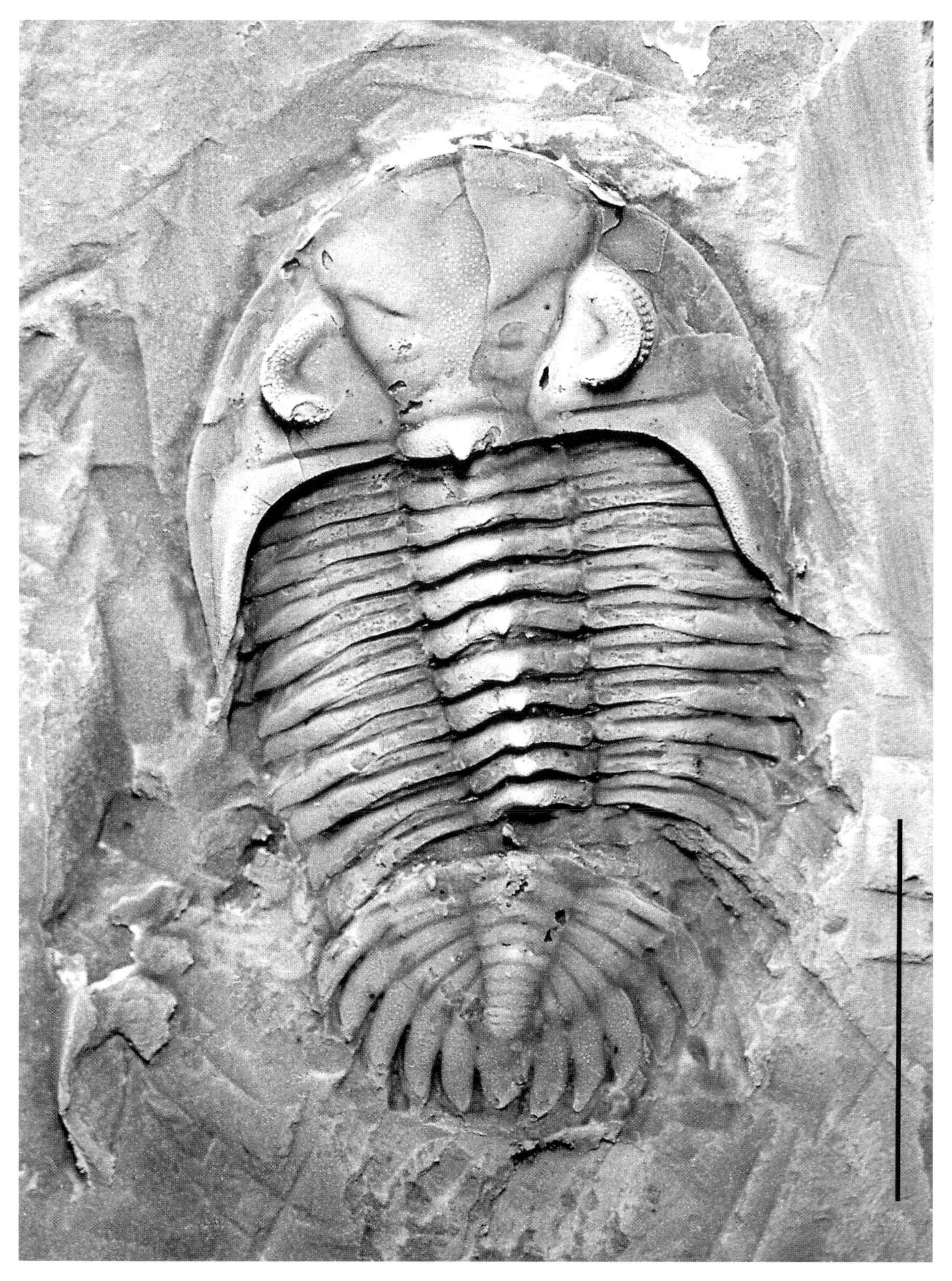

PLATE 56. *Greenops* species from the Middle Devonian Kashong Shale, Hamilton Group. The specimen is 24 mm long. In the collection of Gerald Kloc.

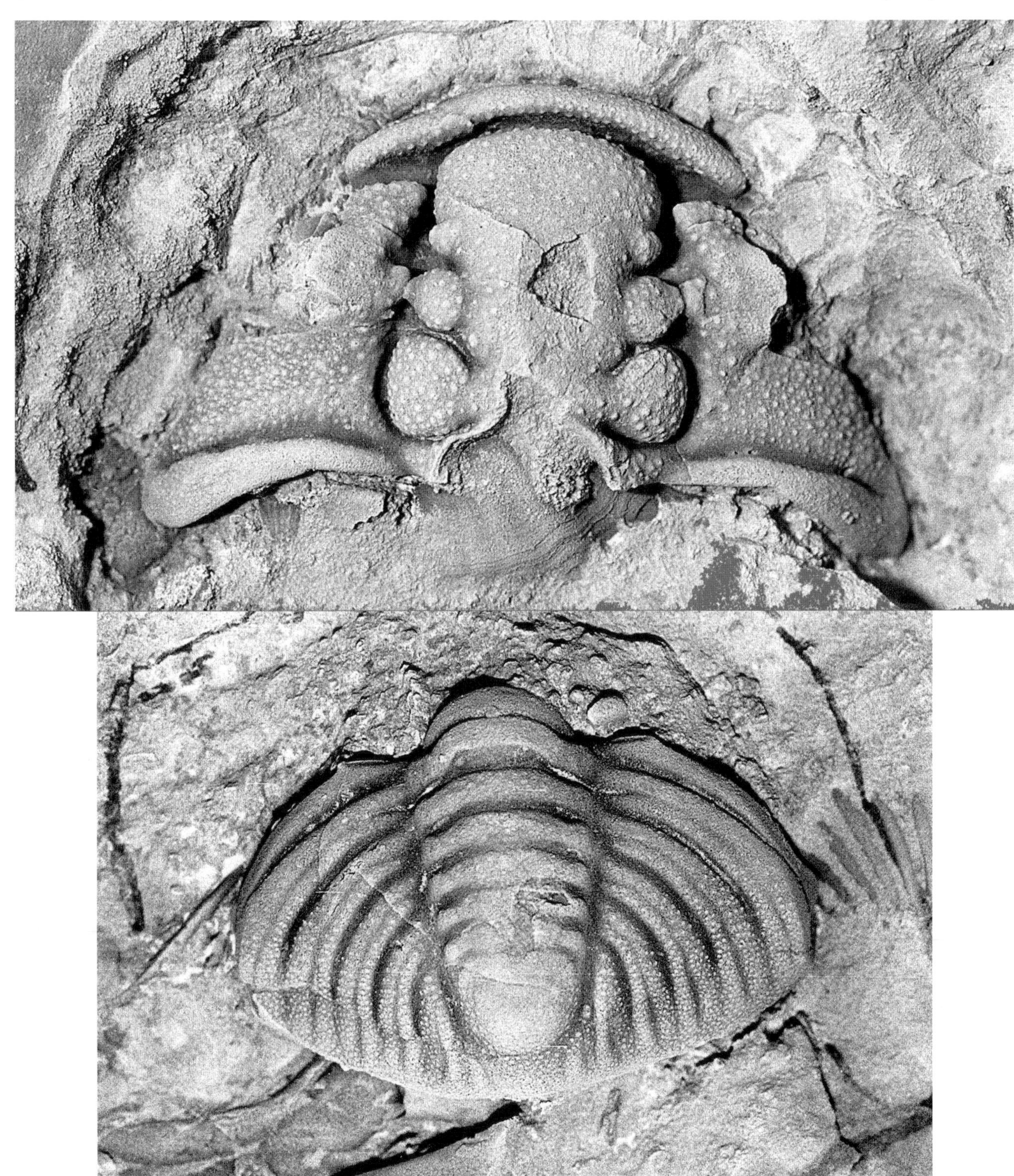

PLATE 57. *Calymene camerata* from the Upper Silurian Cobleskill Limestone at Jerusalem Hill, Herkimer County. This calymenid is distinguished by the inward curve of the anterior cheek toward the glabella. PRI 49624.

PLATE 58. *Calymene niagarensis* from the Lower Silurian Rochester Shale near Middleport, Orleans County. The specimen is 36 mm long. NYSM 16795.

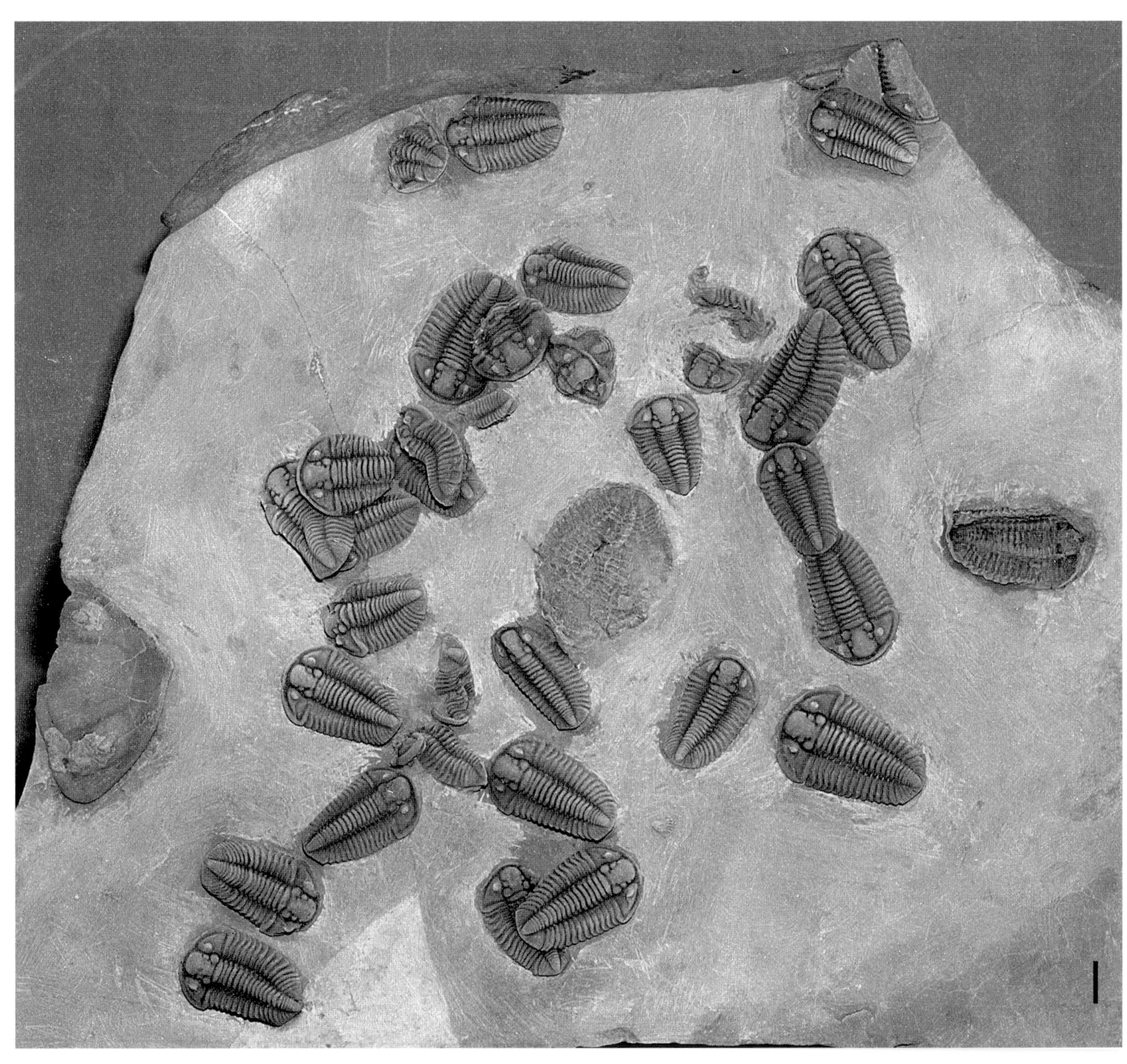

PLATE 59. *Calymene niagarensis* from the Lower Silurian Rochester Shale near Middleport, Orleans County. The plate is an exceptional cluster of trilobites. The largest is 34 mm long. In the collection of Gregory Jennings.

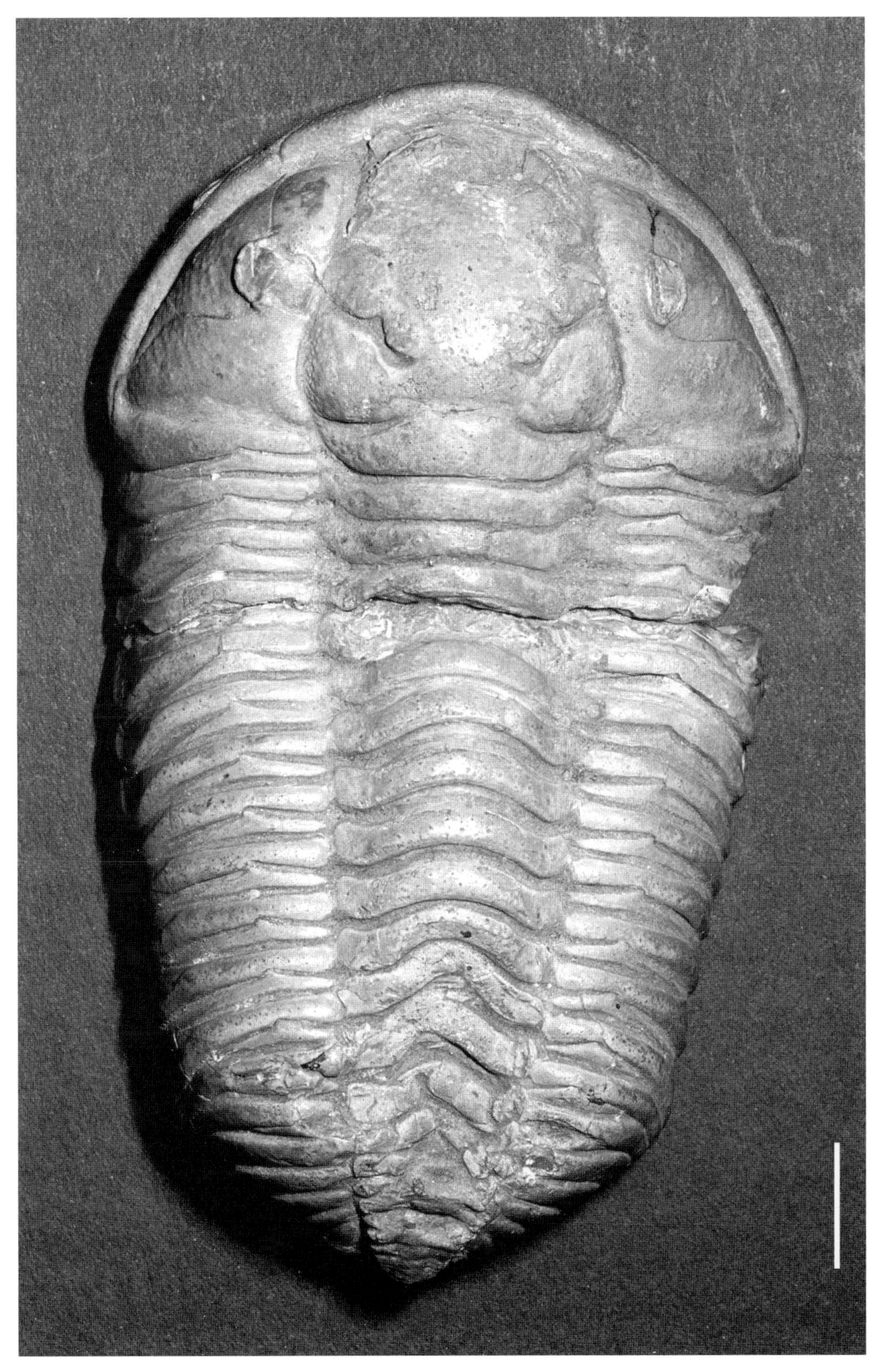

PLATE 60. *Calymene platys* from the Middle Devonian Onondaga Limestone in Hagersville, Ontario, Canada. This species is reported from New York. The specimen is 95 mm long. USNM 380869.

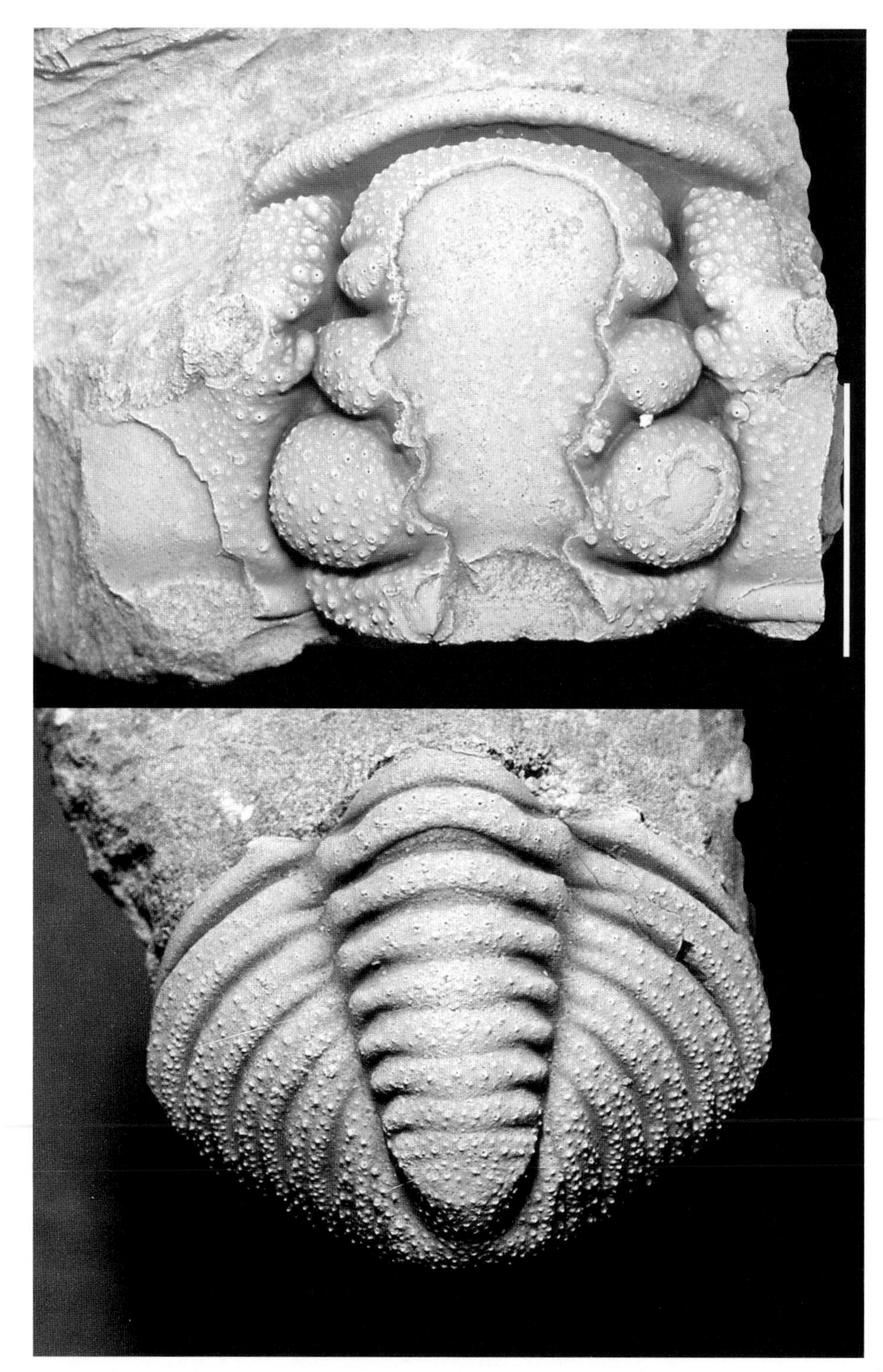

PLATE 61. *Calymene singularis* from the Upper Silurian Oak Orchard Member of the Lockport Group. The figured cephalon is 20 mm long. USNM 488139 (paratype).

PLATE 62. *Calymene* species from the Lower Silurian Rochester Shale on an unnamed creek near Sodus, Wayne County. The specimen is 28 mm long. In the collection of Sam Insalaco.

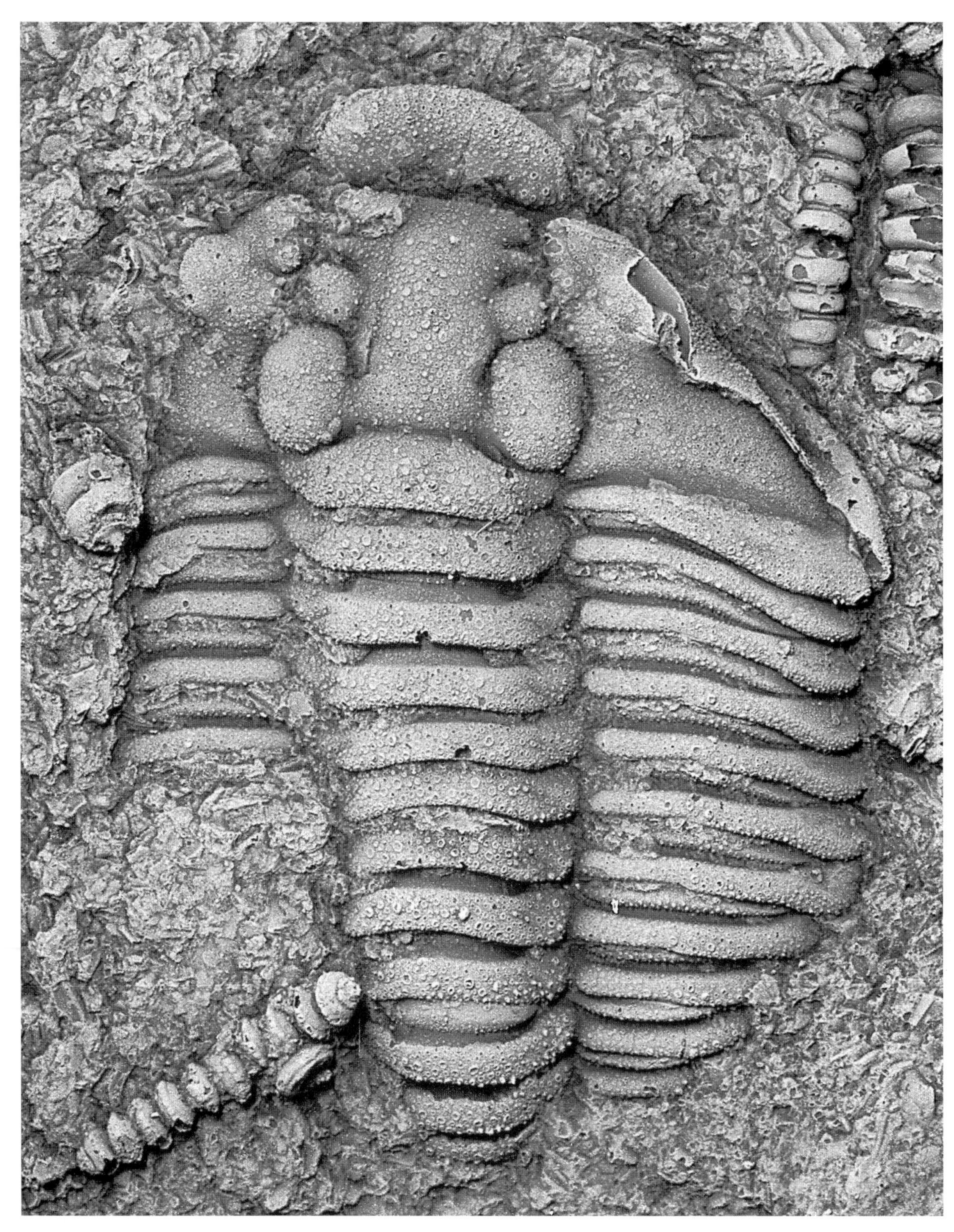

PLATE 63. Calymenid from the Upper Ordovician Pulaski Member, Lorraine Group, at Burke Creek, Oneida County. The latex pull figured here is very pustulose. PRI 49641.

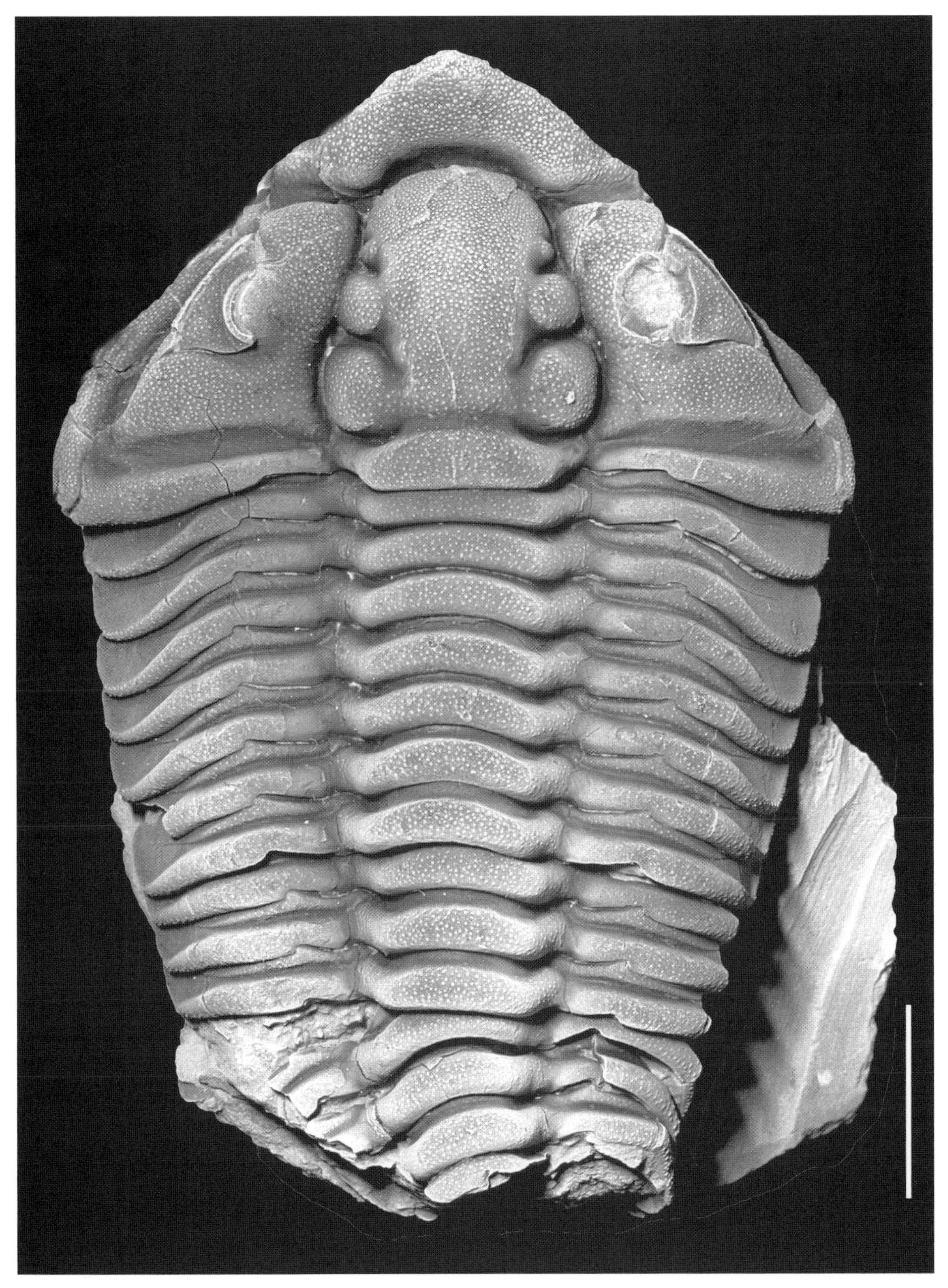

PLATE 64. *Diacalymene* species from the Lower Silurian Rochester Shale at Densmore Creek, Rochester, Monroe County. The specimen, though incomplete, is strikingly different from the Rochester Shale calymenids found to the west of Rochester. The specimen is 60 mm long. NYSM 16799.

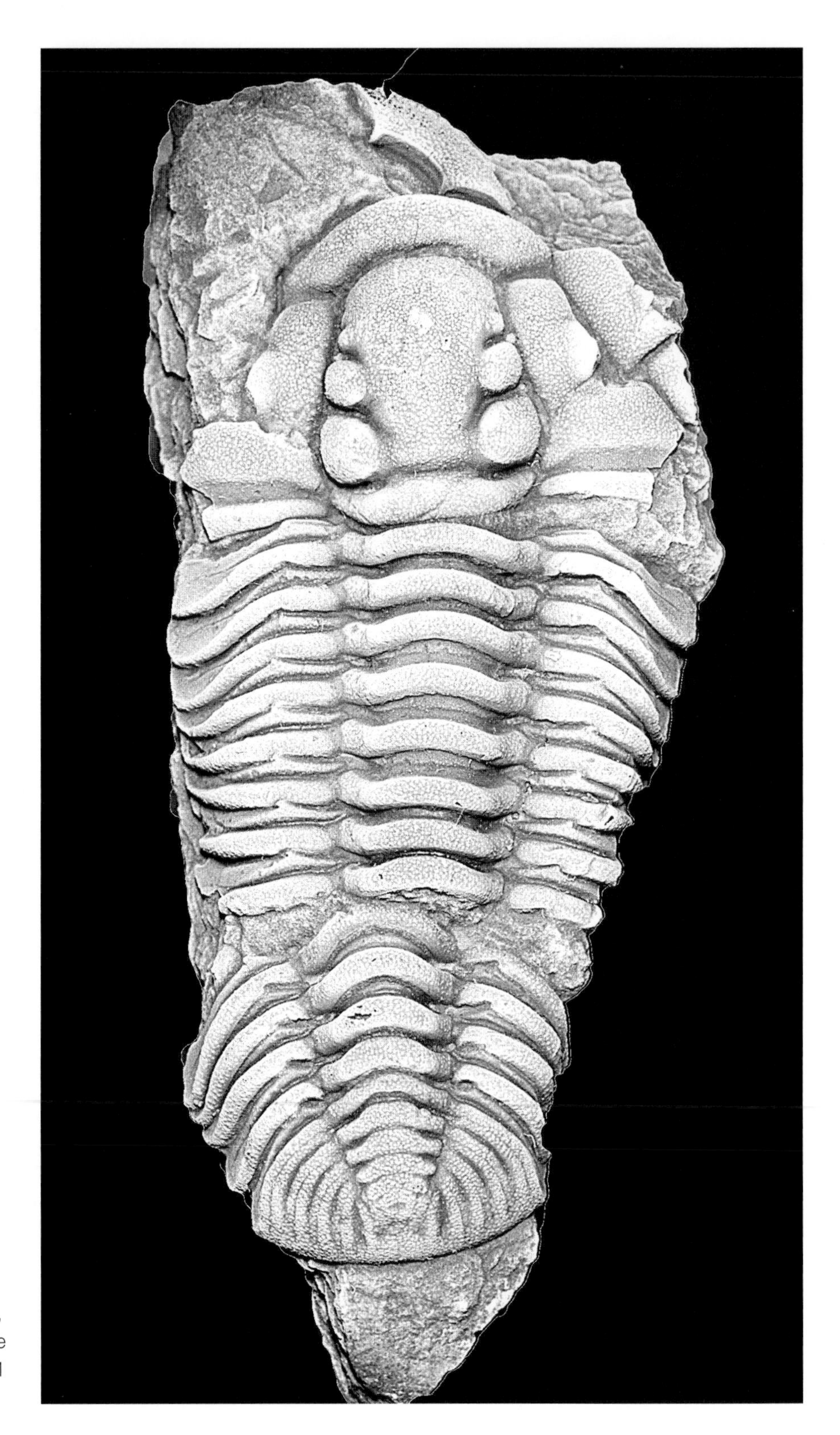

PLATE 65. *Diacalymene* species from the Lower Silurian Rochester Shale in Wayne County. This trilobite is different, primarily in the anterior of the cephalon, from the *Diacalymene* specimen in Plate 64. The specimen is 45 mm long. NYSM 16793.

PLATE 66. *Flexicalymene meeki* from the Upper Ordovician shales at Cincinnati, Ohio. This specimen is figured because Upper Ordovician calymenids in New York have been identified as this species. The trilobite is 55 mm long. PRI 49642.

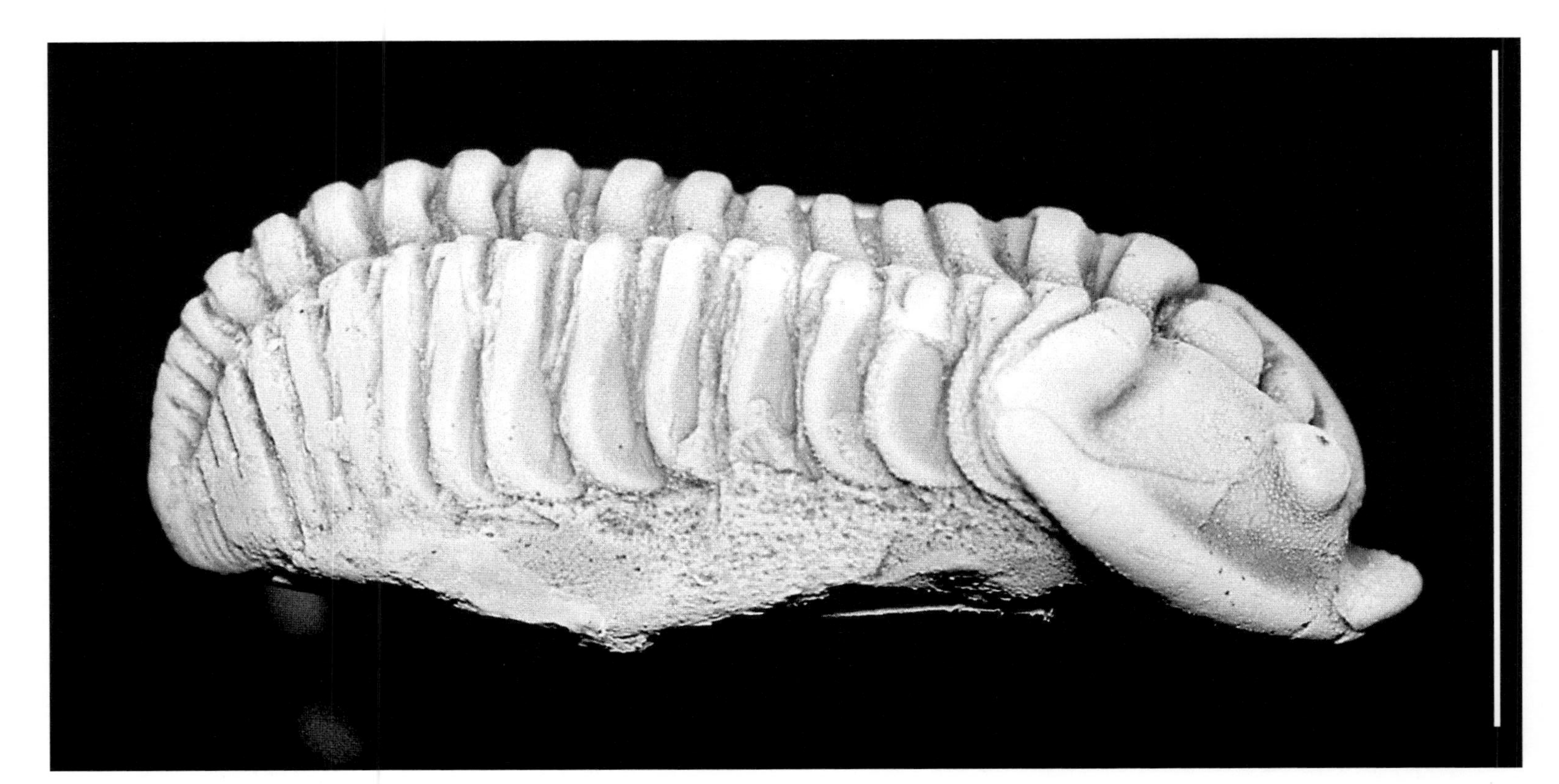

PLATE 67. *Flexicalymene senaria* from the Middle Ordovician limestones at Middleville, Herkimer County. The original type for this species is lost, and this specimen, figured by Hall, was chosen as the neotype. The specimen is 24.5 mm long. AMNH 29474 (neotype).

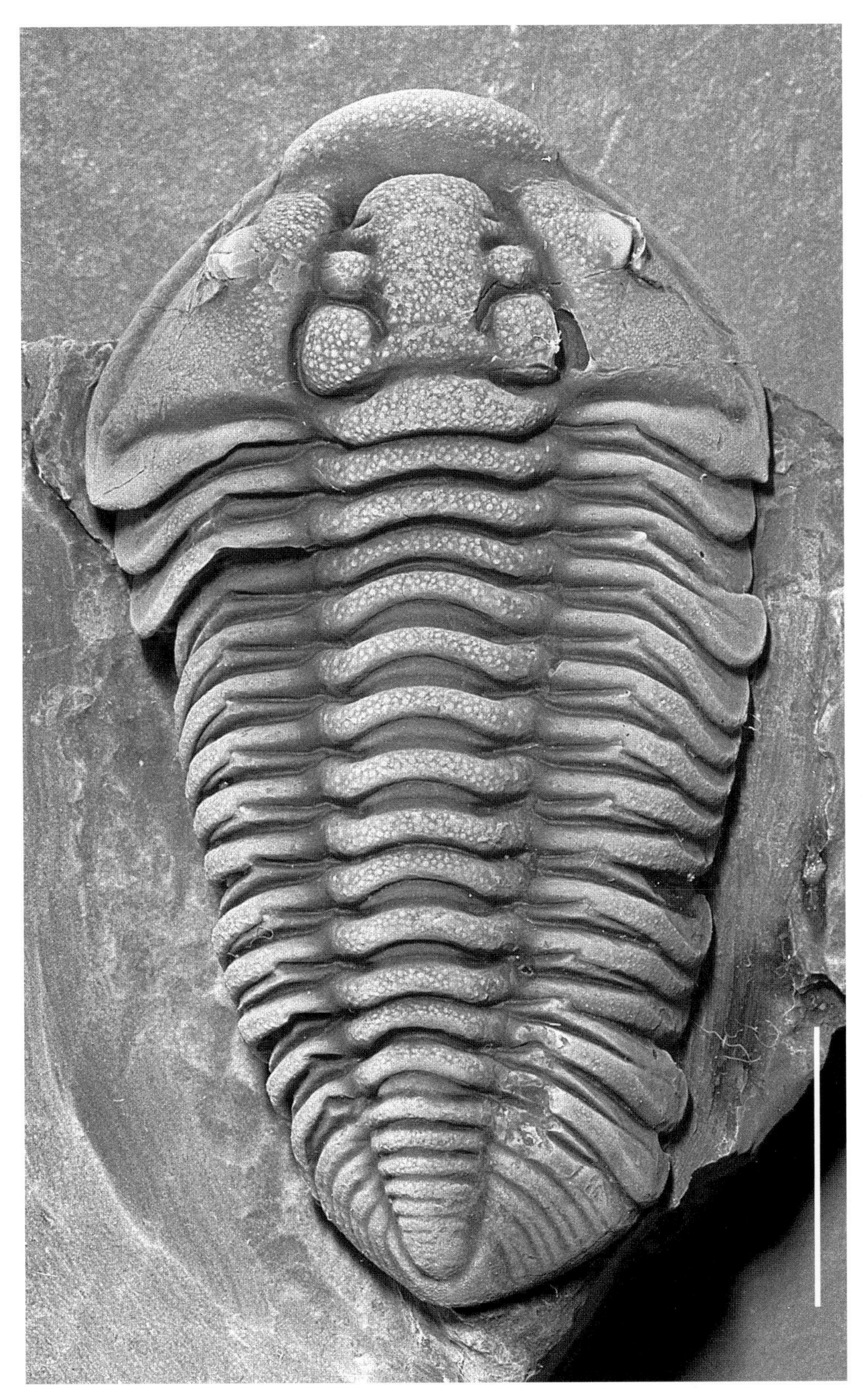

PLATE 68. *Flexicalymene senaria* from the Middle Ordovician Poland Member of the Trenton Group at Trenton Falls, Herkimer County. This specimen is characteristic of the middle and lower Trenton *F. senaria*. The trilobite is 43 mm long. In the collection of Gerald Kloc.

PLATE 69. *Flexicalymene* cf. *F. senaria* from the Middle Ordovician Rust Formation, Trenton Group, Walcott-Rust Quarry. This trilobite is somewhat different from many *F. senaria* specimens in that it is more pustulose and has a spinelike, acute genal angle. The specimen is 28 mm long. MCZ 111710.

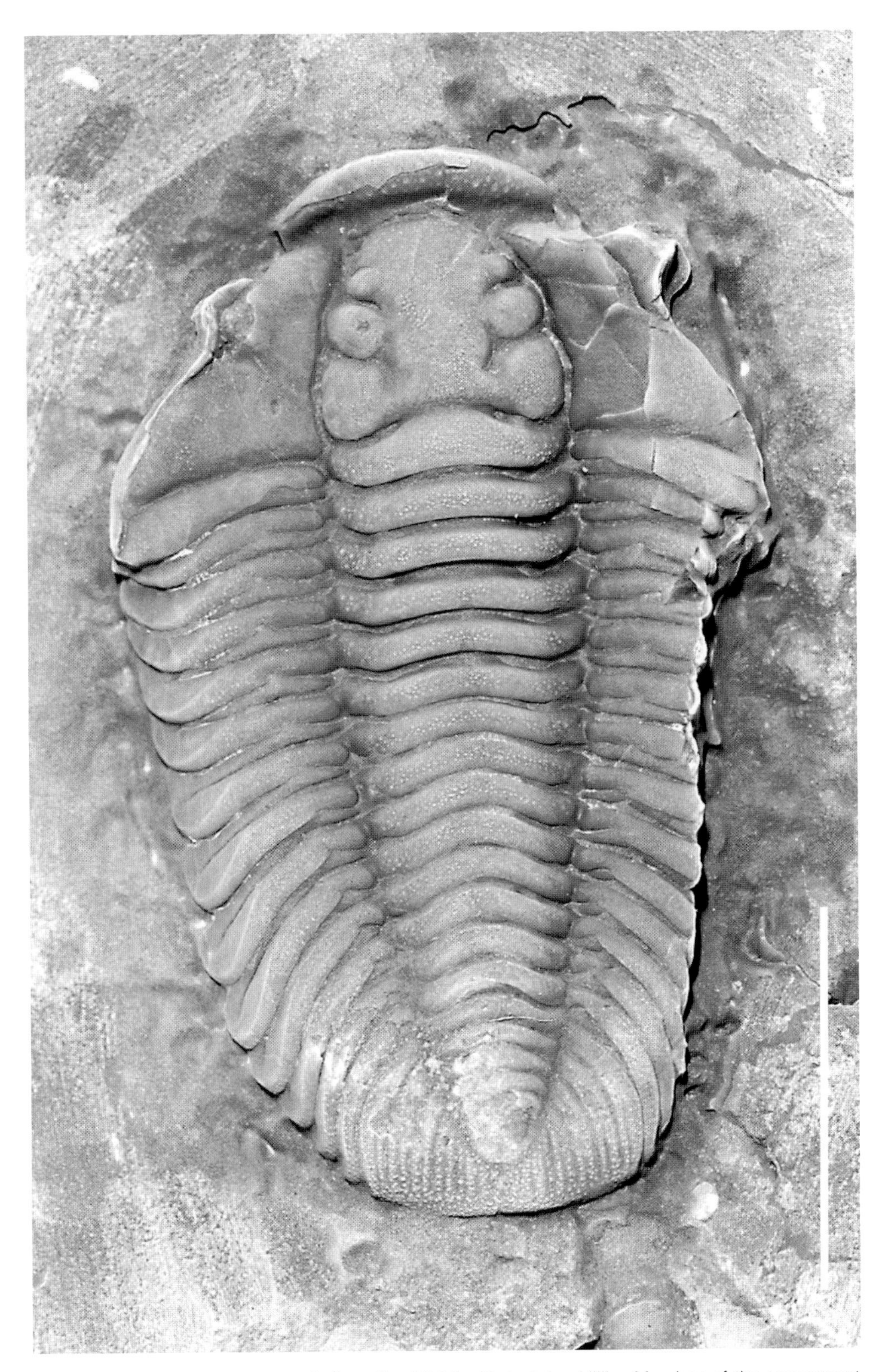

PLATE 70. *Flexicalymene senaria* from the Middle Ordovician Hillier Member of the uppermost Trenton Group in a quarry off Rte. 3 in Jefferson County. The trilobite is flattened because it was on a bedding plane, but it is apparently much less pustulose than the trilobite in Plate 69. The specimen is 28 mm long. PRI 49643.

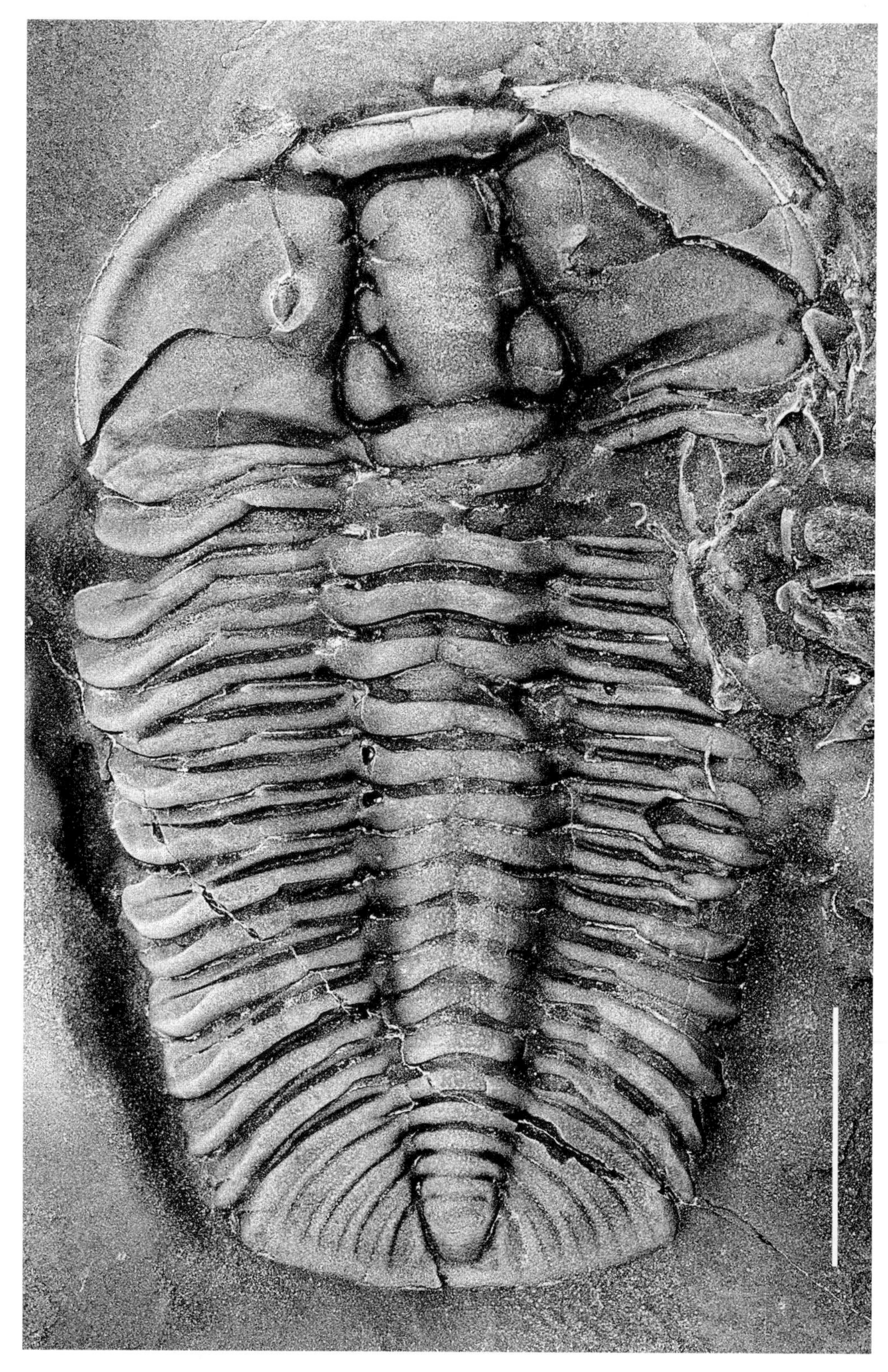

PLATE 71. *Gravicalymene magnotuberculata* from the Middle Ordovician Sugar River Formation, Trenton Group, North Creek, Herkimer County. Note the very differently shaped glabella compared to the one in *F. senaria* in Plate 70. The trilobite is 45 mm long. PRI 49644.

PLATE 72. *Gravicalymene magnotuberculata* from the Middle Ordovician Sugar River Formation, Trenton Group, on Allen Road, Fulton County. The ventral exoskeletal anatomy of this trilobite is shown. This is a molt as there is no sign of the hypostome. The specimen is 40 mm long. PRI 49645.

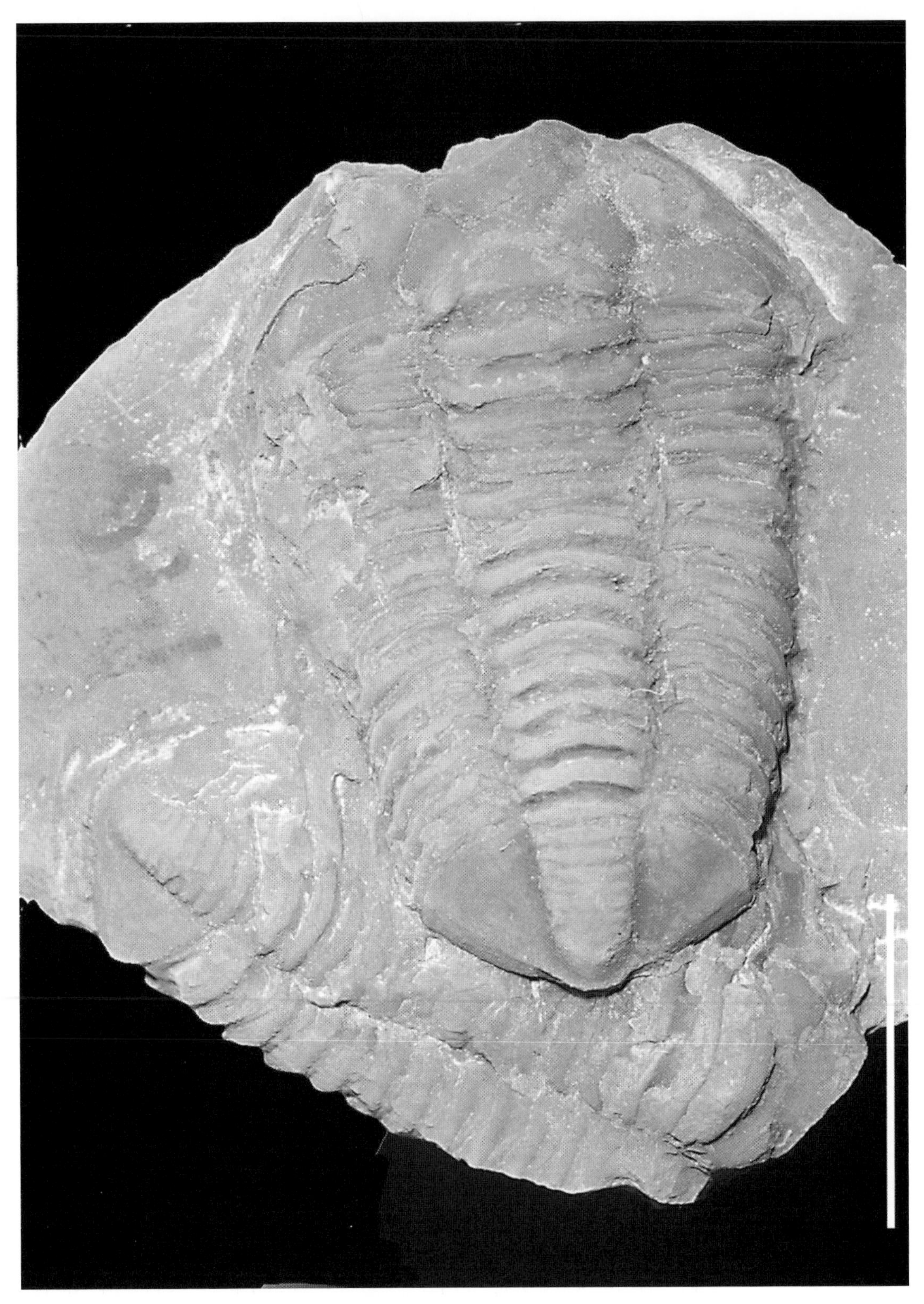

PLATE 73. *Liocalymene clintoni* from the Lower Silurian Clinton Group, Clinton, Oneida County. Most of the *Liocalymene* specimens one sees are internal molds like this one. The specimen is 26 mm long. PRI 49625.

PLATE 74. *Ceraurinella* cf. *C. scofieldi* from the Middle Ordovician Watertown Member, Black River Group, Buck's Quarry, Poland, Herkimer County. This trilobite is usually found in the equivalent rocks in the Midwest. The cranidium is 12.5 mm wide. MCZ 13.

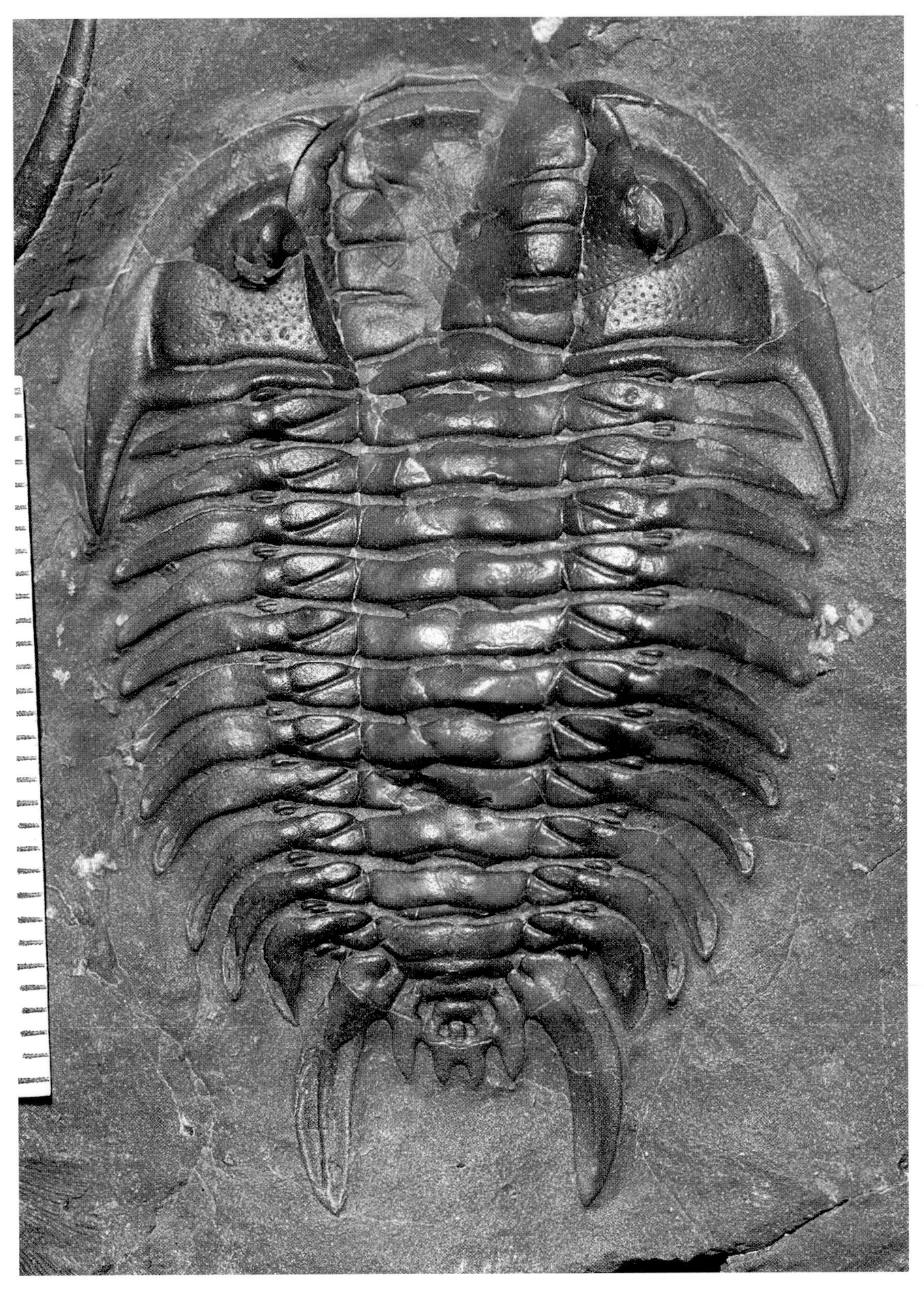

PLATE 75. *Ceraurinus marginatus* from the Middle Ordovician Lindsay Formation, Trenton Group, Colborne Quarry, Colborne, Ontario, Canada. This trilobite is known from New York but is uncommon. In the collection of Kevin Brett.

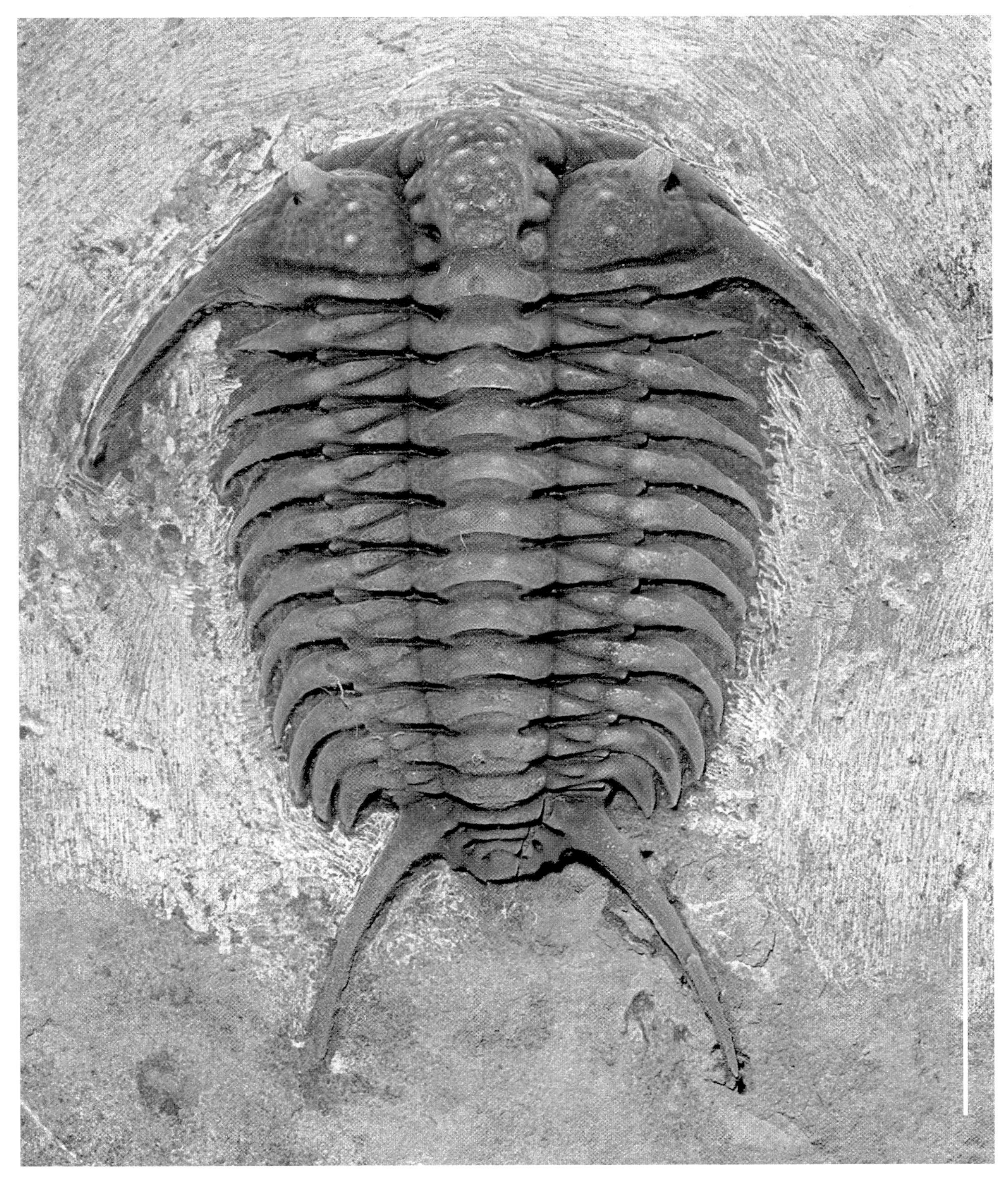

PLATE 76. *Ceraurus pleurexanthemus* from the Middle Ordovician Rust Formation, Trenton Group, Walcott-Rust Quarry, Herkimer County. This specimen is 35 mm long. MCZ 111708.

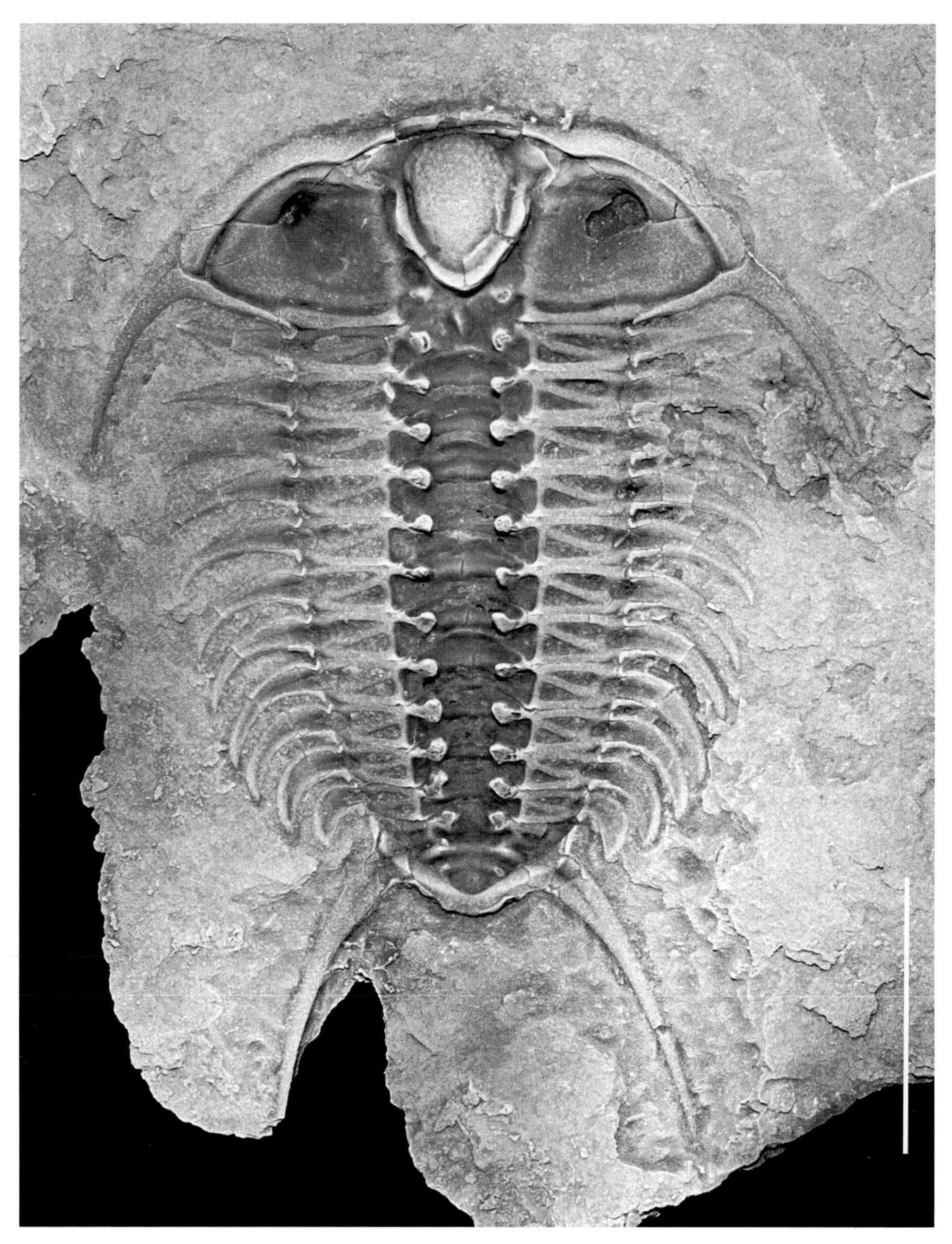

PLATE 77. *Ceraurus pleurexanthemus* from the Middle Ordovician Rust Formation, Trenton Group, Walcott-Rust Quarry, Herkimer County. This specimen shows the ventral exoskeletal anatomy. It is 29 mm long. MCZ 111715.

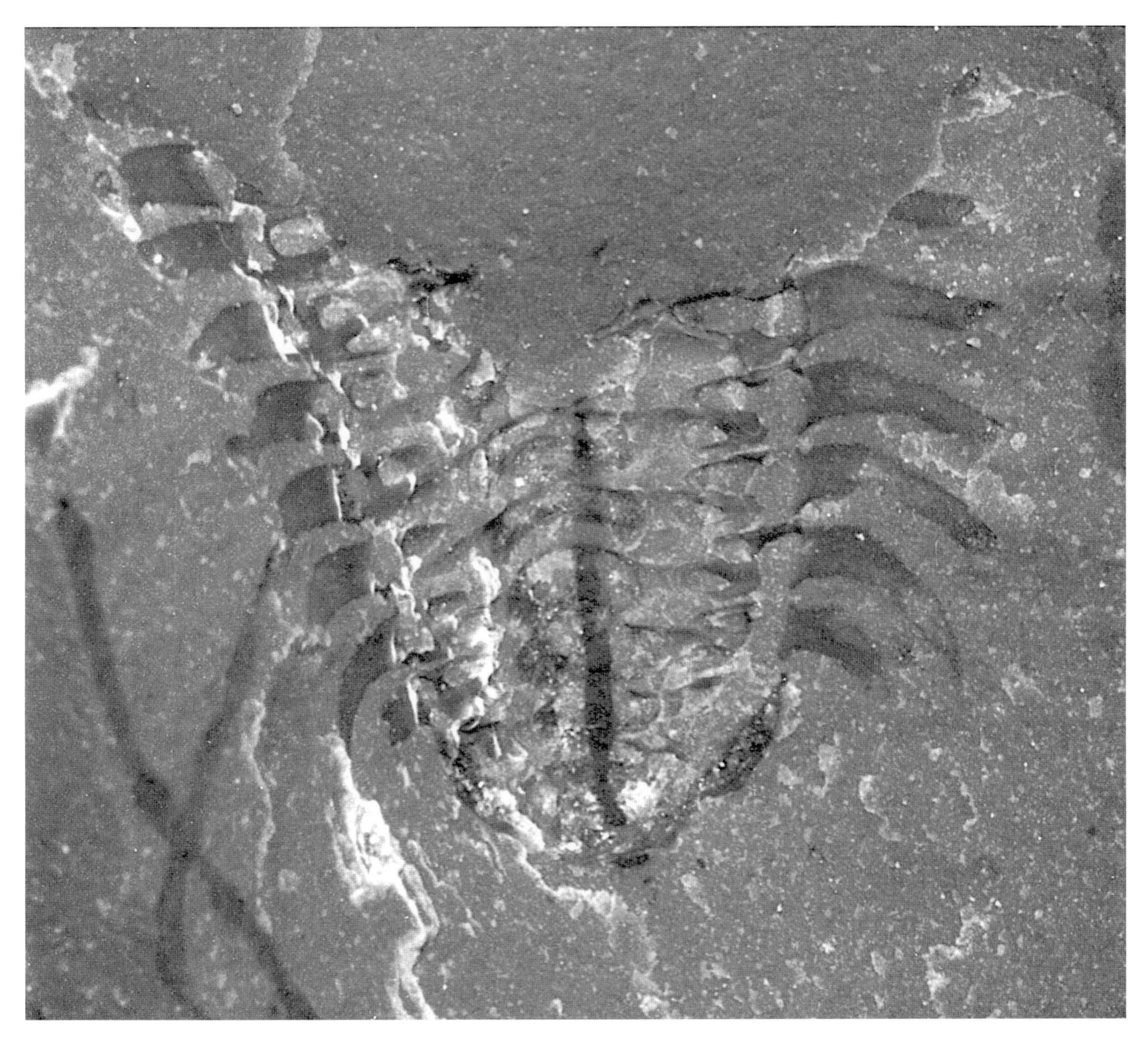

PLATE 78. *Ceraurus pleurexanthemus* from the Middle Ordovician Rust Formation, Trenton Group, Walcott-Rust Quarry, Herkimer County. This specimen from inside a limestone shows a ferruginous trace along the axis, which is interpreted to be a trace of the gut. MCZ 111716.

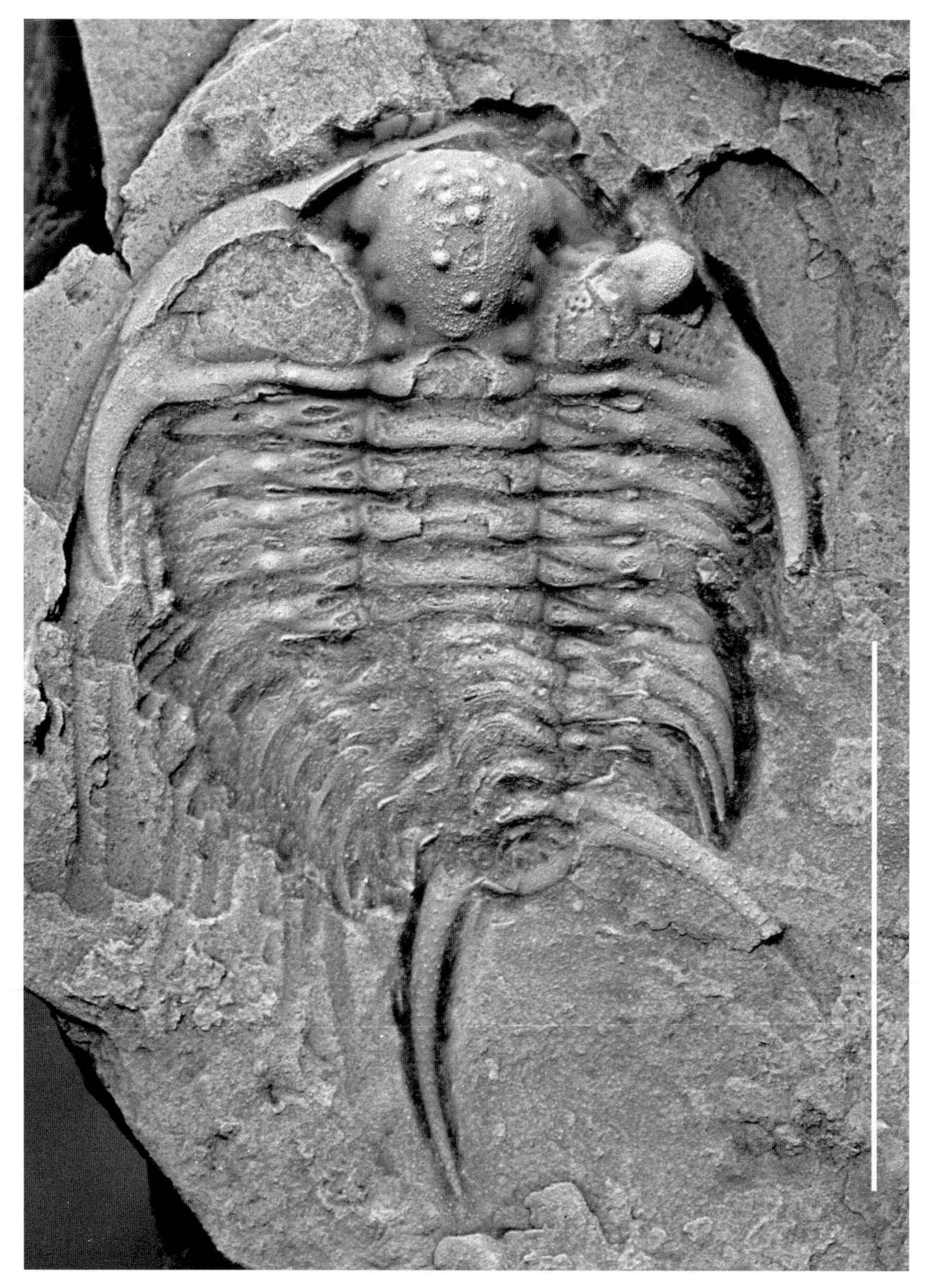

PLATE 79. *Ceraurus* species from interface of the Middle and Upper Ordovician at a flooding surface between the Hillier Member of the Trenton Group and the Utica Shale on Gulf Stream, Rodman, Jefferson County. This trilobite shows differences from the more common *C. pleurexanthemus*. The specimen is 13 mm long. NYSM 17000.

PLATE 80. *Ceraurus* species from the Middle Ordovician Trenton Group in a Quarry near Watertown, Jefferson County. Although similar to the specimen in Plate 79, there are differences from *C. pleurexanthemus*. The trilobite is 35 mm long. In the collection of Gerald Kloc.

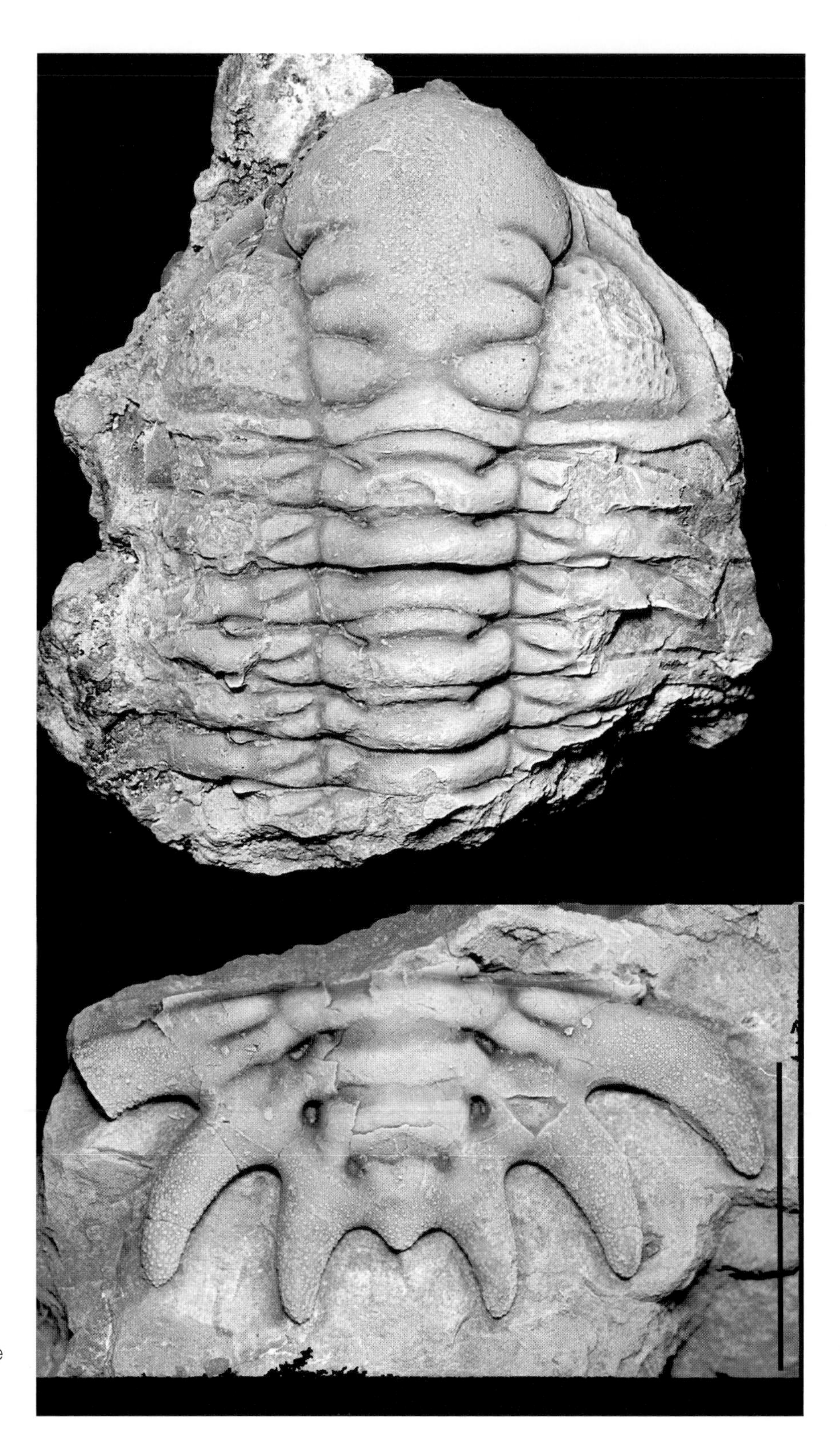

PLATE 81. *Cheirurus* cf. *C. niagarensis* from the Lower Silurian Irondequoit Limestone in Monroe County. The cephalothorax is USNM 489751, and the pygidium, which is 23 mm wide. NYSM 17022.

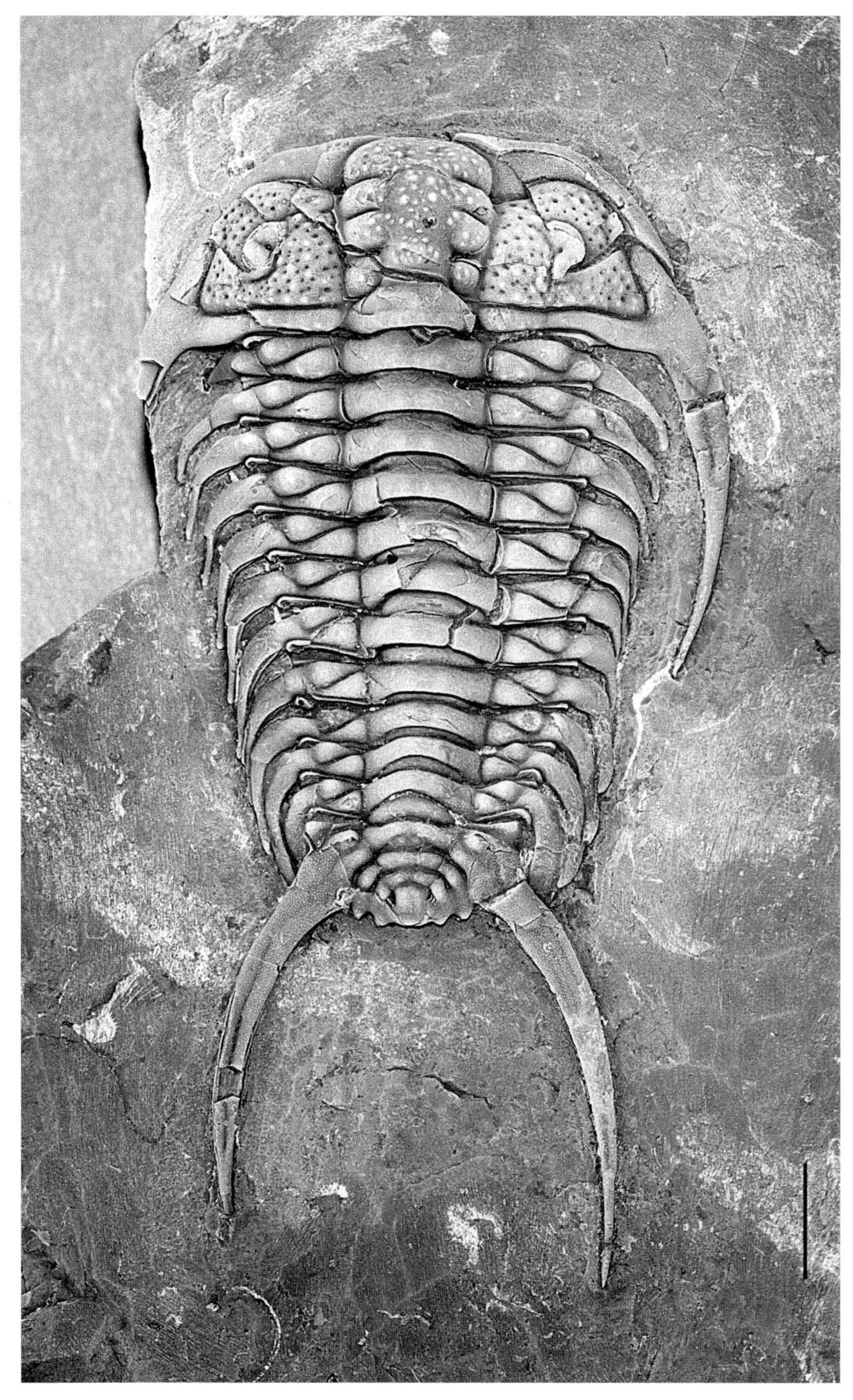

PLATE 82. *Gabriceraurus dentatus* from the Middle Ordovician Kings Falls Limestone, Trenton Group, Ingham Mills, Fulton County. This fairly three-dimensional specimen is 70 mm long. In the collection of Gerald Kloc.

PLATE 83. *Gabriceraurus dentatus* from the Middle Ordovician Lindsay Formation, Trenton Group, Colborne Quarry, Colborne, Ontario, Canada. This specimen is 69 mm long. In the collection of Kevin Brett.

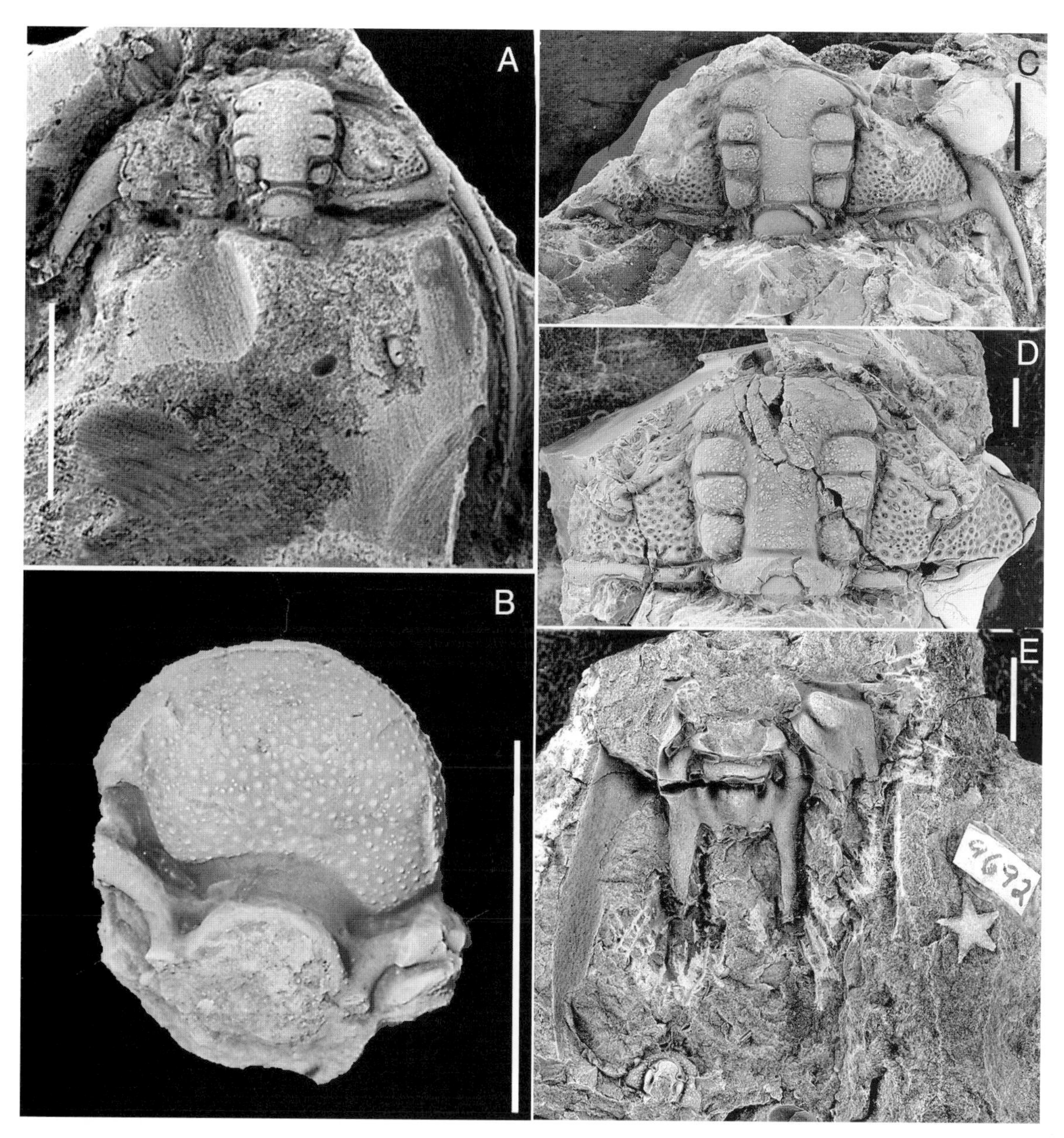

PLATE 84. Cheirurids from New York for which articulated specimens are not known. A. *Gabriceraurus hudsoni* from the Middle Ordovician Chazy Group, Valcour Island, Clinton County. YPM 23296. B. *Deiphon* species from the Lower Silurian, possibly at Lockport, Niagara County. The figured glabella is 10 mm wide. PRI 49626. C–E. *Cerauropeltis ruedemanni* from the Chazy Group limestones near Chazy, Clinton County. C. MCZ 7701. D. MCZ 1612. E. NYSM 9692 (holotype). The negatives for A and C–E were supplied by Dr. Fredrick Shaw, Lehmann University.

PLATE 85. *Sphaerocoryphe robusta* from the Middle Ordovician Rust Formation, Trenton Group, Walcott-Rust Quarry, Herkimer County. This small but interesting trilobite generally is rarely found, but a fair number were found at the Walcott-Rust Quarry, probably because of the area quarried. This specimen is 13 mm long. MCZ 110901.

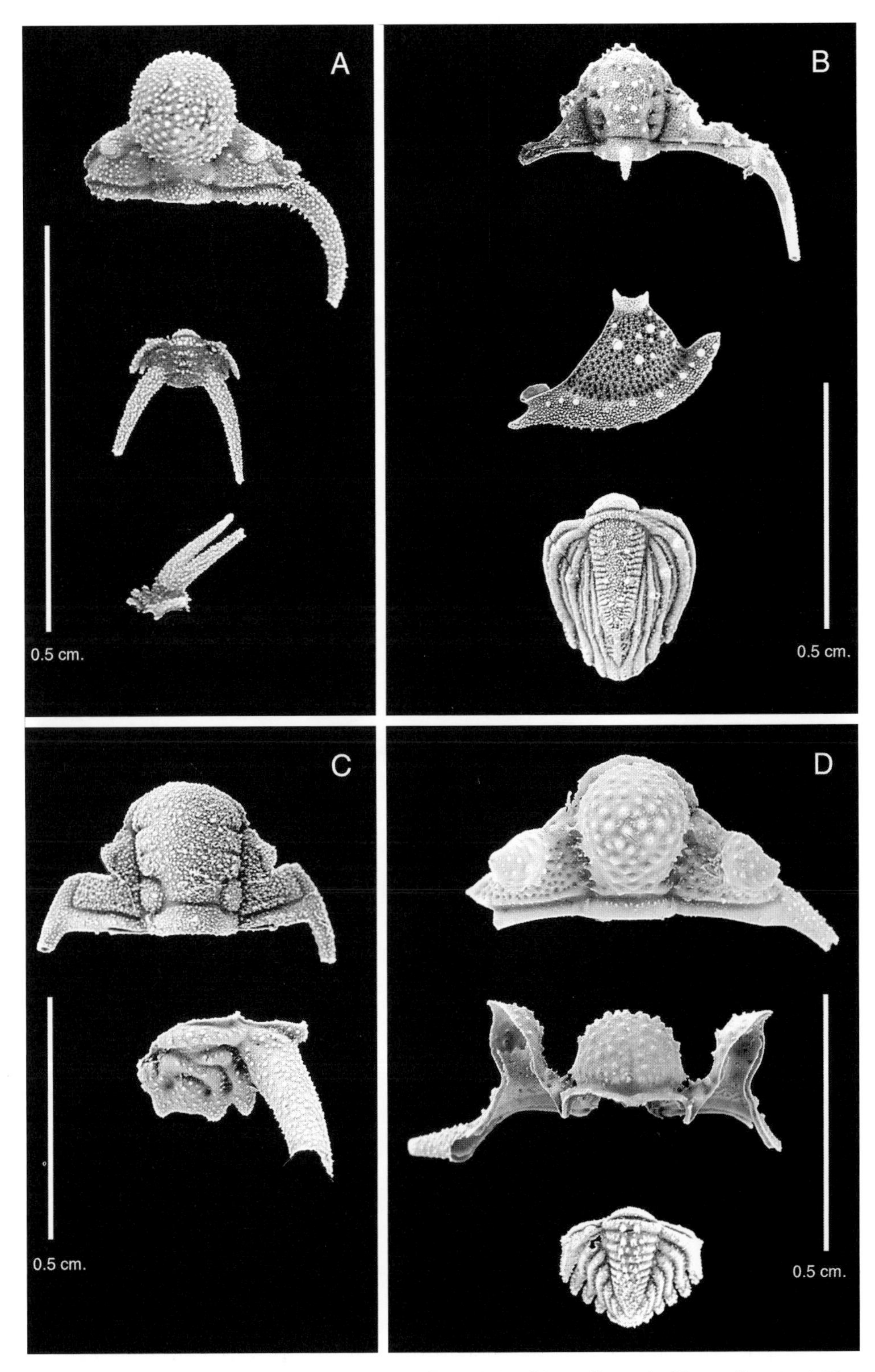

PLATE 86. Silicified trilobites from the Middle Ordovician Chazy Group in Clinton County. A. The cephalon (NYSM 12404) and two views of the pygidium (NYSM 12402) of *Sphaerocoryphe goodnovi* from Chazy. B. The cranidium (NYSM 12317), free cheek (NYSM 12325), and pygidium (NYSM 12318) of *Cybeloides prima* from the Crown Point Formation, Valcour Island. C. The cranidium (NYSM 12442) and pygidium (NYSM 12450) of *Ceraurinella latipyga* from the Crown Point Formation, Valcour Island. D. Two views of the cranidium (NYSM 12331) and one of the pygidium (NYSM 12330) of *Physemataspis insularis* from the Crown Point Formation, Valcour Island. The negatives for all these pictures were supplied by Dr. Fred Shaw, Lehmann University.

PLATE 87. *Dalmanites aspinosus* from the Silurian Keyser Formation in Flintstone Maryland. This trilobite is also reported in New York. The pygidium is 30 mm long. In the collection of Gerald Kloc.

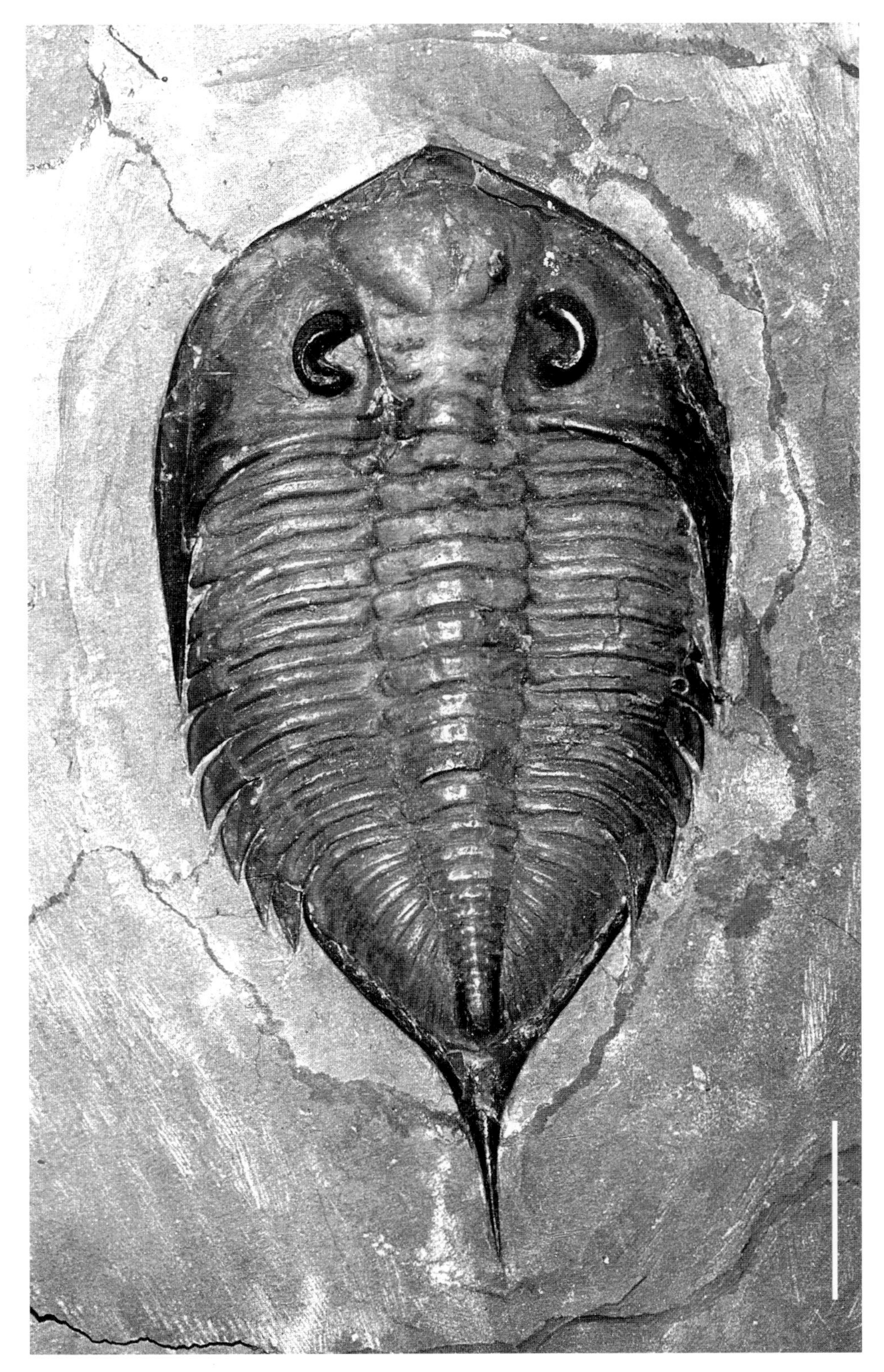

PLATE 88. *Dalmanites limulurus* from the Lower Silurian Rochester Shale in Brockport, Monroe County. The actual specimen is stained red by iron minerals. It is 60 mm long. In the collection of Gerald Kloc.

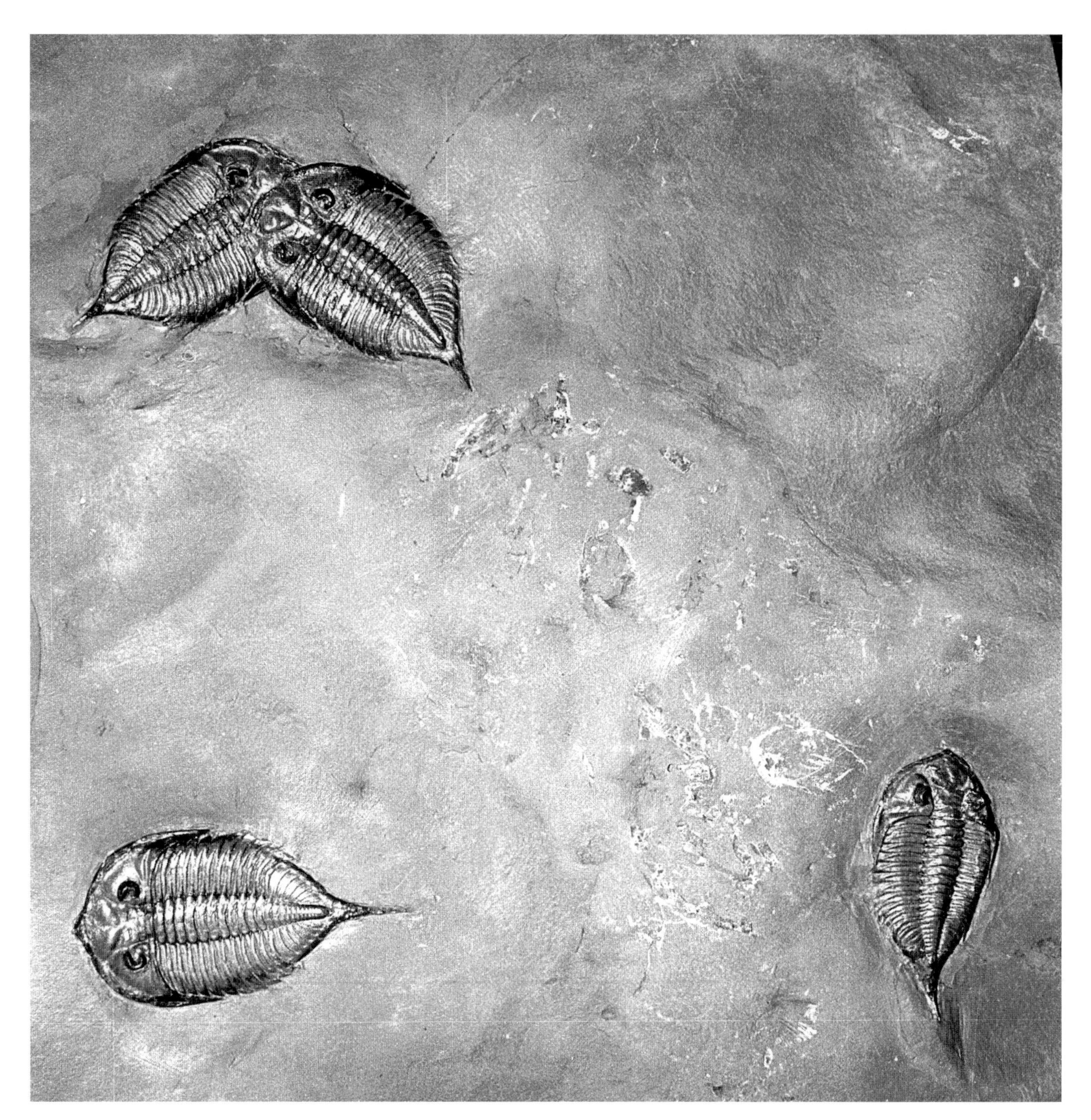

PLATE 89. Four specimens of *Dalmanites limulurus* from the Lower Silurian Rochester Shale in a commercial quarry near Middleport, Orleans County. In the collection of Kent Smith.

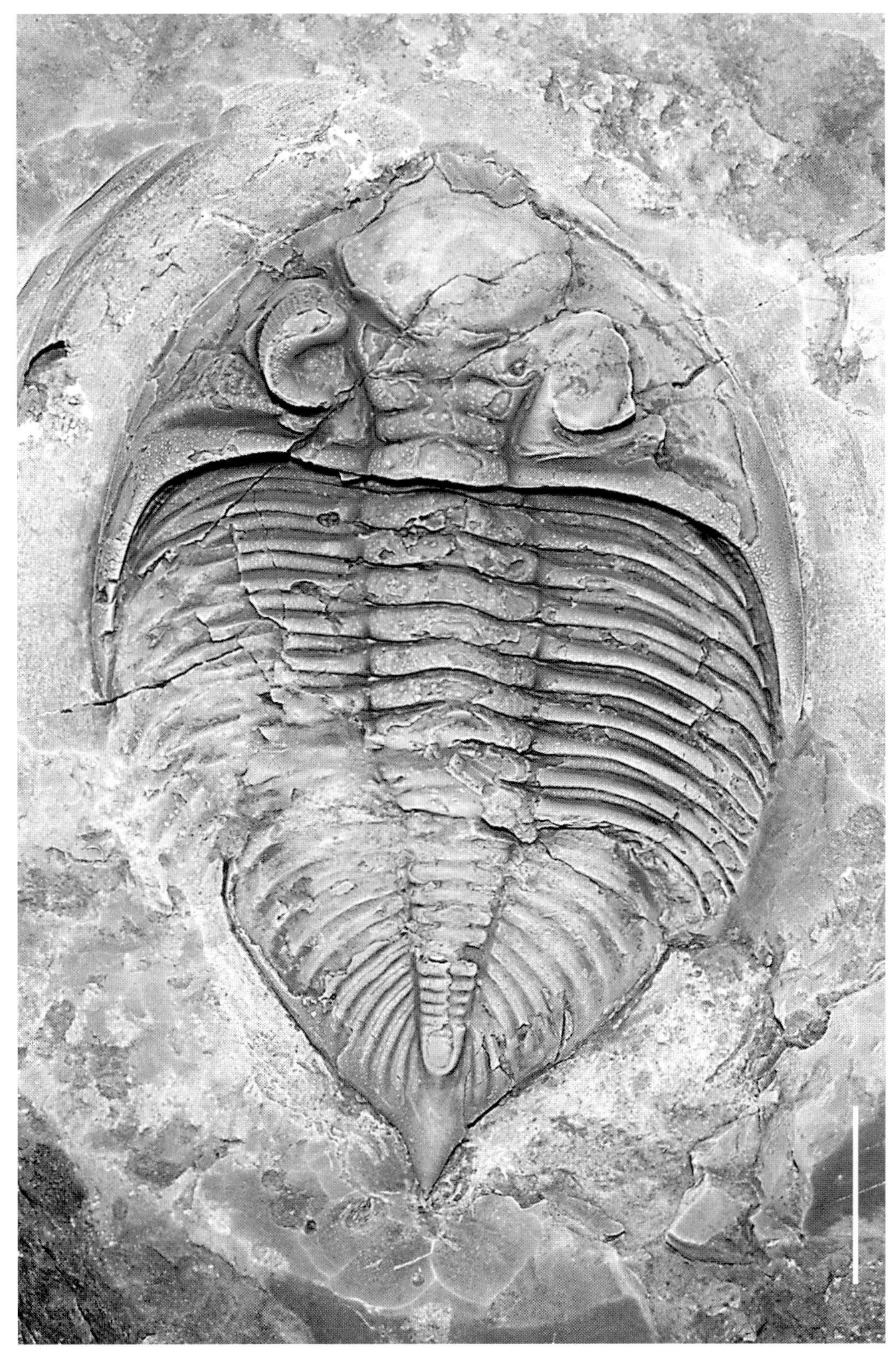

PLATE 90. *Dalmanites pleuroptyx* from the Lower Devonian Helderberg Group at Slingerlands near Clarksville, Albany County. The specimen is 60 mm long. NYSM 17003.

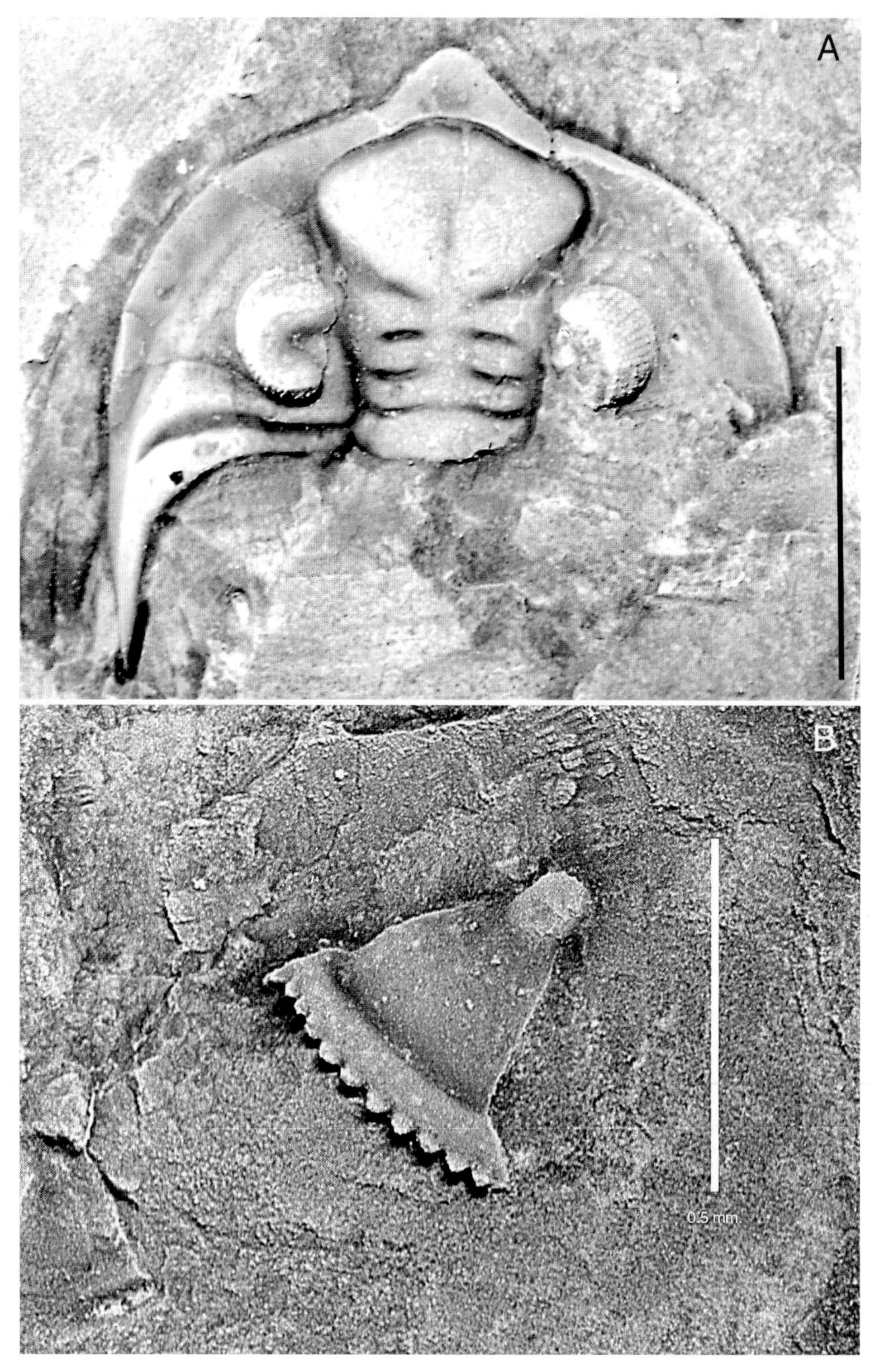

PLATE 91. A. *Dalmanites* species from the Upper Silurian Lockport Group on an unnamed creek near Sodus, Wayne County. This unusual cephalon is 21 mm wide. In the collection of Gerald Kloc. B. *Staurocephalus* species from the Lower Silurian Rochester Shale on an unnamed creek near Sodus, Wayne County. This trilobite is rare, possibly due to the small size of the sclerites. In the collection of Gerald Kloc.

PLATE 92. Pygidium of *Odontochile micrurus* from the Lower Devonian at a Jerusalem Hill quarry on Albany Road, Herkimer County. PRI 49627.

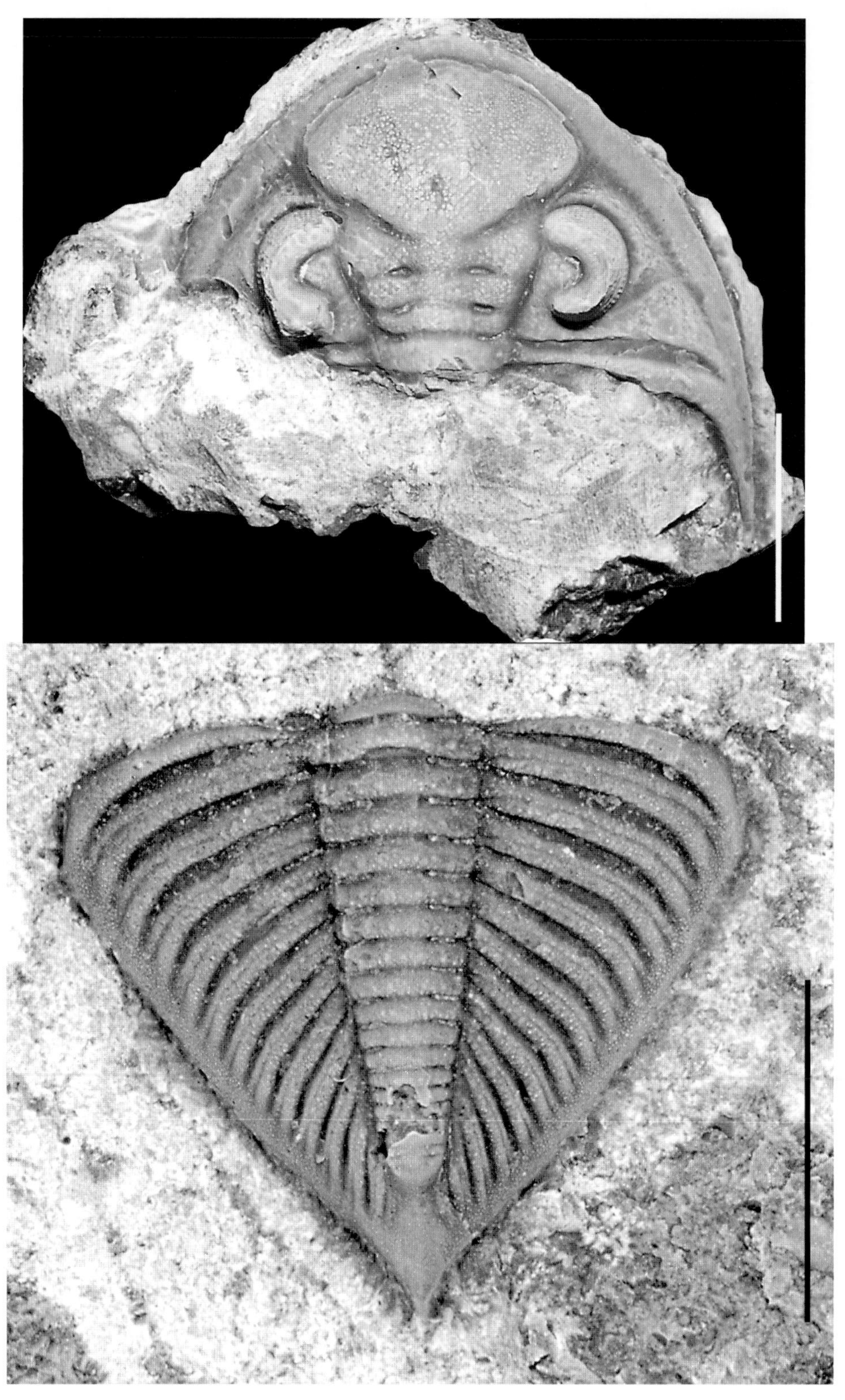

PLATE 93. A cephalon and pygidium from an odontochilid from the Lower Devonian, upper Manlius Limestone, Helderberg Group, Cobleskill, Schoharie County. The cephalon is 16 mm long and the pygidium is 18 mm long. Both in the collection of Gerald Kloc.

PLATE 94. *Erratencrinurus vigilans* from the Middle Ordovician at the Brechin Quarry, Brechin, Ontario. This trilobite is known from the lower Trenton Group in New York, but this is a particularly good specimen. This specimen is 14 mm long. RMSC 2001.46.1.

PLATE 95. *Encrinurus* cf. *E. raybesti* from the Lower Silurian Reynales Limestone at Reynolds Basin, Orleans County. This specimen is 31 mm long. PRI 49628.

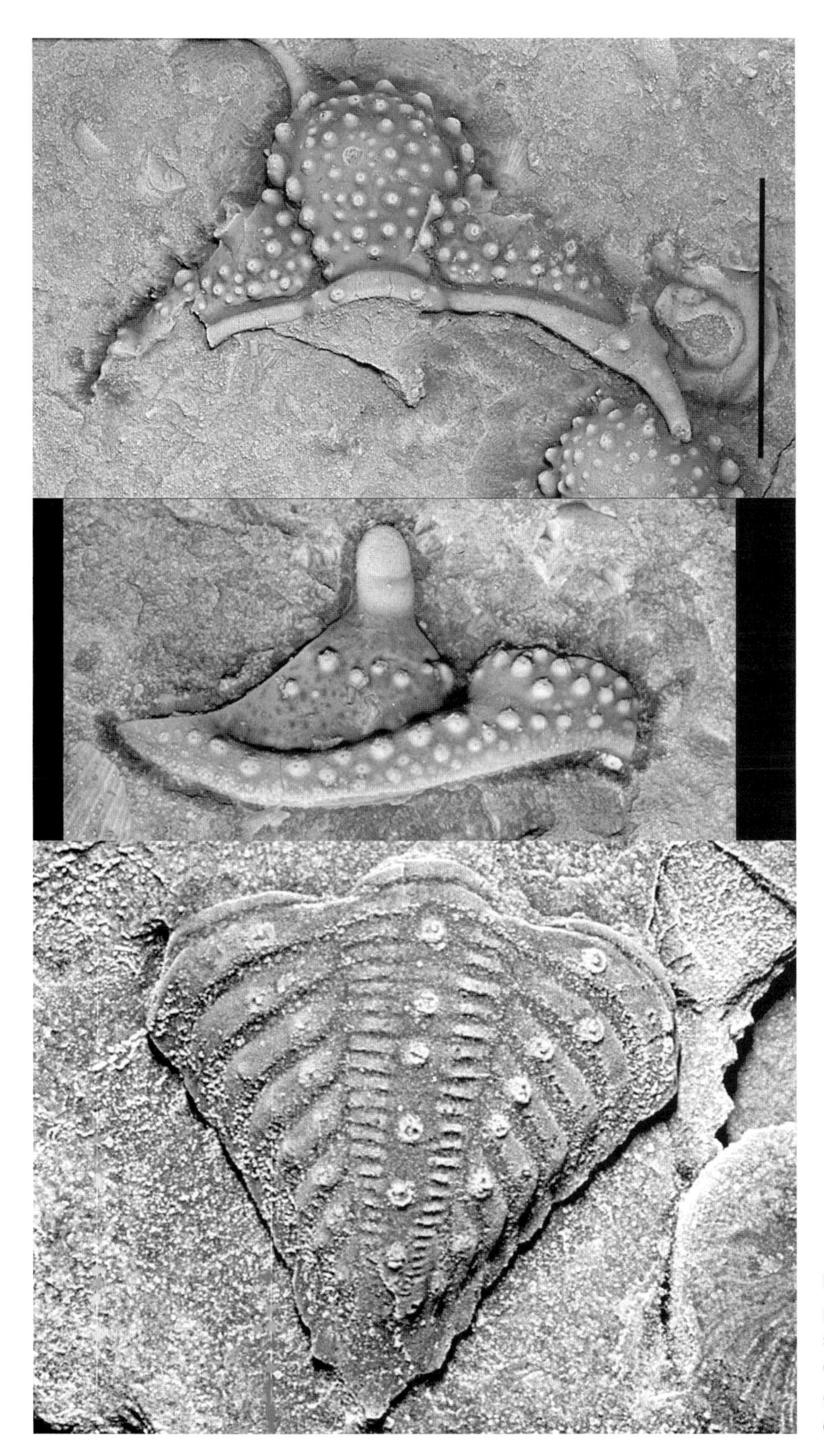

PLATE 96. A cranidium, free cheek, and pygidium of an unidentified *Encrinurus* species from the Upper Silurian Lockport Group in Wayne County. The figured glabella is 7 mm long. All in the collection of Gerald Kloc.

PLATE 97. *Dipleura dekayi* from the Kashong Shale, Hamilton Group, Livingston County. The trilobite shows healed damage to the lower right thorax, possibly due to a predator. This large trilobite is 115 mm long. In the collection of Fred Barber.

PLATE 98. *Dipleura dekayi* from the Middle Devonian Hamilton Group, Interlaken Beach, Seneca County. This large specimen has some exoskeleton broken on the posterior thorax. The broken edges show clearly that the apparent pits on the dorsal surface actually go all the way through the exoskeleton and are holes. The cephalon is 119 mm wide. PRI 49629.

PLATE 99. A *Trimerus delphinocephalus* and a *Dalmanites limulurus* together on one slab from the Lower Silurian Rochester Shale at a commercial quarry near Middleport, Orleans County. The large *T. delphinocephalus* is 172 mm long, and *D. limulurus* is 60 mm long. In the collection of Kent Smith.

PLATE 100. *Eldredgeops crassituberculata* from the Middle Devonian Silica Formation, Silica, Ohio. This trilobite is well known in the Midwest where this specimen was found. It is also reported from New York but is considered uncommon. This specimen is 47 mm long. PRI 49646.

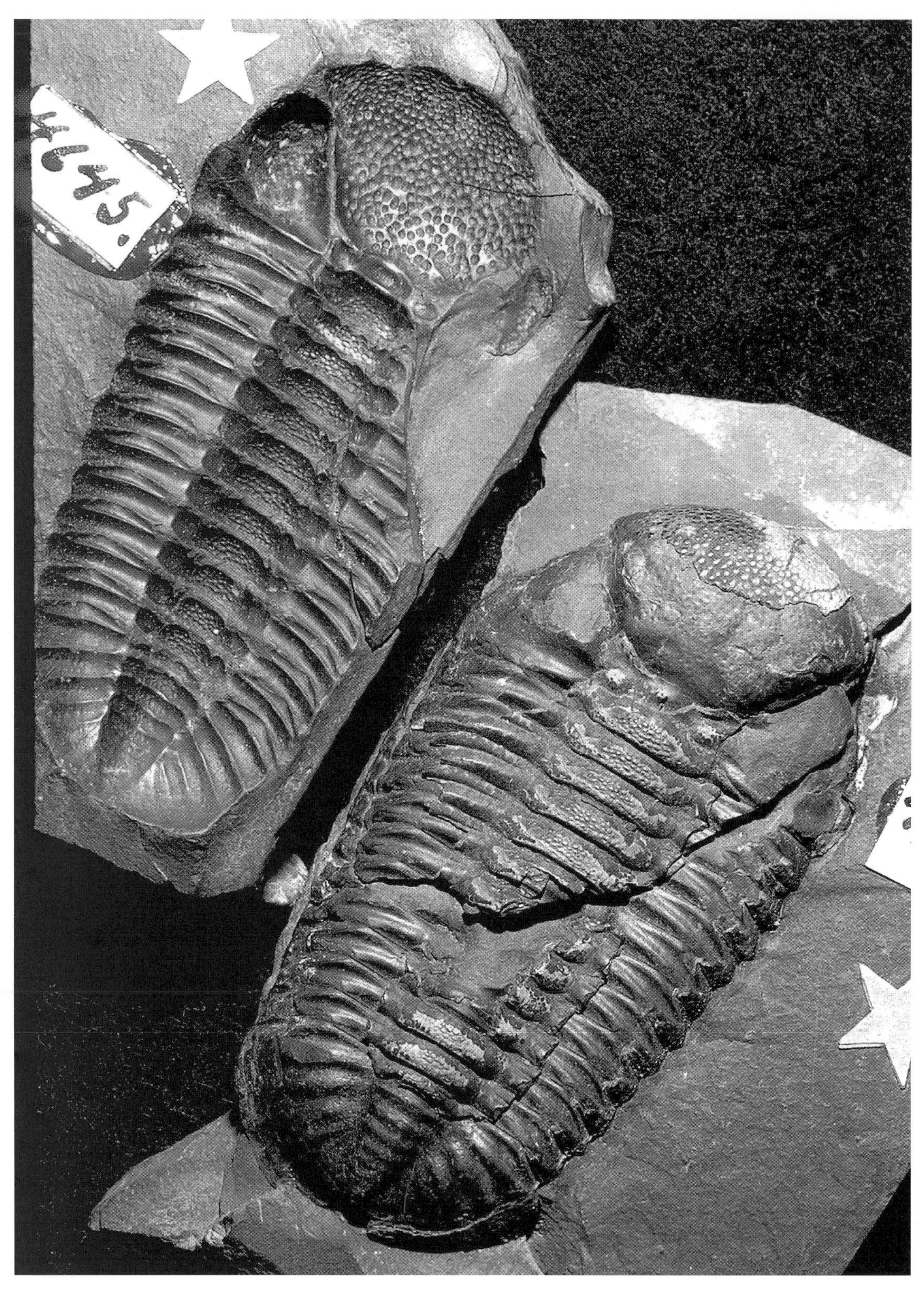

PLATE 101. *Eldredgeops rana* from the Middle Devonian Hamilton Group in Seneca, Ontario County. This specimen has both the part and the counterpart, with the mold to the upper left. NYSM 4645 (holotype).

PLATE 102. *Eldredgeops rana* from the Middle Devonian Windom Member, Hamilton Group, Penn-Dixie Quarry, Hamburg, Erie County. This site is noted for the occasional finding of clusters of trilobites such as this. The *Eldredgeops* specimens in this cluster are relatively small, as are most of them from this site. USNM 403874.

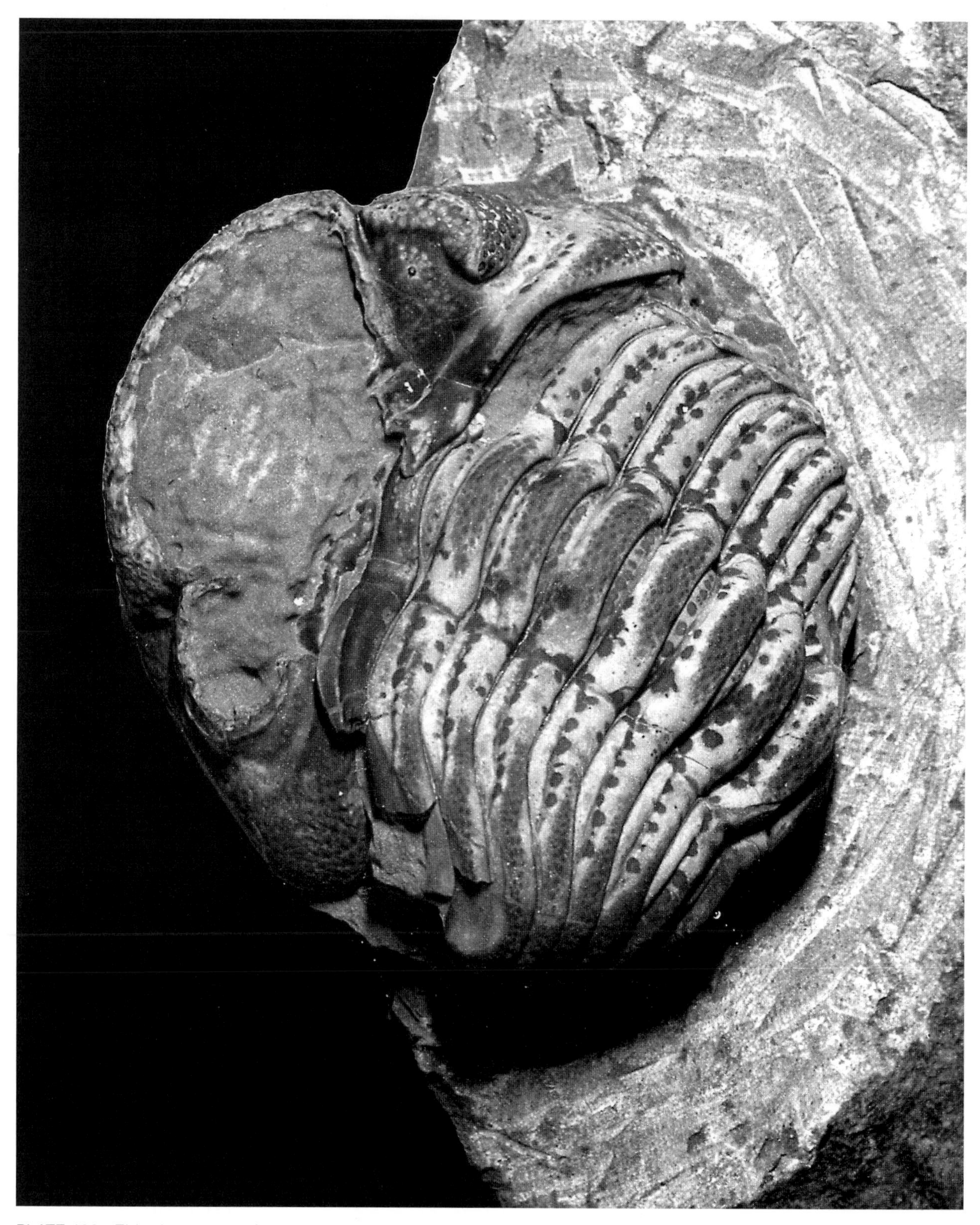

PLATE 103. *Eldredgeops rana* from the Middle Devonian Wanakah Member, Hamilton Group, Darien, Genesee County. This specimen exhibits color markings in non-random patterns on the dorsal exoskeleton. These markings have been interpreted as the positions of internal muscle attachment. PRI 49647.

PLATE 104. *Eldredgeops rana* from the Middle Devonian Wanakah Member, Hamilton Group, Darien, Genesee County. A good view of the dorsal exoskeleton. The specimen is 22 mm long. PRI 49648.

PLATE 105. *Eophacops? trisulcatus* from the Lower Silurian Williamson Member of the Clinton Group at Densmore Creek, Rochester, Monroe County. The preservation of exoskeletal material is poor in these rocks. USNM 79137.

PLATE 106. *Eophacops? trisulcatus* from the Lower Silurian Sodus Shale Member of the Clinton Group at the Genesee Gorge, Rochester, Monroe County. This is a latex mold from a not-too-good mold in shale. It does show a little more detail than Plate 105. NYSM 17041.

PLATE 107. *Paciphacops logani* from the Lower Devonian New Scotland Member of the Helderberg Group at a roadcut on Interstate 88 in Schoharie County. The specimen is 41 mm long. PRI 49649.

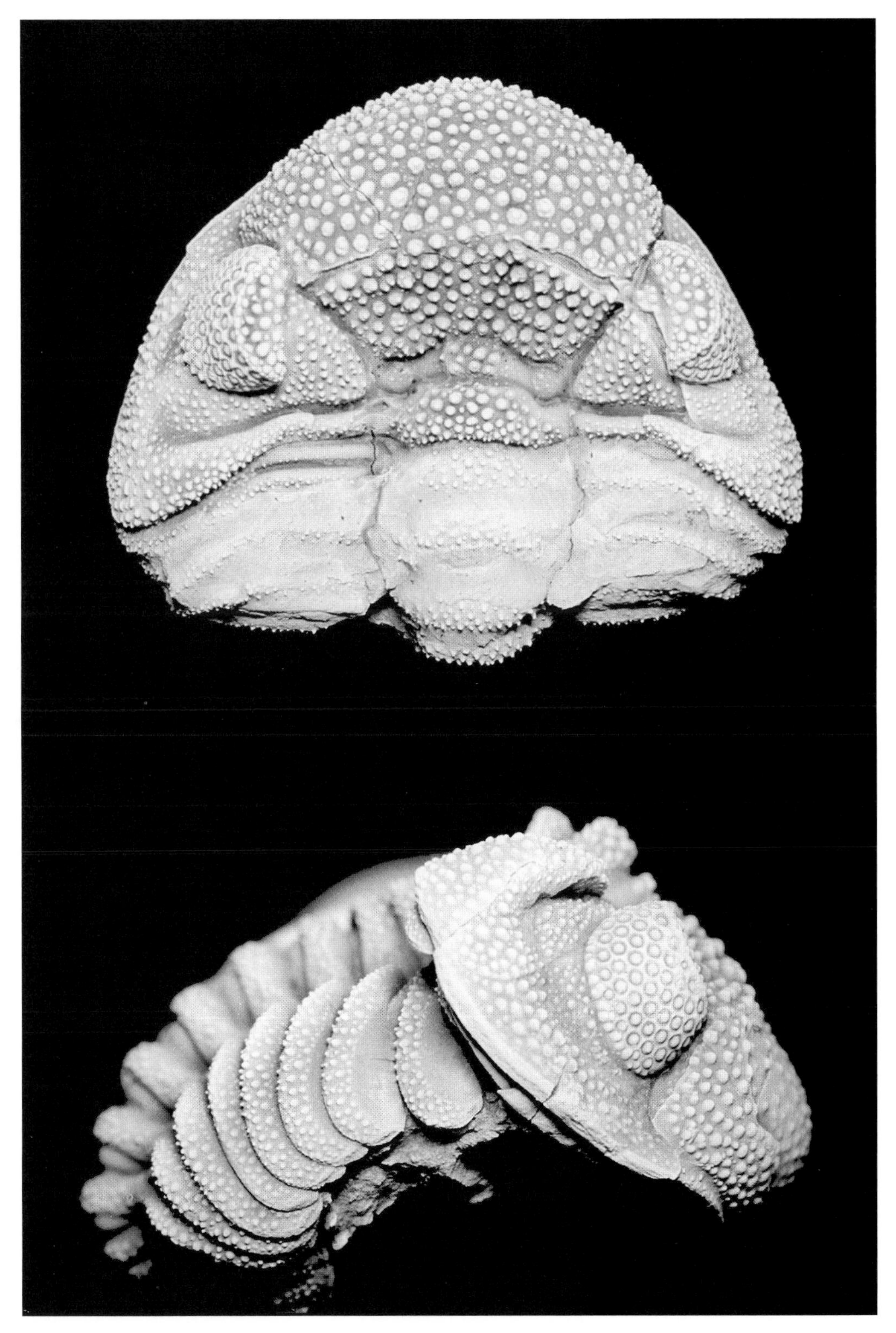

PLATE 108. *Phacops? iowensis* from the Middle Devonian Taunton Beds of the Windom Member, Hamilton Group, Livingston County. This trilobite has only been reported from the upper Windom in New York. In the collection of Steve Pavelsky.

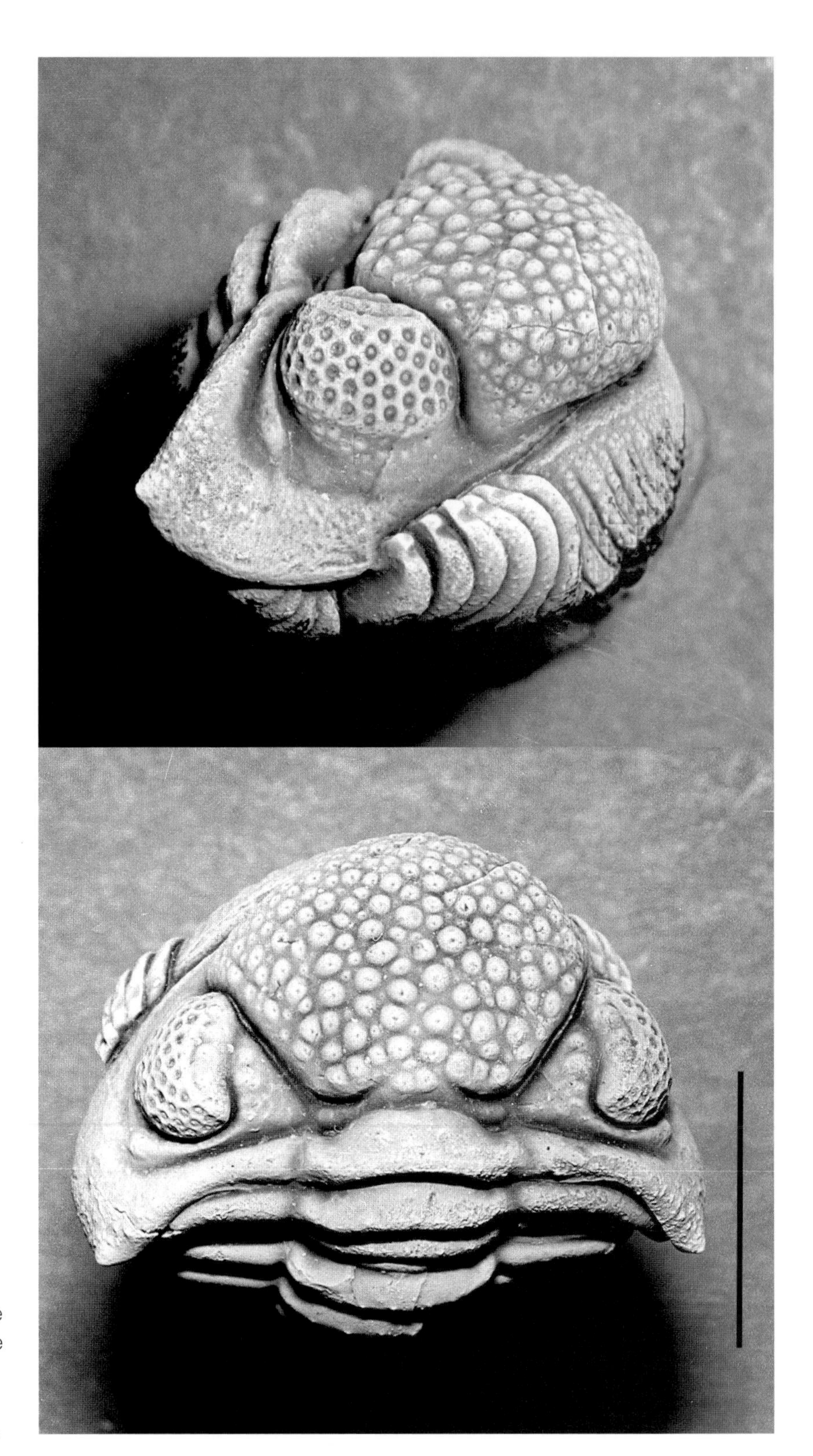

PLATE 109. *Viaphacops bombifrons* from the Middle Devonian Moorehouse Member of the Onondaga Limestone in the Honeyoe Falls Quarry, Monroe County. The specimen is 24 mm wide. In the collection of Gerald Kloc.

A *Paciphacops logani*
B *Viaphacops bombifrons*
C *Eldredgeops rana*
D *Eldredgeops crassituberculata*
E *Eldredgeops norwoodensis*
F *Eldredgeops milleri*
G *Phacops (Eldredgeops?) iowensis*

PLATE 110. A composite plate showing the characteristics of the eyes of seven phacopid trilobites found in New York.

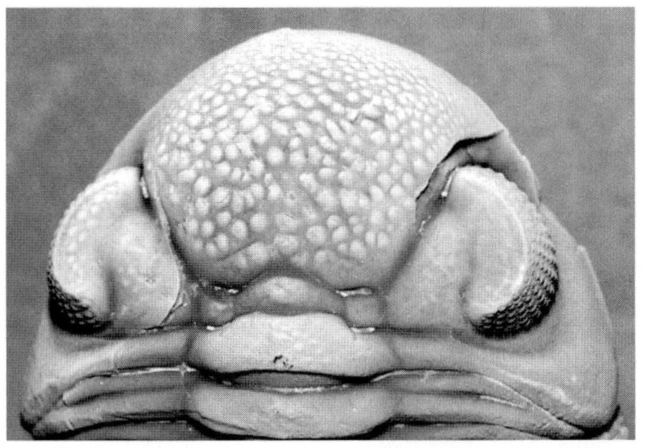

Eldredgeops milleri

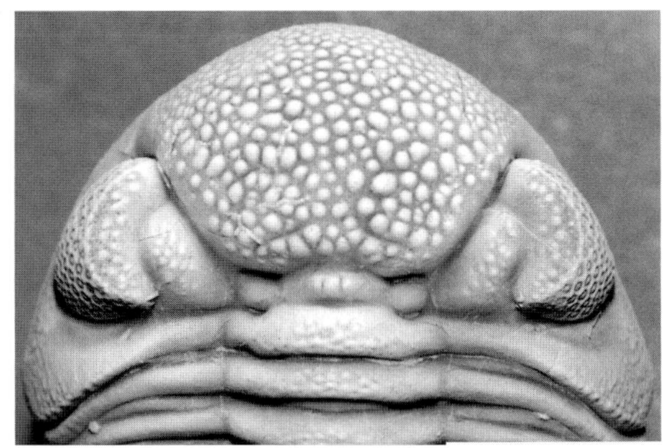

Eldredgeops crassituberculata

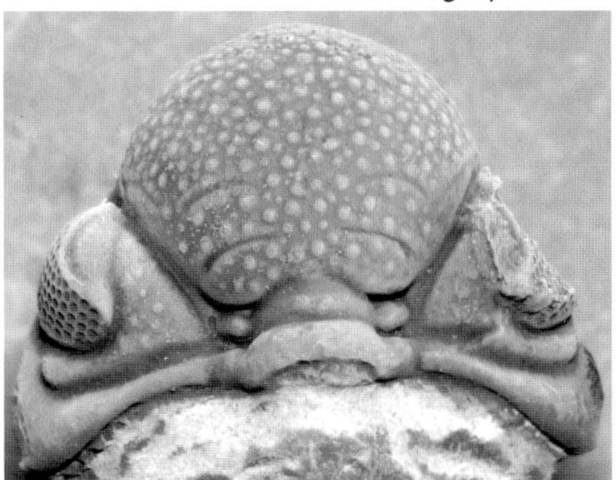

Paciphacops logani

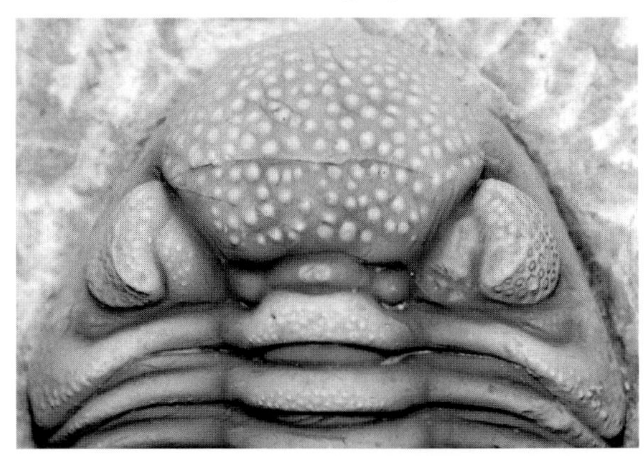

Eldredgeops norwoodensis

Phacops (Eldredgeops?) iowensis

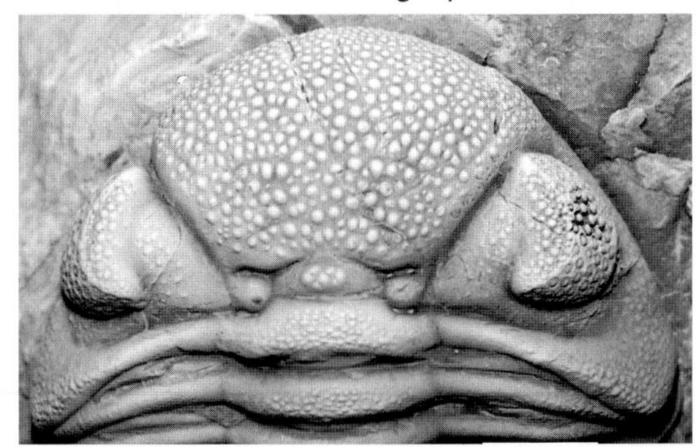

Eldredgeops rana

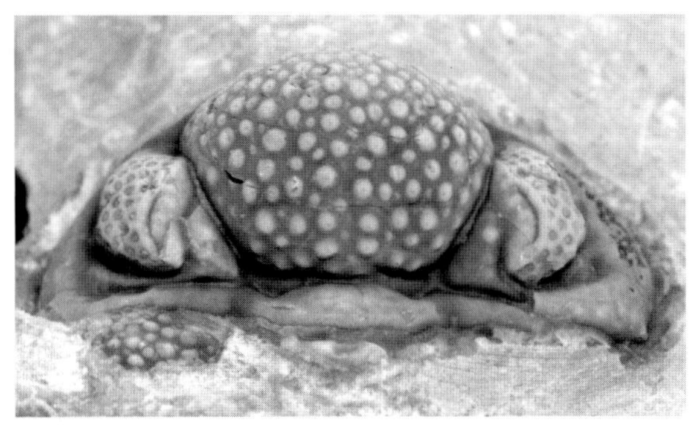

Viaphacops species

Viaphacops bombifrons

PLATE 111. A composite plate showing some dorsal characteristics of New York phacopids. The combination of eye structure and dorsal cephalic structure are often diagnostic to species.

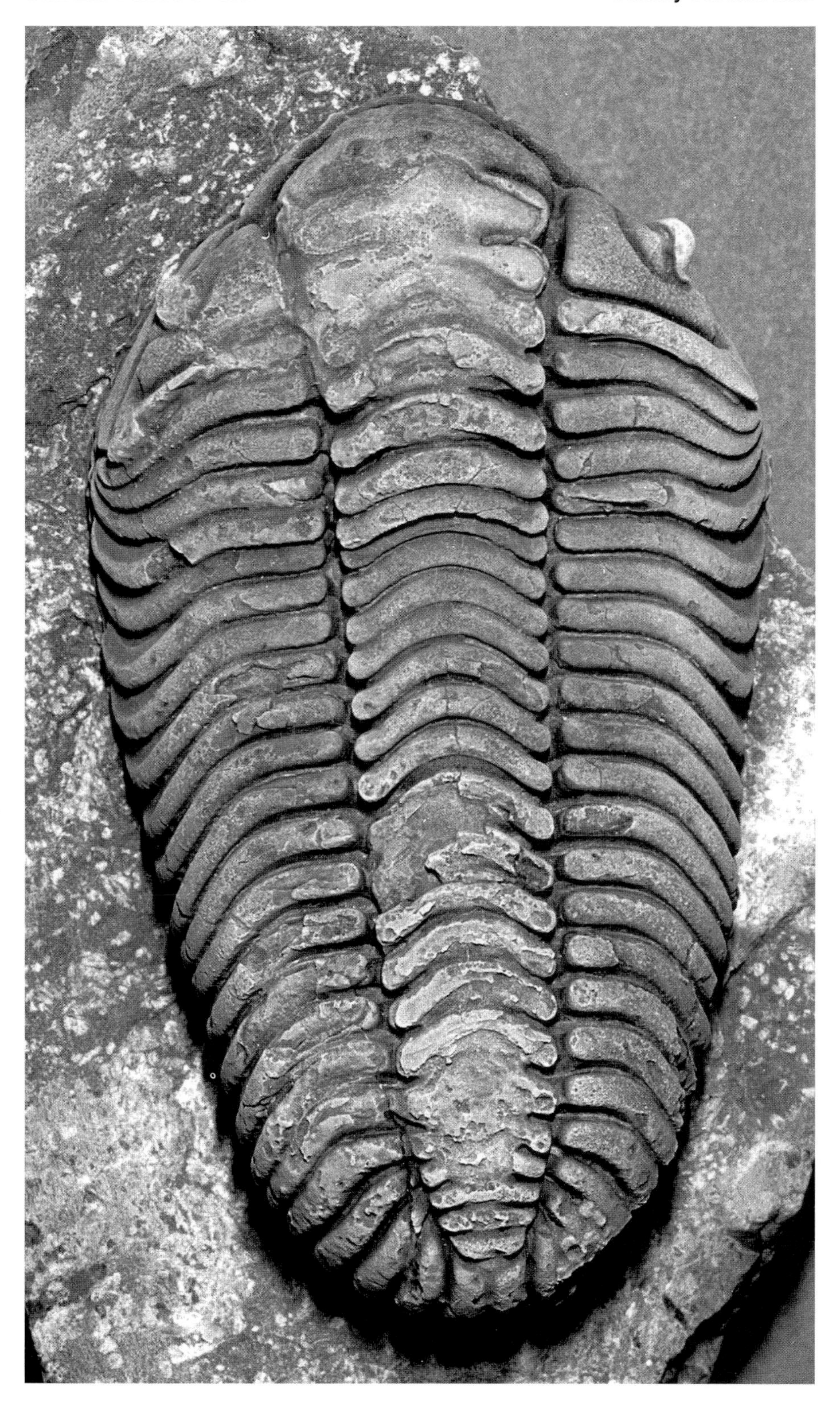

PLATE 112. *Pliomerops canadensis* from the Middle Ordovician Chazy limestones in Clinton County. Articulated trilobites are rare in the Chazy because of the high-energy conditions of these limestones. MCZ 6920.

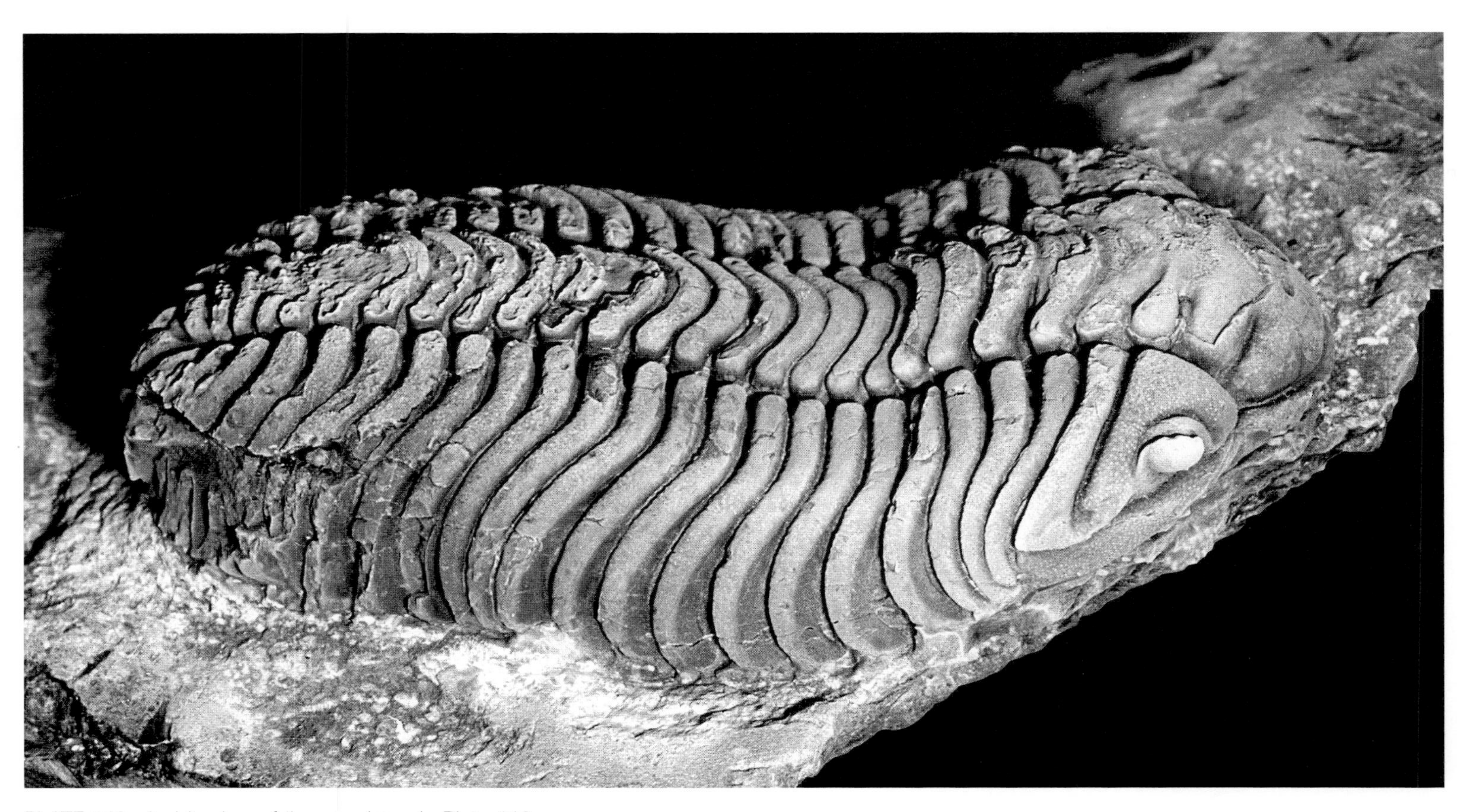

PLATE 113. A side view of the specimen in Plate 112.

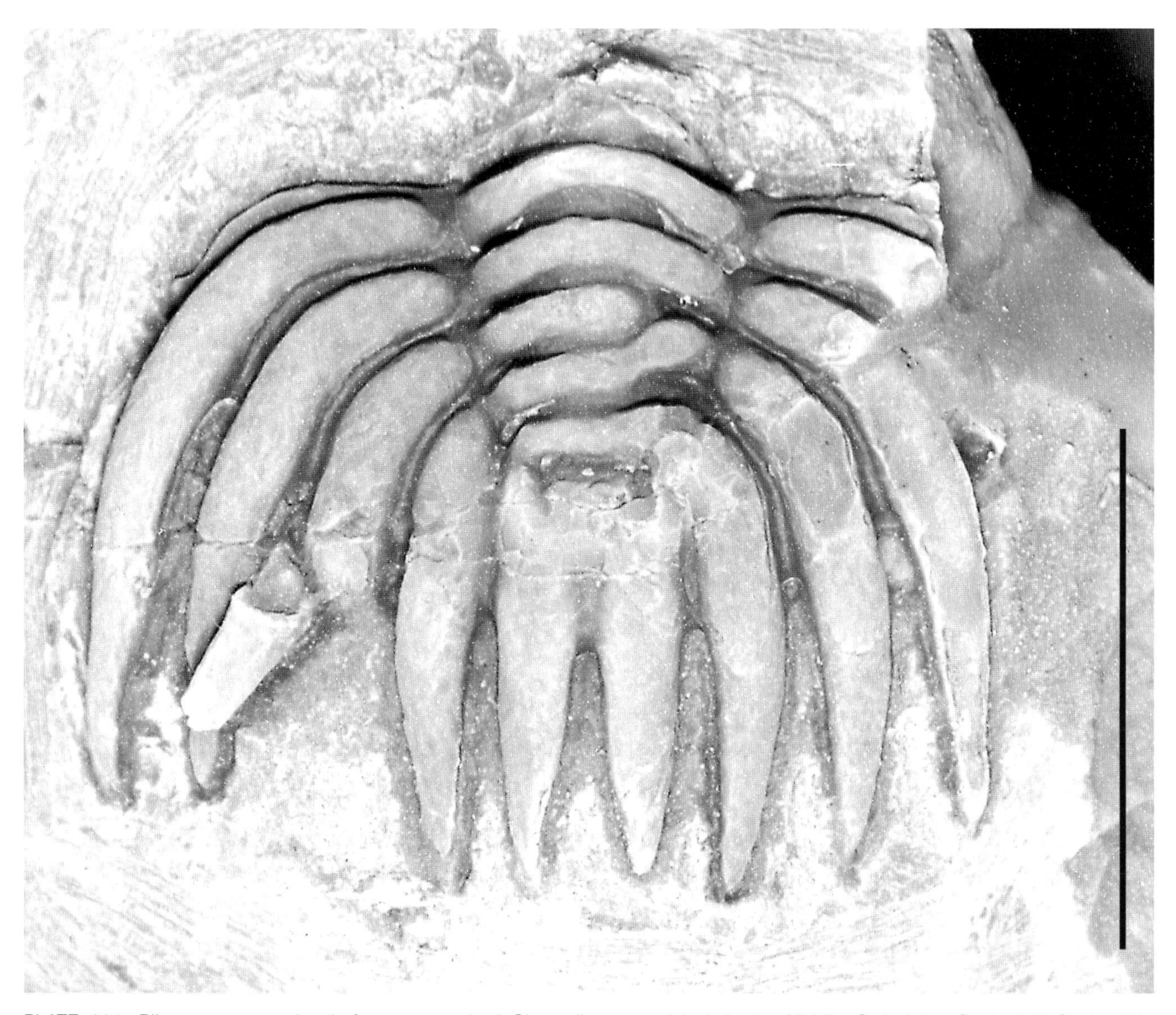

PLATE 114. *Pliomerops canadensis* from a reworked Chazy limestone block in the Middle Ordovician Snake Hill Shale. This pygidium shows an unusual displacement of one of the axial rings, perhaps a healed injury. In the collection of Gerald Kloc.

PLATE 115. *Achatella achates* from the Middle Ordovician Trenton rocks at Trenton Falls, Herkimer County. This specimen is 21 mm long. PRI 49630.

PLATE 116. *Calyptaulax callicephalus* from the Middle Ordovician Rust Formation, Trenton Group, Walcott-Rust Quarry, Herkimer County. This specimen is 28 mm long. MCZ 111712.

PLATE 117. *Calyptaulax callicephalus* from the Middle Ordovician Rust Formation, Trenton Group, Walcott-Rust Quarry, Herkimer County. This specimen shows the ventral exoskeletal anatomy of the trilobite. The specimen is 22 mm long. MCZ 111713.

PLATE 118. *Chasmops? bebryx* from the Middle Ordovician Trenton Group at Jacksonburg, Herkimer County. This trilobite was used by Billings (1863) to illustrate the species, although he was naming a Canadian specimen. It is 39 mm long. PRI 49631 (type).

PLATE 119. *Eomonorachus convexus* from the Middle Ordovician Glens Falls Limestone, Trenton Group, Amsterdam, Montgomery County. The holotype is reported from Trenton Falls, but it is a very uncommon trilobite in the Trenton rocks. PRI 49632.

PLATE 120. Cephalon of *Anchiopella anchiops* from the Lower Devonian Helderberg Group. NYSM 4262.

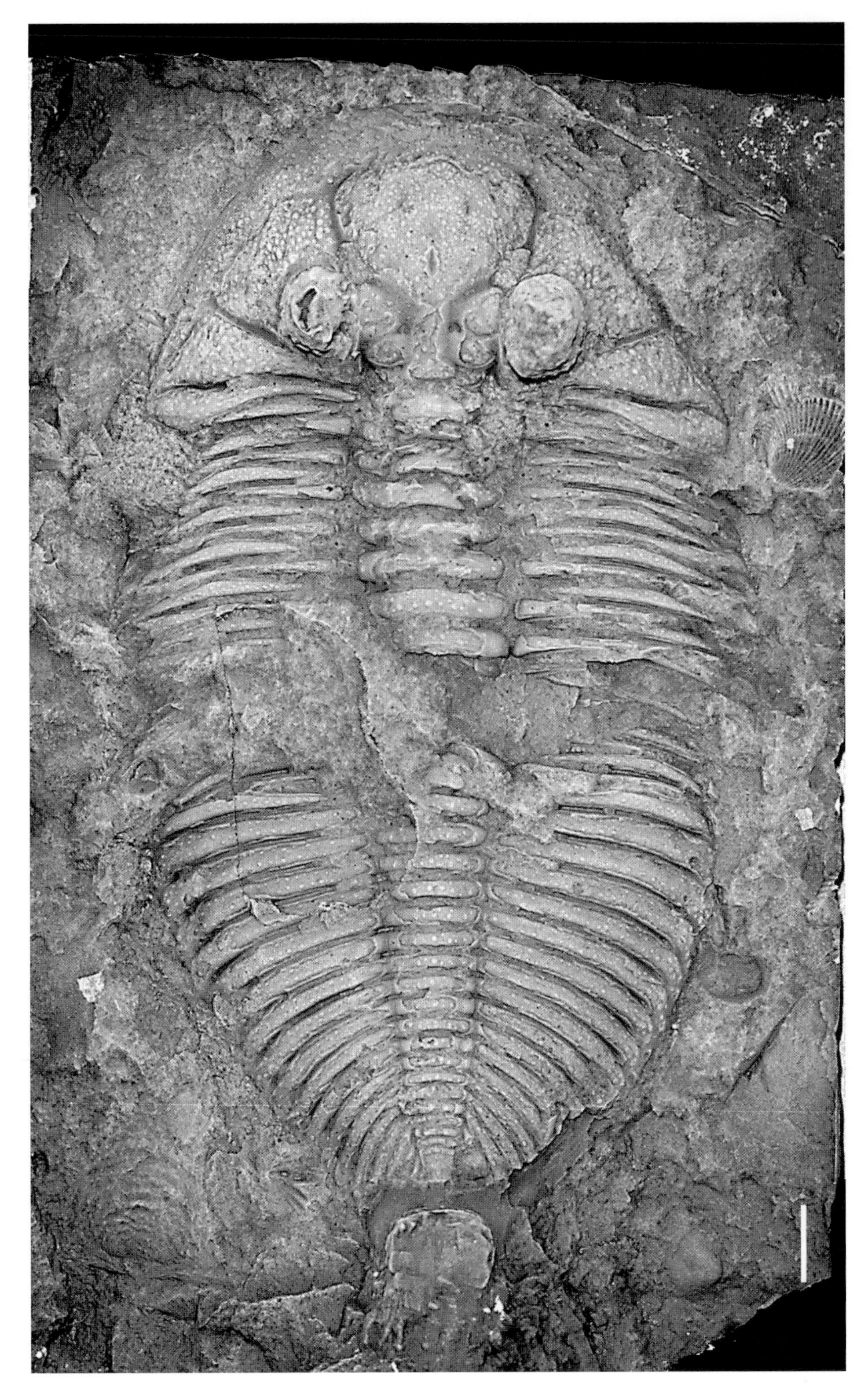

PLATE 121. *Coronura aspectans* from the Middle Devonian Onondaga Limestone at LeRoy, Genesee County. This latex cast of the external mold, or counterpart, is in better condition than the internal mold of the trilobite. The trilobite is 145 mm long. NYSM 4316 (hypotype).

PLATE 122. *Coronura aspectans* from the Middle Devonian Moorehouse Member of the Onondaga Limestone at LeRoy, Genesee County. As one can see from the 1-cm scale bars, the cephalon and pygidium are very different sizes. Both in the collection of Gerald Kloc.

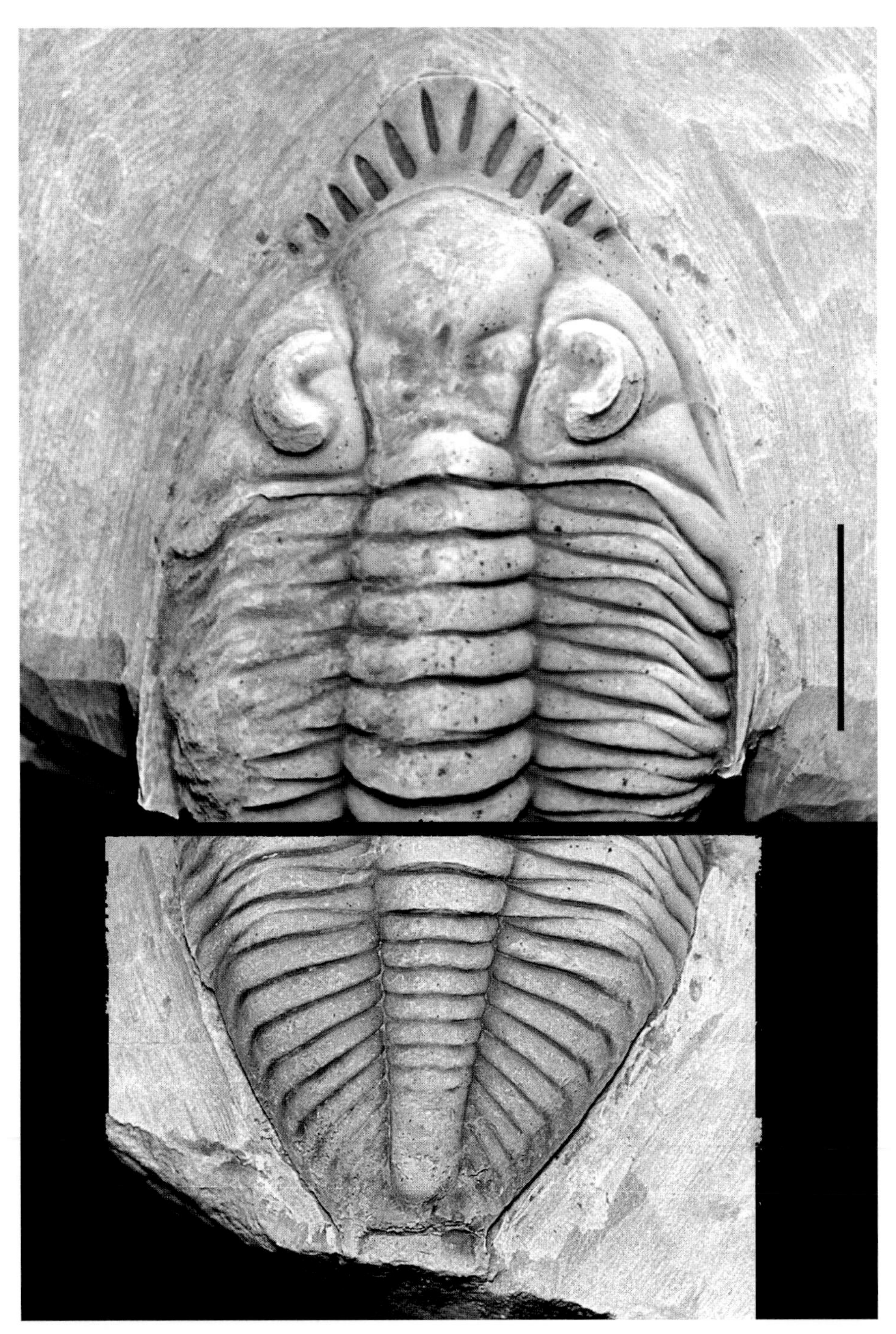

PLATE 123. *Odontocephalus aegeria* from the Middle Devonian Needmore Shale at Cumberland, Maryland. This same species is reported in New York. The plate is two views of the same specimen, as it is semicoiled. The trilobite is 30 mm long. In the collection of Gerald Kloc.

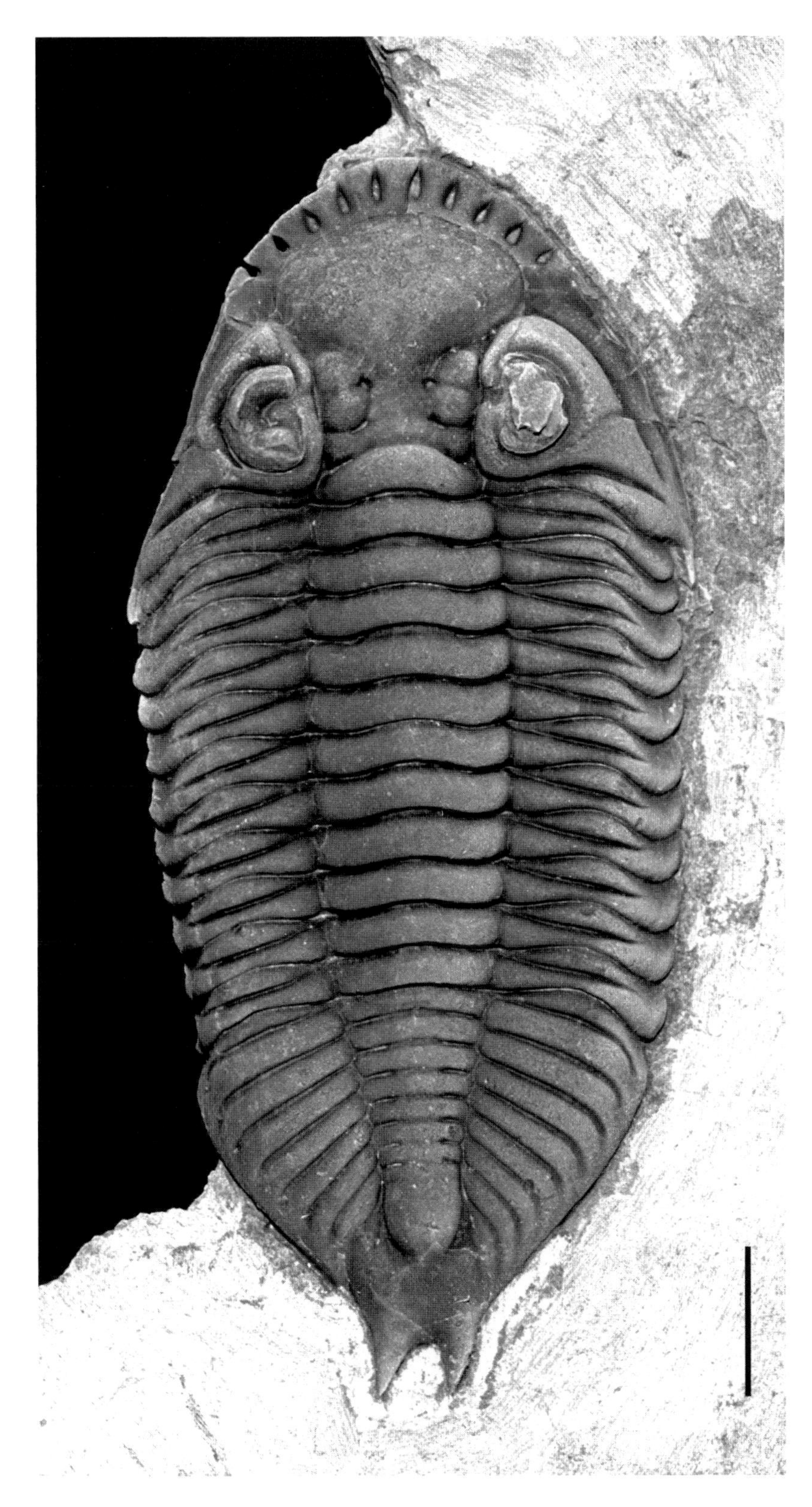

PLATE 124. *Odontocephalus bifidus* from the Middle Devonian Moorehouse Member of the Onondaga Limestone at the LeRoy Quarry, LeRoy, Genesee County. The specimen is 85 mm long. In the collection of Gerald Kloc.

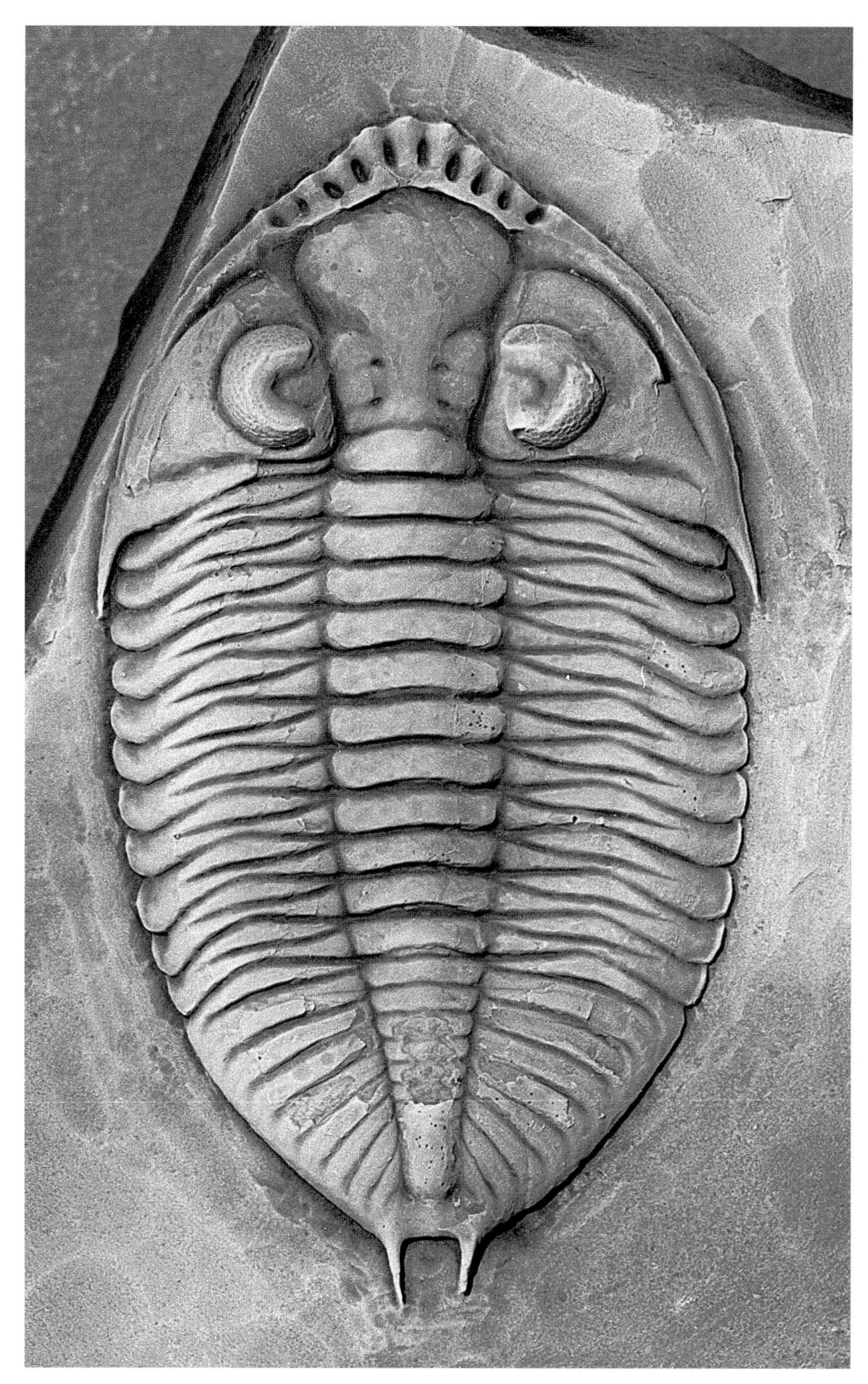

PLATE 125. *Odontocephalus selenurus* from the Middle Devonian Moorehouse Member of the Onondaga Limestone at the Seneca Stone Quarry, Seneca, Seneca County. The trilobite is 41 mm long. In the collection of Gerald Kloc.

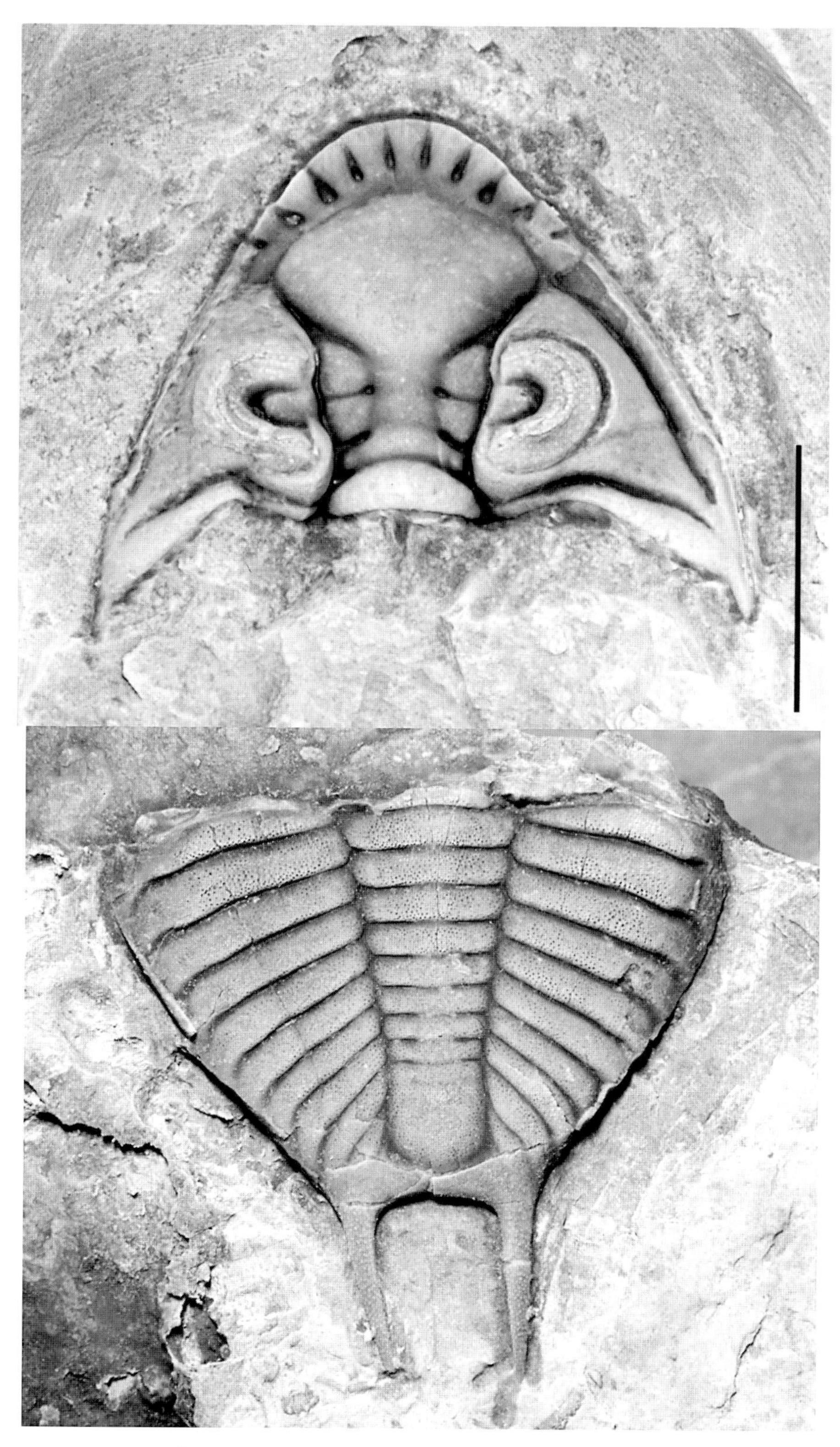

PLATE 126. *Odontocephalus* species from the Middle Devonian Moorehouse Member of the Onondaga Limestone in the LeRoy Quarry, LeRoy, Genesee County. There are enough differences in these specimens from the other *Odontocephalus* species to question whether these are of known species. The cephalon is 25 mm wide. In the collection of Gerald Kloc.

PLATE 127. A. *Trypaulites erinus* found in a loose block, possibly Lower Devonian Bois Blanc Limestone, at Mumford, Monroe County. This pygidium is 17 mm wide. B. *Coronura helena* from the Middle Devonian Moorehouse Member of the Onondaga Limestone. The pygidium was found in a construction excavation at Clarence, Erie County. Both in the collection of Gerald Kloc.

PLATE 128. *Harpidella craspedota* from the Middle Devonian Centerfield Limestone, Hamilton Group, Browns Creek, York, Livingston County. These are small trilobites, about 15 mm long. NYSM 16798.

PLATE 129. *Harpidella spinafrons* from the late Middle Devonian, Tully Limestone at Moravia, Cayuga County. This specimen is 16 mm long. USNM 89980.

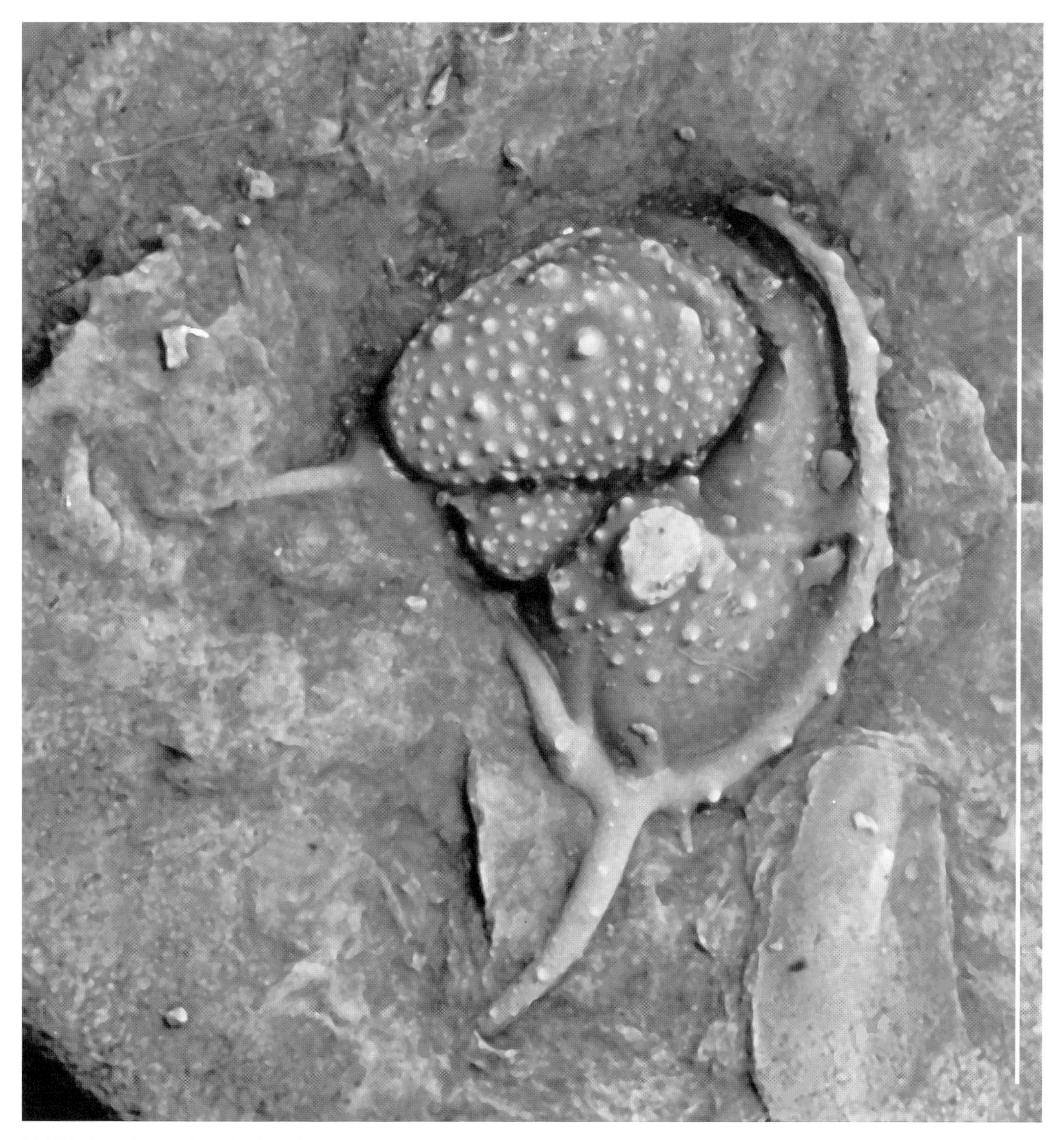

PLATE 130. *Harpidella spinafrons* from the late Middle Devonian, Tully Limestone on West Brook, Sherburne, Chenango County. The original is an external mold and this is a gutta-percha cast. The longest dimension is 8 mm. USNM 516589.

PLATE 131. *Harpidella* species from the Middle Devonian Onondaga Limestone in the Honeoye Falls Quarry, Honeoye Falls, Monroe County. This is a previously unreported species. RMSC 2001.48.1.

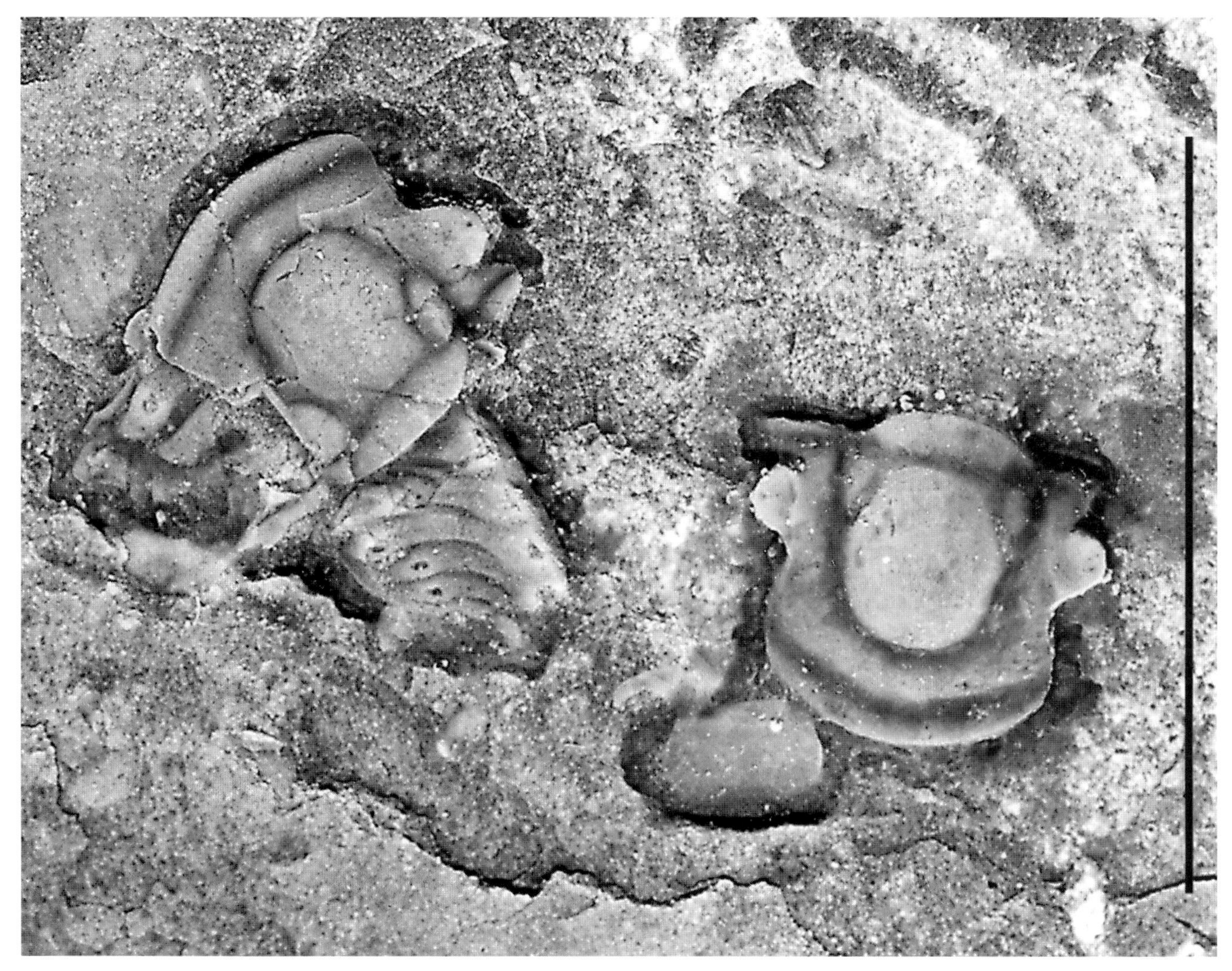

PLATE 132. *Maurotarion* species from the Upper Silurian Lockport Group on an unnamed creek near Sodus, Wayne County. The cranidia are 5 mm wide. In the collection of Gerald Kloc.

PLATE 133. *Bathyurus extans* from the Middle Ordovician Black River Group, Black River, Great Bend, Jefferson County. This specimen is 23 mm long. USNM 92844.

PLATE 134. *Australosutura gemmaea* from the Middle Devonian Windom Member on Fall Brook, Geneseo, Livingston County. This trilobite is rare in articulated form. The specimen is 12mm long. In the collection of James Scatterday.

PLATE 135. *Radnoria* species from the Lower Silurian Rochester Shale. This small trilobite is usually found associated with bryozoan colonies. The specimen is 14 mm long. NYSM 16792.

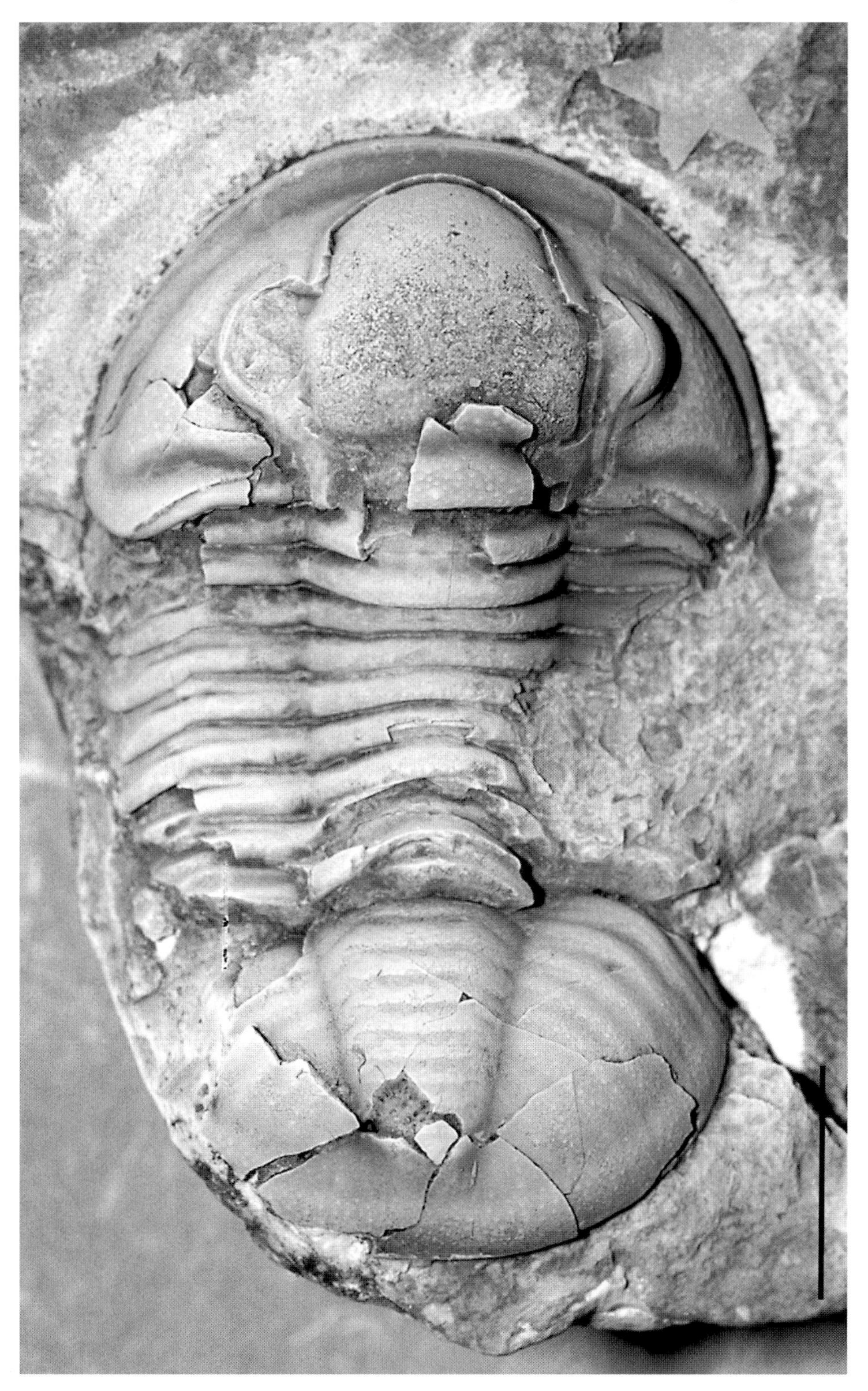

PLATE 136. *Coniproetus folliceps* from the Middle Devonian Onondaga Limestone at LeRoy, Genesee County. This specimen is 48 mm long. NYSM 4722 (lectotype).

PLATE 137. *Cyphoproetus?* cf. *C. wilsonae* from the Middle Ordovician, lower Trenton exposures at Sacketts Harbor, Jefferson County. This rare trilobite is 11.5 mm long. PRI 49633.

PLATE 138. *Dechenella haldemani* from the Middle Devonian Union Springs Member in Cherry Valley, Otsego County. NYSM 6489.

PLATE 139. *Decoroproetus corycoeus* from the Lower Silurian Rochester Shale at Brockport, Monroe County. This uncommon trilobite is 29 mm long. In the collection of Tod Clements.

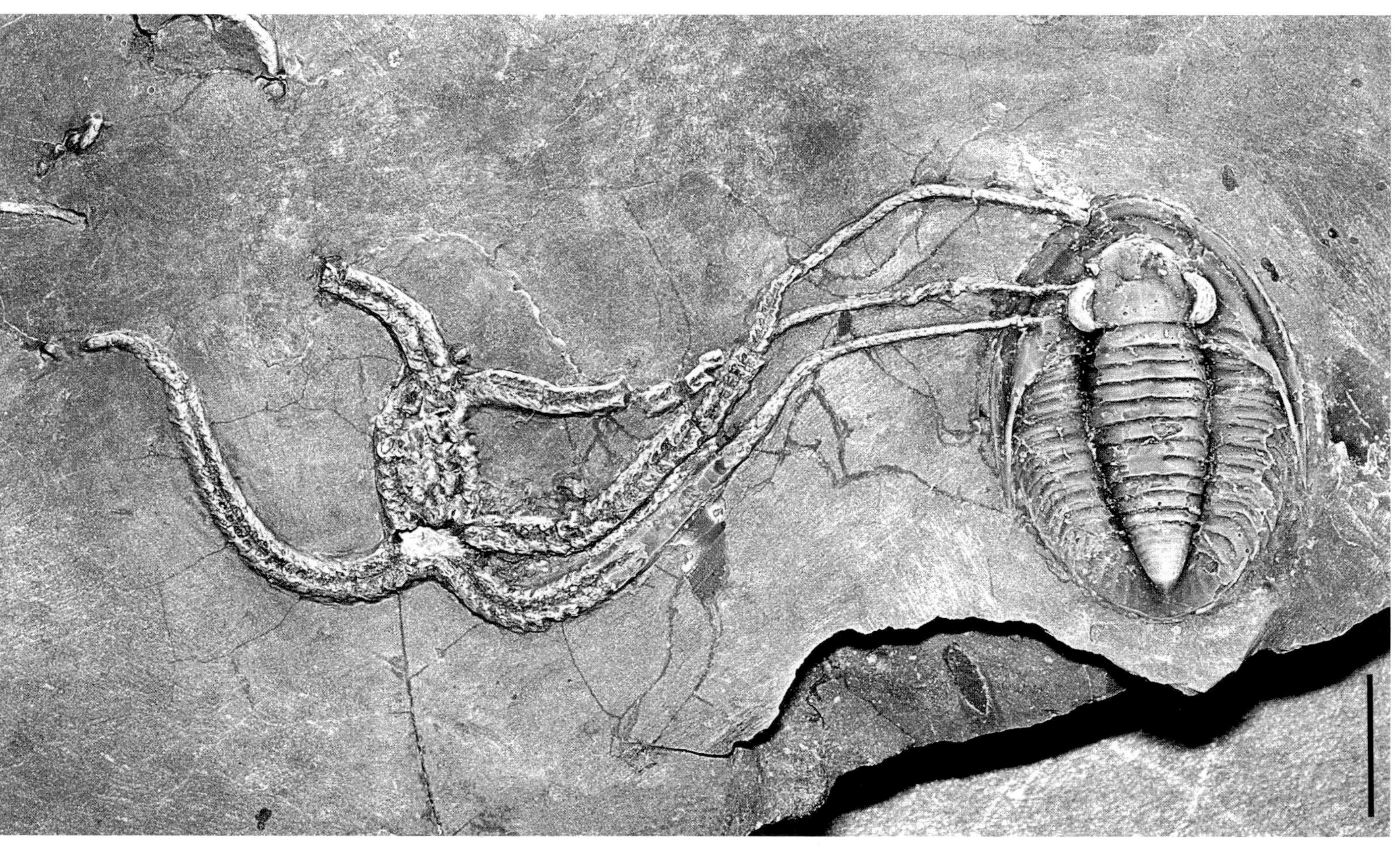

PLATE 140. The same specimen as in Plate 139, showing the associated ophiuroid (brittle star).

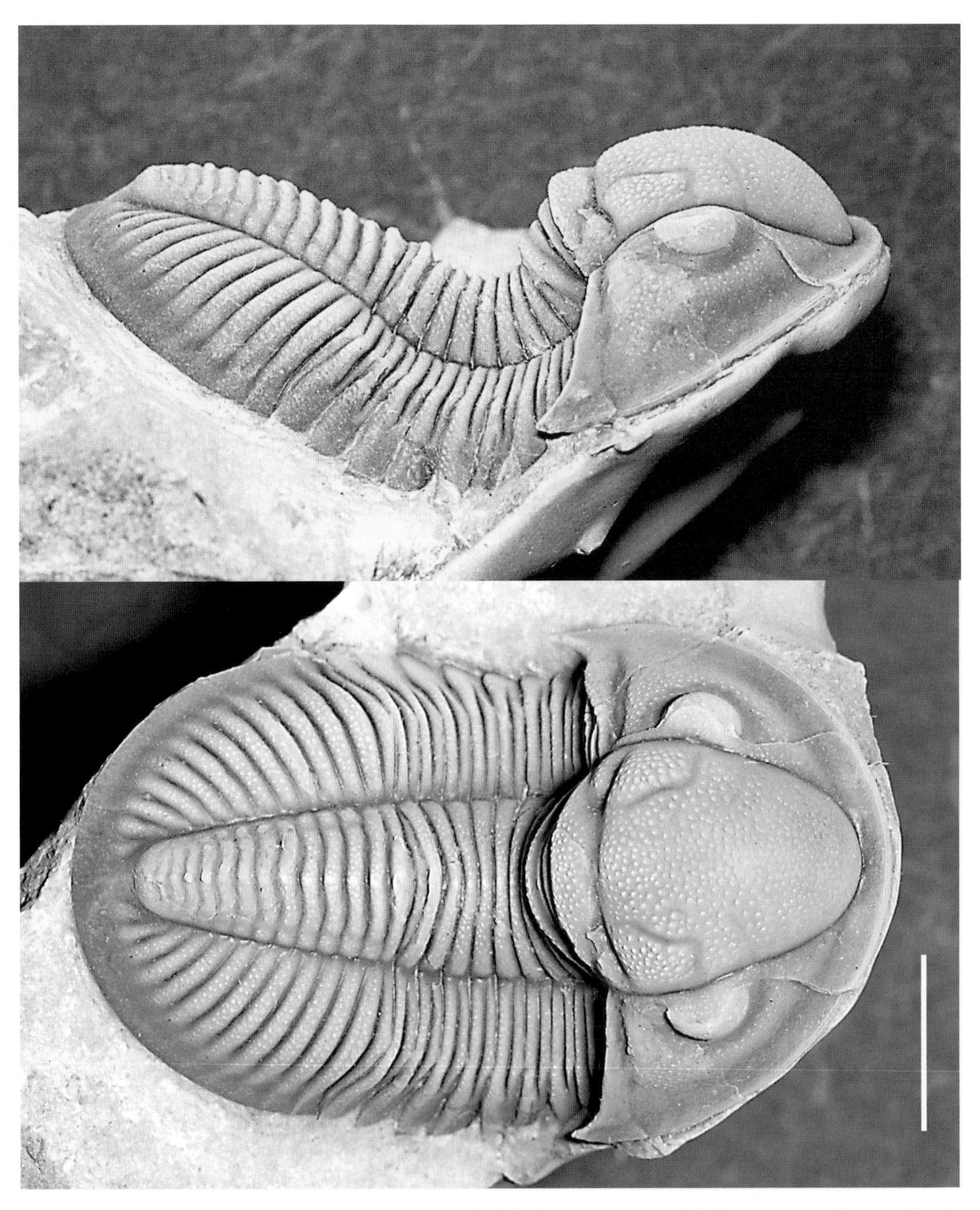

PLATE 141. *Monodechenella macrocephala* from the Middle Devonian Centerfield Limestone at Browns Creek, York, Livingston County. This trilobite is often found curved upward in these beds, and the two views give a good perspective on this preservation. The specimen is 29 mm long. In the collection of Douglas DeRosear.

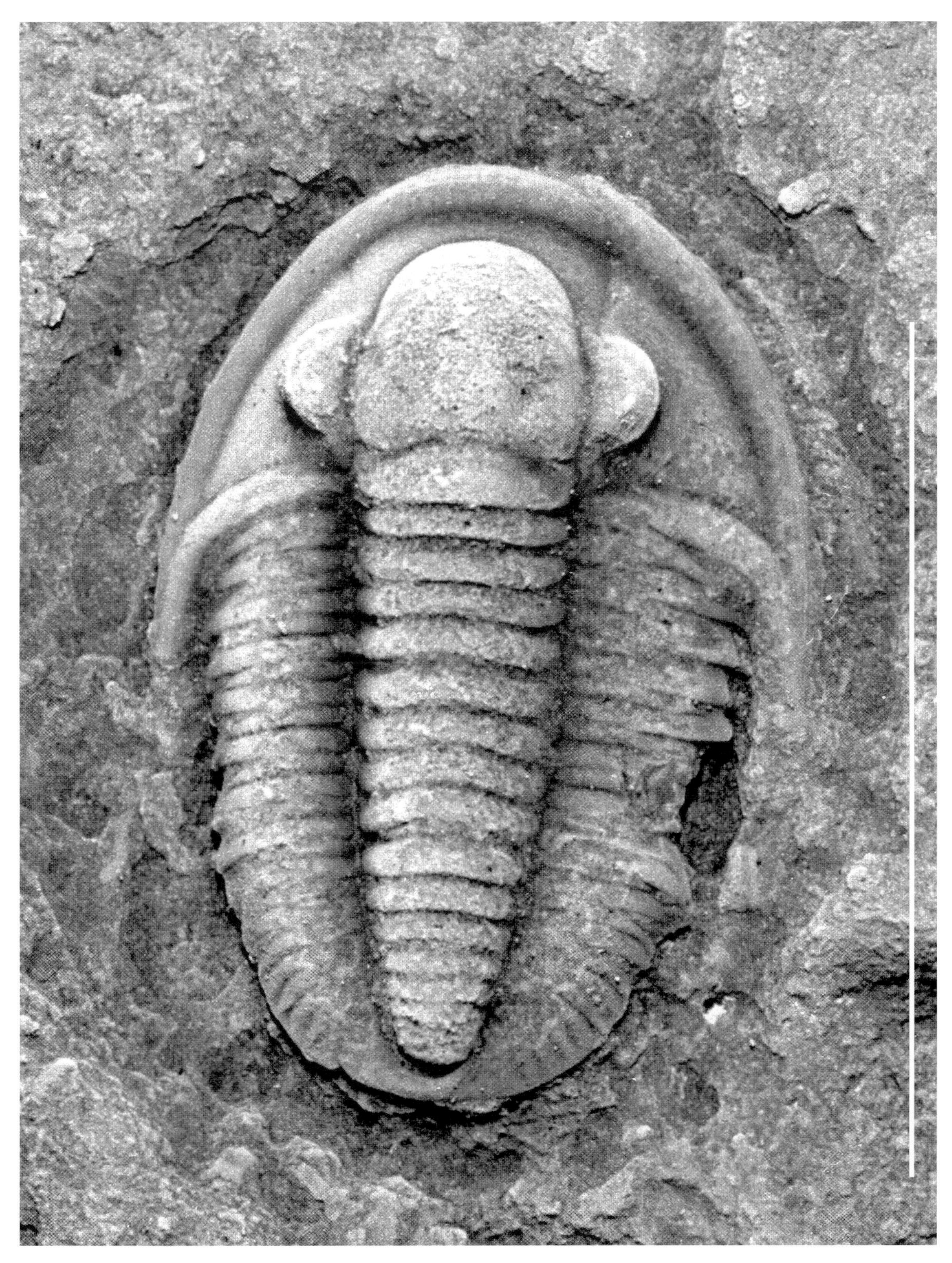

PLATE 142. Proetid from the Middle Ordovician Rust Formation, Trenton Group, Walcott-Rust Quarry, Herkimer County. Proetids are very uncommon in the New York Trenton. This specimen is 11 mm long. MCZ 111714.

PLATE 143. *Pseudodechenella clara* from the Middle Devonian Moorehouse Member of the Onondaga Limestone, Honeoye Falls Quarry, Honeoye, Monroe County. This is a nice specimen of an uncommon trilobite. It is 31 mm long. RMSC 2001.47.1.

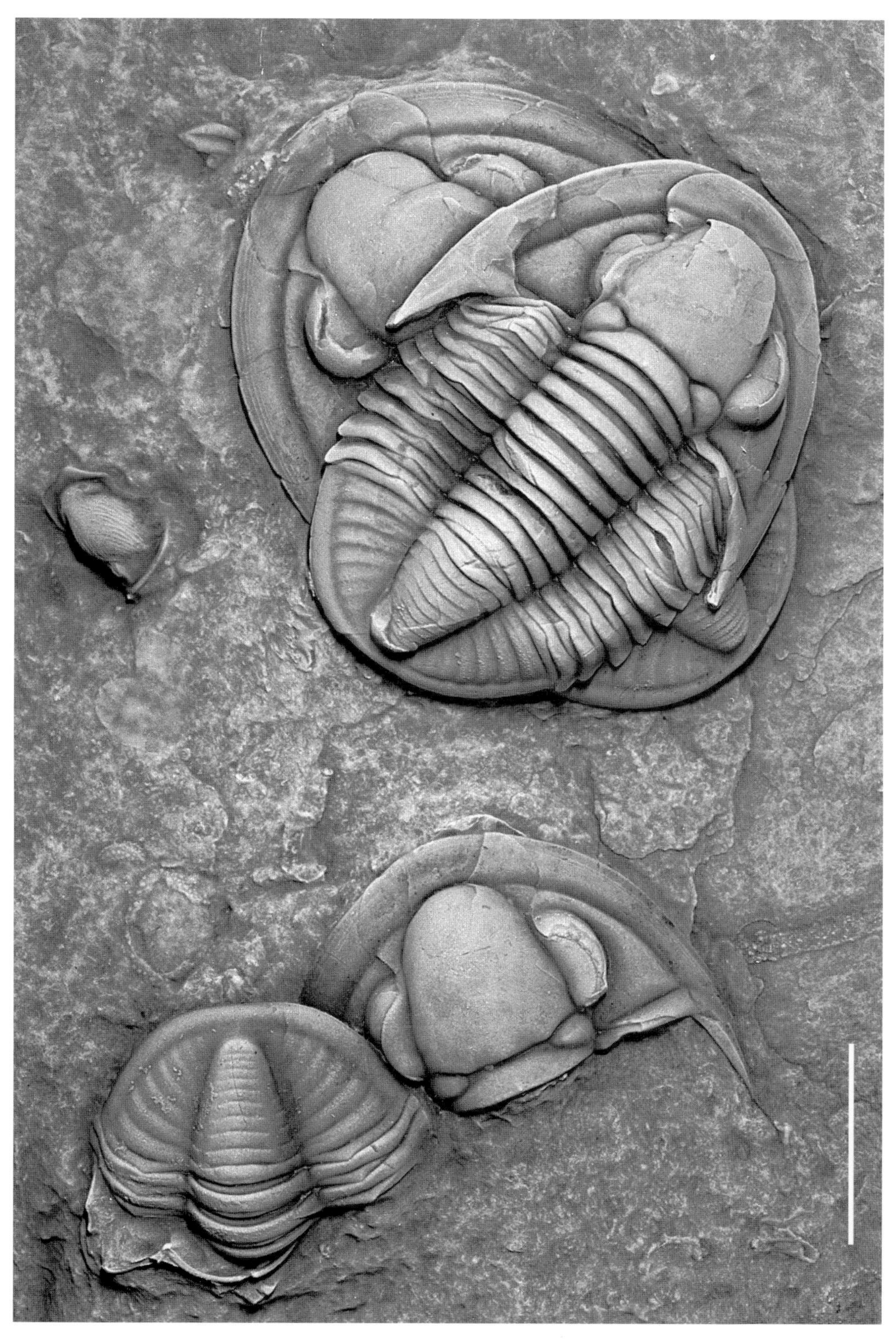

PLATE 144. *Pseudodechenella rowi* from the Middle Devonian Windom Member, Hamilton Group, Penn-Dixie Quarry, Hamburg, Erie County. Shown are two specimens with the molt remains of a third. The largest trilobite is 31 mm long. In the collection of Gregory Jennings.

PLATE 145. *Pseudodechenella rowi* along with *Bellacartwrightia whiteleyi* from the Middle Devonian Wanakah Member, Hamilton Group, Darien, Genesee County. In addition to these trilobites, one can see the pygidium of *Eldredgeops rana*, which shows the richness of these beds. PRI 49634.

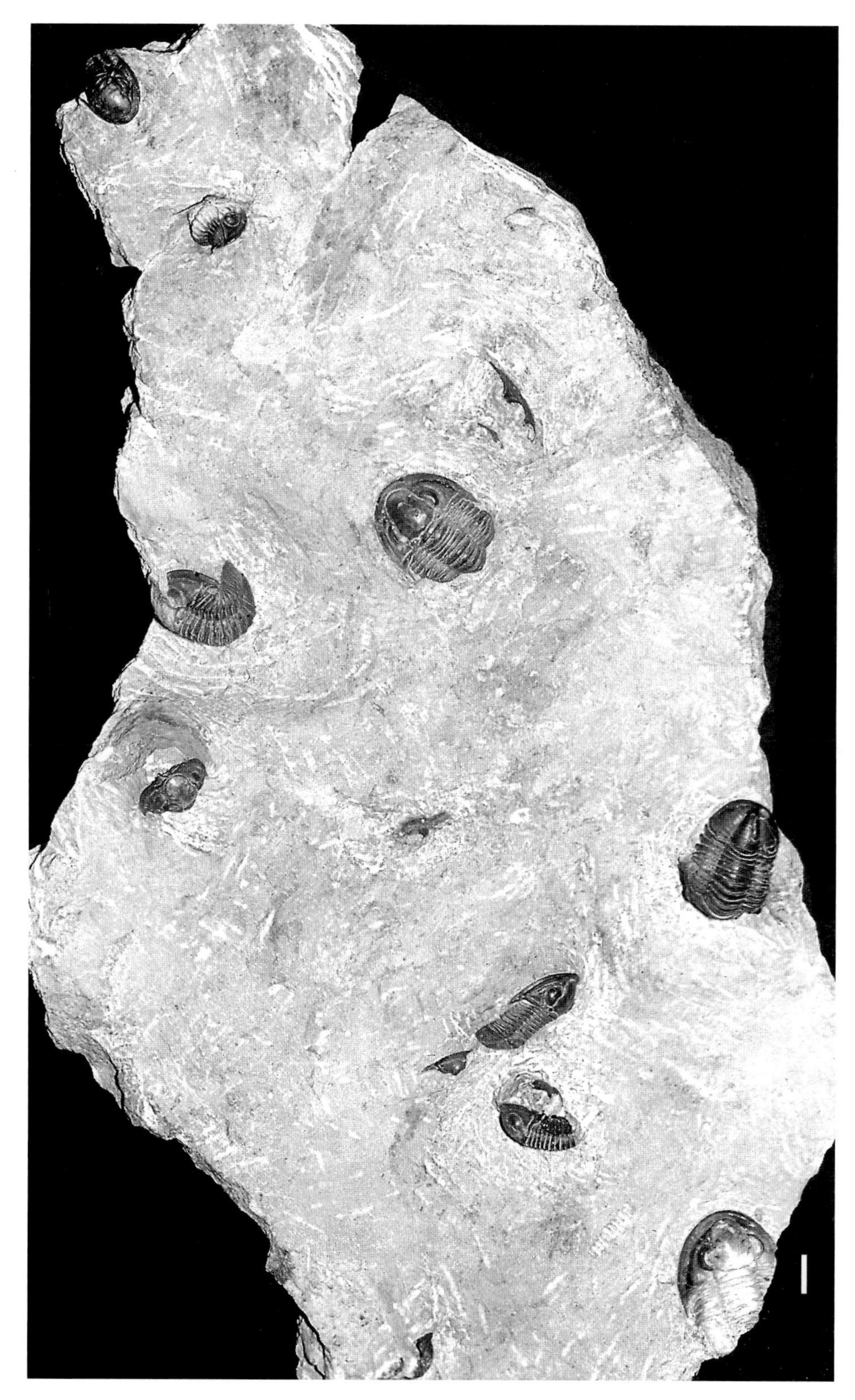

PLATE 146. *Pseudodechenella rowi* from the Middle Devonian Centerfield Limestone, Hamilton Group, Bethany, Genesee County. Although this is not a close cluster, it does record an association of these trilobites before the burial event. The widest cephalon is 24 mm. In the collection of Gerald Kloc.

PLATE 147. *Pseudodechenella rowi* from the upper Middle Devonian, Tully Limestone at Moravia, Cayuga County. This trilobite is recorded over a long time range in the Middle Devonian, ending with the Tully. The largest specimen is 34 mm long. USNM 26767.

PLATE 148. *? Ectenaspis homalonotoides* from the Middle Ordovician Bobcaygeon Formation, Trenton Group, in the Brechin Quarry, Brechin, Ontario, Canada. This trilobite is reported in New York but is very uncommon. In the collection of Kevin Brett.

PLATE 149. *Isoteloides canalis* from the Lower Ordovician Fort Cassin Formation, Crown Point, Essex County. *Isoteloides canalis* is the most common trilobite in these rocks, but articulated material is rare. CM 1433. The negative for this plate was supplied by Dr. Stephen Westrop, Oklahoma State Museum.

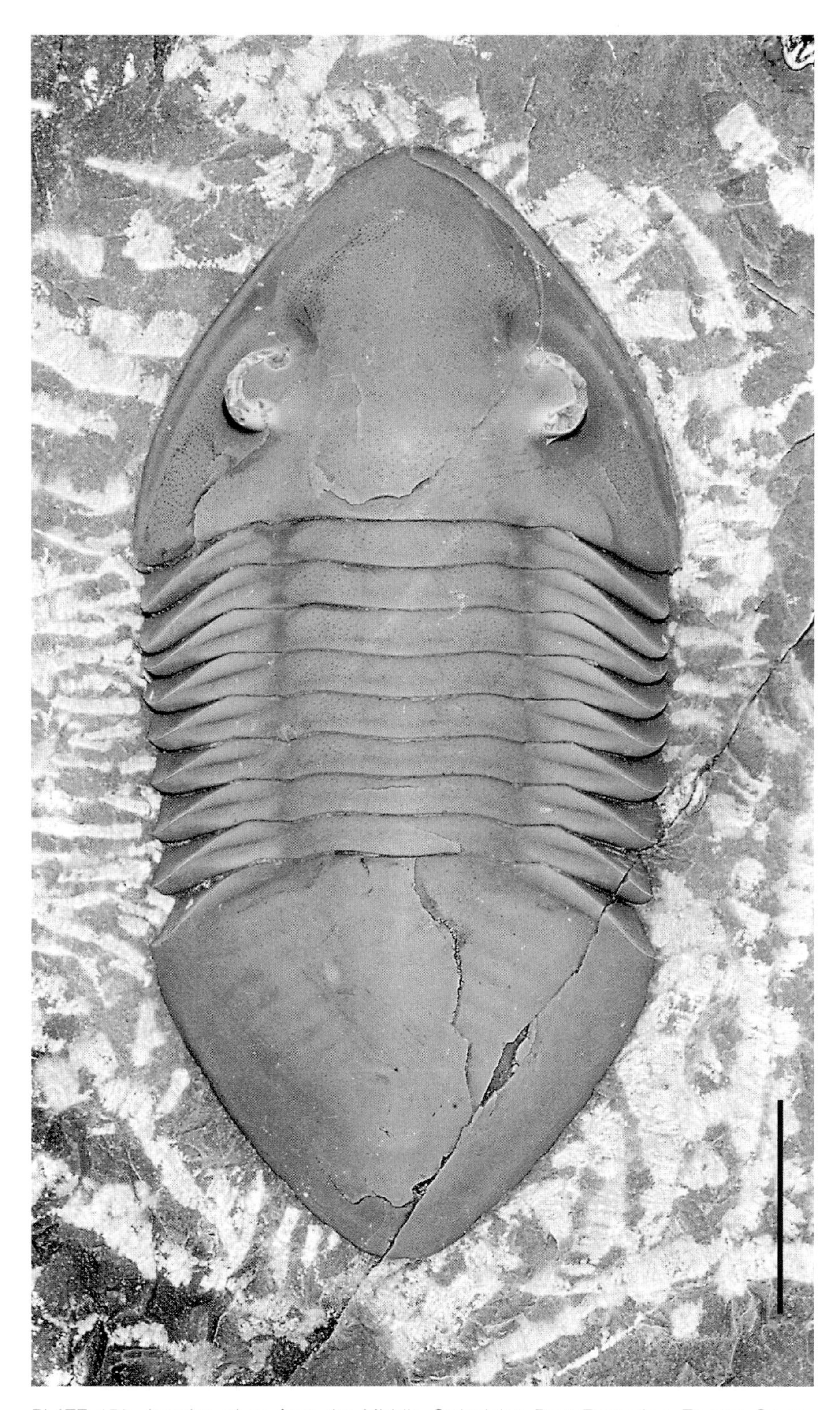

PLATE 150. *Isotelus gigas* from the Middle Ordovician Rust Formation, Trenton Group, Walcott-Rust Quarry, Herkimer County. This specimen was elected as the neotype, as the original type is lost. *Isotelus gigas* is characterized by the small pits abundant on the cephalon. This trilobite is 52 mm long. MCZ 100938 (neotype).

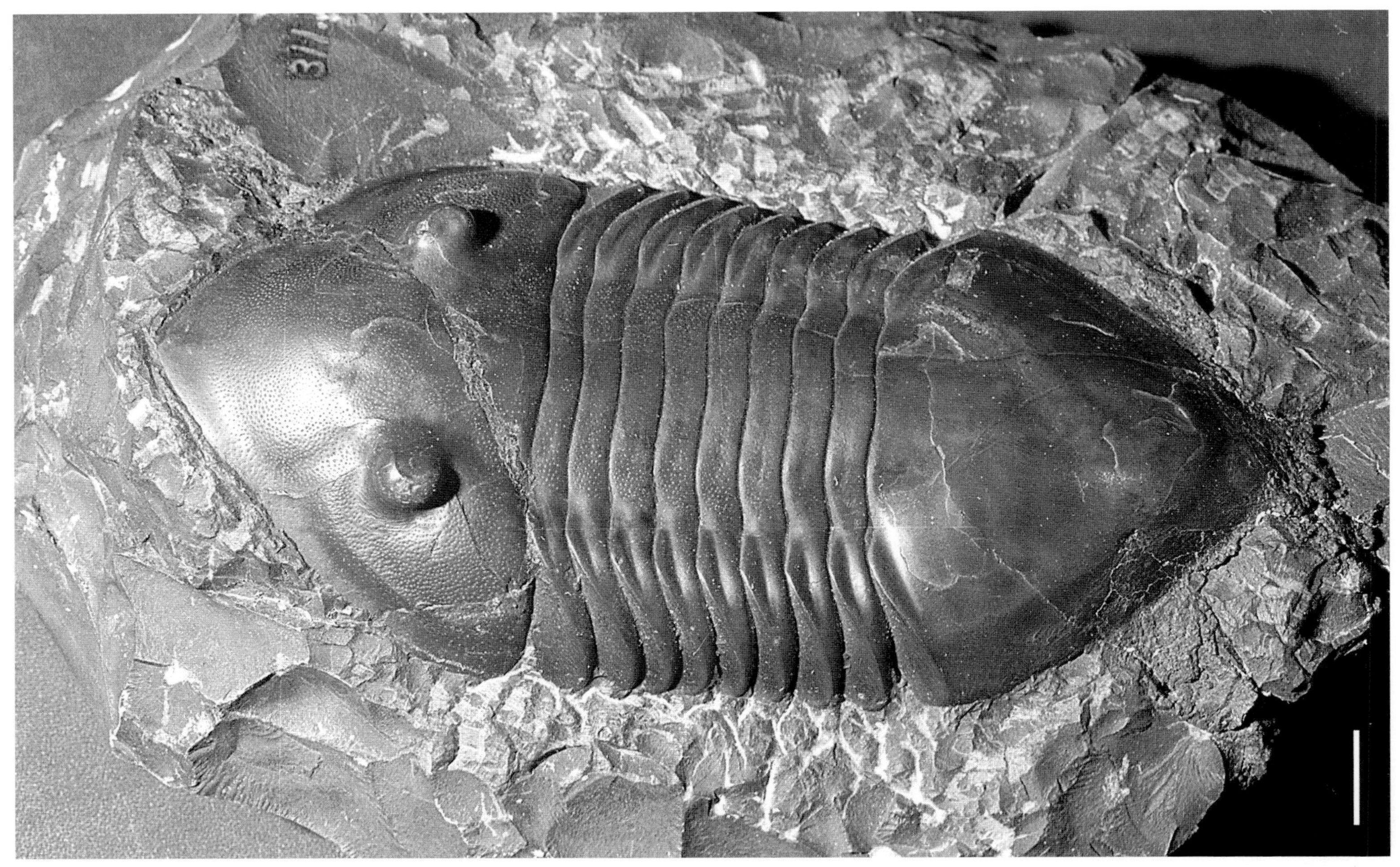

PLATE 151. *Isotelus gigas* from the Middle Ordovician Rust Formation, Trenton Group, Walcott-Rust Quarry, Herkimer County. This large specimen is more characteristic of the large cephalic parts and pygidia common in the Trenton rocks. The trilobite is 122 mm long. MCZ 311.

PLATE 152. *Isotelus gigas* from the Middle Ordovician Rust Formation, Trenton Group, at Trenton Falls, Herkimer County. Many *I. gigas* specimens found in museums came from one bed at Trenton Falls. This is an unusual cluster from this area. USNM 139617.

PLATE 153. *Isotelus* cf. *I. gigas* from the Middle Ordovician Grandine Member of the Neuville Formation in a quarry in Quebec City, Quebec, Canada. This is a view of the ventral exoskeletal anatomy. In the collection of Kevin Brett.

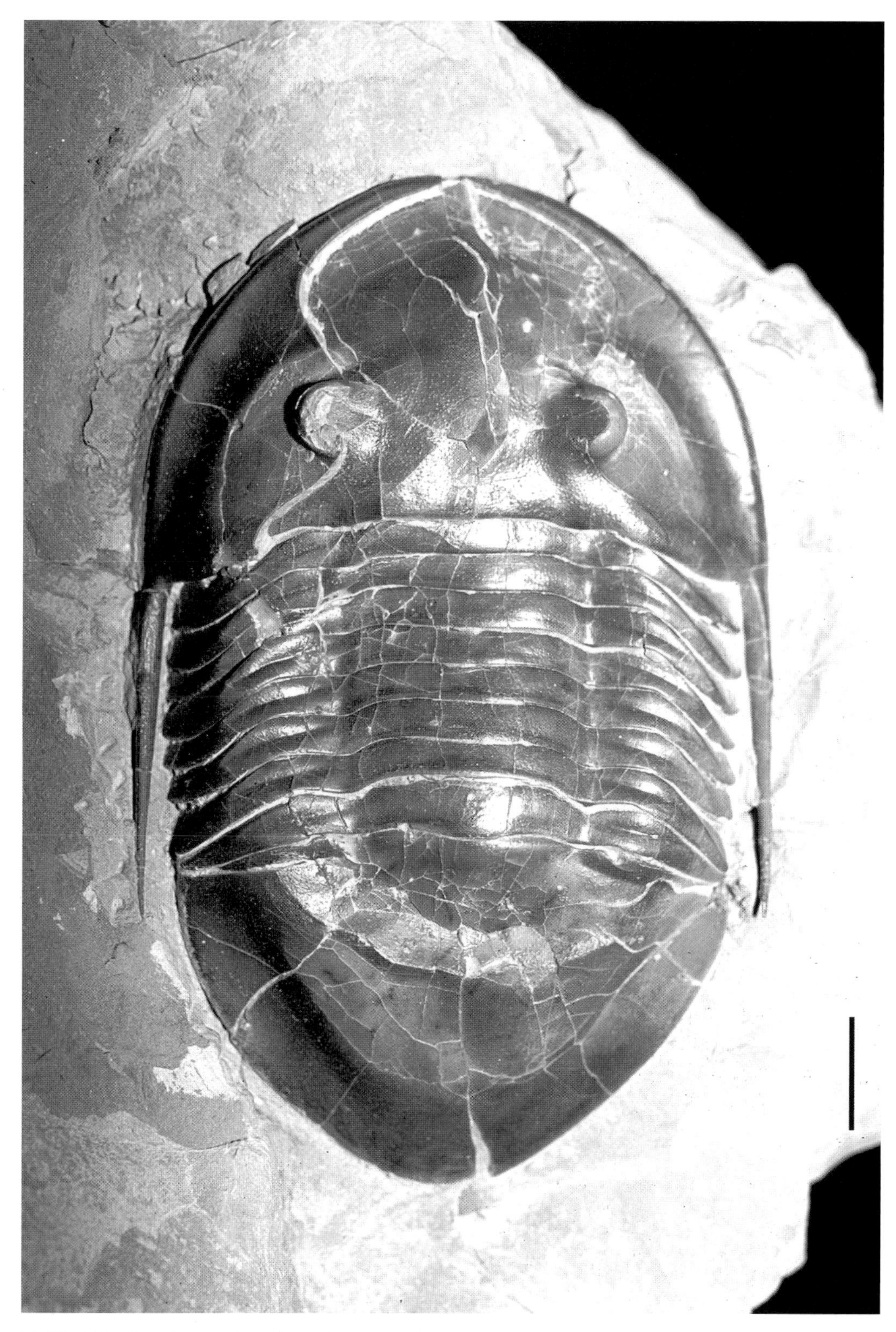

PLATE 154. *Isotelus maximus* from the Upper Ordovician Fort Ancient Member of the Waynesville Formation at Mount Orab, Ohio. *Isotelus maximus* is reported from the Upper Ordovician of New York, but this needs confirmation. This specimen is 34 mm long. PRI 49650.

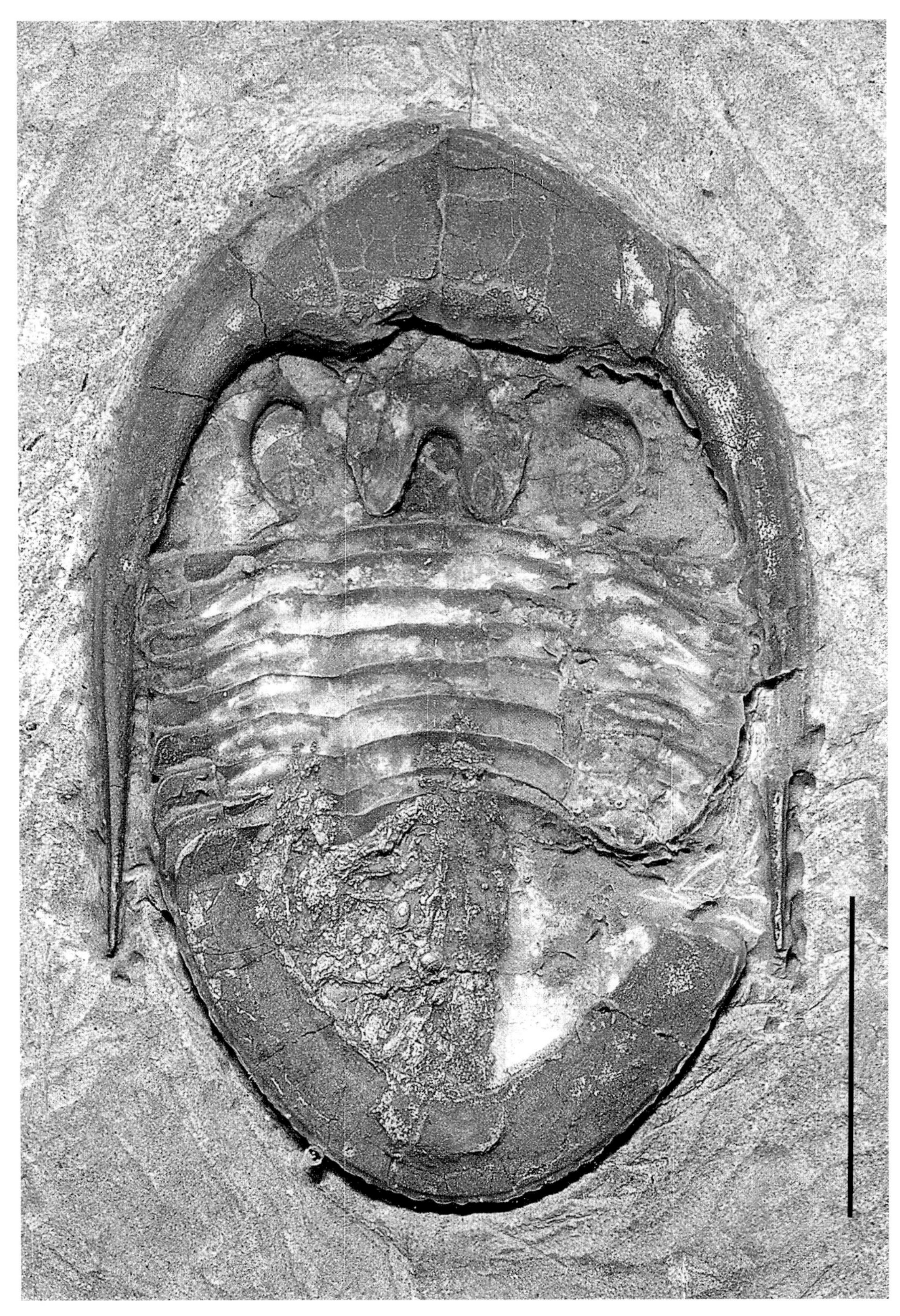

PLATE 155. *Isotelus maximus* from the Upper Ordovician at Cincinnati, Ohio. This is a ventral view of a heavily pyritized exoskeleton. The specimen is 87 mm long. PRI 49651.

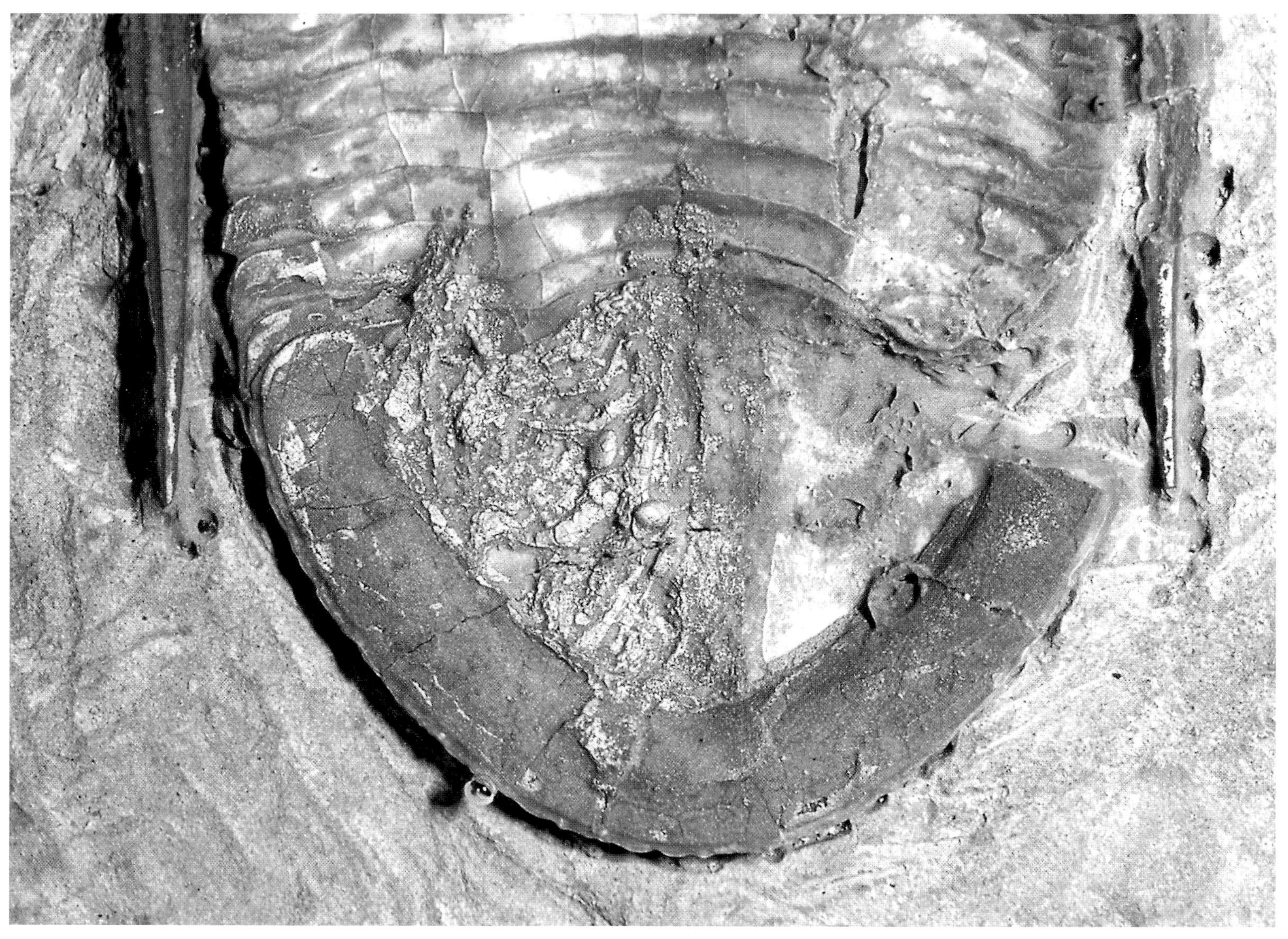

PLATE 156. A closer view of the pygidium from the trilobite in Plate 155. One can make out evidence of pyritized appendages on the left.

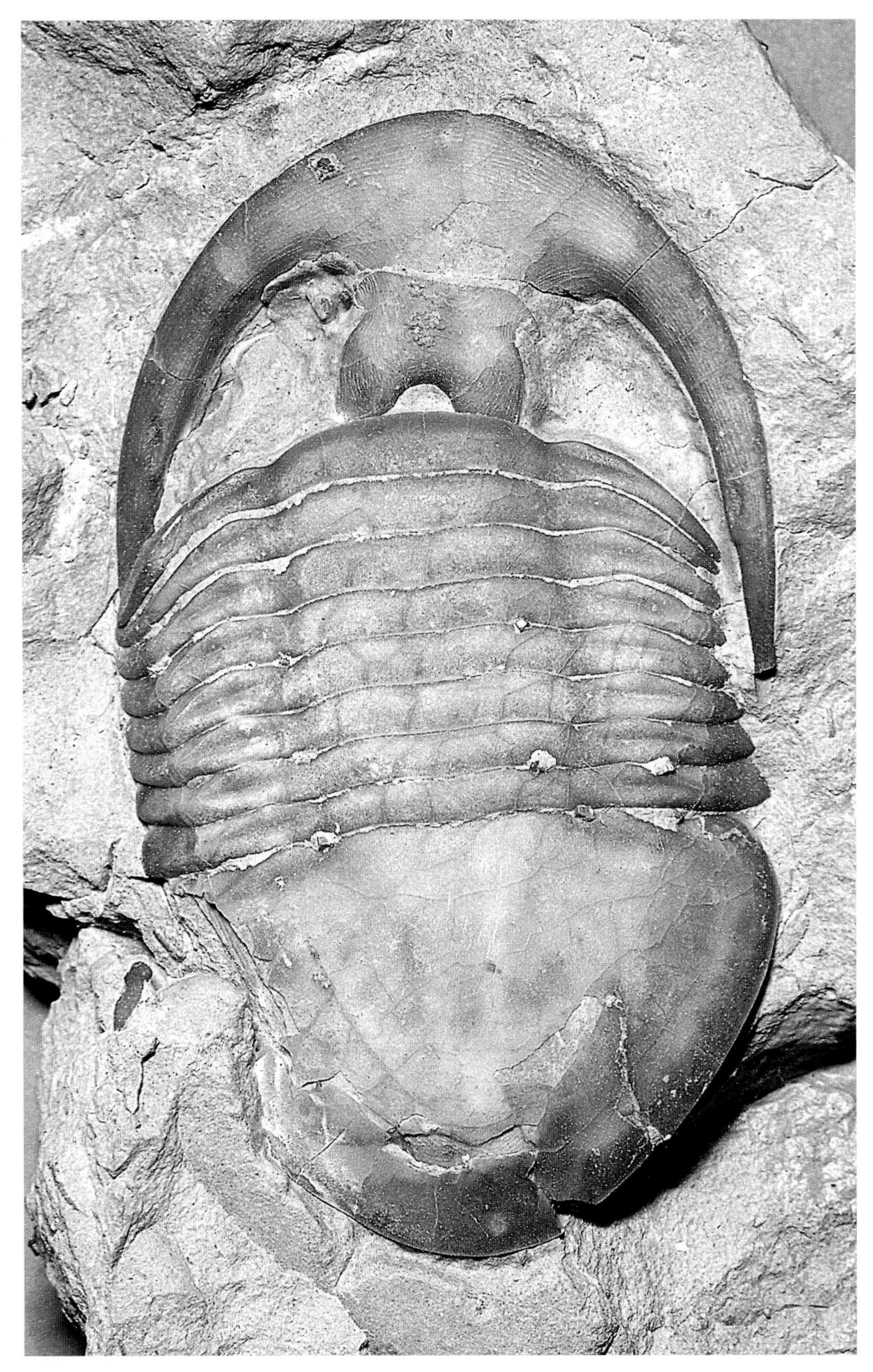

PLATE 157. *Isotelus maximus* from the Upper Ordovician at Cincinnati, Ohio. This molt assemblage shows the turnover of the cranidium from the molting process. The specimen is 55 mm long. PRI 49652.

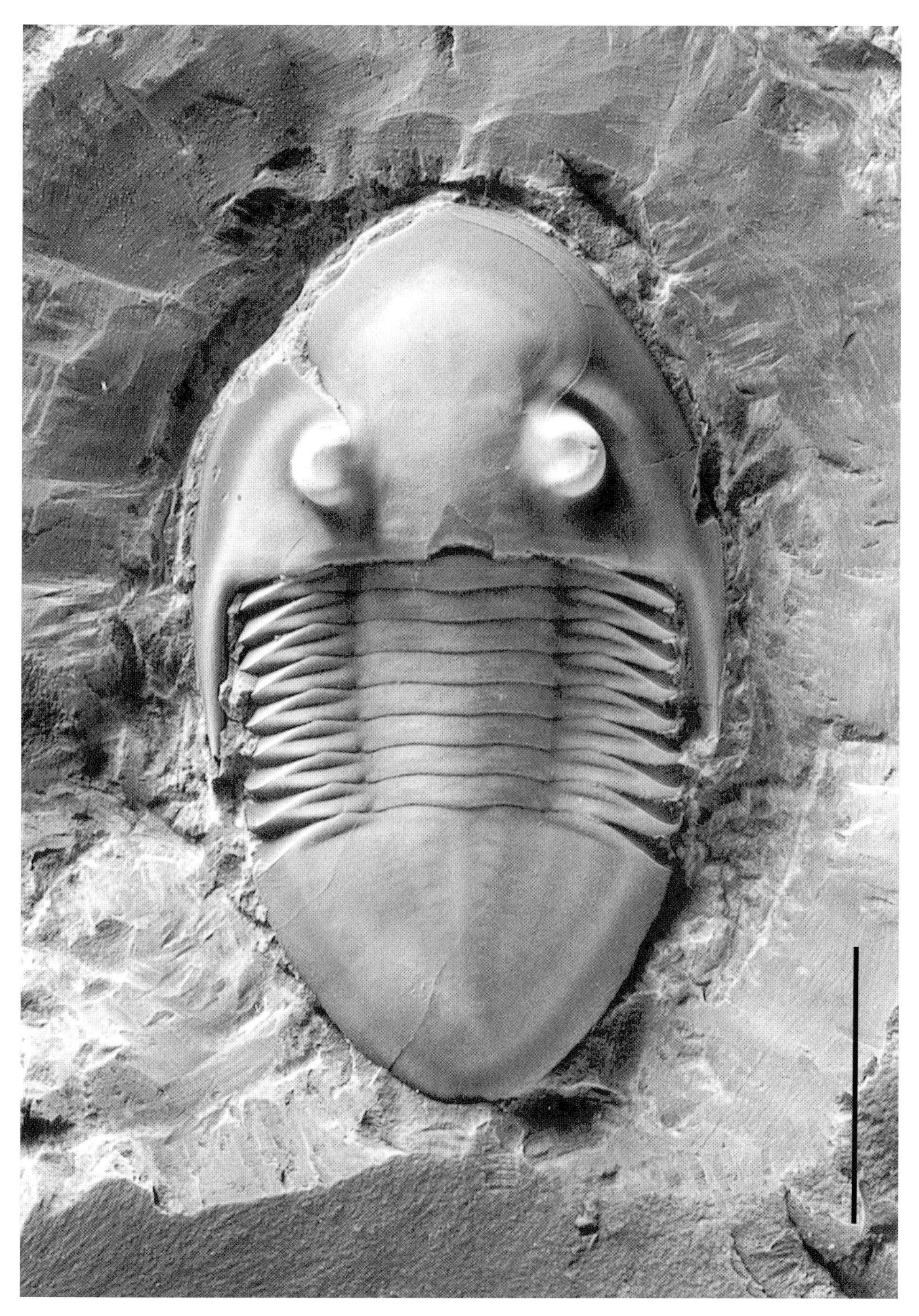

PLATE 158. *Isotelus walcotti* from the Middle Ordovician Rust Formation, Trenton Group, Walcott-Rust Quarry, Herkimer County. This trilobite is only reported from the Rust Formation in the immediate Trenton Falls vicinity. Although the illustrated specimen is small, one has been found at the Walcott-Rust Quarry measuring 120 mm across the width of the cephalon. The specimen is 32 mm long. USNM 61261a (holotype).

PLATE 159. *Pseudogygites latimarginatus* from the Middle/Upper Ordovician Utica Shale at Alder Creek, Oneida County. This trilobite is only found in New York in the uppermost Trenton, Hillier Member, and the Middle/Upper Ordovician transition shales. It is common in the Collingwood Shale in Ontario, Canada. This specimen is 82 mm long. NYSM 17005.

PLATE 160. *Hypodicranotus striatulus* from the Middle Ordovician Rust Formation, Trenton Group, Walcott-Rust Quarry, Herkimer County. This specimen prepared from inside a limestone layer shows the streamlining, suggesting that the species is free swimming. The tiny granules on the surface are not "dirt" but silica sand granules that could not be removed without causing damage. The specimen is 26 mm long. MCZ 115910.

PLATE 161. *Hypodicranotus striatulus* from the Middle Ordovician Rust Formation, Trenton Group, Walcott-Rust Quarry, Herkimer County. This specimen was found inside limestone and reveals the remarkable hypostome, which is almost the full length of the thorax. The specimen is 26 mm long. In the collection of Thomas Whiteley.

PLATE 162. *Hypodicranotus striatulus* from the Middle Ordovician Trenton Group at Trenton Falls, Herkimer County. This specimen from a bedding plain shows the unusual, split genal spine and the very small pygidium. PRI 49635.

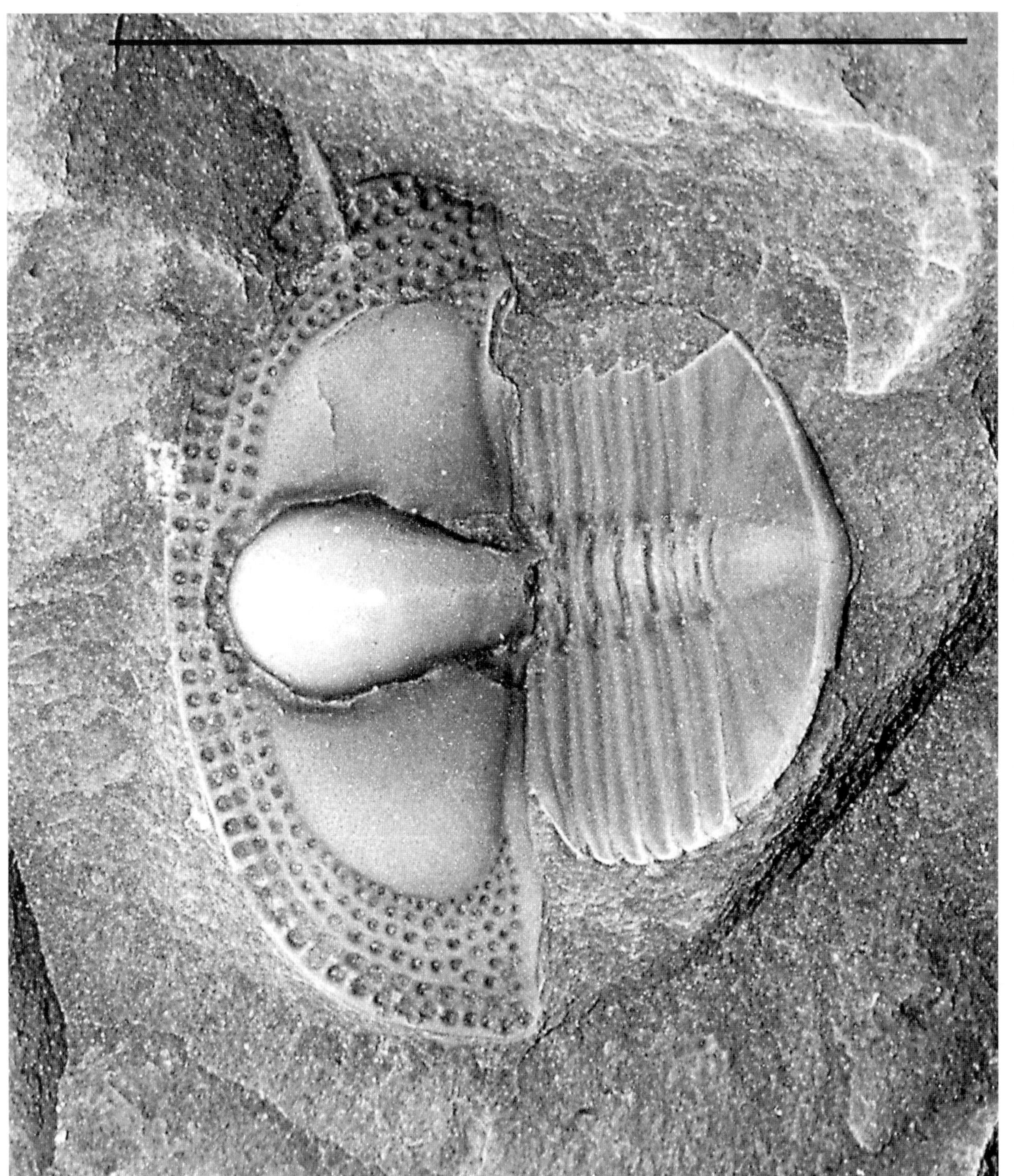

PLATE 163. *Cryptolithus bellulus* from the Upper Ordovician Whetstone Gulf Shale, Lorraine Group, on Burke Creek, Oneida County. The original exoskeletal material is not well preserved in these shales, but there are good external molds. This specimen does not have the long genal spine characteristic of *Cryptolithus*, which indicates it is a molt. *Cryptolithus* molted by losing the lower lamella of the ventral cephalon and exiting the exoskeleton by crawling forward. This is a latex cast from such a mold. The specimen is 8 mm long. PRI 49654.

PLATE 164. *Cryptolithus lorettensis* from the Middle Ordovician Sugar River Formation, Trenton Group, at Mill Creek off Rte. 3, Jefferson County. This cephalon shows the reticulation on the glabella reported for this species. The occipital spine is also nicely preserved. NYSM 17006.

PLATE 165. *Cryptolithus lorettensis* from the Middle Ordovician Sugar River Formation, Trenton Group, at Ingham Mills, Fulton County. Articulated specimens of the Middle Ordovician *Cryptolithus* are very rare, as these trilobites lived in fairly high-energy conditions and the thorax is very fragile. The specimen is a molt and is 19 mm long. NYSM 17007.

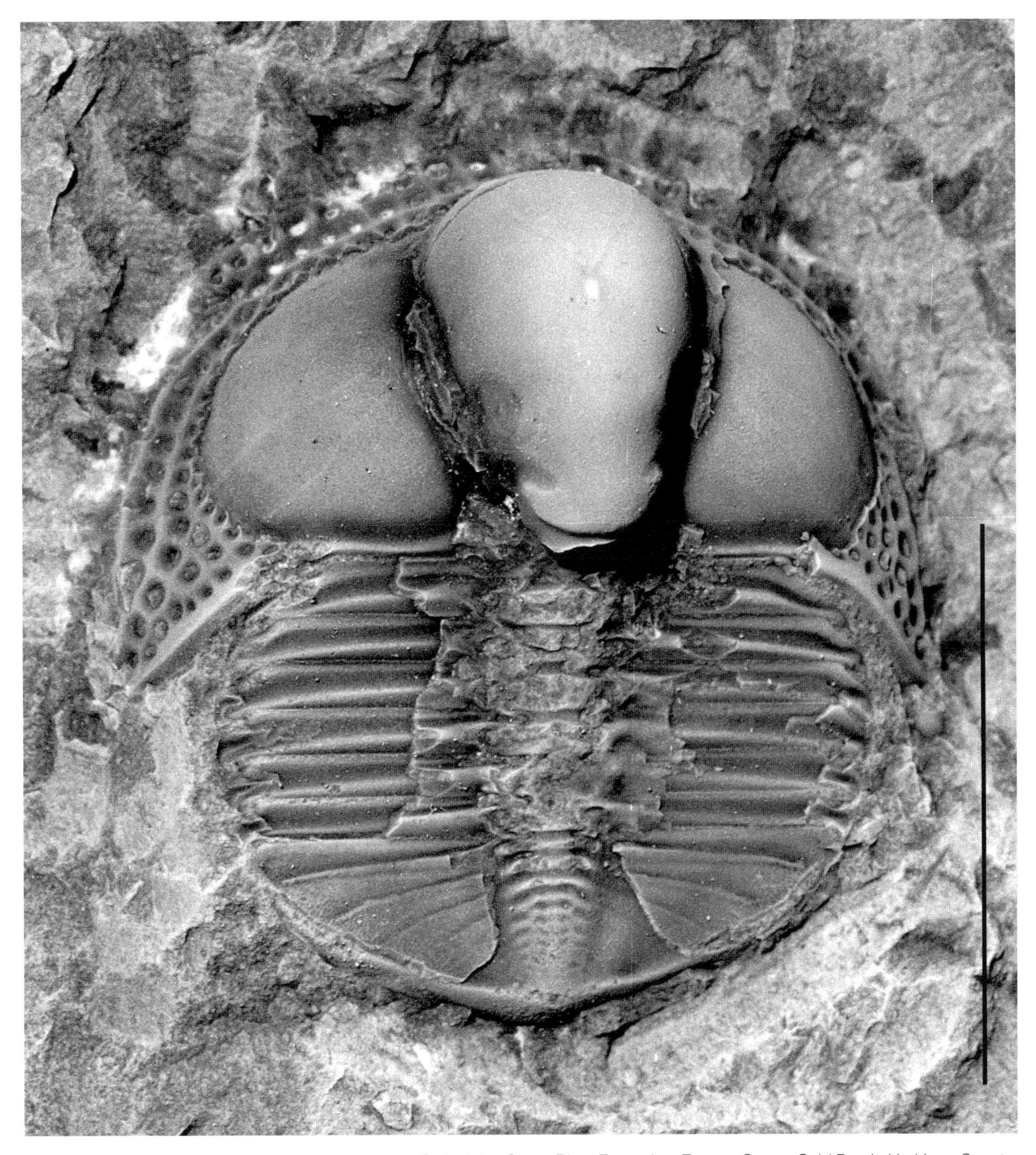

PLATE 166. *Cryptolithus tessellatus* from the Middle Ordovician Sugar River Formation, Trenton Group, Cold Brook, Herkimer County. Although the cephala of this trilobite may be found in large numbers in beds of the Sugar River Formation, articulated specimens are rare. This trilobite is 15 mm long. MCZ 921.

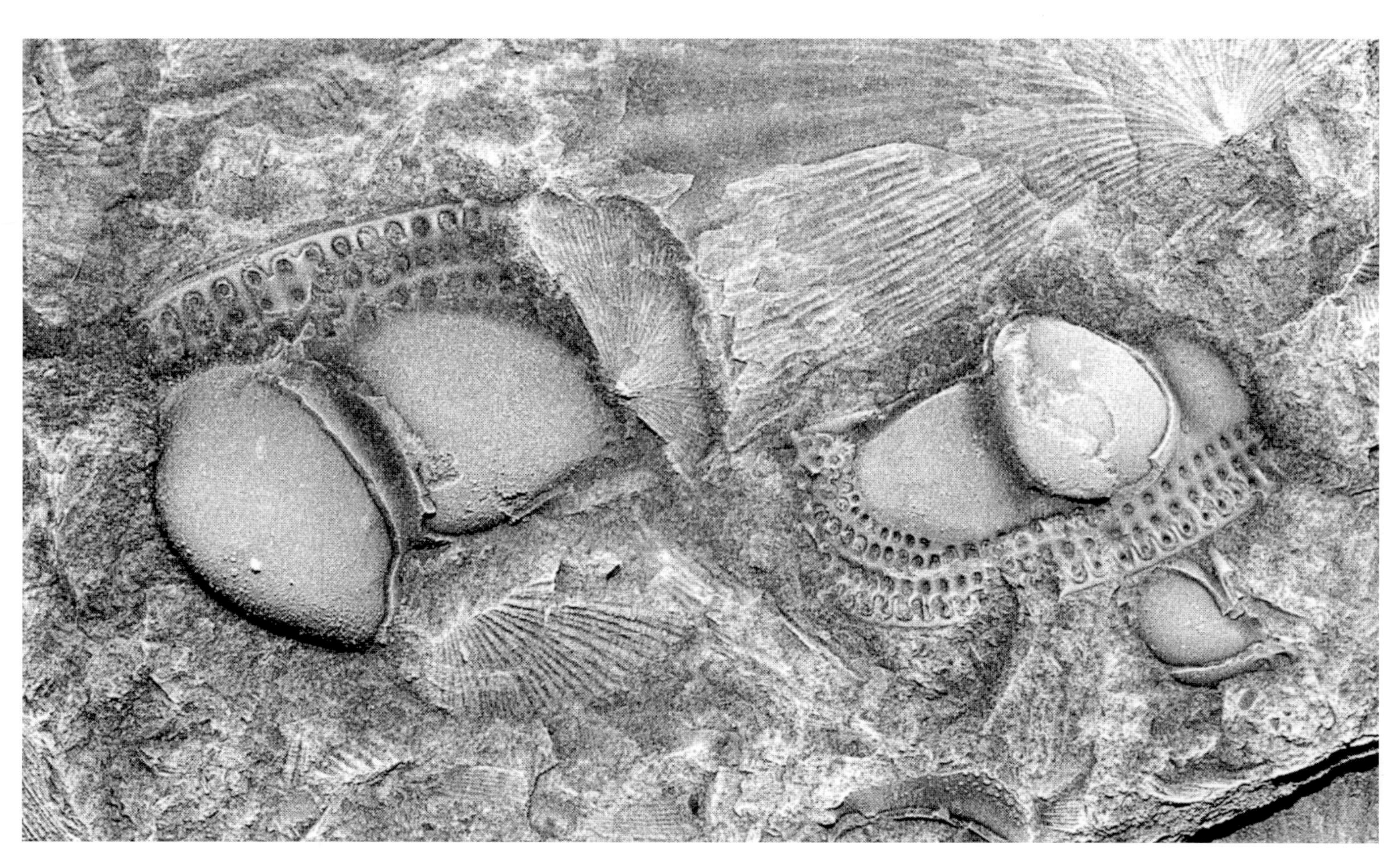

PLATE 167. *Cryptolithus tessellatus* from the Middle Ordovician Sugar River Formation, Trenton Group, at Trenton Falls, Oneida County. This plate shows the pit structure on the cephalic brim, which differentiates the *Cryptolithus* species. *Cryptolithus tessellatus* has three rows of pit immediately anterior to the cheek area. PRI 49636.

PLATE 168. Glaphurids from the Middle Ordovician Chazy Group. A. *Glaphurus pustulosus* from near Chazy, Clinton County. CM 1274. B. *Glaphurina lamottensis* from Isle La Motte, Vermont. This trilobite is also found in the New York Chazy. In the collection of the University of Vermont Museum, specimen 2–63. The negatives for this plate were supplied by Dr. Fredrick Shaw, Lehmann University.

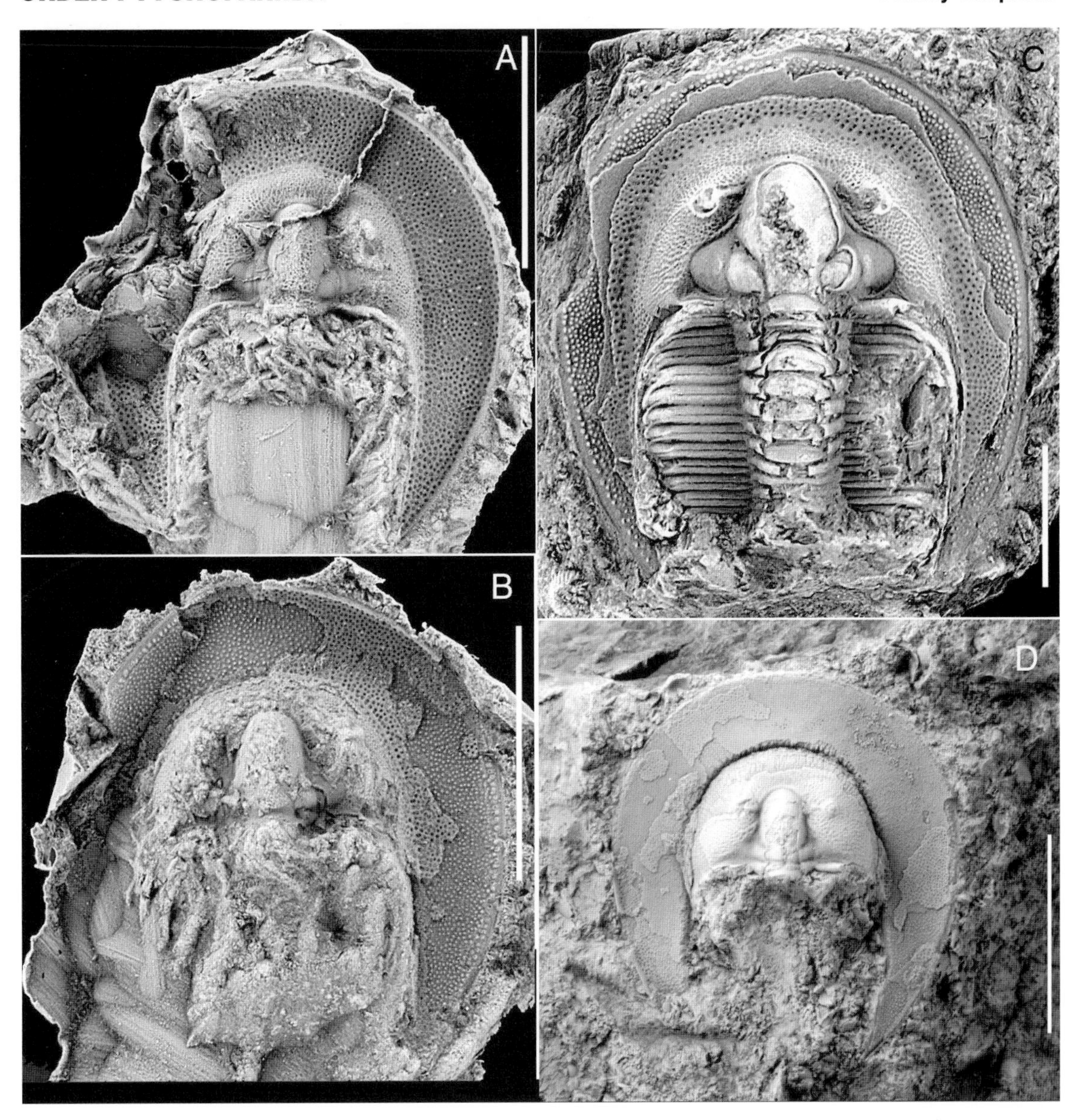

PLATE 169. Trilobite from the family Harpidae. A. *Hibbertia valcourensis* from the Middle Ordovician Day Point Formation, Chazy Group, at Day Point, Clinton County. This is a rubber cast of the original external mold of the trilobite. NYSM 12293 (holotype). B. *Hibbertia valcourensis* from the Middle Ordovician Day Point Formation, Chazy Group, at Day Point, Clinton County. This is a rubber cast of the original external mold of the trilobite. NYSM 12292. C. *Hibbertia ottawaensis* from the Middle Ordovician of Canada. This trilobite is also found in New York. The figured specimen is a metal cast of the original. GSC 329 (holotype). D. *Scotoharpes cassinensis* from the Lower Ordovician Fort Cassin Formation, Fort Cassin Point, Vermont. This trilobite is believed to occur in New York. USNM 35817 (holotype). The negatives for A–C were supplied by Dr. Fredrick Shaw, Lehmann University. The negative for D was supplied by Dr. Stephen Westrop, Oklahoma State Museum.

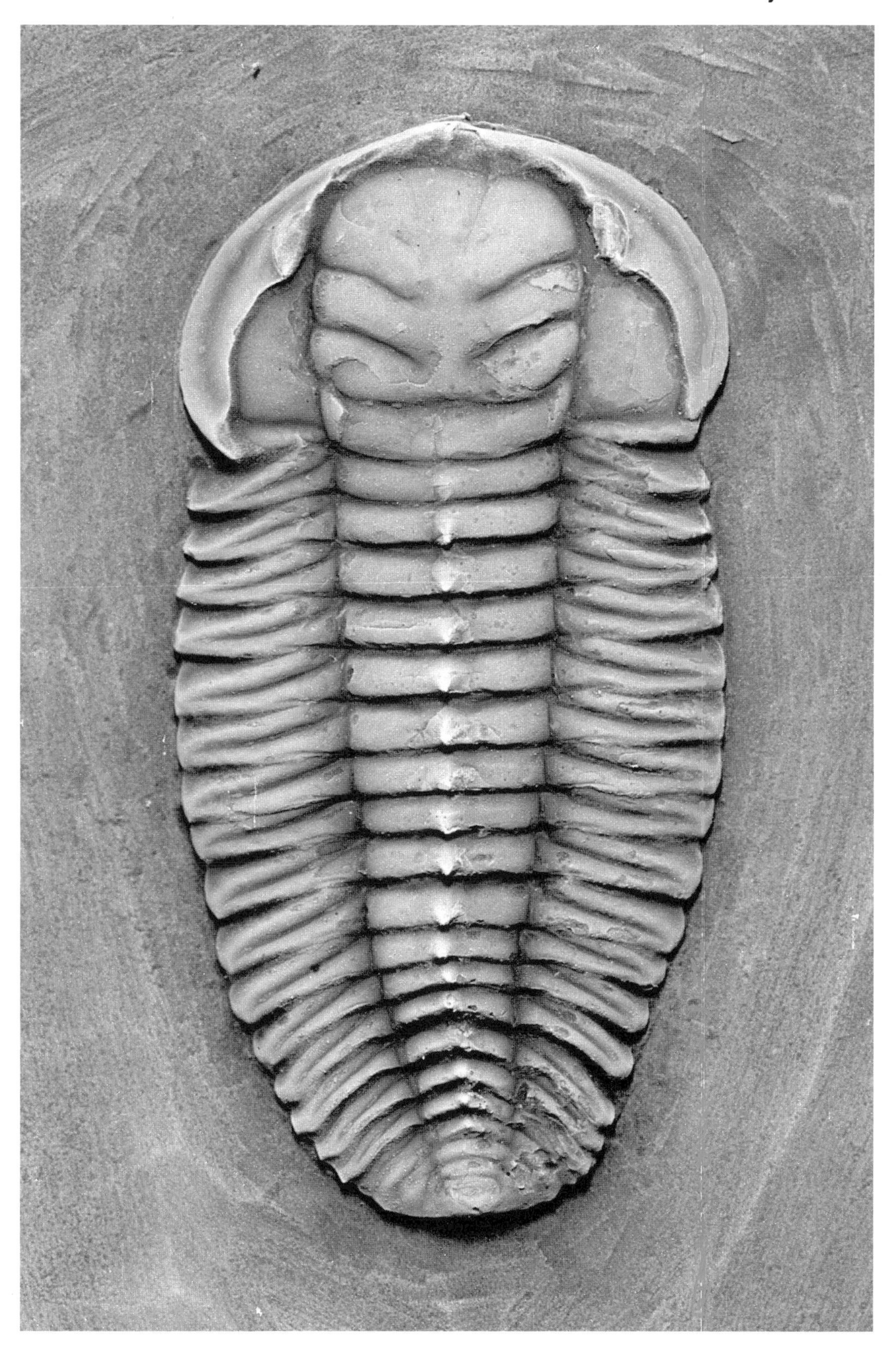

PLATE 170. *Triarthrus beckii* from the Middle Ordovician Utica Shale at Dolgeville, Herkimer County. This specimen shows the yoked, connected, free cheeks of this trilobite. They are slightly displaced here. In the collection of Fred Wessman.

PLATE 171. *Triarthrus* cf. *T. eatoni* from the Upper Ordovician Whetstone Gulf Member, Lorraine Group, on Burke Creek, Oneida County. This specimen is unusual in that there are no axial nodes on the anterior thoracic segments. The specimen is 21 mm long. PRI 49655.

PLATE 172. *Triarthrus eatoni* from the Upper Ordovician Frankfort Member, Lorraine Group, on Six Mile Creek north of Rome, Oneida County. This specimen is from the world renowned Beecher Trilobite Bed, which is the most prolific known source of trilobites with appendage information. This trilobite is 40 mm long. YPM 228.

PLATE 173. *Triarthrus spinosus* from the Upper Ordovician Utica Shale on Nine Mile Creek, Oneida County. This is a rare trilobite in New York but is often found in Canada. The width across the posterior aspect of the cephalon is 15 mm. USNM 96235 (holotype).

PLATE 174. *Triarthrus spinosus* from the Upper Ordovician at Ottawa, Ontario, Canada. This specimen shows the characteristic genal spines and thoracic spines of the species. The trilobite is 15 mm long. In the collection of Kevin Brett.

PLATE 175. Trilobites from the Middle Cambrian limestones of New York. A. *Prosaukia hartti* from the Hoyt Limestone, Greenfield railroad cut, Saratoga County. NYSM 14079. B. *Saratogia calcifera* from the Hoyt Limestone in the Hoyt Quarry at Lester Park, Saratoga County. NYSM 14081. C. *Plethopeltis saratogensis* from the Briarcliff Dolostone at Poughkeepsie, Dutchess County. NYSM 14147. The negatives for A and B were supplied by Dr. Stephen Westrop, Oklahoma State Museum.